Springer Series in
CHEMICAL PHYSICS

65

W0245880

Springer
Berlin
Heidelberg
New York
Barcelona
Hong Kong
London
Milan
Paris
Singapore
Tokyo

Physics and Astronomy

ONLINE LIBRARY

http://www.springer.de/phys/

Springer Series in
CHEMICAL PHYSICS

Series Editors: F. P. Schäfer   J. P. Toennies   W. Zinth

The purpose of this series is to provide comprehensive up-to-date monographs in both well established disciplines and emerging research areas within the broad fields of chemical physics and physical chemistry. The books deal with both fundamental science and applications, and may have either a theoretical or an experimental emphasis. They are aimed primarily at researchers and graduate students in chemical physics and related fields.

Series homepage – http://www.springer.de/phys/books/chemical-physics/

R. Rigler   E.S. Elson

Fluorescence Correlation Spectroscopy

Theory and Applications

With 174 Figures

 Springer

Professor Rudolf Rigler

Karolinska Institutet, Department of Medical Biophysics
17177 Stockholm, Sweden
e-mail: rudolf.rigler@mbb.ki.se

Professor Elliot S. Elson

Washington University, Department of Biochemistry and Molecular Biophysics, School of Medicine
660 S. Euclid Avenue, St. Louis, MO 63110–1093, USA
e-mail: elson@biochem.wustl.edu

Series Editors:

Professor F.P. Schäfer

Max-Planck-Institut für Biophysikalische Chemie
D-37077 Göttingen-Nikolausberg, Germany

Professor J.P. Toennies

Max-Planck-Institut für Strömungsforschung
Bunsenstrasse 10
D-37073 Göttingen, Germany

Professor W. Zinth

Universität München,
Institut für Medizinische Optik
Öttingerstr. 67
D-80538 München, Germany

ISSN 0172-6218
ISBN-13: 978-3-642-64018-6   e-ISBN-13: 978-3-642-59542-4
DOI: 10.1007/978-3-642-59542-4

Library of Congress Cataloging-in-Publication Data

Rigler, Rudolf.
Fluorescence correlation spectroscopy : theory and applications / R. Rigler, E.S. Elson. p. cm. – (Springer series in chemical physics, ISSN 0172-6218 ; 65)
Includes bibliographical references and index.
ISBN-13: 978-3-642-64018-6
1. Fluorexcence spectroscopy. I. Elson, Elliot, 1937 - II. Title. III. Series.

QD96.F56 R54 2001
543'.08584–dc21   00-069229

Springer-Verlag Berlin Heidelberg New York
a member of BertelsmannSpringer Science+Business Media GmbH

http://www.springer.de

Typesetting: PTP–Berlin, Stefan Sossna
Cover concept: eStudio Calamar Steinen
Cover production: *design & production* GmbH, Heidelberg

SPIN: 10759774   57/3020Rw - 5 4 3 2 1 0

Preface

Fluorescence correlation spectroscopy (FCS) was developed in order to characterize the dynamics of molecular processes in systems in thermodynamic equilibrium. FCS determines transport and chemical reaction rates from measurements of spontaneous microscopic thermally driven molecular concentration fluctuations. Since its inception, and particularly in recent years, technical and conceptual advances have extended the range of practical applicability and the information obtainable from FCS measurements. Improvements in microscopy, data acquisition, and data processing have greatly shortened the time required for FCS measurements. FCS can now be routinely applied to labile systems such as cells, and for the acquisition of large volumes of data as required for high-throughput screening. Cross correlation methods provide a powerful tool for characterizing interactions among different molecular species. Analysis of the amplitude of concentration fluctuations can provide a wealth of information about aggregation/polymerization process and the compositions of mixtures.

Furthermore, FCS provides a bridge between conventional measurements of dynamic processes on a macroscopic concentration scale and the currently developing field of single molecule measurements. Both FCS and single molecule approaches measure directly stochastic fluctuations in molecular properties, and so must be analyzed by statistical methods to yield conventional phenomenological parameters. As commonly practiced, FCS yields these phenomenological parameters, e.g., diffusion coefficients and chemical rate constants, directly in terms of a fluorescence fluctuation autocorrelation function. Single molecule measurements are most interesting when they provide a view of differences in behavior among the molecules in a population. However, when properly averaged, these different behaviors must all be consistent with conventional ensemble parameters as measured by FCS or older techniques.

As a result of these developments, FCS and related methods now provide a powerful arsenal that can be used to attack diverse problems over a wide range of fields. This book presents a representative selection of basic and biotechnological applications of FCS to measurements in chemically defined systems, complex mixtures and cells. It provides a progress report on the current state of development of a rapidly evolving field of study with broad ramifications. Our aim is to provide a source which students and lab-

VI   Preface

oratory scientists can survey current FCS concepts, experimental methods, and results, and can assess the potential utility of FCS methods in their own research.

The editors would like to express their gratitude to all the authors who have contributed to this collection, and to Ms. Adelheid Duhm for her excellent editorial work, without which this volume would not have been finished or published.

Stockholm, St. Louis (Missouri)   *Rudolf Rigler*
May 2000   *Elliot Elson*

Contents

Part III   Applications in Biotechnology, Drug Screening, and Diagnostics

9   Dual-Color Confocal Fluorescence Spectroscopy and its Application in Biotechnology

Andre Koltermann*, Ulrich Kettling, Jens Stephan, Thorsten Winkler,
and Manfred Eigen 187

List of Contributors

van den Berg, Petra A.W.
MicroSpectroscopy Centre
Department of Biomolecular
Sciences
Laboratory of Biochemistry
Wageningen University and Research
Centre
Dreijenlaan 3, 6703 HA Wageningen
The Netherlands

Brinkmeier, Michael
Abteilung Biochemische Kinetik und
Experimentelle Biophysik
Max-Planck-Institut für Bio-
physikalische Chemie
Am Fassberg 11
D-37077 Göttingen, Germany

Brock, Roland
Abteilung Molecular Biologie
Max-Planck-Institut für Bio-
physikalische Chemie
Am Fassberg 11
D-37077 Göttingen, Germany

Buffle, Jacques
CABE (Analytical and Biphysical
Environmental Chemistry)
University of Geneva, Sciences II
30, quai E. Ansermet
CH-1211 Geneva 4, Switzerland

Chen, Yan
School of Physics and Astronomy
University of Minnesota
Ninneapolis, MN 55455, USA

Edman, L.
Department of Medical Biophysics
Karolinska Institutet
S-171 77 Stockholm, Sweden

Eigen, Manfred
Abteilung Biochemische Kinetik und
Experimentelle Biophysik
Max-Planck-Institut für Bio-
physikalische Chemie
Am Fassberg 11
D-37077 Göttingen, Germany

Elson, Elliot
Department of Biochemistry and
Molecular Biophysics
Washington University School of
Medicine
St. Louis, MO 63110
USA

Földes-Papp, Zeno
Department of Medical Biophysics
Karolinska Institutet
S-171 77 Stockholm, Sweden

Gratton, Enrico
Laboratory for Fluorescence
Dynamics
University of Illinois at Urbana
-Champaign
Urbana, IL 61801, USA

Hink, Mark A.
MicroSpectroscopy Centre
Department of Biomolecular
Sciences
Laboratory of Biochemistry
Wageningen University and Research
Centre
Dreijenlaan 3, 6703 HA Wageningen
The Netherlands

Janka, Reinhard
Carl Zeiss Jena
Promenade 10
D-07740 Jena
Germany

Jankowski, Tilo
Carl Zeiss Jena
Promenade 10
D-07740 Jena
Germany

Jovin, Thomas M.
Abteilung Molecular Biologie
Max-Planck-Institut für Bio-
physikalische Chemie
Am Fassberg 11
D-37077 Göttingen, Germany

Kask, Peet
EVOTEC BioSystems AG
Schnackenburgallee 114
Hamburg, Germany

Kettling, Ulrich
Abteilung Biochemische Kinetik und
Experimentelle Biophysik
Max-Planck-Institut für Bio-
physikalische Chemie
Am Fassberg 11
D-37077 Göttingen, Germany

Kinjo, Masataka
Research Institute for Electronic
Sciences
Hokkaido University, Sapporo, 060,
Japan

Koltermann, Andre
DIREVO Biotech AG
Nattermannallee 1
D-50829 Köln, Germany

Matayoshi, Edmund
Department of Structural Biology
Pharmaceutical Products Division,
Discovery Research
Abbott Laboratories
Abbott Park, Illinois 60064-6114
USA

Mets, Ülo
Institute of Chemical Physics and
Biophysics
Rävala 10
EE0001 Tallinn
Estonia

Meyer-Almes, F.J.
EVOTEC BioSystems AG
Schnackenburgallee 114
Hamburg, Germany

Mitchell, Jennifer L.
University of North Carolina
Departmen of Chemistry
Campus Box 3290
Chapel Hill, NC
USA 27599-3290

Müller, Joachim D.
University of Illinois
Department of Physics,
Loomis Laboratory of Physics
1110 West Green Street
Urbana, IL 61801-3080, USA

Palo, Kaupo
EVOTEC BioSystems AG
Schnackenburgallee 114
Hamburg, Germany

Petersen, Nils O.
Department of Chemistry
The University of Western Ontario
London, Ontario
Canada, N6K 3E5

Petushkov, Valentin N.
MicroSpectroscopy Centre
Department of Biomolecular
Sciences
Laboratory of Biochemistry
Wageningen University and Research
Centre
Dreijenlaan 3, 6703 HA Wageningen
The Netherlands

Pramanik, Aladdin
Department of Medical Biophysics
Karolinska Institutet
S-171 77 Stockholm, Sweden

Qian, Hong
Department of Applied Mathematics
University of Washington
Seattle, WA 98195
USA

Riesner, Detlev
Institute für Physikalische Biologie
und
Biologisch-Medizinisches
Forschungszentrum
Heinrich-Heine-Universität
Düsseldorf
D-40225 Düsseldorf, Germany

Rigler, Rudolf
Department of Medical Biophysics
Karolinska Institutet
S-171 77 Stockholm, Sweden

Schwille, Petra
Experimental Biophysics Group
Max-Planck-Institute for Biophysical
Chemistry
Am Fassberg 11
D-37077 Göttingen, Germany

Starchev, Konstantin
CABE (Analytical and Biphysical
Environmental Chemistry)
University of Geneva, Sciences II
30, quai E. Ansermet
CH-1211 Geneva 4, Switzerland

Stephan, Jens
Abteilung Biochemische Kinetik und
Experimentelle Biophysik
Max-Planck-Institut für Bio-
physikalische Chemie
Am Fassberg 11
D-37077 Göttingen, Germany

Swift, Kerry
Department of Structural Biology
Pharmaceutical Products Division,
Discovery Research
Abbott Laboratories
Abbott Park, Illinois 60064-6114
USA

Thompson, Nancy L.
University of North Carolina
Department of Chemistry
Campus Box 3290
Chapel Hill, NC
USA 27599-3290

Visser, Antonie J.W.G.
MicroSpectroscopy Centre
Department of Biomolecular
Sciences
Laboratory of Biochemistry
Wageningen University and Research
Centre
Dreijenlaan 3, 6703 HA Wageningen
The Netherlands

Webb, Watt W.
Cornell University
School of Applied and Engineering
Physics
23 Clark Hall
Ithaca, NY 14853-2501, USA

Wennmalm, S.
Department of Medical Biophysics
Karolinska Institutet
S-171 77 Stockholm, Sweden

Widengren, Jerker
CABE (Analytical and Biphysical
Environmental Chemistry)
University of Geneva, Sciences II

30, quai E. Ansermet
CH-1211 Geneva 4, Switzerland

Wilkinson, Kevin J.
Department of Medical Biophysics
Karolinska Institutet
S-171 77 Stockholm, Sweden

Winkler, Thorsten
Abteilung Biochemische Kinetik und
Experimentelle Biophysik
Max-Planck-Institut für Bio-
physikalische Chemie
Am Fassberg 11
D-37077 Göttingen, Germany

1 Introduction

E. Elson

Fluorescence correlation spectroscopy (FCS) is one of a family of fluctuation correlation methods that can measure the dynamics of molecular processes from observations of spontaneous microscopic fluctuations in molecular position or number density. Principally, these measurements are carried out on systems in thermodynamic equilibrium; but are also applicable to non-equilibrium steady states, an area that could merit future study. The driving forces for spontaneous fluctuations arise from thermal energy and the fluctuations have a stochastic variation of amplitude and time course. Therefore, to derive the macroscopic phenomenological parameters by which we conventionally measure transport and chemical reaction rates, e.g. diffusion coefficients and chemical rate constants, it is necessary to perform a statistical analysis on a large number of measurements of individual fluctuations. Typically, this statistical analysis takes the form of a correlation function, which yields information about both the time courses and amplitudes of the molecular processes taking place in the system.

FCS is closely related to an earlier optical fluctuation method, quasi-elastic light scattering, or dynamic light scattering (DLS). From measurements of scattered coherent laser light, DLS characterizes molecular motion in terms of optical interference effects [1.1]. DLS provides an excellent method for studying molecular transport in highly resolved and fairly concentrated systems, or in mixtures in which one species, e.g., erythrocytes in blood, dominates the scattering. The polarizability, which determines the extent of scattering, varies weakly with molecular structure. Therefore DLS is not useful for measuring the progress of chemical reactions or for following the motion of a specific type of molecule at high dilution or in the presence of large concentrations of other molecular species. The high sensitivity and chemical specificity of fluorescence makes these areas accessible to FCS. In contrast to scattered laser light, fluorescence is incoherent due to the random time delay between the absorption of the excitation photon and the emission of the fluorescence photon. Therefore, the "optical mixing" techniques used in laser light scattering experiments cannot be applied to fluorescence. In a typical FCS experiment fluctuations in the numbers of fluorescent molecules in a small, usually open, observation region of the solution are registered as corresponding fluctuations in detected fluorescence.

There is a renaissance of interest in FCS due to improvements in measurement techniques and a growing recognition of the wide range of problems to which it can be applied productively. Chapter 14 describes the development of FCS from its origins to the present. An original goal was to provide a method for characterizing chemical kinetics, which couples the specificity of relaxation methods [1.2] with the advantages of the fluctuation approach. An advantage of fluctuation measurements relative to conventional relaxation kinetics is that no perturbation, e.g., temperature jump or rapid mixing, is required to place the system in a non-equilibrium state. (However, temperature- and pressure-jump measurements do provide reaction enthalpies and volume changes, which are not as readily obtained from FCS.) FCS derives conventional rate parameters for transport, e.g., diffusion coefficients, and for reaction processes, e.g., chemical rate constants, from measurements of spontaneous fluctuations of the numbers of molecules of specified types measured by their specific fluorescence emission. For example, a simple FCS measurement of diffusion registers fluctuations in the number of fluorescent molecules diffusing into and out of the observation region, defined by a focused laser beam, in terms of the corresponding fluctuations of emitted light.

Technical experimental challenges result from the reliance of FCS measurements on spontaneous microscopic fluctuations. The small size of the fluorescence fluctuations make them difficult to measure with high precision. Even if the individual fluctuations could be measured with infinite precision, their stochastic character demands that many fluctuations be sampled to obtain an accurate estimate of macroscopic phenomenological transport coefficients and chemical rate constants. For a molecule with a diffusion coefficient $D = 10^{-8} \, \mathrm{cm}^2/\mathrm{s}$ and using a beam radius of $2 \, \mu\mathrm{m}$, a typical fluctuation would last about $1 \, \mathrm{s}$. On average, this would be the time needed for a molecule to diffuse out of the observation region from some point near its center. Then it would be necessary to integrate the correlation function for about $1000 \, \mathrm{s}$ to observe 1000 fluctuations. The measurement system must be stable over the time required to collect the required amount of fluctuation data.

Fluorescence photobleaching recovery (FPR, also called fluorescence recovery after photobleaching, FRAP) was first developed to circumvent the stability requirements imposed by the long duration of the initial FCS measurements and thereby to enable measurements of dynamic processes in and on living cells [1.3]. In an FPR experiment, a laser irreversibly photobleaches a fraction of the fluorophores in the observation region. The recovery of fluorescence then characterizes rates of transport and chemical reaction involving the fluorescence-detected molecules in the observation region. Hence, the fundamental difference between FCS and FPR is the measurement in the latter of a macroscopic concentration change (as in a conventional relaxation kinetics experiment) in contrast to a microscopic fluctuation in FCS. In contrast to the need for many fluctuations in an FCS measurement, a single macroscopic transient, which requires approximately the same time as one fluctuation, yields the phenomenological transport coefficients and chemical rate constants to the limit of accuracy of the fluorescence recovery measurement. FCS and FPR are very closely related in both theory and experimental approach [1.4,1.5]. One disadvantage of FPR is its requirement for a photobleaching process, which typically involves uncharacterized photochemistry. FCS does not require this process. Nevertheless, photobleaching during measurement presents a serious problem for both FCS and FPR (see Chap. 13 for a discussion of the balance between suppression of photobleaching by the elimination of O_2 and the resulting enhancement of the triplet state population, which reduces emission intensity).

A crucial technical advance for FCS stems from minimizing the observation region and thereby accelerating the rate of diffusional fluctuations. Shortening the average duration of the fluctuations correspondingly reduces the measurement time and thereby substantially increases the applicability of the method to repetitive and routine measurements such as screening pharmaceuticals [1.6,1.7]. Experiments that once required FPR can now be done using FCS including studies on living cells [1.8–1.11] and as discussed in Chaps. 6 and 7. Reduction of the sample volume was facilitated by the development of more sensitive photon detectors such as avalanche photodiodes, that allowed acquisition of a fluorescence signal from a smaller number of fluorophores. Corresponding improvements in time resolution and correlation function computation have further simplified FCS measurements and extended their range. The commercial availability of an integrated instrument for FCS also substantially adds to the accessibility of this method to a wide range of laboratories (Chap. 15).

Compared to conventional methods for measuring molecular transport and chemical kinetics, FCS offers several important advantages which have contributed to the recent upsurge of interest in this method. Some of these include:

1. Changes in molecular concentration or in chemical states, e.g., conformations of biological macromolecules, arise from spontaneous fluctuations and are therefore very small. This property is potentially advantageous, for example, in resolving transitions in complex systems with states closely spaced in free energy, e.g., folding of proteins and untwisting of DNA (Chap. 4). It is also useful in simplifying the analysis of highly nonlinear systems, which for such small changes can be linearized safely. These considerations suggest the potential utility of FCS to study highly nonlinear processes such as phase changes.

2. Due to the use of laser excitation and optical microscopy systems, the region of the system sampled by the measurement is very small, limited only by the optical resolution limit of the laser excitation to $\approx \lambda/2$ ($\lambda =$ wavelength). This is especially advantageous for the analysis of small systems such as cells.

3. Fluorescence detection provides molecular specificity, which enables the observation of chemical reaction progress and the motions of labeled molecules in the presence of a vast quantity of unlabeled molecules. How-

ever, even in the absence of fluorescence changes, reaction progress in favorable systems can be measured by changes in diffusion coefficient (Chap. 11). Fluorescence also provides the possibility of measuring fluctuations in molecular or group proximity using fluorescence resonance energy transfer and of rotational mobility via polarization (Chaps. 5, 16).

4. FCS has an exceptionally wide dynamic range, covering processes with rates from $\langle\mu s^{-1}$ to $\rangle s^{-1}$. This is shown strikingly in applications of FCS to fluorophore photophysics (Chap. 13). At the fast end of the range, measurements are limited by the rate at which photons can be excited from the system; at the slow end, the limits are determined by the stability of the system being measured and the patience of the experimenter.

5. The amplitudes of fluorescence fluctuations can provide unique information about the composition of molecular systems. An interesting recent example is the use of FCS to determine the concentration of a GFP-labeled signal transducing protein in individual bacterial cells [1.12]. Higher order fluorescence fluctuation autocorrelation functions as well as photocount distribution or histogram analysis are useful for characterizing molecular interactions including aggregations and heterogeneous mixtures of fluorescent molecules (Chaps. 8, 19–21).

6. Again taking advantage of the molecular specificity of fluorescence labeling and the possibility of using two or more labels with distinct excitation and/or emission properties, the recent introduction of cross-correlation techniques provides a powerful approach to studying molecular interactions (Chaps. 9, 17 and 18).

FCS provides one avenue for entry into the field of single molecule studies [1.13,1.14] (Chaps. 4 and 22). Single molecule measurements as well as FCS center on spontaneous fluctuations of molecular states driven by thermal energy. Therefore, as for FCS, the stochastic character of measured fluorescence changes from single molecules must be analyzed in statistical terms. The essential similarity between FCS and single molecule studies is reflected in the formal similarity of the theoretical analyses of the two kinds of experiments. For example, the equations defining the correlation of concentration fluctuations in the standard FCS theory [1.15] are identical formally to those describing the statistical behavior of individual molecular conformational fluctuations (Chap. 4).

With the technical improvements introduced during the last few years (Chap. 22) it has been feasible to apply FCS to a wide range of basic and applied problems [1.16,1.17]. This volume samples a range of these applications. Földes–Papp and Kinjo (Chap. 3) provide an extensive survey of applications of FCS that can supplement or replace more conventional methods to study interactions among nucleic acid molecules. Matayoshi and Swift (Chap. 5) provide a comparison of conventional FCS and fluorescence polarization measurements to study protein-ligand interactions in solution. The two types of measurements conveniently complement each other in their advantages and disadvantages. In Chap. 2, Visser and co-workers demonstrate that FCS can be used to characterize photophysical and structural properties of photosensitive flavoproteins.

During the early days of the development of the FCS method, it was clear that the method was not yet suitable for routine biotechnological applications or for environmental analysis and monitoring. Progress in simplifying and speeding up the method and extending its range of applicability has made FCS a potentially important tool in a variety of biotechnological areas (see Chaps. 3, 9, and 10). One limitation of FCS stems from the relative insensitivity of translational diffusion measurements to the molecular weight of the diffusing molecule. Mayer-Almes (Chap. 10) describes a way of circumventing this problem for the analysis of ligand binding using nanoparticles to slow down the diffusion of one of the interacting species. However, Riesner (Chap. 11) demonstrates that conventional measurements of diffusion rate by FCS can provide useful information about protein aggregation processes, cases associated with Alzheimer's and prion diseases. In his discussion of the advantages of FCS for environmental applications, Starchev (Chap. 12) lists its high sensitivity and selectivity for specifically labeled molecules and provides examples of FCS data obtained with samples of natural organic matter. A potential disadvantage is the polydispersity of natural colloidal systems. However, this might provide a favorable area for the application of higher order autocorrelation functions and intensity distribution (histogram) analysis.

As the chapters discussed above demonstrate, the renaissance of FCS seen during the last few years has encouraged applications to a wide range of basic and applied problems. Current work is extending further the capability and range of FCS and there are many ideas for further developments. Mets (Chap. 16) provides a summary of the current status of the FCS analysis of antibunching and rotational diffusion. Although the required measurements in the nanosecond time domain are difficult, FCS has clear advantages over conventional fluorescence polarization methods, especially to the analysis of slow rotational diffusion. Finally, Webb (Chap. 14) concludes with some suggestions about future directions and challenges for FCS, including two-photon excitation, further reduction of the observation volume, and positional cross-correlation.

References

1.1 B.J. Berne and R. Pecora: Dynamic Light Scattering with Applications to Chemistry, Biology, and Physics (Wiley, New York. 1976)

1.2 M. Eigen and L.C. De Maeyer: Techniques of Organic Chemistry, edited by S.L. Freiss, E.S. Lewis, A. Weissberger (Interscience, New York, 1963) P. 895

1.3 D. Axelrod, D.E. Koppel, J. Schlessinger, E. Elson, and W.W. Webb: Biophys. J. **16**(9), 1055–1069 (1976)

1.4 E.L. Elson: Ann. Rev. Phys. Chem. **36**, 379–406 (1985)

1.5 D.E. Koppel, D. Axelrod, J. Schlessinger, E.L. Elson, and W.W. Webb: Biophys. J. **16**(11), 1315–1329 (1976)

1.6 A. Koltermann, U. Kettling, J. Bieschke, J. Winkler, and M. Eigen: Proc. Natl. Acad. Sci. USA **95**(4), 1421–1426 (1998)

1.7 S.A. Sundberg: High-throughput and ultra-high-throughput screening: solution- and cell-based approaches, Curr. Opin. Biotechnol. **11**(1), 47–53 (2000)

1.8 R. Brock, M.A. Hink, and T.M. Jovin: Fluorescence correlation microscopy of cells in the presence of autofluorescence, Biophys. J. **75**(5) 2547–57 (1998)

1.9 R. Brock and T.M. Jovin: Fluorescence correlation microscopy (FCM)-fluorescence correlation spectroscopy (FCS) taken into the cell, Cell. Mol. Biol. (Noisy-le-grand), **44**(5), 847–56 (1998)

1.10 J.C. Politz, E.S. Browne, D.E. Wolf, and T. Pederson: Intranuclear diffusion and hybridization state of oligonucleotides measured by fluorescence correlation spectroscopy in living cells, Proc. Natl. Acad. Sci. USA **95**(11), 6043–48 (1998)

1.11 P. Schwille, U. Haupts, S. Maiti, and W.W. Webb: Molecular dynamics in living cells observed by fluorescence correlation spectroscopy with one- and two-photon excitation, Biophys. J. **77**(4), 2251–65 (1999)

1.12 P. Cluzel, M. Surette, and S. Leibler: An ultrasensitive bacterial motor revealed by monitoring signaling proteins in single cells, Science **287**(5458), 1652–55 (2000)

1.13 L. Edman, Ü. Mets, and R. Rigler: Conformational transitions monitored for single molecules in solution, Proc. Natl. Acad. Sci. USA **93**(13), 6710–15 (1996)

1.14 S. Wennmalm, L. Edman, and R. Rigler: Conformational fluctuations in single DNA molecules, Proc. Natl. Acad. Sci. USA **94**(20), 10641–46 (1997)

1.15 E. Elson and D. Magde: Fluorescence Correlation Spectroscopy, I: Conceptual Basis and Theory. Biopolymers, **13**, 1–27 (1974)

1.16 M. Eigen and R. Rigler: Sorting single molecules: application to diagnostics and evolutionary biotechnology, Proc. Natl. Acad. Sci. USA **91**(13), 5740–47 (1994)

1.17 S. Maiti, U. Haupts, and W.W. Webb: Fluorescence correlation spectroscopy: diagnostics for sparse molecules, Proc. Natl. Acad. Sci. USA **94**(22), 11753–57 (1997)

Part I

FCS in the Analysis of Molecular Interactions

2 Fluorescence Correlation Spectroscopy of Flavins and Flavoproteins

Antonie J.W.G. Visser, Petra A.W. van den Berg, Mark A. Hink, and Valentin N. Petushkov

2.1 Introduction

The remarkable revival of fluorescence correlation spectroscopy (FCS), stimulated by major technical improvements, has led to novel insights into dynamical processes occurring in biological macromolecules over a large time span [2.1–2.3]. The application of FCS has enabled the retrieval of information on the properties of single biomolecules as distinguished from the collective properties of a whole molecular ensemble [2.4,2.5]. The FCS technique is not confined to clear solutions of single purified biomolecules, since the next challenging step to apply the technique to living cells has already been taken [2.6,2.7]. To date, most FCS experiments have been restricted to fluorescent dye molecules, which possess the ability to emit a large number of fluorescence photons before being destroyed. The application of FCS to natural fluorophores has been extremely limited, since the fluorescent moiety usually does not have favorable properties for FCS. The famous green fluorescent protein (GFP) [2.8] was the first example where FCS provided very specific dynamic information on a natural fluorophore [2.9,2.10]. However, in the latter two references the fluorophore in GFP was engineered to yield more brightness and a higher absorption cross section in the blue-green spectral range. The natural occurrence of GFP is limited to a few species of marine organisms.

In this chapter, we report on the initial results of applying FCS to another class of natural, biological fluorophores, namely flavins, which are the important redox-prosthetic groups in flavoproteins (Fig. 2.1). They are ubiquitous in nature and are involved in metabolic oxidation-reduction and in biological electron transport processes. Flavins in the excited state play an important role in photobiology, such as light dependent reactions in blue-light photoreception [2.11] and photorepair [2.12,2.13].

Our first aim was, not surprisingly (see below), to investigate a particular class of flavoproteins functioning as antenna proteins in bacterial bioluminescence. Bioluminescence organisms provide a rich source of fluorescent proteins. Antenna proteins participate in the bioluminescence reaction along with the flavoenzyme luciferase, thereby forming a protein-protein complex. It is proposed that antenna proteins function as acceptors of excited state

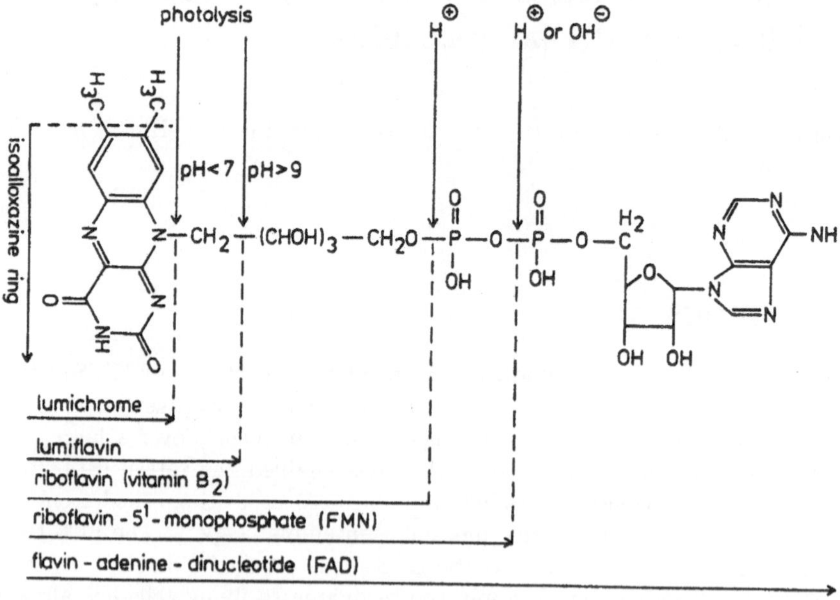

Fig. 2.1. Structure of FAD, FMN and riboflavin

energy generated in a flavin intermediate during the luciferase oxidation reaction [2.14]. The antenna proteins then emit light with high quantum yield.
The earlier quoted GFP has exactly the same antenna function in the bioluminescence emission of a number of coelenterates. In some bacterial bioluminescent bacteria the fluorescent prosthetic group is a flavin. In other photobacteria another luminophore called lumazine is abundantly present [2.15].
One flavoprotein with the same antenna function is the so-called yellow fluorescence protein (YFP). This protein is the most fluorescent flavoprotein,
having a fluorescence quantum yield of approximately 0.5–0.6 [2.16]. YFP
contains FMN as the prosthetic group and can be isolated from the bioluminescent bacterium *Vibrio fischeri*, strain Y1. From the same bacterial strain
another brightly fluorescent protein has been isolated, which has been identified with blue fluorescence protein (BFP), having a fluorescence quantum
yield of 0.5–0.6, as well [2.17]. The prosthetic group in BFP is lumazine, the
precursor metabolite in riboflavin biosynthesis [2.15]. In BFP, lumazine can
be exchanged for riboflavin [2.18].

Returning to our comparison with GFP, the distinct advantage of the
GFP-fluorophore is its covalent attachment in the central helix of the protein.
In contrast to the fluorescent group in GFP, the prosthetic groups of antenna
proteins of photobacteria are non-covalently bound to the protein matrix. In
general, most flavoproteins have the flavin prosthetic group non-covalently
bound. For antenna proteins of photobacteria this has large consequences for

the applicability of FCS. In order to reach a sufficiently high amplitude in the autocorrelation function, the antenna proteins must be diluted to nanomolar concentration, which evidently results in dissocation of the prosthetic group having an equilibrium dissociation constant on the order of 40–400 nM, depending on the type of antenna protein studied [2.15,2.16,2.19].

During our investigations, it turned out that a systematic investigation of the FCS properties of free flavins is essential to understand the photophysics of the protein-bound chromophores. In this chapter, we will give a brief phenomenological account of what we have observed. This chapter concludes with the description of some FCS experiments on bioluminescent antenna proteins.

2.2   Materials and Methods

The preparation of the antenna proteins has been described elsewhere [2.15–2.17], and the replacement of lumazine by riboflavin has been described recently [2.18]. The free flavins FMN, FAD and riboflavin were of the highest purity grade available. FAD was additionally purified from traces of other flavin contaminants by polyacrylamide size-exclusion chromatography using Biogel P2. Samples were prepared in 50 mM sodium phosphate buffer at pH 7.0.

Full details concerning the FCS experimental setup currently used by our research group can be found in [2.20,2.21]. The 488 nm Ar-ion laser line was used for excitation and the "fluorescein" filter set selected the emission wavelength range. The laser power was measured using a power meter, which was mounted in one of the objective holders in a carousel. Approximately 200 μl of sample was measured in an eight-chambered cover glass with a 0.13 mm borosilicate bottom. In later experiments, 96-well microtiter plates were used with ∼ 50 μl of sample in a well.

The fluorescence autocorrelation functions were analyzed using a fluorescence correlator data processor [2.22]. The analysis was performed using a global approach, where several autocorrelation traces were fitted simultaneously. Standard operations such as parameter fixing and constraining were implemented in the program. The quality of the fit was judged by the χ^2 criterion and by visual inspection of the residuals. The confidence intervals for each parameter can be obtained by the exhaustive search method. The built-in experimental database allowed for easy storage and management of the experimental data. The software was developed on the basis of a modular, object oriented architecture, allowing for easy extension of the model library. The data presented in this chapter has been analyzed with the "Triplet-State" model (for details, see [2.3,2.23]):

$$G(t) = 1 + \frac{(1 - F_{\text{trip}} + F_{\text{trip}}e^{-t/T_{\text{trip}}})}{N} \left(\sum_i \frac{F_i}{\left(1 + \frac{t}{T_i}\right)\sqrt{1 + \frac{t}{a^2 T_i}}} \right), \quad (2.1)$$

in which N is the average number of fluorescent molecules in the detection volume, F_{trip} and T_{trip} are, respectively, the fractional population and decay time of the triplet state (these parameters are, to a first approximation, assumed equal for all components), F_i and T_i are, respectively, the contribution and translational diffusion time of the i-th fluorescent component, and a is the "structural" parameter of the instrumental setup.

2.3   Results and Discussion

2.3.1   FCS on FMN and FAD

We observed considerable photobleaching when an FCS experiment on either riboflavin or FMN is performed. In Fig. 2.2, an example of the number of fluorescence photons emitted by FMN is shown during 30 s of constant illumination with $\sim 800\,\mu\text{W}$ of 488 nm laser light. For this reason, the applied laser power should be chosen to be as low as possible. On the other hand, the number of fluorescence photons emitted by flavins is considerably lower than that of the frequently used rhodamine dyes, implying a much longer collection time. We therefore adopted another strategy. A repetitive cycle of short FCS experiments (20 s duration) was collected and selected autocorrelation traces were globally analyzed. The selected curves were those where no loss in fluorescence intensity was observed during the experiments, indicating an apparent steady state of photobleaching and replenishment of fresh molecules.

A typical example of the aforementioned global analysis of five autocorrelation traces of FMN (out of 10) is presented in Fig. 2.3 showing clearly the amplitude of the autocorrelation function increased during the course of the experiment. This suggests that the number of molecules in the detection volume decreased due to photodestruction. The results of the analysis are collected in Table 2.1, and data from the reference dye rhodamine 6G is listed for comparison. The diffusion time T_{FMN} was recovered with remarkable precision ($40.3 \pm 0.6\,\mu\text{s}$). A distinct discrepancy exists in the number of FMN molecules in the detection volume ($N = 9$): the corresponding concentration is 20 nM, which is much lower than the initial concentration of 100 nM in this sample. The major reason is that FMN is photobleached efficiently, taking into consideration that the triplet yield is rather high (on the order of 50%), although this is not apparent from the analysis yielding a much smaller value (Table 2.1). A systematic study of this phenomenon will be presented elsewhere [2.24].

FAD is much less susceptible to photobleaching. The fluorescence of FAD is strongly quenched (the relative fluorescence quantum yield of FAD is one tenth that of FMN). The quenching mechanism of the flavin in FAD involves both dynamic and static interactions with the adenine group [2.25,2.26]. Because of this quenching, the triplet state is expectedly much less populated, resulting in reduced photobleaching. Five autocorrelation traces of FAD are

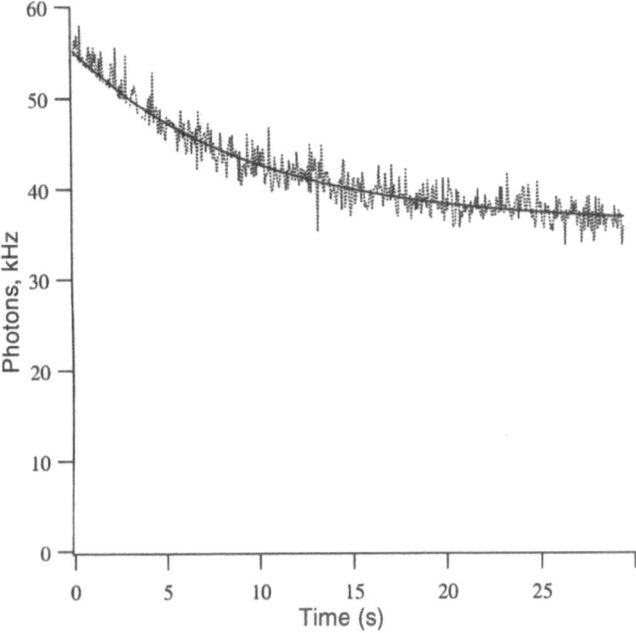

Fig. 2.2. Decrease in number of fluorescence photons emitted by FMN (input concentration 100 nM) upon illumination with the 488 nm Ar-ion laser line at 800 μW of power. The *drawn line* is a fit to the function $\{a + b \cdot \exp(-k_{\text{bleach}}t)\}$ yielding the photobleaching rate constant $k_{\text{bleach}} = 0.105 \pm 0.004\text{s}^{-1}$

Table 2.1. Parameters obtained after global analysis of FCS experiments on 100 nM FMN and FAD

Sample	Diffusion time $T, \mu s$	Triplet lifetime $T_{\text{trip}}, \mu s$	Triplet fraction F_{trip}	Number of molecules, N
FMN	40.3 ± 0.6	3.9 ± 1.3	0.088 ± 0.011	9.0 ± 0.3
FAD	54 ± 12	2.0 ± 1.3	0.196 ± 0.070	5.2 ± 1.1
Rhodamine 6G	52 ± 2	2.0 ± 0.1	0.163 ± 0.010	1.1 ± 0.1

All parameters except the number of molecules (N) are linked in a global analysis of 10 experiments of flavins. This has been repeated on two other sets of 10 experiments. All experiments presented were conducted directly after each other using the same experimental configuration. The same symbols as in (2.1) are used. The standard errors are obtained from the final results of the three sets of 10 experiments. N and the standard error of N are determined from six initial experiments of flavins and 10 experiments of rhodamine 6G in water. The structure parameter was determined from the rhodamine 6G data and fixed in the global analysis: $a = 7.6$. The detection volume element $V = 0.67 \text{ fL}$.

14   A.J.W.G. Visser et al.

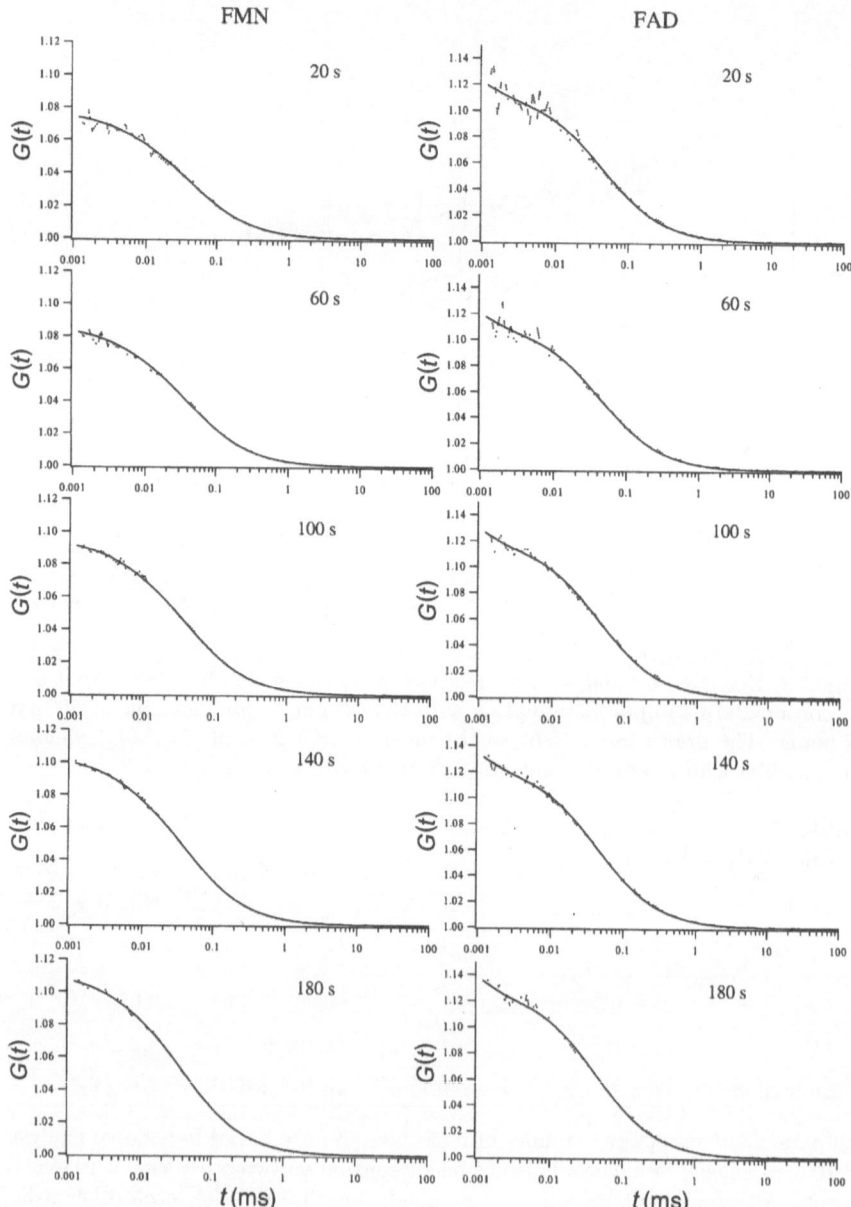

Fig. 2.3. *Left panel,* Example of global analysis of five experimental and fitted autocorrelation functions (out of 10) of FMN (input concentration 100 nM) collected sequentially over 20 s each at given time intervals. *Right panel,* the same traces but now for FAD (input concentration 100 nM). The laser power at 488 nm was 340 μW

shown in Fig. 2.3 as well, for comparison with the results of FMN. The traces are collected at the same time intervals. The amplitude remains almost unaffected, illustrating that during the course of the experiments no significant photobleaching took place. Again, the number of FAD molecules in the detection volume ($N = 5$, which is equivalent to 11 nM) is less than expected from the concentration of FAD used (100 nM). The recovered diffusion time of FAD, T_{FAD}, is longer (and less precise) than found for FMN ($T_{FAD} = 54 \pm 12\,\mu s$). Taking the differences in molecular mass (FMN = 456 Da, FAD = 830 Da) into account, one would expect a diffusion time $T_{FAD} = 49\,\mu s$, which is close to the observed value.

Another observation gleaned from Fig. 2.3 is that the apparent triplet fraction is significantly higher than that found for FMN and, in addition, the apparent triplet lifetime is shorter. In view of the argument presented above, an enhanced triplet yield and shorter triplet lifetime are not very likely for FAD. Therefore, another dynamic process must play a role. Since FAD exists for $\sim 80\%$ of the time as an intra-molecular complex between flavin and adenine [2.25,2.26], this dynamic process is most likely the stacking and unstacking of the two aromatic moieties. This is schematically depicted in Fig. 2.4.

An interesting related FCS investigation of the kinetics of conformational fluctuations in DNA hairpin-loops has been recently reported [2.27]. In analogy to the situation in FAD, the opening and closing of the loop was studied by using a so-called DNA molecular beacon, which consisted of a fluorescent probe at one end and a quencher attached to the other end of the loop. The autocorrelation function of the beacon (G_{beacon}) contained contributions of both diffusion and chemical kinetics. A DNA hairpin without quencher was used as a control sample, for which the correlation function (G_{control}) consisted of diffusion only. The ratio $G_{\text{beacon}}(t)/G_{\text{control}}(t)$ isolated the kinetics

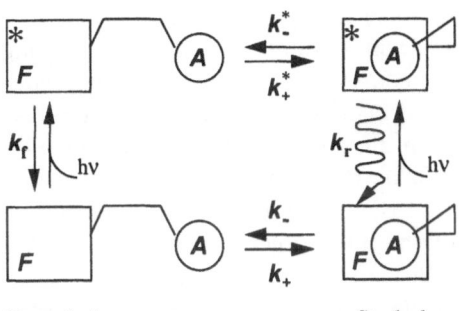

Unstacked   Stacked

Fig. 2.4. Schematic view of stacking (rate constant k_+) and unstacking (rate constant k_-) of the flavin (*rectangle*) and adenine (*circle*) parts of FAD. In the open conformation the flavin moiety is fluorescent (fluorescent rate constant k_f), whereas the flavin fluorescence is strongly quenched in the closed conformation (non-radiative rate constant k_r)

part, and yielded the chemical reaction rate $1/\tau_{\text{reaction}} = k_- + k_+$, where k_- is the opening and k_+ the closing rate constant of the loop Fig. 2.4 depicts the analogous situation in FAD. Another (bulk) fluorescence experiment gave the equilibrium constant $K = k_-/k_+$, enabling both rate constants to be determined separately.

Exactly the same procedure has been applied to the ratio of autocorrelation functions of FAD and FMN (control) and the results are presented in Fig. 2.5a, b. It is very clear from the trace in Fig. 2.5b that two kinetic processes are present: a very fast one in the sub-microsecond time scale and a slower one in the microsecond time scale. It is very probable that the apparently slow process is due to the slightly different diffusion times of FAD and FMN and some changes in triplet population and kinetics. This conclusion is supported via simulations of the ratio of the autocorrelation functions of FAD and FMN differing in diffusion time and triplet yield (several examples are shown in Fig. 2.5c). Therefore, the rapid kinetic process most likely reflects the stacking and unstacking of FAD. In principle, the fast part of the ratio can be fitted to an exponential function:

$$G_{\text{ratio}}(t) = G_{\text{FAD}}(t)/G_{\text{FMN}}(t) = A + B\exp(-t/\tau_{\text{reaction}}). \qquad (2.2)$$

However, it is important to note that the FCS experiment was carried out with a relatively slow correlator. The decay of the autocorrelation function only encompasses two points in the time scale: the first point is at 200 ns and the second point is at 400 ns, when the fast decay is over. In order to recover a more reliable τ_{reaction}, the experiments must be repeated with a faster correlator. Nonetheless, we can set an upper limit of the reaction time constant: $\tau_{\text{reaction}} \approx 200$ ns. Since the equilibrium constant of the intramolecular FAD complex is known ($K = 0.2$) [2.25], the individual rate constants can be directly determined: $k_- \approx 0.8\,(\mu\text{s})^{-1}$ and $k_+ \approx 4.2\,(\mu\text{s})^{-1}$.

The latter rate constants are larger than those found for the opening and closing rates of the larger DNA hairpin-loops [2.27]. From time-resolved fluorescence measurements on FMN and FAD one obtains the dynamic quenching constant k_+^*. This rate constant can be conceived of as the rate of collisions between the adenine and flavin moiety during the lifetime of the excited state leading to ultra-rapid quenching of flavin fluorescence (the superscript * indicates an excited-state process) [2.25,2.26]. By assuming the same rate constant in the ground state (k_+) and from the known equilibrium constant, it is possible to estimate the lifetime of the intra-molecular complex in FAD ($1/k_-$), which turns out to be only ~ 30 ns from dynamic fluorescence measurements. From FCS experiments at higher time resolution the reaction time constant (τ_{reaction}) can be obtained directly, from which the lifetime of the complex can be obtained. Using the crude assumptions of this experiment the lifetime of the complex is now much longer, $\sim 1.2\mu\text{s}$.

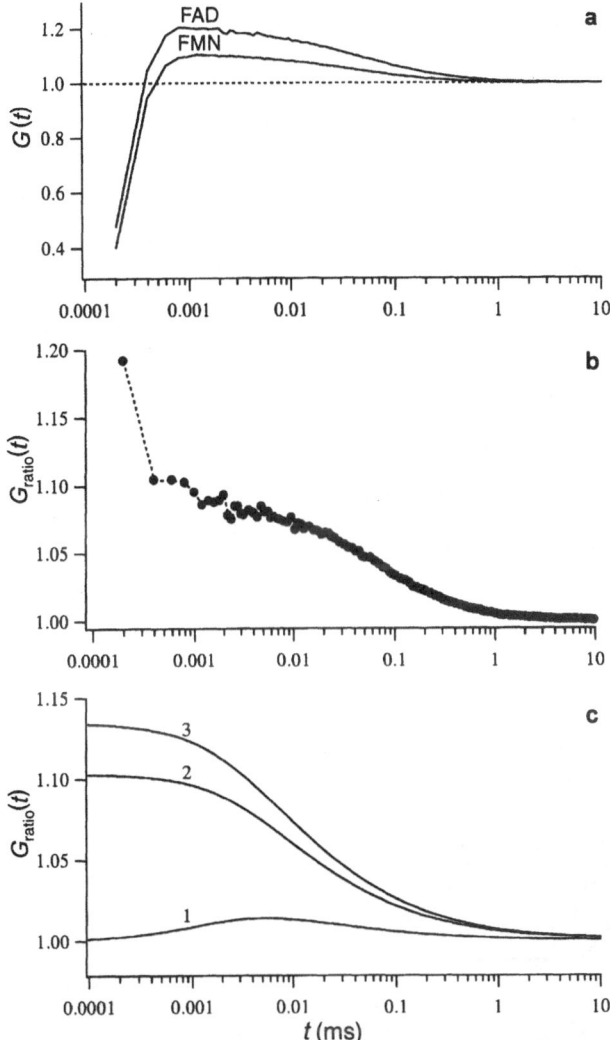

Fig. 2.5. a Autocorrelation functions (G) of FMN and FAD (the average of 20 experimental traces), and **b** the ratio of both autocorrelation functions $G_{\text{ratio}} = G_{\text{FAD}}/G_{\text{FMN}}$. **c** Simulations of the ratio of autocorrelation curves for FAD and FMN (G_{ratio}). The curves were created using the parameters obtained from the experimental data as listed in Table 2.1. *Curve 1* $N = 5.2$, $T = 54\,\mu\text{s}$ for FAD; $N = 5.2$, $T = 40\,\mu\text{s}$ for FMN; $T_{\text{trip}} = 2\,\mu\text{s}$, $F_{\text{trip}} = 0.196$ for both. *Curve 2* same FAD and triplet parameters as for curve 1; $N = 9$, $T = 40\,\mu\text{s}$ for FMN. *Curve 3* same FAD parameters as for curve 1 and same FMN parameters as for curve 2; $T_{\text{trip}} = 3.9\,\mu\text{s}$, $F_{\text{trip}} = 0.088$ for both. As can be clearly seen from the simulations, different N values for FAD and FMN have the largest impact on the shape of the curve

2.3.2   FCS on YFP and BFP

The absorption and fluorescence emission spectra of YFP and its prosthetic group FMN are presented in Fig. 2.6. There are significant shifts in absorption and emission maxima of free and bound FMN. YFP is very efficient at absorbing 488 nm laser excitation wavelength. In the FCS experiment we used a relatively high YFP concentration of 250 nM to ensure that a measurable fraction of FMN was bound. One example of experimental and fitted autocorrelation functions is shown in Fig. 2.7. As in the free flavin experiments, the correct YFP concentration is not recovered ($N = 18$ is equivalent to 27 nM in that experiment). A global analysis of three experiments resulted in an apparent diffusion time $T_{\text{apparent}} = 60\,\mu\text{s}$, which is longer than the 53 μs of free FMN, but much shorter than the diffusion time of $\sim 150\,\mu\text{s}$ expected for a 20 kDa protein. A two-component analysis with fixed diffusion times of 53 μs and 150 μs showed indeed that $88 \pm 7\%$ belongs to free FMN. The reason for this apparent anomaly is explained below (see experiments on BFP). Since YFP is less stable than BFP [2.18] and less readily available, more elaborate FCS studies were performed with BFP.

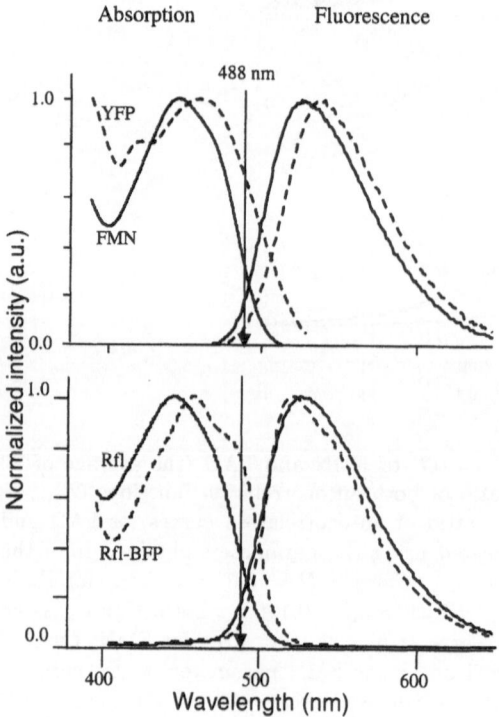

Fig. 2.6. Visible absorption and technical fluorescence spectra of YFP and FMN and of Rfl-BFP and riboflavin (Rfl). The *arrow* denotes the 488 nm Ar-ion laser line

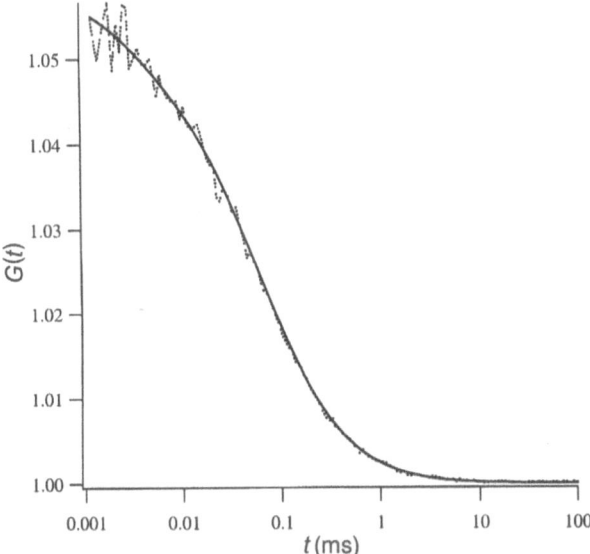

Fig. 2.7. Experimental (average of three traces) and fitted autocorrelation functions (G) of YFP (input concentration 250 nM). A two-component fit to the experimental data is shown (for details see text). The laser power at 488 nM was 800 μW

A systematic FCS investigation has been performed on BFP, which has been recombined with riboflavin (Rfl-BFP). The laser intensity and the protein concentration were varied. The absorption and fluorescence spectra are shown in Fig. 2.6. As with YFP, Rfl-BFP efficiently absorbs at 488 nm. The first result was that experiments using relatively low laser intensity (in combination with rather high protein concentration) lead to the recovery of the correct diffusion time and the correct concentration (Table 2.2). Relatively long data acquisition times (180 s) were required, but global analysis of several autocorrelation traces gave consistent results.

Three other FCS experiments are presented in Fig. 2.8. The first experiment (Fig. 2.8a) shows that the analysis of the variation of the free riboflavin concentration in combination with relatively high laser power yielded the correct diffusion time ($T_{\text{riboflavin}} = 52 \pm 1\,\mu\text{s}$), but failed to recover the correct number of molecules (Table 2.2). The diffusion time of the dye Oregon Green measured under comparable conditions was $72 \pm 1\,\mu\text{s}$. The experiments suggest that riboflavin-like FMN either undergoes photodestruction or is trapped in the triplet state leading to an apparently lower concentration [2.23].

The second experiment (Fig. 2.8b) shows the effect of Rfl-BFP concentration at relatively low laser power. The correct concentrations can now be retrieved and a two-component fit indeed shows the dissociation of riboflavin at lower protein concentration (Table 2.2). The third experiment (Fig. 2.8c)

Table 2.2. Parameters obtained after global analysis of FCS experiments on riboflavin (Rfl) and BFP recombined with riboflavin (Rfl-BFP)

Sample	F_1 (%)	T_1 (μs)	F_2 (%)	T_2 (μs)	N	Meas. conc. (nM)	Input conc. (nM)	Laser power (μW)
Rfl	100	52 (51–53)	-	-	2.6 ± 0.1	3.2	60	800
Rfl	100	52 (51–53)	-	-	1.7 ± 0.1	2.1	30	800
Rfl	100	52 (51–53)	-	-	1.2 ± 0.1	1.5	10	800
Rfl-BFP	100	180 (92–357)	-	-	190 ± 13	243	250	15
Rfl-BFP	100	180 (92–357)	-	-	95 ± 4	121	120	15
Rfl-BFP	100	180	-	-	62 ± 7	79	60	15
Rfl-BFP	52 ± 5	52	48 ± 5	180	82 ± 1	105	100	50
Rfl-BFP	65 ± 20	52	35 ± 20	180	49 ± 1	63	60	50
Rfl-BFP	79 ± 1	52	21 ± 1	180	27 ± 1	34	30	50
Rfl-BFP	100	52	-	180	7.8 ± 0.2	10	100	800
Rfl-BFP	86 ± 4	52	14 ± 4	180	44 ± 1	56	100	160
Rfl-BFP	57 ± 2	52	43 ± 2	180	81 ± 1	103	100	50

All parameters, except the number of molecules (N) and the relative fractions F_1 and F_2, are linked in a global analysis of six experiments (each three consecutive rows). The same symbols as in (2.1) are used. The values in parentheses are obtained after a detailed error analysis at the 67% confidence level. N and the standard error of N, as well as the fractions F_1 and F_2, are determined from two experiments (each single row). The latter six entries in the table arose from a two-component analysis with fixed diffusion times. The structure parameter was determined from data of the dye Oregon Green and fixed in the global analysis: $a = 9.1$. The detection volume element $V = 1.3$ fl.

shows the effect of laser intensity on a 100 nM sample of Rfl-BFP. At relatively high laser power the remarkable result is that only the diffusion time of free riboflavin is seen, whereas at relatively low laser intensity mainly bound riboflavin is observed (Table 2.2). Therefore, at high laser intensity the riboflavin apparently dissociates from the protein and does not recombine. It

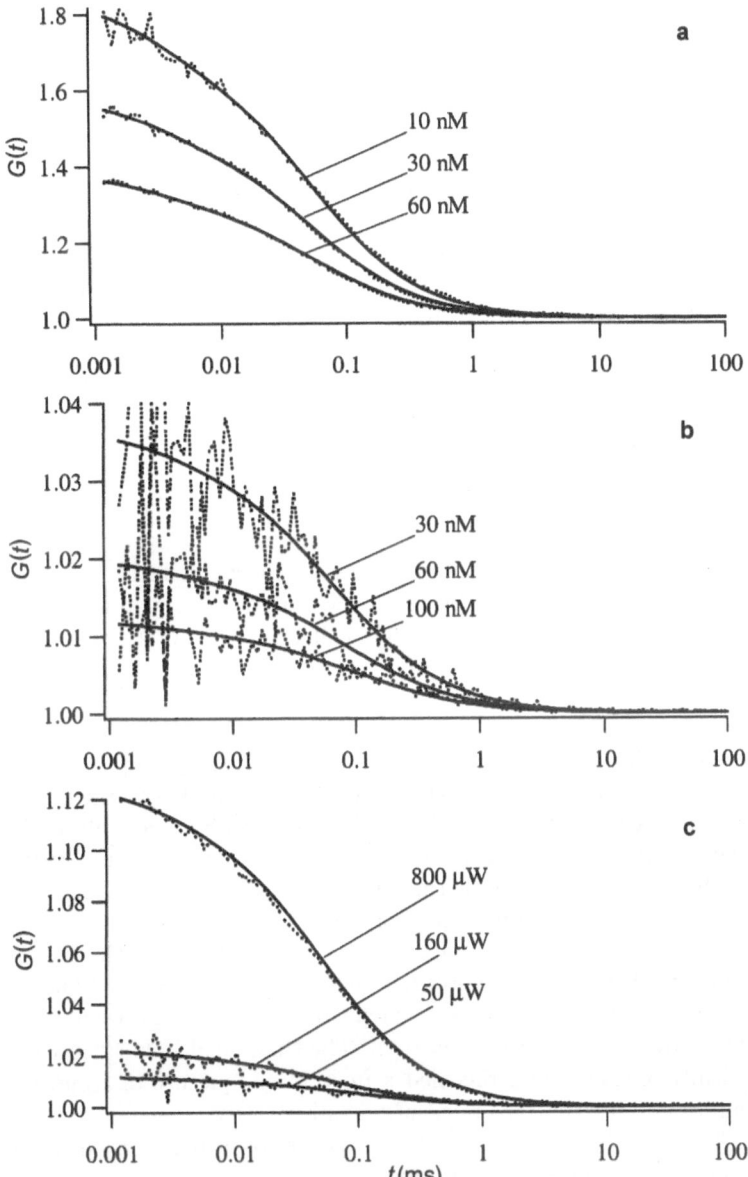

Fig. 2.8. Three FCS experiments on riboflavin (Rfl) and Rfl-BFP with laser excitation at 488 nm. **a** Three different concentrations of Rfl using 800 µW laser power. **b** Three different concentrations of Rfl-BFP using 50 µW of laser power. **c** The effect of laser power on the autocorrelation function of 100 nM Rfl-BFP. All parameter values are collected in Table 2.2

is known that, in contrast to flavoproteins with a specific redox function, reduced flavins do not bind to antenna proteins. Therefore the following sequence of events is proposed for the photo-induced dissociation of riboflavin and FMN from the proteins:

$$\text{Rfl-BFP} \rightarrow \text{Rfl}^*\text{-BFP} \rightarrow \text{Rfl}^T\text{-BFP} \rightarrow \text{RflH} \ldots$$
$$^+\text{BFP} \rightarrow \text{RflH} + {}^+\text{BFP} \rightarrow \text{RFl} + {}^+\text{BFP}$$

Step one is excitation in the first singlet state, step two is triplet formation, step three is hydrogen abstraction in the protein to form reduced riboflavin, and the final step is dissociation of reduced riboflavin from the protein followed by immediate oxidation to normal riboflavin. The modified oxidized protein is not able to rebind riboflavin any longer. In the experiment using YFP apparently this photodissociation has taken place, since relatively high laser power ($800\,\mu\text{W}$) was used.

2.4   Conclusions

The set of experiments described here provides prospects for further research on flavins and flavoproteins. The direct observation of the stacking and unstacking of both aromatic moieties in FAD will be a very interesting topic, provided that the experimental conditions are further improved. The conditions to enable observation of FCS of antenna proteins need to be optimized. The availability of shorter wavelength laser excitation wavelengths are very suitable for the excitation of lumazine proteins. An alternative option would be two-photon FCS. The obvious next step will be to use FCS to investigate the interactions of the antenna proteins with luciferase. Such an interacting system is then composed of a natural reporter group. It would be interesting to compare these FCS results with the single molecule fluorescence experiments conducted with the flavoprotein cholesterol oxidase [2.28]. In the latter experiments the flavoprotein was trapped in a rigid matrix and the fluorescence photons of FAD of the immobilized enzyme were monitored. The fluorescence showed stochastic blinking behavior during enzymatic turnover as FAD is reduced by the substrate cholesterol. The single-molecule trajectories were statistically analyzed and revealed a microscopic picture of conformational dynamics and fluctuations in reaction rate.

In the FCS experiments described here the flavoproteins are diffusing freely in solution, and any enzymatic turnover events must be distinguished from those arising from diffusion. However, their peculiar photochemical properties make single-molecule detection of flavoproteins during turnover conditions with FCS difficult. In the latter case one must develop other (labeling) strategies to enable observation of single-flavoenzyme dynamics. One approach is to label a flavoprotein at a specific site with a donor fluorophore having a large absorption cross section and which can sense the flavin chromophore via fluorescence resonance energy transfer (FRET). In this measurement setup the detected donor fluorescence will reach a maximum number of emission photons during turnover conditions, because FRET will not occur when the flavin is reduced, since it does not have absorption spectral overlap with the donor fluorescence. The FCS experiment will therefore give information on the turnover rate of an individual enzyme in action.

Acknowledgments

The support of a fellowship from The Netherlands Organization of Scientific Research (to V.N.P.) is gratefully acknowledged. We thank John Lee (University of Georgia) for stimulating discussions.

References

2.1 M. Eigen and R. Rigler: Proc. Natl. Acad. Sci. USA **91**, 5740–5747, (1994)

2.2 S. Maiti, U. Haupts, and W.W. Webb: Proc. Natl. Acad. Sci. USA **94**, 11753–11757 (1997)

2.3 J. Widengren and R. Rigler: Cell. Mol. Biol. **44**, 857–879 (1998)

2.4 L. Edman, U. Mets and R. Rigler: Proc. Natl. Acad. Sci. USA **93**, 6710–6715 (1996)

2.5 S. Wennmalm, L. Edman, and R. Rigler: Proc. Natl. Acad. Sci. USA **94**, 10641–10646 (1997)

2.6 J.C. Politz, E.S. Browne, D.E. Wolf, and T. Pederson: Proc. Natl. Acad. Sci. USA **95**, 6043–6048 (1998)

2.7 R. Brock, M.A. Hink, and T.M. Jovin: Biophys. J. **75**, 2547–2557 (1998)

2.8 R.Y. Tsien: Ann. Rev. Biochem. **67**, 509–544 (1998)

2.9 U. Haupts, S. Maiti, P. Schwille, and W.W Webb: Proc. Natl. Acad. Sci. USA **95**, 13573–13578 (1998)

2.10 A.J.W.G. Visser and M.A. Hink: J. Fluoresc. **9**, 81–87 (1999)

2.11 C. Lin, D.E. Robertson, M. Ahmad, A.A. Raibekas, M.S. Jorns, P.L. Dutton, and A.R. Cashmore: Science **269**, 968–970 (1995)

2.12 S.P. Jordan and M.S. Jorns: Biochemistry **27**, 8915–8923 (1988)

2.13 S.T. Kim, P.F. Heelis, T. Okamura, Y. Hirata, N. Mataga, and A. Sancar: Biochemistry **30**, 11262–11270 (1991)

2.14 J. Lee: Biophys. Chem. **48**, 149–158 (1993)

2.15 J. Lee, I.B.C. Matheson, F. Müller, D.J. O'Kane, J. Vervoort, and A.J.W.G. Visser: In: Chemistry and Biochemistry of Flavoenzymes, Vol. II. (F. Müller (ed.)), 109–151. (CRC Press, Boca Raton, Fl. 1991)

2.16 A.J.W.G. Visser, A. van Hoek, N.V. Visser, Y. Lee, and S. Ghisla: Photochem. Photobiol. **65**, 570–575 (1997)

2.17 V.N. Petushkov and J. Lee: Eur. J. Biochem. **245**, 790–796 (1997)

2.18 V.N. Petushkov, B.G. Gibson, A.J.W.G. Visser, and J. Lee: Methods Enzymol. **305**, 64–180 (2000)

2.19 A.J.W.G. Visser and J. Lee: Biochemistry **19**, 4366–4372 (1980)

2.20 M. Hink and A.J.W.G. Visser: In: W. Rettig, B. Strehmel, S. Schrader, and H. Seifert (eds.) Applied Fluorescence in Chemistry, Biology and Medicine. 101–118 (Springer-Verlag, Berlin, 1998)

2.21 M.A. Hink, A. van Hoek, and A.J.W.G. Visser: Langmuir **15**, 992–997 (1999)

2.22 A.V. Digris, V.V. Skakoun, E.G. Novikov, V.V. Apanasovich, M.A. Hink, and A.J.W.G. Visser: Manuscript in proparation

2.23 J. Widengren, Ü. Mets, and R. Rigler: J. Phys. Chem. **99**, 13368–13379 (1995)

2.24 P.A.W. van den Berg, J. Wildengren, M.A Hink, R. Rigler, and A.J.W.G. Visser, submitted

2.25 R.D. Spencer and G. Weber: In: Å. Åkeson and A. Ehrenberg (eds.) Structure and Function of Oxidation Reduction Enzymes. 393–399 (Pergamon Press, Oxford, 1972)

2.26 A.J.W.G. Visser: Photochem. Photobiol. **40**, 703–706 (1984)

2.27 G. Bonnet, O. Krichevsky, and A. Libchaber: Proc. Natl. Acad. Sci. USA **95**, 8602–8606 (1998)

2.28 H.P. Lu, L. Xun, and X.S. Xie: Science **282**, 1877–1882 (1998)

3 Fluorescence Correlation Spectroscopy in Nucleic Acid Analysis

Zeno Földes-Papp* and Masataka Kinjo

3.1 Introduction

Laser induced fluorescence correlation spectroscopy (FCS) is a new, developing branch in nucleic acid analysis. Interest in FCS has increased enormously over the past few years [3.1–3.3] as a result of the introduction of the confocal volume of detection [3.4–3.6]. Fluorescent molecules are observed in a very small volume element of $\approx 1\,\mathrm{fl}$ ($10^{-15}\,\mathrm{l}$) excited by a well-focused laser beam. The femtoliter "cavity" defined by the laser is either stationary or scans through the solution or matrix [3.7–3.13]. The emitted fluorescence bursts are detected from individual molecules. This enables a single fluorescent molecule in solution or on a matrix at room temperature to be measured with a negligible background. The principles of the identification nucleic acids target sites are linked closely to the concept of thermal fluctuations [3.14–3.17]. Due to thermodynamic fluctuations the number of molecules, for example, will fluctuate in the observed volume element. For small numbers N of molecules, Poisson statistics are valid and the statistical parameters, mean value, variance and standard deviation, can be determined. The variance is directly related to the number of molecules. In other words, examination of the thermal fluctuations is sufficient to obtain the absolute number of molecules without any calibration. If the biochemical relaxation time τ_{chem} is much larger than the characteristic diffusion time τ_{Diff}, then the biochemical reaction is not equilibrated during diffusion through the three-dimensional volume element. Under this condition the normalized correlation function of the intensity fluctuations contains other information regarding the biochemical kinetics of the system. We obtain immediately the size of the components in terms of their diffusion times of the molecules that occupy the volume element and their relative fraction during the time course of the reaction.

Beyond the thermal motion of fluorescent molecules and their reaction kinetics, FCS is a powerful tool for studying conformational fluctuations of single nucleic acid molecules immobilized on a surface or in a matrix [3.7–3.12,3.18–3.20]. Most advances in nucleic acid analysis were related to the

* Corresponding author. Present address: Clinical Immunology and Jean Dausset Laboratory, Graz University M.S. and Hospital, Auenbruggerplatz 8, A-8036 Graz, LKH, Austria, e-mail: Zeno.Foldes-Papp@kfunigraz.ac.at

development of new techniques. As we shall demonstrate in the following sections, there has been outstanding progress in the development of FCS-based methods and techniques that focus on nucleic acid analysis.

The original techniques in nucleic acid analysis such as Southern blots [3.21] (DNA analysis) and Northern blots [3.22] (RNA analysis) led to a myriad of related hybridization and quantitation approaches. Edwin M. Southern combined the restriction enzyme technology [3.23,3.24], the resolution of gel electrophoresis assays, and the binding capacity of nitro-cellulose membranes for DNA to provide a sensitive and simple means for detecting specific DNA sequences. Although enzyme immunoassays, chemiluminescence-based detection probes, and hybridization probes have almost completely replaced Southern blot and slot-blot methods as diagnostic tools, the rationale for gel transfers, colony and plaque lifts is still of great importance for analyzing recombinants as well as clinically relevant viruses, for example, as an adjunct procedure to polymerase chain reaction (PCR).

In today's drug research, the way to discover bioactive compounds with improved or tailor made properties is to synthesize a large series of derivates (libraries) with high diversity by combinatorial approaches. All of the compounds are screened in microarrays by fluorescence-based assays. These assays can be hybridization and binding assays, enzyme preparations, or in vitro cell cultures. Laboratory procedures are applied to microstructures that are based on the principles of microelectronics [3.25]. By going to microscopic dimensions the miniaturization of bioanalytical procedures and detection strategies leads to new horizons. For example, DNA sequencing, that is the determination of the linear pattern of nucleotides, is one of the key analytical tools of molecular biology and consequently databases and gene discoveries are growing exponentially.

In the field of the human genome about three billion nucleotides are being determined to create the complete human genetic map. For widespread biomedical applications such as medical diagnostics, individual comprehensive human gene maps, that are cross referenced, of all genes are mandatory. At present, the sequence variability of the human genome individuals is beyond the scope of classic as well as existing large scale sequencing technologies. Developing new, faster sequencing technologies is one of the most important challenges of modern biotechnology and molecular biology. The deconstructing of single, high density fluorescence labeled DNA molecules by an exonuclease has the potential to speed up the whole sequencing process [3.26]. Labeling of DNA with one or more fluorescent groups can be achieved by different approaches. The labeling strategies are quite varied depending on the particular usage of the DNA. Exonuclease degradation, restriction enzyme cutting and DNA sizing provide core capabilities of molecular biology. Key to these approaches is the development of new techniques for detection and quantitation. Nowadays, PCR-based methods are the dominating tools for gene quantification [3.27]. Techniques and methodologies developed during the past few years have expedited virtually every area of life sciences, including clinical diagnostics, drug discovery, forensics, and evolutionary biology. These areas appear to be the main fields for mutation-function analysis by FCS.

3.2   Oligonucleotide-Target Interactions

Hybridization is based on the physico-chemical properties of nucleic acids. The interaction of a DNA/oligonucleotide-duplex includes electrostatics, solvent forces, base stacking, and base pairing [3.28]. Electrostatic contributions such as dipole-dipole and London-dispersion forces are reasonable well understood, both theoretically and experimentally. Bulk solvent contributions are not so pronounced. The contributions of base stacking and pairing to oligonucleotide duplex formation have been extensively studied on small as well as large sets of oligonucleotides. They depend on oligonucleotide length, sequence, base composition, base mismatches, concentration and hybridization solvent. From these measurements, kinetic and thermodynamic parameters of oligohybridization can be obtained by FCS, such as association and dissociation rate constants as well as melting temperatures. By contrast, there are many empirical equations for calculating thermodynamic properties in nucleic acid analysis. These formulae are more or less applicable to special experimental situations. The temperature dependent behavior of the DNA/oligonucleotide-duplex structure can be predicted from a complete thermodynamic library of all 10 Watson–Crick DNA nearest neighbor interactions [3.29]. It is largely these neighbor interactions that determine the local three-dimensional structure imparted by the sequence [3.30].

The dependence of the rate of re-association on the concentration of two strands was measured by FCS. The fluorescence autocorrelation function in this situation is generally of the form [3.31]

$$G_g(\tau) = \frac{1}{N[(1-y)+yQ]^2}[(1-y)\text{Diff}_{\text{free}} + yQ^2\text{Diff}_{\text{bound}}] + 1\,, \qquad (3.1)$$

with

$$\text{Diff} = \left(\frac{1}{1+(\tau/\tau_g)}\right)\sqrt{\frac{1}{1+(\omega_{x,y}/\omega_z)^2(\tau/\tau_g)}}\,.$$

Q corrects for the quantum yield difference of a free dye and a bound dye. For example, a weighted average of the photochemical lifetimes of an internally incorporated dye was calculated and decreased to less than a third of its free nucleotide dye value [3.32]. The redox potential between rhodamine dye molecules and purine, as well as pyrimidine bases, can sustain an electron transfer from, for example, a guanine to the dye [3.8,3.9,3.33,3.34]. Therefore, fluorescence emitted from tetramethylrhodamine residues is quenched in probes where the dye is located, such that the secondary structure of the probe brings the guanosine nucleotide into the vicinity of the dye. Obviously, this effect is independent of the nucleotide length of dye-oligonucleotide or dye-DNA, but in fact it depends on the base sequence and on the number of atoms by which the dye is linked to the probe.

Rapid and versatile methods have been developed for the synthesis of oligonucleotides which contain an aliphatic a the 5'-end via a linker. Six-atom carbon linkers are used in standard chemical synthesis of dye-oligonucleotides. This amino group reacts specifically with a variety of electrophiles allowing fluorescent dyes to be attached to the oligonucleotides. The modified oligonucleotides are highly fluorescent and retain their ability to prime specifically DNA. For example, an 18-mer oligodeoxyribonucleotide directed against single stranded M13mp18(+) DNA was 5'-tagged with fluorescent dyes [3.35]. Here, the quantum yield difference between a free oligonucleotide-dye and a bound oligonucleotide-dye was negligible and Q becomes 1 in (3.1). For this case, Fig. 3.1 a shows the change in the autocorrelation function between a free and a bound fluorescent labeled oligonculeotide hybridized to the single stranded template DNA. Since the difference in translational diffusion times between a free and a bound oligonucleotide was well identified, the fraction of the hybridized oligonucleotide can be quantified by a two-component model of autocorrelation function, as shown in Fig. 3.1 b. The Cot-analysis revealed the rate constants k for the re-associations at different hybridization temperatures as well as the stoichiometry of the hybridization process (Fig. 3.1 b). The activation energy was also evaluated from the temperature dependence of k.

Growing interest in gene targeting led to many nucleic acid analogs which can be used as small hybridization probes. In peptide nucleic acids (PNA) the entire sugar-phosphate backbone has been replaced by the flexible chain of peptide-like N-(2-aminoethyl) glycine units. PNAs can sequence-specifically recognize both DNA and RNA [3.36]. Kinetic FCS measurements were performed with the DNA-mimic PNA [3.37]. The FCS study gave insight into the thermodynamic features and mechanical aspects of the ligand reaction. Reaction pathways were identified and the time scale of the processes was determined. The hybridization was about 100-fold faster than the oligonucleotide displacement at room temperature measured with the single stranded DNA target M13mp18^{+} and a fluorescently labeled 18-mer PNA, a sequence that was identical to the sequencing primer of the template DNA.

In nucleic acid research, hybridization methods are used in order to detect specific DNA and RNA sequences. A specific probe (oligonucleotide, DNA or RNA) recognizes its complementary sequence in the target DNA and/or RNA. Fluorescently labeled DNA oligonucleotides (19mer–23mer) were used as probes for the HIV virus target RNA (101 base) [3.38]. Although the changes in the diffusion times of free primers (0.13–0.2 ms) and bound primers (0.37–0.50 ms) were not so large, characteristic time-dependent properties were observed in the primers. Antisense oligonucleotides can be designed to hybridize to specific mRNAs, thereby inhibiting protein expression.

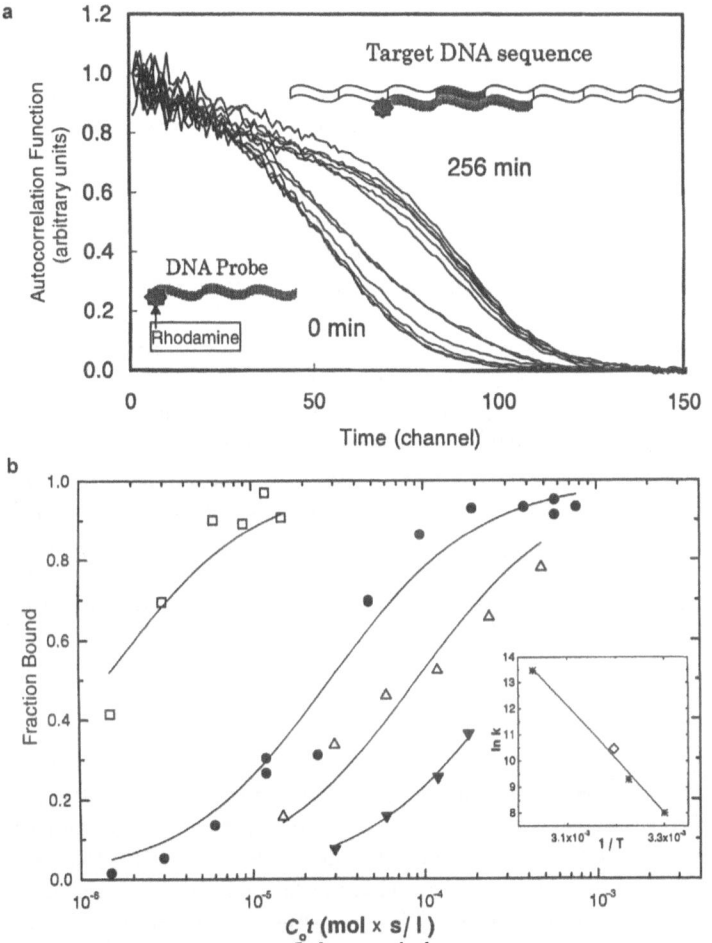

Fig. 3.1. a The increasing of the translational diffusion time in the case of the hybridization process of probe DNA associated to the target DNA sequence. **b** Time course of hybridization between the universal sequencing primer (50 nM) and ssM13 DNA (50 nM) in 0.18 M NaCl. The reaction was carried out with the Bodipy-labeled primer at 40 °C (*solid circle*), with the tetramethylrhodamine labeled primer at 57 °C (*open square*), 37 °C (*open up triangle*) and 30 °C (*solid up triangle*). A *solid line* indicates the second order fitting versus time using $1 - [1/(k \cdot C_0 t + 1)]$, where k = rate constant for association, C_0 = initial concentration of single strand primer and single strand template DNA, respectively. The association rate constants obtained at 30 °C, 37 °C, 40 °C and 57 °C were 3×10^3, 11×10^3, 35×10^3 and 716×10^3 $M^{-1} s^{-1}$, respectively. The activation energy (38.8 kcal/mol) was calculated from an Arrhenius plot (insert) taking the experimentally determined k values for the sample with the tetramethylrhodamine labeled primer (*asterisk*) as well as the sample with the Bodipy labeled primer (*open diamond*). Reproduced from [3.35]

Once hybridized to their RNA target, the oligonucleotide can block access and binding of various factors such as ribosomes [3.39,3.40]. The heteroduplex formed can be degraded by RNase H. The approach enables a level of selectivity that is not available with traditional drugs and is working in several animal models [3.41]. Antisense act as anticancer and antiviral agents. However, the low capacity for passive diffusion across cell membranes is one of the limitations with antisense oligonucleotides as therapeutics [3.42]. There are several promising strategies to circumvent both permeability and stability problems [3.43,3.44]. For example, nanoparticles can be used as carriers for these molecules.

Hybridization reactions are commonly analyzed by electrophoresis or liquid chromatography techniques. Electrophoretic separations in agarose and polyacrylamide gels have several drawbacks such as poor resolution and lower sensitivity for long, natural DNAs. Capillary electrophoresis and capillary zone electrophoresis have been applied to separate and quantify hybridization mixes, but it remains a time-consuming task of optimizing the separation conditions. Presently, expression monitoring provides the best quantitative results for practical applications. However, all of the approaches face competition from FCS. Recently developed FCS allows detection and quantification of hybridization products without the necessity of bound-free physical separation by gel electrophoresis, filters, or column chromatography. Optical separation by FCS relies upon a reasonable difference in diffusion times between the free and bound dye-oligonucleotides. An approach is described for performing multi-component analysis using a combination of capillary electrophoresis and FCS [3.45]. The mixture of analytes flowed continuously through a microcapillary in the presence of an electric field. The transition times measured by FCS depended on the electrophoretic mobilities of the analytes as well as on the applied field. This method will be useful for probe-target binding assays in which the diffusion times of the free and bound probe molecules are not significantly different [3.45]. The sensitivity of FCS given by confocal volume detection leads to an enormous reduction in sample consumption and analysis time, and enables high-throughput screening in microcapillaries, microarrays or biochips.

The possibility of detecting DNA target sequences at the single-copy level without amplification can overcome PCR (polymerase chain reaction) limitations resulting from non-specific amplification at initially very low template numbers. Two oligonucleotide probes fluorescently labeled in two different colors, with each probe carrying one color, can be hybridized to a complementary target gene (DNA). When the target is present in the sample, both oligonucleotides bind to the same target. The sample can then be analyzed by two color cross-correlation FCS. Brownian motions of probe molecules that are not bound to the same target are not correlated to each other. The DNA carrying both probes on the same molecule is detected, even when there is an excess of molecules in one color. The cross-correlated signals due to the presence of the target molecule will simultaneously appear in time and space and emerge from the background of non-bound or non-specifically bound probe molecules. Two color confocal fluorescence coincidence analysis [3.46] could have the potential to indicate the presence of a target molecule in the sample. Stable, yet discriminating, hybridization has already been a demand in molecular biology for oligonucleotides serving as diagnostics or therapeutics.

Oliogonucleotide-target interactions can be found in vivo. Hybridization reactions are involved, for example, in replication (DNA primer), transcription (DNA-RNA complex), and translation (codon-anticodon interaction). Fluorescently labeled oligo-dT as well as oligo-dA were applied to living cells by the culture medium [3.47]. The intra-nuclear diffusion and the hybridization state of the oligonucleotides were measured by FCS. The diffusion properties of the small molecules (oligo-dT) and the complex with polyA in the cytosol and in the nucleus revealed two phases. The slow part may represent the bound complex with mRNA. Interestingly, the fast-moving parts separated are almost the same as for oligo-dT and oligo-dA in aqueous solution. The observation of hybridization reactions in living cells is fascinating for cell biologists and cell molecular biologists because the translation and/or transcription state can be monitored and the physiological conditions of the cytosol and the nucleus can be determined without crucial damage to the cell. The combination of FCS with high sensitivity imaging, three-dimensional micro-positioning and micro-manipulation enables the characterization of the localization and topology of proteins [3.104] in living cells and the visualization of lipid bilayers [3.105]. It also offers the possibility to analyze gene expression profiling in vivo.

3.3   DNA Analysis by "Going Micro"

Excited state lifetime measurements of DNA molecules in solution, where the DNA was linked to tetramethylrhodamine, indicated different conformational states of the DNA [3.8,3.18]. A spectrum of reaction rates between an open and a closed conformation was found by FCS measurements on many molecules in solution [3.8]. Studying the DNA molecular beacon with a fluorophore and a quencher by FCS revealed the kinetics between the opening and the closing of the hairpin for different sizes and sequences of the loop and for various salt concentrations [3.19]. Assuming that the two conformational states have the same diffusion times, a distribution of reaction rates was not reported for one DNA molecule under the same experimental conditions of measurements. However, the geometry of the open, random polymer coil and the closed, quenched state of the molecular beacon is significantly different. Conformational fluctuations of DNA attached to a microsphere were quantitatively analyzed by single particle tracking with FCS [3.20]. The behavior of polymer molecules can be described from weak bending to flexible circles. Studies of individual DNA molecules immobilized on a streptavidin-coated surface gave detailed insight into the distributed relaxation behavior of conformational states [3.9,3.12], the heterogeneous-homogeneous behavior of many single molecules as the observed fraction of time is changed during the evaluation time [3.10], and the total number of detected photons before decomposition of single DNA molecules as well as their survival times [3.11].

Synthetic single strands of DNA are being used for a variety of applications, e.g., as gene fragments, primer, as antisense and antigene oligonucleotides for tools of both molecular biology research and clinical diagnostics, as well as for potential therapeutic uses. Nucleotide building blocks are chemically joined together in vitro to form, usually, short oligonucleotides (up to about 40-mers) that have been designed to recognize and bind to specific genes. Coupled with the solid phase strategy for oligonucleotide libraries (for a general review of chemical libraries, see [3.48]) the chemical methods are now the basis for technologies to produce such tools on a large scale. In particular, the quantitative analysis of synthetic oligonucleotide libraries is a demand for modern molecular biology applying rational and irrational biotechnological approaches, but the experimental determination of the number and weight fraction of individual product molecules with different sequences has, so far, not been reached by conventional state-of-the-art analysis. The quantitation methods for synthesis predominantly in use, although well-studied, have limitations in many respects [3.49]. However, arrays ("microstructures") are a systematic way to read out all components. The array technologies become tools of molecular biology and clinical diagnostics in order to survey DNA and RNA variations [3.44]. For example, in today's DNA array technology for genotyping arrays, the complete set of 4^N polydeoxynucleotides of length N, or any subset, can be produced in light directed combinatorial chemical synthesis on high density synthetic oligonucleotide arrays [3.50]. Each position in the array contains a defined, short DNA fragment of possible combinations of all possible oligonucleotides N. The oligonucleotides are designed, for example, from all the open-reading frames of a known, whole genome sequence. In gene chips for examining gene expression, the array positions can be occupied by synthetic or complementary DNAs or PCR products (longer DNA fragments). Genotyping arrays and gene expression arrays are probed with the "unknown", fluorescently labeled target DNA sample.

Laser confocal fluorescence scanning systems presently dominate the detection devices [3.51]. They are similar to the line scanning detector systems for DNA sequencing instruments. Integrated microanalysis systems are described with microfabricated fluidic channels, heaters, temperature sensors and fluorescence detectors [3.52], and with sizing and sorting of DNA molecules [3.53]. In processing microarrays through hybridization and reading, more sensitive detection methods will extend the range of application beyond analysis of DNA variation and comparative expression analysis [3.54]. Alternative detection methods can be applied, such as fluorescence resonance energy transfer, fluorescence quenching, fluorescence polarization or time-resolved fluorescence. In a recent development of two color cross-correlation FCS [3.2,3.55–3.60], it was shown that sensitivity and specificity of optical detection and separation of biomolecules could be further enhanced by combining two fluorophores in two different colors. The possibility of using two color cross-correlation FCS for high throughput screening is described [3.58–3.60].

Devices based on microstructure chips and confocal fluorescence detection have been developed [3.61–3.63]. In the ongoing research to develop high throughput screening and selection of biomolecules [3.63], loss-poor detection of single molecules and hydrodynamic flow are performed in silicon-based microstructures (microchannels). For example, a Gaussian line focus is achieved by a cylindrical lens stretching the detection volume element in one direction and focusing it in the other. The direction of the flow from the drain to the collection branch is changed by a flow switch operating in the microsecond range. In principle, biological libraries such as a library of a mutant enzyme generated by a random mutagenesis of amino acids residues, or a library of a repertoire of diverse immunoglobuline specificities (phage antibody display system), or a ribosome display system can be screened by single molecule selection instead of biopanning with higher discrimination and selection efficiencies.

The development of single molecule analysis in solution in the Los Alamos laboratory [3.64–3.68] as well as at the Karolinska Institute [3.1,3.4–3.6,3.17] has furnished two key technologies to follow up the idea of rapid exonucleolytic single molecule DNA sequencing with identification of fluorescent dye nucleotides. The Keller group has realized the possibility of concentric flow using time gated single molecule detection [3.69–3.72]. Alternatively, microchannel flow using confocal single molecule detection is proposed for DNA sequencing [3.1,3.61–3.63,3.73,3.74]. Without any doubt, rapid DNA sequencing at the single-molecule level will have a profound impact on molecular biology and molecular medicine.

3.4 Incorporation of Dye Nucleotides into DNA

3.4.1 Low-Density Labeling

During PCR fluorescently labeled deoxynucleoside triphosphates such as fluorescein-11-dUTP (FluoroGreen from Amersham Pharmacia Biotech), Cy3-dUTP (FluoroLink from Amersham Pharmacia Biotech), tetramethylrhodamine-4-dUTP (FluoroRed from Amersham Pharmacia Biotech), Oregon-Green-5-dUTP (from Molecular Probe), etc. are incorporated somehow into the formed DNA. The PCR product is randomly labeled by fluorescent nucleotides after thermal cycling. Using a molar ratio Fluorescein-11-dUTP [Flu-dUTP/(dTTP+Flu-dUTP)] of 0.1, 0.2, 0.3, 0.4 and 0.5, $\approx$ 4000-bp DNA was synthesized by the poll type DNA polymerase. An increase in the number of PCR product molecules was detected, in this case after 10 cycles,

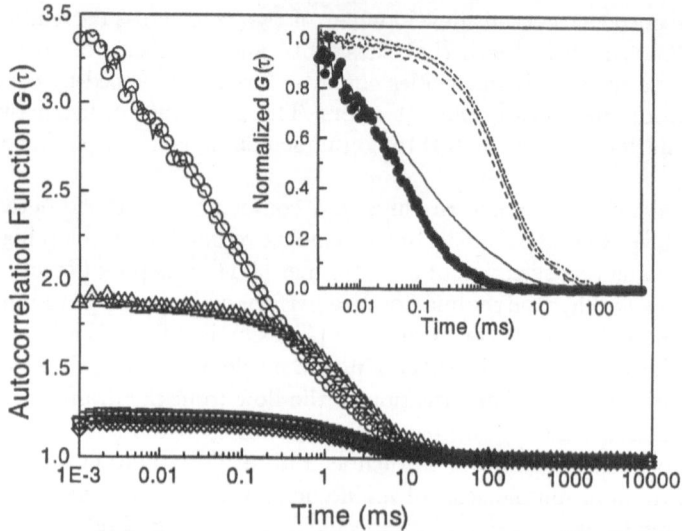

Fig. 3.2. The shift of the autocorrelation curves in the time course of the PCR reaction. *open circle*: 10 cycles; *up triangle*: 20 cycles; *down triangle*: 30 cycles; *diamond squares*: 40 cycles; *open square*: 50 cycles. The autocorrelation function was normalized to zero delay (*inset*). In this plot, the curve of Flu-dUTP is indicated for comparison (*solid circle*). *solid line*: 10 cycles; *dashed line*: 20 cycles; *dotted line*: 30 cycles; *dash-dotted line*: 40 cycles; *dash-dot-dotted line*: 50 cycles. Reproduced from [3.101]

as shown in Fig. 3.2. For the FCS measurements, non-incorporated labeled nucleotides have to be removed, for example, by size-exclusion spin column chromatography. The yields of the labeled PCR product depended on the fluorescent dye-nucleotides, the DNA polymerase as well as on the conditions of thermal cycling used.

In lightly labeled DNA an approximately linear relationship was found between the photon counts per molecule and second and the number of incorporated tetramethylrhodamine fluorphores [3.32]. At high dye nucleotide concentrations in the labeling reactions (for example, above $50\,\mu$M), non-covalently attached dye molecules tend to stick together or form aggregates together with the covalently incorporated ones. Thus, an approach to purify the labeled DNA under these experimental conditions is the extraction with organic solvents [3.81]. If n green and m red labels are incorporated into the same target (or from the viewpoint of the FCS model: "bound to" the same target), the fluctuations in the number of molecules can be analyzed by two color cross-correlation FCS. The parameters of the normalized cross-correlation function

$$G_{\mathrm{gr}}(\tau) = 1 + \frac{1}{N}\left(\frac{1}{1+(\tau/\tau_{\mathrm{gr}})}\right)\left(\frac{1}{1+(\hat{\omega}_{\mathrm{x,y}}/\hat{\omega}_{\mathrm{z}})^2(\tau/\tau_{\mathrm{gr}})}\right)^{1/2}, \qquad (3.2)$$

are the cross-correlation diffusion time τ_{gr} and the size of the cross-correlated volume element $V_{gr} = (\pi/2)^{3/2}(\omega_g^2 + \omega_r^2)(z_g^2 + z_r^2)^{1/2}$ [3.55]. The diffusion time of the two color molecule is the weighted green and red autocorrelation diffusion times $\tau_{gr} = (\tau_g + \tau_r)/2$, since $\tau_{gr} = \hat{\omega}^2/4D$ and $\hat{\omega}^2 = (\omega_g^2 + \omega_r^2)/2$ [3.55]. It can be obtained directly from the logarithmic time axis (correlation time) and is related to the size of the two color DNA by the diffusion constant D. The amplitude of the cross-correlation term is interpreted in terms of the theoretical model [3.57]

$$\lim_{\tau \to 0} \frac{\langle \delta I_g(0) \delta I_r(\tau) \rangle}{\langle I_g \rangle \langle I_r \rangle} = \frac{1}{N} , \tag{3.3}$$

where

$$\frac{1}{N} = \frac{N_g(Q_B^R/Q_R^R) + N_{gr}[n(n(Q_B^R/Q_R^R) + m)]}{[N_g + nN_{gr}][N_r + N_g(Q_B^R/Q_R^R) + N_{gr}(n(Q_B^R/Q_R^R) + m)]} . \tag{3.4}$$

By defining the emission rates per molecule as $Q_{\lambda\text{excitation}}^{\lambda\text{emission}}$ [3.56], (3.4) applies to the cross talk in one direction from green to red. The other case from red to green can be disregarded. A small portion of photons emitted by the green fluorophore passes the fluorescence filters and dichroic mirrors or beam splitter for the red fluorophore. This occurs no matter how excellent the optical properties of fluorescence filters dichroic mirrors or beam splitter are. In our optical setup (Sect. 3.7) the cross talk from green to red is about 2% of the photon counts per molecule and second at blue excitation and green emission for Rhodamine Green. In the case that $n(Q_B^R/Q_R^R) < 1$, we obtain from (3.4) [3.57]

$$\frac{1}{N} = \frac{nmN_{gr}}{(N_g + nN_{gr})(N_r + mN_{gr})} . \tag{3.5}$$

Equation (3.5) is exact if a green emission in the red channel is negligible. It represents a signal amplification in which the two color target itself provides an enhanced cross-correlated fluorescence signal by the factor nm. Equation (3.4) found by converting the physical ideas into mathematical form is the expansion of (3.11) for n green and m red labels.

Results are shown in Table 3.1 for the two color "signal amplification" approach via tetramethylrhodamine-4-dUTP (FluoroRed, Amersham Pharmacia Biotech, UK) and Cy-5-dUTP (Amersham Pharmacia Biotech) in the presence of 167 μM of each natural dNTP with $n = 11$ and $m = 2$ (Table 3.1 A) and for the two color "target" amplification via 5'-fluorescently tagged primers with $n = m = 1$ (Table 3.1 B). The "signal amplification" approach is an example for interactions of two ligands with different colors at multiple positions of the same target. The amplitude of the cross-correlated fluorescence signals increases proportionally with the number of incorporated green (n) and red (m) labels under the condition of a constant level of molecules in one color over two color molecules. Furthermore,

Table 3.1. Parameters of cross-correlation measurements for the amplified target double stranded DNAs. The specific target DNAs were amplified (**A**) in the presence of 30 μM tetramethylrhodamine-4-dUTP and 30 μM Cy5-dCTP and (**B**) in the presence of 42 nM of each of the amplification primers 5′-tagged with Rhodamine Green or Cy5. In (**A**) the excess of green and red fluorescent nucleotides was adjusted after completion of PCR followed by spin column chromatography. All calculations were corrected for the different size of the autocorrelated and cross-correlated volume elements, and performed for the undiluted PCR mixes

Initial template concentration in PCR [M]	τ_{gr}^{a} [ms]	N	$N_{r,total}$	$N_{g,total}$	N_{gr}
A					
217-bp target DNA					
$1.4 \times^{-14}$	1.64^{b}	12.4	0.7	4.3	0.01
B					
217-bp target DNA					
7×10^{-13}	1.88	289	17.3	33.1	2.00
389-bp target DNA					
7×10^{-13}	2.14	1030	22.8	51.4	1.14

[a] The green and red autocorrelation diffusion times of two color DNA measured were: $\tau_g = 1.7$ ms and $\tau_r = 2.6$ ms, averaged values for 389-bp nucleotide length; $\tau_g = 1.3$ ms and $\tau_r = 2.2$ ms, averaged values for 217-bp nucleotide length. Data analysis was perfomed using the Marquardt nonlinear least squares algorithm.

[b] The behavior of the fluctuations of the molecule number in two color cross-correlation measurements analyzed under the effect of $n = 11$ green and $m = 2$ red labels covalently incorporated into the two color target: signal amplification of the target sequence.

the signal-to-noise ratio S/N in correlation functions increases proportionally to the product of n and m, in agreement with Koppel [3.16]. Special PCR conditions for thermal cycling were applied to suppress PCR artifacts and to ensure higher specific product yield [3.57]. However, the "signal" amplification of a target sequence led to two color target molecules together with target sequences in one color. Post PCR purification of the reaction mixes was necessary. The target yield was much lower in comparison with the "target" amplification via 5′-fluorescently tagged primers.

3.4.2 Nick Translation

Fluorescence in situ hybridization (FISH) is bridging the gap between sub-chromosomal, molecular genetic analyses and chromosome level studies. A cloned segment of "probe DNA" is labeled with a fluorescent compound, denatured (made single stranded), and annealed (hybridized) to the denatured (single stranded) DNA of chromosomes fixed in situ to the microscope slide [3.75]. The labeled probe DNA hybridizes to regions where the chromosomal DNA sequence is identical or very similar to the probe DNA sequence.

A major application of FISH is the detection of microdeletions in a number of syndromes [3.76]. The microdeletion is frequently not detectable by conventional, high resolution chromosome analyses. The detection of microdeletion by FISH provides the conformation of a clinical diagnosis [3.77]. Entire chromosomes or chromosome regions can also be identified by FISH. Painting probes consisting of DNA from an entire human chromosome are used in situations where the origin of chromosome regions with abnormal banding patterns cannot be determined by conventional cytogenetics. Conventional cytogenetics are highly dependent on the quality of metaphase chromosomes and FISH is frequently not, therefore, FISH is the method of choice in analyzing chromosomes of neoplastic cells [3.78].

Numerous methods exist for labeling probe DNA in FISH. Nick translation is the most direct method of labeling hybridization probes. The double stranded DNA that is to be used as a probe is "nicked" by DNase I E. coli DNA polymerase I, for example, three unlabeled dNTPs, and a mixture of one labeled dNTP with its unlabeled analog are added. The $5' \rightarrow 3'$ nuclease activity of the DNA polymerase I removes nucleotides beginning at the $5'$-end of each nick while the $3' \rightarrow 5'$ polymerase activity of the enzyme fills in the enlarging gap with free dNTPs, some of which are fluorescently labeled. Since the DNA polymerase I is a poorly processive enzyme, the length of DNA polymerized on any run is about 400 nucleotides. Among the parameters determining maximal incorporation of free tetramethylrhodamine-4-dUTP, the incubation time for DNase I is essential. Incubation beyond the point of maximal incorporation will result in probe degradation.

3.4.3 Linear Primer Extension Reactions

For many years there has been a particular interest in the property of DNA polymerases and reverse transcriptases to incorporate deoxynucleoside triphosphate analogs into DNA. Experiments have been reported [3.79] that measure the ability of wild type enzymes and mutants to incorporate fluorescent dye-deoxynucleoside triphosphates into DNA substrates (DNA cassettes). What is different to other assays is that (1) one of the naturally occurring dNTPs was completely replaced by a fluorescent dye-deoxynucleoside triphosphate derivative and (2) up to 40 subsequent elongation steps could be analyzed in detail in a defined template-primer system by denaturing polyacrylamide gel electrophoresis at high spatial resolution. Starting from a binding site for a specificity primer (20-mer) $5'$-labeled with a reporter group, 1 times all 4 nucleotides (cassette 1), 2 times all 4 nucleotides (cassette 2), 3 times all 4 nucleotides (cassette 3), and so forth up to template cassette

10 were chemically synthesized the template-directed synthesis of short DNA fragments in primer extension reactions with fluorophore-bearing dNTPs, the experimental method named "cassette technique" was developed by measuring the probability density of stop and pause sites of DNA polymerases and reverse transcriptases [3.79]. For example, the dTTP was replaced by its fluorescent analog tetramethylrhodamine-6-dUTP (from Boehringer Mannheim) in the presence of all other natural dNTPs. In the incorporation assay the mass increase of the resulting enzymatic polymerization products led to multiband electrophoretic patterns of nonlinear mobility shifts for each template cassette.

The different DNA species (erroneous sequences) formed during the template directed polymerization were caused by stop and pause sites of the enzyme [3.79]. The relative nucleotide length of the target sequences were given by comparison with the relative migration distances of the enzymatically polymerized cassettes 1 to 10. The target sequences were identified as the longe sequences at which the fluorescence signals could be detected at 532 nm laser excitation at and emission at 605 nm. We studied Taq-DNA polymerase, Klenow polymerase, Pwo polymerase, AMV and HIV reverse transcriptases, Vent polymerase, and polymerase x (a purified polymerase from a hitherto uncharacterized bacterium, Boehringer Mannheim). The ability to incorporate tetramethylrhodamine-6-dUTP opposite to A in the template DNA varied significantly among the polymerases and reverse transcriptases tested. Taq-DNA polymerase, Klenow polymerase, AMV reverse transcriptase, Vent polymerase, and polymerase x were able to polymerize the target length for cassette 1, 5, and 10, which resulted in 4, 20 and 40 elongation steps. However, these polymerases differed in their tendency to abort DNA synthesis after the incorporation of the fluorophore-bearing nucleotide. The stops for Vent polymerase were more pronounced at the end of the target sequence. The intrinsic $3' \rightarrow 5'$ exonuclease activity of Vent polymerase might remove initially misincorporated nucleotides, but tetramethylrhodamine-6-dUTP also had an inhibitory effect on the $3' \rightarrow 5'$ exonuclease activity of this enzyme. In the case of AMV reverse transcriptase more stops were found at the beginning of the polymerization process. For HIV reverse transcriptase the incorporation of one tetramethylrhodamine-6-dUTP was detected in cassette 1, whereas the primer annealed with cassette 5 and 10 as templates was not elongated to the maximum nucleotide lengths. AMV reverse transcriptase did not reveal a better acceptance of the modified nucleotide compared to DNA polymerases of families A (Taq-DNA polymerase and Klenow polymerase) and B (Vent polymerase). In the presence of tetramethylrhodamine-6-dUTP, the probability density of stop and pause sites (erroneous sequences) was lower for polymerase x in comparison with that for Vent polymerase and AMV reverse transcriptase. In contrast, Taq-DNA polymerase, Klenow polymerase, and AMV reverse transcriptase produced more stops and pause sites in the presence of tetramethylrhodamine-6-dUTP than in the presence of the naturally occurring dTTP. The probability densities of stop and pause sites for Pwo polymerase and the phylogenetically related Vent polymerase were almost the same in the presence of tetramethylrhodamine-6-dUTP.

In order to compare different fluorescent and non-fluorescent modifications, linkers, spacers and dyes at different positions, this incorporation assay [3.79] was also useful for looking at two, three or even four deoxynucleoside triphosphate derivatives together, substituting completely for their natural analogs in one reaction mix [3.80]. A modification of this incorporation assay was also used which consists of a defined template-primer system with consecutive nucleotides of the same base, for example, 18 A residues in a row. This looks more like a sensitivity assay for fluorescent dye-deoxynucleoside triphosphates and non-fluorescent deoxynucleoside triphosphate derivatives.

3.4.4 High-Density Labeling

It has been reported that complete labeling of DNA with fluorophore-bearing derivatives of the dNTPs is obtained by a "tailor made", $3' \rightarrow 5'$ exonuclease deficient mutant of the mesophilic T4-DNA polymerase (polymerase family B) [3.72]. An extensive screening study was performed at Roche Diagnostics GmbH [3.80] which led to the selection of a $3' \rightarrow 5'$ exonuclease deficient mutant of the thermophilic Tgo-DNA polymerase by means of the incorporation assay developed [3.79] and of its modification, similar to that incorporation assay. The principle of the incorporation assay [3.79] is briefly described in Sect. 3.4.3. The Rhodamine Green-X-dUTP was substituted completely for the normal nucleotide; X stands for a total distance of 12 atoms in a chain between the fluorophore and base. It was found that the $3' \rightarrow 5'$ exonuclease deficient mutant of the thermophilic Tgo-DNA polymerase (polymerase family B) is able to incorporate Rhodamine Green-X-dUTP into a 217-b "natural" (model) DNA with a determined labeling efficiency of 0.89–0.96 [3.81].

By means of the novel technologies of mobility shift gel electrophoresis, reversion-PCR (re-amplification of the shifted band) and re-sequencing that were developed for the first time in [3.81,3.83], the correct error free covalent incorporation of dye-nucleotides into the DNA was demonstrated. Within this context the determined labeling efficiency of 0.89–0.96 indicates that almost all possible incorporation sites of a thymine contained a fluorescently modified uracil, but a fraction of the substrate analog did not bear the fluorophore. The absence of any non-fluorescent substrate in the preparation of the fluorescent substrate analog can improve the statistics of labeling. The maximum number of 92 incorporations of the fluorescent U* label was given by the sequence of the DNA fragment. Moreover, the data indicated that labeled DNA has to be carefully purified to prevent the adsorption of non-incorporated dye-dNTPs. The purification procedure depends on the type and concentration of dye-nucleotides used in the labeling reaction. It can involve denaturing conditions (e.g., preparative denaturing polyacrylamide gel electrophoresis) or extraction with organic solvents on solid-phase as well as in solution [3.81,3.83,3.84].

The results of the full, error free incorporation of Rhodamine Green-X-dUTP into a 217-bp "natural" (model) DNA demonstrated that the product formation of high density labeled DNA can be detected and quantified in terms of absolute numbers of molecules by fluorescence correlation spectroscopy [3.81]. However, the green fluorescence intensity was reduced substantially ($\approx$ 98% is quenched). The second label, Cy5, was introduced via the 5'-end of the amplification primer in order to identify the two color product. In the case of high density labeled, green DNA we have to consider the interaction of the green fluorophores with their local surroundings. For this reason, two quantities characterizing this kind of interaction were first introduced [3.81]: the relative fluorescence detection efficiencies $\tilde{Q}_B^G$ at blue (green) excitation and green emission, and $\tilde{Q}_B^R$ at blue (green) excitation and red emission. These parameters are in addition to the relative fluorescence detection efficiencies for the free dyes Q_B^G and Q_B^R introduced in [3.56,3.57], quantifying the cross talk from green to red. The amplitude (minus one) of the cross-correlated signals then takes the form

$$\frac{1}{N} = \frac{N_g(Q_B^R/Q_R^R) + N_{gr}[n\tilde{Q}_B^G(1 + n(\tilde{Q}_B^R/Q_R^R))]}{[N_g + N_{gr}n\tilde{Q}_B^G][N_r + N_g(Q_B^R/Q_R^R) + N_{gr}(1 + n(\tilde{Q}_B^R/Q_R^R))]} . \tag{3.6}$$

Here, the cross-correlation results from intensity fluctuations for many n green labels (high density labeling) and one red label incorporated into the same two color target [3.81]. The apparent (measured) number of the low-fluorescence two color DNA heavily green quenched is $N_{gr}n\tilde{Q}_B^G = \tilde{N}_{gr}$ and N_{gr} is the number equivalent of two color DNA in the case of negligible interactions (quenching) between the internal dyes and their surroundings. In the realistic case $(Q_B^R/Q_R^R) \ll 1$ and $n(\tilde{Q}_B^R/Q_R^R) < 1$ for the optical setup, we obtain from (3.6) the following result

$$\frac{1}{N} = \frac{N_{gr}n\tilde{Q}_B^G}{(N_g + N_{gr}n\tilde{Q}_B^G)(N_r + N_{gr})} . \tag{3.7}$$

Under the experimental conditions that (1) in red autocorrelation $N_{gr} \ll N_r$, and (2) $\tilde{Q}_B^R \ll \tilde{Q}_B^G$, (3.7) yields directly the number of heavily green labeled two color target molecules

$$\tilde{N}_{gr} = \frac{N_r N_g}{N - N_r} . \tag{3.8}$$

In the limit $n\tilde{Q}_B^G = 1$ for large numbers n of internally incorporated labels, the result is exact: $\tilde{N}_{gr} = N_{gr}$. The apparent number of high density green labeled two color DNA is decreased and equals the real number N_{gr} of two color target molecules. It is then possible to calculate the number of fluorophores that are covalently incorporated into the DNA. Experimentally, we performed the complete degradation of the high density green labeled two color DNA by an exonuclease after exhaustive purification [3.81]. FCS green autocorrelation curves measured before and after complete degradation were used, with a best fit procedure, to obtain the number of released fluorophore-bearing nucleotides. The FCS-based method enables very accurate determination, but the green fluorescence autocorrelation function measured before degradation must be corrected for the relative difference in the quantum yield between high density, green labeled DNA and free, green dye molecules (see equation (3.1)) [3.81]. The number n of incorporated green labels per one molecule of two color DNA is then given by

$$n = \frac{N_{g,dye}}{\tilde{N}_{gr}} , \tag{3.9}$$

where $N_{g,dye}$ is the number of internal green labels released after complete enzymatic degradation of the exhaustive purified DNA and $N_{gr} \cong \tilde{N}_{gr}$. All numbers of molecules are specified for the size of the green illuminated volume V_g, i.e., they are corrected for the different size of the autocorrelated and cross-correlated volume elements. The models (equations (3.6–3.9) are confirmed by a series of measurements [3.81]. The sequence integrity after high density labeling (complete substitution of dTTP by its fluorescently tagged analog) was measured by the newly developed biotechnologies of mobility shift gel electrophoresis, reversion-PCR and re-sequencing [3.81]. These measurements were in agreement with the results of the models indicating error free substitution for a normal nucleotide. Our models enable a whole series of problems to be tackled.

The complete, base-specific labeling of the "natural" model DNA fragment with the fluorophore-bearing dTTP is one of the enzymatic prerequisites for rapid DNA sequencing based on the deconstruction of fluorescent DNA by an exonuclease in a microchannel flow or in a microcapillary using single molecule detection [3.26] (Sect. 3.3). Next, one has to prove that base-specific, fluorescent labeling can be obtained by the DNA polymerase with two different type of nucleotides. A set of two types of nucleotides labeled in two colors (green and red) enables the complete DNA sequence to be read out within the concept of single-molecule sequencing.

3.5   Exonuclease Degradation

The determination of exonuclease activities in homogeneous solutions has been of interest not only for enzymological reasons but also for basic experiments within the context of ultrafast sequencing of single DNA molecules (Sect. 3.3). The T7 exonuclease ($3' \rightarrow 5'$ exonuclease activities of the bacteriophage T7 DNA polymerase) is a processive enzyme. It digests DNA from the $3'$-end by cutting off nucleoside $5'$-phosphate (dNMP, mononucleotide). In analyzing the degradation of double stranded DNA that was internally labeled with tetramethylrhodamine-4-dUTP (FluroRed, Amersham Pharmacia

Biotech, UK) by low density labeling (random way), the reaction was carried out in $10\,\mu l$ volume and continuously monitored by FCS [3.85]. The increase in the number of fluorescent molecules and the decrease in the diffusion time were observed by the fluorescence correlogram (Fig. 3.3). The autocorrelation function was analyzed in detail by a fluorescence intensity weighted derived model. It was assumed that the overall reaction can be described by two different fluorescent components: cleaved fluorescent monomers (TMR-4-dUMPs) and remaining polymeric product neglecting its size distribution. The analysis yielded the averaged size of the remaining polymer fragment as a function of the reaction time, the incorporation efficiency for tetramethylrhodamine-4-labels under the labeling conditions, and the ratio of the quantum yield between DNA and free rhodamine label. The analysis of the reaction mix was performed by optical separation without any physical separation procedure such as column chromatography, electrophoresis or membrane binding assays.

A universally applicable assay (method) for kinetic analysis of double stranded and/or single stranded DNA was developed and established using a red fluorescent reported dye (Cy-5) attached to a 5'-end of the DNA via the amplification primer [3.86]. The reporter dye enabled us (1) to prove complete enzymatic degradation of the DNA substrate and (2) laser excitation at a wavelength (633 nm) at which no absorption of internally incorporated fluorescent dye-deoxynucleotides occured. We incorporated internally

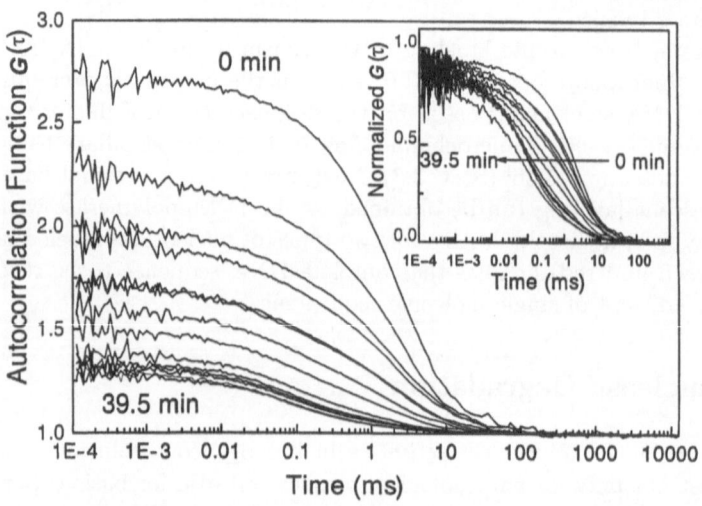

Fig. 3.3. Changes in the autocorrelation function caused by the cleavage of double stranded 500-bp DNA internally labeled with tetramethylrhodamine-4-dUTP (low density labeling). The degradation was performed by $3' \rightarrow 5'$ exonuclease activities of the T7 DNA polymerase (wild type). Each experimental curve was obtained by a one minute measurement

tetramethylrhodamine-4-dUTP (Fluoro-Red, Amersham Pharmacia Biotech, UK) or Rhodamine Green-5-dUTP (Molecular Probes, Netherlands) at different labeling densities [3.32,3.81,3.83,3.86]. The double stranded, 217-bp DNAs were used as substrates in the degradation assay.

The biochemical reaction parameters were chosen in such a way that only full length DNA, that was 5'-labeled with the reporter dye Cy5, and free Cy5-dye (the latter comes from the chemical oligonucleotide preparation) were detetctable at the beginning of the reaction. As the reaction proceeded, free Cy5-mononucleotide appeared. The time course of the enzymatic degradation was recorded by FCS (Fig. 3.4). In the autocorrelation curves measured at red laser excitation (633 nm) and red emission (685 nm, peak intensity value), a decrease in the number of full length DNA and an increase in the number of free Cy5-mononucleotides was found. The fraction of free Cy5 dye was always constant. The $3' \rightarrow 5'$ exonuclease of the T7 DNA polymerase resides in the gene 5 protein (80 kDa) and has a high activity on single stranded DNA and a weak activity on double stranded DNA. The processivity factor thioredoxin, forming a strong 1:1 complex with the gene 5 protein, increases the $3' \rightarrow 5'$ exonuclease activity on double stranded DNA several hundred-fold, but it has little effect on the $3' \rightarrow 5'$ exonuclease activity, if single stranded DNA is the substrate [3.87,3.88]. Thioredoxin stabilizes the gene 5 protein-(primer)-template complex. The reconstitution of the gene 5 protein-thioredoxin complex with an excess of reduced thioredoxin is indistinguishable from the complex assembled in vivo.

FCS measurements analyzed a series consisting of sets of 10 single 15 s measurements. The simultaneous analysis of multiple measurements covered almost the complete reaction time (for example, $\approx$ 30 min up to one hour). This was obtained by choosing data in each set with, for example, a constant and sufficiently large difference in time between subsequent measurements. Thus, multiple measurements were simultaneously analyzed several times but never within identical sets. In this way, sliding mean values and standard deviations were calculated for all the fractions of the full length DNA substrate at any reaction time. Software for the simultaneous analysis of multiple measurements was developed [3.83] which can also be performed as a global parameter analysis. We obtained the initial velocity for the enzymatic reactions, providing access to molecular parameters such as the turnover numbers of molecules [3.89]. A synopsis of the results obtained is shown in Table 3.2.

3.6   Restriction Enzyme Cutting and DNA Sizing

Analyses of restriction endonuclease size distribution, restriction fragment length polymorphisms (RFLP) or DNA fingerprinting by a variety of means are examples of indirect structural comparative approaches [3.90]. Type-II restriction endonucleases recognize short double stranded (or single stranded) nucleotide stretches (4 base pairs or more) in a genome, and cut the DNA

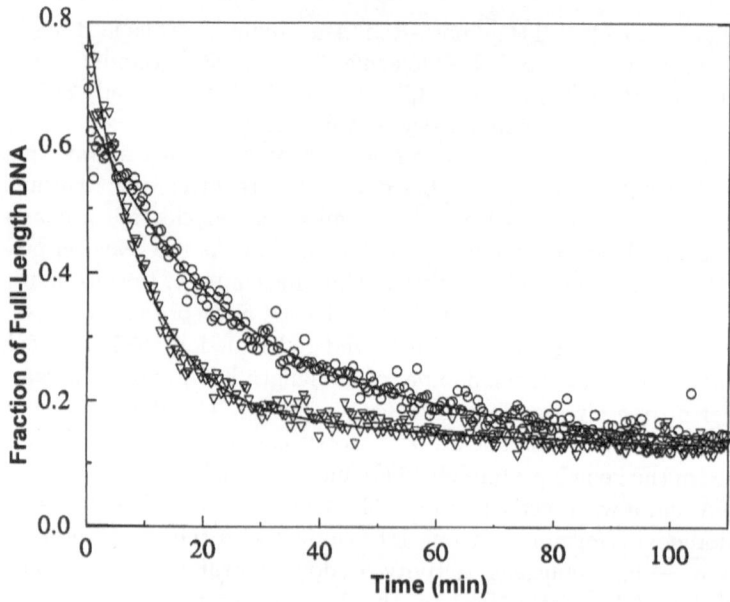

Fig. 3.4. Time course of the degradation of double-stranded, "natural" 217-bp DNAs by $3' \to 5'$ exonuclease activities of the T7 DNA polymerase (wild type). The DNAs were labeled at the $5'$-end with the reporter dye Cy5 via the amplification primer. The FCS measurements were carried out at the laser excitation wavelength of 633 nm (red) and the emission was detected at 685 nm (red peak intensity value). *Open circle*: 217-bp DNA internally labeled with 7 Rhodamine Green-5-dUMPs on average (low density labeling). *Open down triangle*: no internal labels. Reproduced from [3.83]

at the recognition sites. Therefore, the enzymatic activity produces cleaved DNA fragments. The pattern of cleaved DNA fragments depends on both the DNA under investigation and the restriction endonuclease used. The approaches are rather convenient when relatively short DNAs (e.g., DNA fragments) are investigated. The methods become much more complicated and time- and labor-consuming as the length increases of the DNA beging investigated. The use of hybridization "tagging" for visualization of primary structure differences in DNA molecules requires long oligonucleotide complementary sequences along a DNA molecule. The longer the length of an oligonucleotide, the less information it produces in mapping experiments. However, standard procedures for producing or detecting hexanucleotide hybrid molecules are not suitable for physical mapping experiments. Thus, the development of physical techniques for the detection of cleaved fragments opens up new opportunities for physical mapping of DNA.

The use of FCS enables the restriction endonuclease reactions to be monitored by counting the increasing number of cleaved DNA fragments in ho-

Table 3.2. Kinetic parameters for complete degradation of low density labeled 217-bp "natural" model DNAs (substrates) by the $3' \rightarrow 5'$ exonuclease activities of the bacteriophage T7 DNA polymerase (wild type) measured in solution

Substrate	Determined number of internal labels	Determined cleavage rate constant k (s^{-1})	Turnover number (k_{cat}) (s^{-1})	k_{cat} value normalized for the internal label (s^{-1})
ds DNA with 5-primed Cy5 reporter dye	no label	5.7×10^{-3}	480	–
ds DNA with 5-primed Cy5 reporter dye, labeling by PCR in the presence of TMR-4-dUTP	11 Tetramethyl-rhodamine-4 labels	3.3×10^{-4}	28	1.5
ds DNA with 5-primed Cy5 reporter dye, labeling by PCR in the presence of Rhodamine Green-5-dUTP	7 Rhodamine Green-5 labels	1.9×10^{-4}	16	0.5
ds DNA with 5-primed Cy5 reporter dye, labeling by PCR in the prescence of Cy5-dCTP	2 Cy5 labels	1.2×10^{-3}	100	1.3

mogeneous solution [3.91]. For this purpose, the M13mp18 target DNA was randomly labeled with tetramethylrhodamine-4-dUTP (FluoroRed, Amersham Pharmacia Biotech) by the Klenow enzyme or by PCR at a molar ratio dye-dUTP / (dye − dUTP + dTTP) of 1:0.5. The cleavage of the fluorescently labeled double stranded M13mp18 DNA (7250-bp) was performed by four different restriction endonucleases: HaeIII, HgaI, BsmAI and BspMI. The experimental autocorrelation functions during cutting of the original double stranded M13m into several fragments are shown in Fig. 3.5. The digestion process results in the decrease of the autocorrelation amplitude (the zero time intercept).

A multi-component analysis of the diffusion of cleaved fragments (time-dependent part of the correlation function) is complicated and has not yet been developed. Therefore, a model of the fluorescence intensity weighted autocorrelation function was introduced for analyzing the number of DNA fragments produced by restriction endonucleases [3.91]. A point mutation in the cutting site inhibits the enzymatic activity of Type-II restriction endonucleases and modifies the number of cleaved products. FCS can monitor the change in the number of cleaved DNA fragments in one step (for example, Fig. 3.6). On the other hand, time-consuming and labor-intensive preparative procedures are involved in conventional gel electrophoresis-based methods in

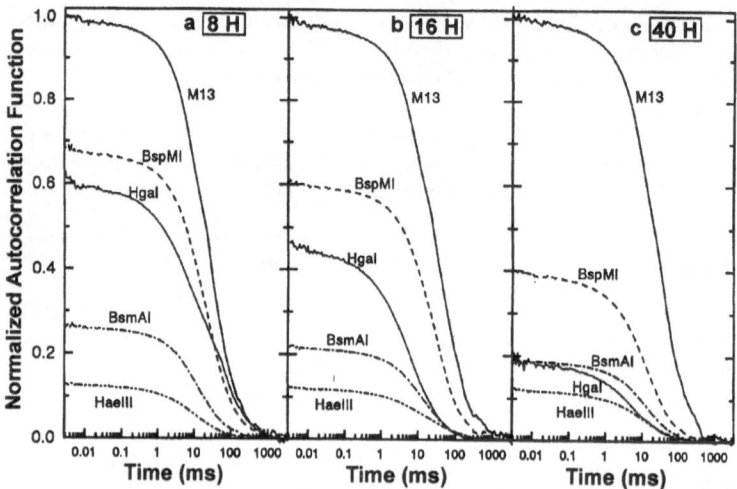

Fig. 3.5. Fluorescence correlation functions of cleaved double stranded M13mp18 DNA using restriction enzymes. The cleavage of fluorescently labeled DNA by restriction enzyme was carried out in 5 μl solution with HaeIII (*dash-dot-dotted line*), HgaI (*dotted line*), BsmAI (*dash-dotted line*) or BspMI (*dashed line*). The reaction solutions contained two units of enzyme with each combination buffer. They were incubated for (**a**) 8 h, (**b**) 16 h, and (**a**) 40 h. The autocorrelation functions were normalized to M13mp18 DNA (*solid line*) measured in the absence of enzyme. The incubation times were the same as in the presence of the enzyme. Reproduced from [3.91]

order to detect point mutations by RFLP [3.92]. The sensitivity and the quantitation of the measurements by FCS can be an important advantage. For example, we used the restriction enzymes HaeIII and SmaI to monitor specific as well as non-specific PCR reactions of a 217-bp target DNA by FCS [3.32]. In principle, the simplicity of the FCS detection is another aspect resulting in short analysis times and high sample throughputs in genetic studies for screening and medical diagnostics.

3.7 Polymerase Chain Reaction

In 1993 Kary B. Mullis' contribution to the field of PCR was recognized by the Nobel Prize in Chemistry. Mullis developed the idea of PCR over 15 years ago. In order to generate an exponential increase in the starting number of copies, Mullis proposed the concept of thermal cycling. In thermal cycling of denaturation, hybridization and extension, the products synthesized in a given cycle serve as templates in the next cycle. Thus, 10 doubling is a 10^3-fold increase, 20 doubling is a 10^6-fold increase, and 30 doubling is a 10^9-fold increase, if everything works efficiently every cycle. In the past 15 years, scientists developed new innovations and applications of PCR. Many of the factors that contribute to efficiency have been discovered, as well as the factors that contribute to the limits of PCR. Among the many innovations are TaqMan PCR detection strategy, hot start PCR, reverse transcription PCR, anchor PCR, alu PCR, multiplex PCR, and in situ PCR. All of the PCR strategies are different permutations of the theme of increasing amplification. Over the past decade there have been many critical improvements. More precisely, the advances are related to new detection strategies [3.27], the search for polymerases with superior properties for special applications, direct amplification with thermostable reverse transcriptases [3.25], hot start dUTP UNG contamination control, as well as thermocyclers. Within this context, FCS provides a number of desirable features that ensures its place at the forefront.

3.7.1   FCS Autocorrelation Analysis: New Detection Methods

FCS can detect the molecular weight changes between fluorescently labeled primers and/or PCR products. When the experimental FCS curves are eval-

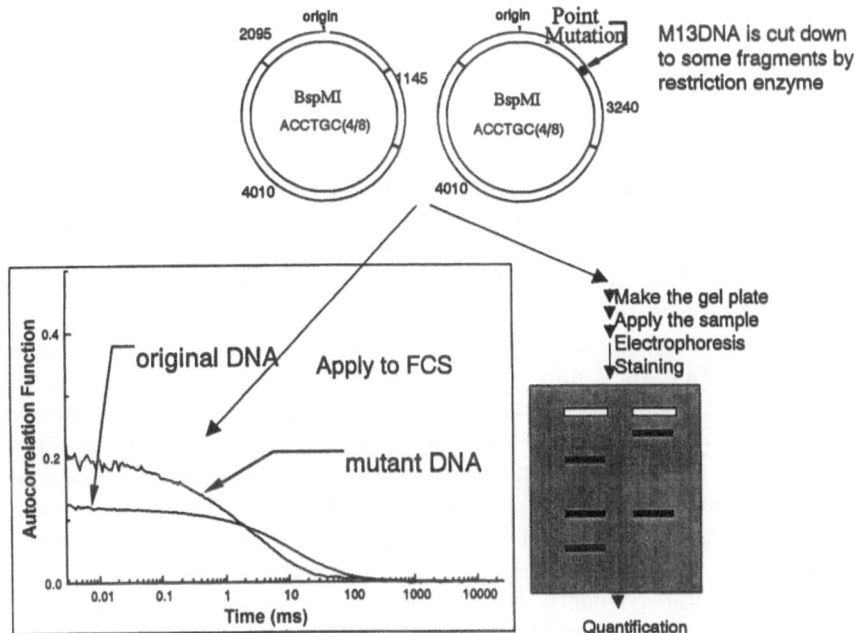

Fig. 3.6. Schematic flow diagram of RFLP (restriction fragment length polymorphisms) analysis carried out by FCS and by gel electrophoresis. The DNA mutation can be detected by the change of the autocorrelation function in FCS measurements performed in one minute. In contrast, many preparative procedures are involved in the detection by conventional gel electrophoresis-based methods. Reproduced from [3.92]

uated by the normalized autocorrelation function (3.1), the translational diffusion times of the sample are obtained, where N is the average number of fluorescent molecules, τ_D the diffusion time of the molecules, and the volume element includes a half-length ω_z and a radius $\omega_{x,y}$. The diffusion times are related to the relative molecular weights of the samples. We developed a FCS calibration curve, where the translational diffusion times of different size DNA fragments were plotted versus the number of base pairs they contain [3.32]. By means of such a calibration curve the molecular weight of DNA products formed can be determined quantitatively.

Non-specific versus Specific Product Formation. It is an enormously frustrating experience in the lab, if one sees there is some amplification of DNA, but without specificity, regarding the intended product. The use of Taq DNA polymerase simplifies the procedure in molecular biology and provides an increase in specificity at the beginning of the setting up of PCR. One of the first modifications of the reaction parameters is to change the annealing stringency. Instead of using $47\,°C$ as the annealing temperature and $72\,°C$ as the extension temperature, the annealing temperature for Taq DNA polymerase can be increased [3.86]. Then, as a result, the non-specific hybridization between primers and templates will fall away, and also the absence of proofreading activity in Taq DNA polymerase will prevent the conversion of mismatched primers to matched primers. While diluting the template copy number, the specific signal can be preserved by increasing the annealing temperature without doing anything else. A higher specific requirement for a perfect match of the primer-template yields an improved discrimination against the extension, which also increases the specificity.

There are a large number of factors that can contribute to PCR specificity. We have demonstrated [3.32] that non-specific product formation is observed for a 217-bp DNA fragment where the reaction was in a sort of standard condition of Taq, Mg^{2+} and primer concentration by diluting Taq using less, reducing Mg^{2+} using less, and reducing the primers using less. This was one of the very first attempts to monitor target DNA by FCS autocorrelation spectroscopy to see if one could detect a low copy sequence under standard PCR conditions. A calibration curve of translational diffusion times versus the number of base pairs of the DNA indicated that the 217-bp target could be detected. We certainly made some of the intended product, but there was a lot of non-specificity despite using the appropriate Taq, Mg^{2+}, and primer concentrations, and $65\,°C$ annealing, coming from the M13mp18 DNA. If everything is set up together in the PCR mix, the enzyme can produce nonspecific extension products under nonstringent reaction conditions. In hot start PCR, the tube is heated up separating one component either enzyme, Mg^{2+}, dNTPs or primers. At elevated temperatures the missing component is mixed to the reaction or becomes activated in the reaction. In order to facilitate hot start in a convenient fashion, an Amplivax barrier or a modified enzyme is used. It is possible to modify the enzyme chemically to provide a thermally activated DNA polymerase. This enzyme (AmpliTaq Gold from Perkin-Elmer, USA) requires high temperatures to reactivate, therefore a small percentage of the enzyme is reactivated at 95 °C in a few minutes, and full reactivation occurs in 10–20 min. There is an enormous increase in specificity and minimization of primer-contamers and other artifacts with hot start. The product is directly quantified at a much higher frequency by FCS autocorrelation and cross-correlation analysis [3.56]. Similarly, looking at low copy numbers via 10 copies or 50 copies, we get a single, specific product [3.56] when hot start is imposed because the primers work mispriming under nonstringent PCR setup conditions. This is in contrast to what is seen with non hot start. In the study, the immediate benefits of increased specificity by FCS detection were observed [3.56].

FA-PCR (fluorescence-labeled primer-based asymmetric PCR) is based on different primer concentrations in PCR [3.93]. Asymmetric primer concentrations were originally used to make single stranded (ss) DNA for direct sequencing or for using ss DNA as hybridization probes [3.76]. However, if the concentration of the fluorescently labeled primer is much lower then the concentration of the non-fluorescently labeled primer, fluorescent double stranded DNA and non-fluorescent single stranded DNA can be expected as products of FA-PCR (Fig. 3.7). From the viewpoint of fluorescent molecules, the short single stranded oligonucleotide (fluorescent primer) is elongated and forms double stranded DNA (amplified, fluorescent DNA) during PCR. Because the diffusion time of the DNA is longer than that of the primer, the analysis of the diffusion yields the DNA size (Fig. 3.8). In FA-PCR only one kind of fluorescent labeled primer is needed. Routine PCR conditions can be modified easily to proper amplification conditions for FA-PCR.

APEX (amplified probe extension) is based on conventional PCR with non-fluorescently labeled outer primers in the μM concentration range. In addition, a fluorescently labeled oligonucleotide probe in the nM concentration range is used and binds to amplification products that have been formed in-between the outer primer binding sites. The fluorescently labeled specificity primer is extended during thermal cycling and yields double stranded DNA. The molecular weight of the specific DNA, and thus the increase in diffusion time, can be monitored directly by FCS [3.94]. APEX does not need physical separation of the non-incorporated probe, therefore quantification is performed in closed reaction vials such as special glass bottom microtiter plates (96, 386 or more). The homogeneous assay format is important for pathogenic samples since all the reactions and FCS measurements can be carried out under non-invasive conditions without any contamination of the sample during thermal cycling and analyses.

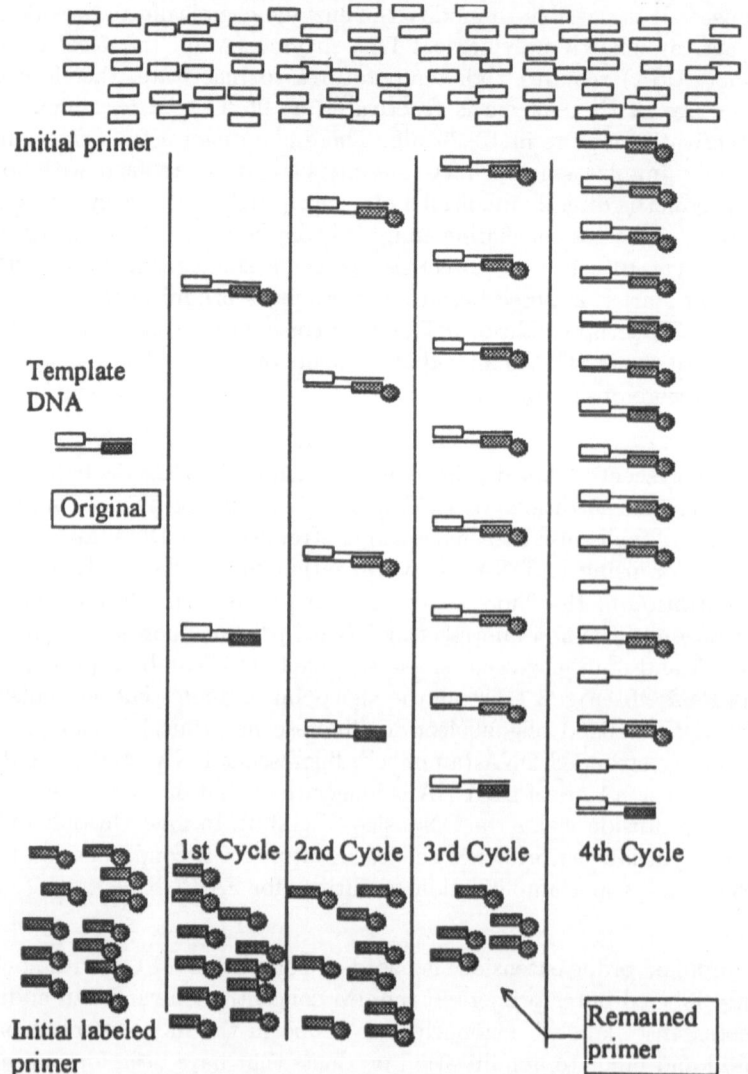

Fig. 3.7. Asymmetric PCR, in which the concentrations of primers are different, was introduced to make single-strand (ss) DNA for direct sequencing or for using ssDNA as hybridization probes. When the concentration of the fluorescently labeled primer (*solid rectangular and star*) in the amplification is lower than that of the non-labeled primer (*open rectangular*), fluorescent dsDNA (*open rectangular with tail,* and *solid rectangular with tail and star*) and non-fluorescent ssDNA (*open rectangular with tail*) can be expected as products of asymmetric PCR. Under ideal conditions, all of the fluorescent primer (low concentration primer) is incorporated into dsDNA after the proper number of PCR cycles. The process of PCR can be analyzed by diffusion constant measurement through FCS

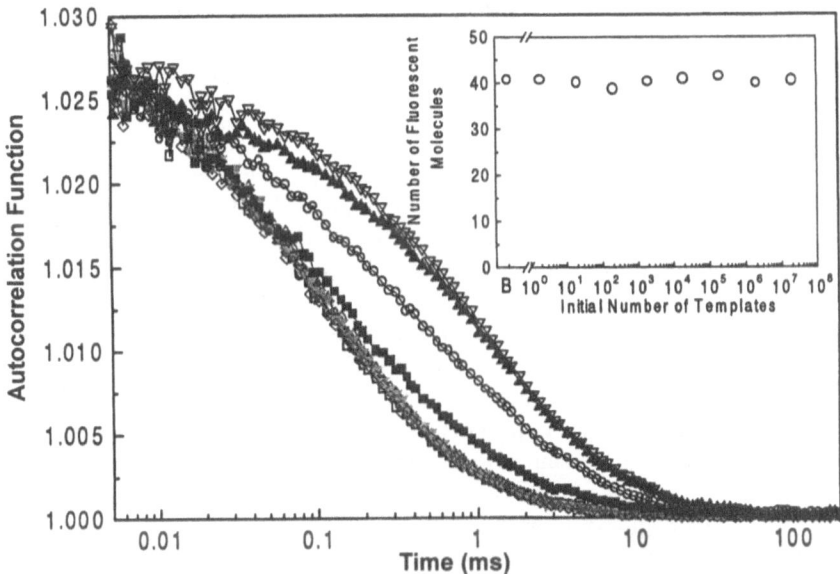

Fig. 3.8. Autocorrelation functions were measured in 10 μl of FA-PCR solution (fluorescence labeled primer based asymmetric PCR) after 30 cycles. Number of template molecules in 25 μl reaction mix: blank (*open square*); 2×10^0 (*solid circle*); 2×10^1 (*open up triangle*); 2×10^2 (*solid down triangle*); 2×10^3 (*open diamond*); 2×10^4 (*solid square*); 2×10^5 (*open circle*); 2×10^6 (*solid up triangle*); 2×10^7 (*open down triangle*). The average number of fluorescent molecules in the volume element is plotted in the inset (*open circle*). Reproduced from [3.93]

RT-PCR. From work in the 1960s, pol1 could be caused to incorporate partially ribonucleotides when activated with manganese [3.95]. In the same way pol1 could slightly read an RNA template. This led to a variety of reverse transcriptases (RT) suitable for PCR. The idea of the single enzyme, single tube, single buffer RT-PCR serves as the basis of all RNA-based virus quantification assays. Generally, templates that many secondary structures do much better with high temperature RT than with low temperature RT [3.75]. The rT-th reverse transcriptase from Perkin-Elmer performs the reverse transcription and the PCR in the same assay. The reversed transcription operates at 70 °C with the "down stream" primer, without using random hexamers or oligo-dT. NASBA (nucleic acid sequence-based amplification, also known as isothermal amplification system, ISA) is an in vitro nucleic acid amplification system with a mixture of RNase H, T7 RNA polymerase and reverse transcriptase (RT). The method avoids heat denaturation (e.g., thermal cycling in PCR); the enzyme alliance amplifies the target sequence via (single stranded) RNAs [3.96]. NASBA is very useful for virus RNA amplification in a small volume. Actually, NASBA was applied to detect the HIV-1 RNA virus by using FCS [3.97]. The combination of NASBA with FCS can be also applied to detect the early stage of infections of other RNA viruses such as hepatitis B and C. This is of particular importance in situations where a time gap (weeks and months) exists between infection and immune response [3.1].

The possibility of using FCS as a quantitation method might open up other applications beyond the scope of present research. For example, SE-LEX pioneering technologies in the early 1990s were used to involve short oligonucleotides that bind to proteins. Aptamers bind to the active site of thermophilic DNA polymerases and shut down the thermophilic DNA polymerases in order to provide a reversible on/off switch of the activity [3.98]. When the aptamer binds the enzyme, the enzyme is inactive. The structure of the aptamer is expected to change the structure of the enzyme. After the aptamer falls off, the enzyme is active. One can work with aptamers in order to improve specificity for manganese-activated RT-PCR [3.25]. In contrast, hot start reactivation with the chemically modified Taq DNA polymerase requires heating the enzyme under certain conditions at 95 °C. These pH conditions are often incompatible with optimum RT activity. Heating RNA at 95 °C is not compatible with optimum detection sensitivity. Using short oligonucleotides that bind and inactivate the RT sustains a circular process. The stringency is imposed because all the aptamers fall inactive by inactivation bonding and then they pull out the inactive enzyme. The aptamer-enzyme complex requires that the enzyme copies the aptamers at elevated temperature and the aptamers again shut down the enzyme at lower temperature. Rounds of that technology involve whole series of aptamers. Thus, aptamers are another way of improving specificity and they may be used in a wide range of PCR applications in combination with FCS as a means to sensitively measure diffusion times of fluorescently labeled nucleic acids.

DNA Polymerases for Special Applications. In addition to specificity enhancement over the last 10 years, there have been varieties of PCR tools, focusing primarily on the design of DNA polymerases. Low resolution structures of the active site of the families of DNA polymerases showed that the C-terminal domain gives the enzyme its name, DNA polymerase [3.95]. The enzymes bind to DNA primer templates and incorporate the deoxynucleotide triphosphates. They cannot bind to RNA templates and incorporate ribonucleotides and modified analogs, like dideoxynucleotides, without termination.

Some of the enzymes may have an inherent proofreading activity. The $3' \rightarrow 5'$ exonuclease activity recognizes single stranded DNA. This can offer a benefit in PCR by providing enhanced fidelity. There are several ways of achieving enhanced fidelity in PCR with proofreading enzymes as well as with non-proofreading enzymes. When the evolution created the 3'-exonucleolytic site of DNA polymerases to have proofreading activity, the site was not designed with a PCR technologist in mind. In evolution the site was designed only to remove a base for an extended primer containing mismatch. This means the primers can be modified during the course of PCR and they cheat back from their 3′-ends. Therefore it is difficult to facilitate mismatch extension discrimination with a proofreading enzyme. In addition, the 3′-end of the strand that is just synthesized is single stranded during the denaturation coming down to the annealing phase of PCR. The last few nucleotides that have been synthesized by the enzyme are the first nucleotides to be removed when the strands are no longer duplex. The N-terminal domain may encode a 5′-nuclease activity. This activity can be a benefit, and led to the hybridization based, real time signal generating TaqMan assay [3.99], in which an energy transfer oligonucleotide not acting as a primer, carries fluorogenic reporter and quencher dyes. The energy transfer oligonucleotide is degraded in a nick translation reaction. The increase of emitted fluorescence after release of the fluorogenic reporter dye nucleotide is recorded and evaluated quantitatively. However, what will be a benefit in certain applications, can be a disadvantage in others and can contribute to product degradation. A recent alternative has been based on two color fluorescence cross-correlation spectroscopy [3.56].

Finding DNA polymerases with superior acceptance and proper incorporation of fluorophore-bearing dNTPs by the $5' \to 3'$ polymerase activity is crucial for high density labeling of natural DNA (Sect. 3.4). We have selected a Tgo DNA polymerase (Thermococcus gorgonarius) for error free substitution of a normal nucleotide by the Rhodamine Green-X-dUTP derivative [3.81]. The enzyme is a $3' \to 5'$ exonuclease deficient mutant. The enzyme belongs to the thermostable type B family of DNA polymerases. Very recently, the 2.5 Å resolution crystal structure of the Tgo-DNA polymerase with $3' \to 5'$ exonuclease activity was described [3.100]. DNA polymerases are clustered into four different structural families (pol A, pol B, pol C, and pol X with pol β and terminal transferase), built from a limited set of subdomains. The common functional polymerase structure among known polymerase families consists of the palm, fingers and thumb domains. The fact that all DNA polymerases, includ Tgo-DNA polymerase, retain a similar topology of the palm domains except polymerase β suggests that during evolution, the tertiary structure of the palm domains changes much more slowly than the structure of the highly diverse fingers and thumb domains. Evolutionary technologies [3.2] would allow the generation of billions of variants of enzymes in order to produce molecular structures of DNA polymerases with further improved labeling characteristics. FCS can detect such tailor made molecules quickly and efficiently by screening.

3.7.2   FCS Cross-Correlation Analysis: A New Concept for PCR

A considerable bulk of experience has been accumulated in our laboratory concerning the fluorescence-based amplification at high template concentrations [3.32,3.79,3.86,3.93,3.101]. However, the amplification of initially low copy numbers of template is still a general problem [3.27]. The reliability of PCR results can be seriously compromised by preparations and analysis conditions. For example, starting with 10^8 down to 10^5 copies of template, the only visible product in gel electrophoresis is that of the expected size [3.102]. When 10^4 starting copies were used, a smaller non-specific amplification product became visible. This product was more prominent when 10^3 copies were used. Using 10^2 copies an equal amount of non-specific and specific product were produced. Sample-to-sample variation of the reaction profiles occured at less than 10^2 template copies, complicating reproducibility below this level [3.102]. "Background" products can also be detected by autocorrelation FCS [3.32].

For many a that only a single species of DNA is formed. Usually, a single species means a single band of DNA on an agarose or polyacrylamide gel, but only sequencing provides the evidence of a homogeneous or nearly homogeneous population of DNA. However, in most cases the non-specific products interfere with the subsequent DNA analysis. Even if a product signal in the negative controls cannot be detected and the product is of the expected size, as well, there is no guarantee that by using non-specific modes of detection that the specific target is present or the correct sequence of the specific target amplicon is produced in the amplification process.

Two color cross-correlation FCS is a specific mode of detection by which the correct amplification product containing, for example, both amplification primers in different colors as specificity probes, is identified and quantified [3.56]. Generally, the specificity oligonucleotides in two colors lie at any position inbetween the target sequence and are bound to the correct target by hybridization. The two color cross-correlation results only from fluorescence intensity fluctuations for which the specificity probes in two colors are incorporated into (or, from the viewpoint of the FCS model: "bound to") the same target molecule. Amplification and cross-correlation analysis are independent and do not interfere with each other. This is superior to the concept of energy transfer primers [3.103] using fluorogenic reporter and quencher dyes (TaqMan PCR detection strategy, see Sect. 3.7).

To make the amplification process quantifiable, no external or internal stand required [3.56]. The amplification is made with all igredients (template DNA, primers, nucleotides, DNA polymerase, buffer, etc.) in the same closed compartment, in which the amplification and measurements are carried out. Postamplification purification steps are avoided. The homogeneous one-tube approach does not depend on fluorescence energy transfer between the fluorogenic dyes. Two color cross-correlation FCS is a very sensitive and specific, as well as a selective and effective biotechnology [3.56]. It allows the continued use of different conditions for amplifying DNA. Although not shown, it is useful for amplification assays in the microchip format to screen for clinically relevant pathogens.

The schematic of the two color fluorescence cross-correlation spectrometer optical setup is shown in Fig. 3.9. It was a prototype built by Carl Zeiss and based on the Confocor device. The main features of the spectrometer are given in [3.56] and [3.57]. The instrument incorporates a confocal opti-

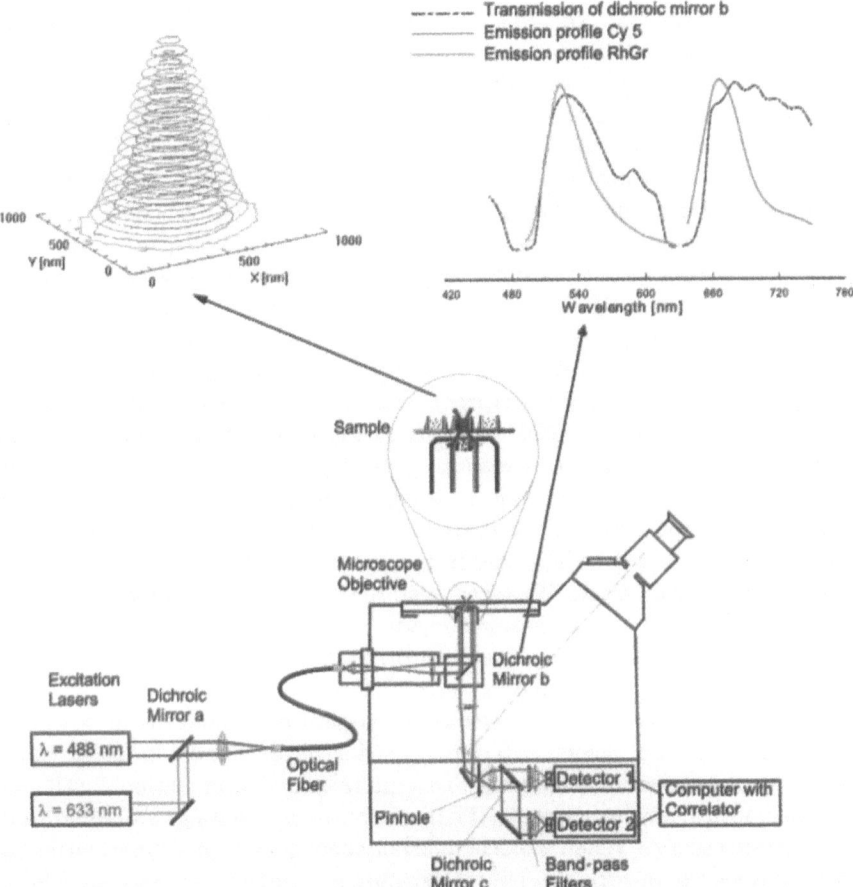

Fig. 3.9. A schematic setup of the fluorescence cross-correlation spectrometer (built by Carl Zeiss). A single line argon ion laser (488 nm) and a helium neon laser (633 nm) are used for the excitation of the sample molecules. The excitation laser light is coupled into the microscope via one single mode fiber. Fluorescence light emitted by the sample molecules is detected in a confocal arrangement. Two single-photon detectors are used. The detector outputs are connected to the hardware correlator. In the *upper right inset* the transmission spectrum of dichroic mirror b (*dashed-dotted line*) is shown together with the emission spectra (*solid lines*) of Rhodamine Green and Cy5 (both dyes were obtained from Amersham Life Science). The wavelengths of the excitation lasers are indicated by *vertical lines*. In the *upper left inset* the intensity distributions of the blue and red laser excitation light in the focal plane of the microscopic object are shown, as measured using a fiber focus scanner. The spatial distance of the two foci is less than 30 nm. See also [3.56]

cal arrangement [3.4–3.6]. The single line argon-ion laser (488 nm, blue) and helium:neon laser (633 nm, red) were co-linearly used to excitate the sample molecules. The emission was detected at 540 nm (green, peak intensity value) and at 685 nm (red, peak intensity value). The laser light was focused by the microscope objective to two superimposed, nearly diffraction-limited spots at the focal plane within the sample. The spatial distance between the blue and red intensity maxima in the focal plane was below 30 nm, as measured with a fiber focus scanner. The observed cross-correlated volume element depended on the intensity profiles around the focal spots of the laser beams at the the optical system. The detection probability profiles are the criteria which determine, to a large extent, the goodness of both the cross-correlation and autocorrelation measurements. The fluorescence emissions were collected by the objective (numerical aperture 1.2) and passed through dichroic mirror b. The dichroic mirror reflects the excitation light (488 nm, 633 nm) in two wavelength regions and transmits the emitted fluorescence (green and red) in two wavelength regions. The transmitted fluorescence light is focused by a lens on a pinhole (30 nm). After passing through a dichroic mirror (dichroic mirror c), residual laser light and Raman scattered light were removed by additional band-pass filters. The green and red fluorescence were focused on two actively quenched avalanche photodiodes (photon counting mode, EG&G). The detector output signals are autocorrelated or cross-correlated by a two channel correlator on a computer (ALV5000).

As the fluorescent two color molecules diffused in and out of the cross-correlated volume element, fluorescence fluctuations arose from stochastic concentration fluctuations about an equilibrium concentration in the well defined optical volume element. The force of the fluctuations is related to the Boltzmann energy. When a biochemical or chemical system is fluctuating, the fluctuations can only be detected if the number of molecules is decreased (e.g., nM or pM concentration range). Poisson statistics describe the trajectories of the two color molecules. Fluorescence intensity fluctuations are compared by using the correlation functions. Figure 3.10 shows that the double stranded target DNA carrying both primers in two colors is formed during the PCR. Its cross-correlated signal emerges from the background of non-incorporated or non-specifically incorporated primers.

We now discus the properties of the two color DNA that can be obtained from the fluctuations in the fluorescence intensities. Figure 3.10 shows a 389-

---

→

Fig. 3.10. a Measured and calculated two-color FCS cross-correlation curves of 389-bp DNA which was 5'-labeled with Rhodamine Green and Cy5 via the amplification primers in PCR. Measured and calculated. **b** red autocorrelation and **c** green autocorrelation curves for the same PCR mix containing the two color 389-bp DNA, the free, labeled primers and the free dyes. For the 389-bp DNA we obtained the following diffusion times: $\tau_{cross} = 2.135$ ms, $\tau_{Red} = 2.599$ ms, $\tau_{Green} = 1.671$ ms

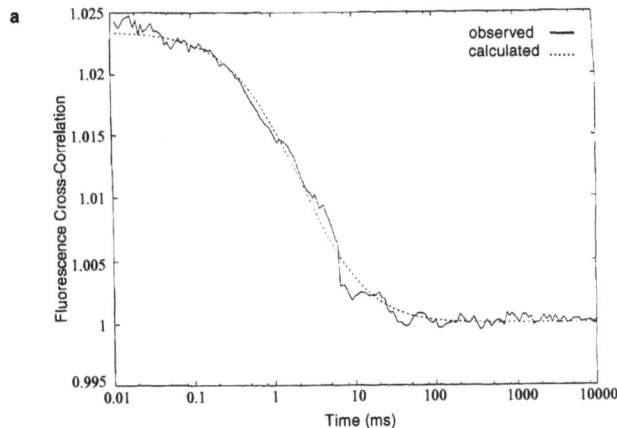

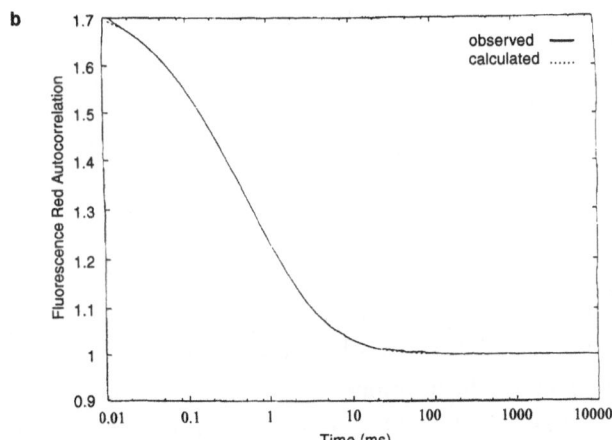

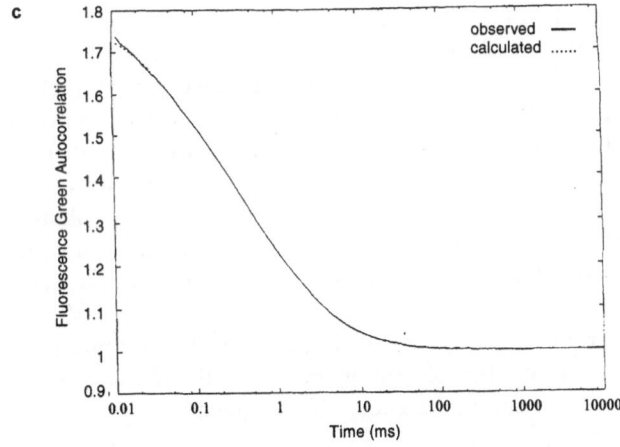

bp two color DNA. The number of two color molecules in the cross-correlated volume element can be calculated in detail. If there is a large excess of two color molecules over molecules containing only one color, then the amplitude of the cross-correlated signals is related to the number of two color molecules by

$$\frac{1}{N} = \frac{1}{N_{gr}}.$$ (3.10)

We have presented this very special case, because the restricted case has been experimentally and theoretically analyzed in detail [3.55,3.58–3.60].

The number of two color molecules in the sample can be changed, for example, by the kinetics of the biochemical process under study. In a situation where it is not possible to neglect the molecules carrying only one color, the motions of the free labels are independent of each other, but the cross-correlation function is due to the species which has both colors in the same molecule. N will then contain correlated and non-correlated quantities [3.55,3.56]. In other words, many molecules in one color give a higher N value (smaller amplitude), which is not identical with the number of two color molecules. The value of N has to be corrected for the number of molecules containing only one label. We have first introduced terms into the number density function (3.11) relating the number of green and red molecules showing no correlation in two colors and the emission rates per molecules $Q_{\lambda\text{excitation}}^{\lambda\text{emission}}$ [3.56]:

$$\frac{1}{N} = \frac{N_g(Q_B^R/Q_R^R) + N_{gr}[1 + (Q_B^R/Q_R^R)]}{[N_g + N_{gr}][N_r + N_g(Q_B^R/Q_R^R) + N_{gr}[1 + (Q_B^R/Q_R^R)]]}.$$ (3.11)

For this case, the expression for fluctuations of the cross-correlated fluorescence intensities due to thermodynamic fluctuations of the number of molecules yields a slightly more compact expression. The cross-correlating particle numbers are calculated by normalizing with the autocorrelation results and the cross talk of the optical setup.

It should be pointed out that (3.11) in fact reflects the cross talk in one direction (from green to red) of the emission rates per molecule $Q_{\lambda\text{excitation}}^{\lambda\text{emission}}$. The cross talk occurs since a small portion of the photons emitted by the green fluorophore is transmitted through the fluorescence filters, dichroic mirrors (or beam splitter) used for the red fluorophore, and detected. In our optical setup (Fig. 3.9), we suppressed the cross talk to about 2% of the photon counts per molecule and second at blue excitation and green emission measured for Rhodamine Green. The other case of a cross talk from red to green can be disregarded.

We discuss the realistic case $(Q_B^R/Q_R^R) \ll 1$. From (3.11), the practical relationship between N, the number of free ligands, N_g and N_r, as well as the number of cross-correlated target molecules N_{gr} is given by

$$\frac{1}{N} = \frac{N_{gr}}{(N_g + N_{gr})(N_r + N_{gr})}.$$ (3.12)

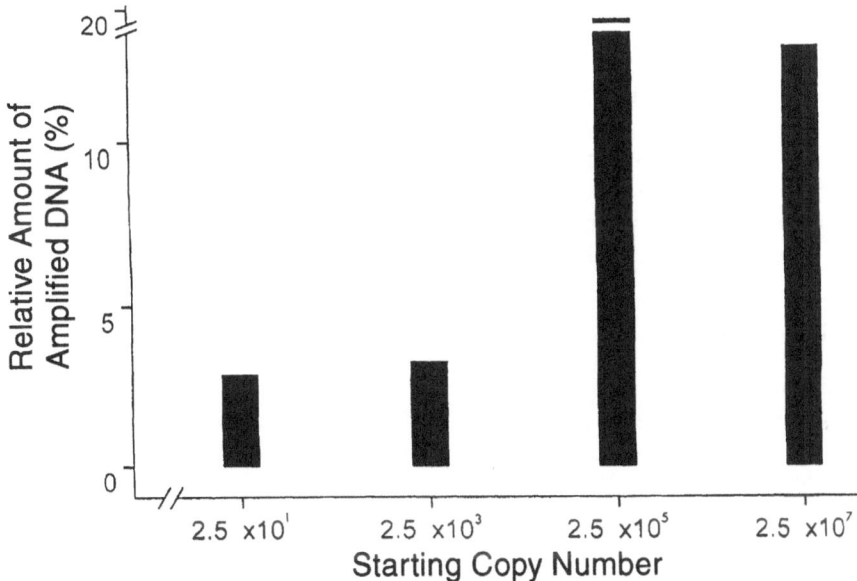

Fig. 3.11. Plot of the relative amount of amplified two color 217-bp DNA versus the initial copy number of the template DNA (single stranded M13mp18$^+$ DNA). The relative amount of amplified DNA was evaluated from the the cross-correlation measurements by taking into account the total number of molecules in the blue, red as well as the cross-correlated volume elements. The corresponding concentrations of the template DNA in the PCR mixes were: 7×10^{-19} M, 7×10^{-17} M, 7×10^{-15} M, and 7×10^{-13} M

The total number of green labeled molecules in V_g is $N_{g,\text{total}} = N_g + N_{gr}$ and the total number of red labeled molecules in V_r is $N_{r,\text{total}} = N_r + N_{gr}$. Equations (3.11) and (3.12) only hold for a single incorporated (or from the viewpoint of the FCS model: "bound to") label of each color [3.56]. Equation (3.12) is exact if no green emission exists in the red channel. This novel method of analysis focuses on the quantification of the number density. The quantity of greatest interest is the number of two color molecules N_{gr} in the cross-correlated volume element. Using this method, it is possible to distinguish the specific interactions caused by molecular species carrying both colors from the molecular species containing only one color. With this approach, the amount of amplified, two color 217-bp DNA is calculated in Fig. 3.11 and depicted for different initial copy numbers of template.

3.8   Summary and Conclusions

The main role of nucleic acid analysis is in monitoring and understanding of various biological processes. This chapter presented a survey on selected areas in nucleic acid analysis in which FCS has been used for particular purposes. FCS is still very much an evolving methodology and by no means a routine method. A recent development is two color cross-correlation FCS. The examples give some idea of the wide range of molecular biological, biochemical and biophysical properties that may be analyzed using FCS. They focus on the important aspects of nucleic acid analysis. We have demonstrated that FCS can contribute to studies on the molecular basis of biological processes that are crucial in molecular medicine. Most topics that have been discussed are of general interest and can be applied to many other applications.

Acknowledgement

Dr.rer.nat. Dr.med. Zeno Földes-Papp thanks Professor Dr. Rudolf Rigler for helpful discussions and critical reading of the manuscript. The work for this article was supported by grants from the Karolinska Institute, Stockholm, Sweden.

References

3.1 R. Rigler: J. Biotechnol. **41**, 177–186 (1995)

3.2 M. Eigen and R. Rigler: Proc. Natl. Acad. Sci. USA **91**, 5740–5747 (1994)

3.3 S. Maiti, U. Haupts, and W.W. Webb: Proc. Natl. Acad. Sci. USA **94**, 11753–11757 (1997)

3.4 R. Rigler and Ü. Mets: SPIE Laser Spectroscopy of Biomolecules **1921**, 239–248 (1992)

3.5 R. Rigler, Ü. Mets, J. Widengren, and P. Kask: Eur. Biophys. J. **22**, 169–175 (1993)

3.6 Ü. Mets and R. Rigler: J. Fluoresc. **4**, 259–264 (1994)

3.7 J. Dapprich, Ü. Mets, W. Simm, M. Eigen, and R. Rigler: Exp. Tech. Phys. **41**, 259–264 (1995)

3.8 L. Edman, Ü. Mets, and R. Rigler: Proc. Natl. Acad. Sci. USA **93**, 6710–6715 (1996)

3.9 S. Wennmalm, L. Edman, and R. Rigler: Proc. Natl. Acad. Sci. USA **94**, 10641–10646 (1997)

3.10 L. Edman, S. Wennmalm, F. Tamsen, and R. Rigler: Chem. Phys. Lett. **63**, 97–109 (1998)

3.11 S. Wennmalm and R. Rigler: J. Phys. Chem. B **103**, 2516–2519 (1998)

3.12 S. Wennmalm, L. Edman, and R. Rigler: Chem. Phys. **247**, 61–67 (1999)

3.13 L. Edman, Z. Földes-Papp, S. Wennmalm, and R. Rigler: Chem. Phys. **247**, 11–22 (1999)

3.14 D. Magde, E.L. Elson, and W.W. Webb: Phys. Rev. Lett. **29**, 705–711 (1972)

3.15 E.L. Elson and D. Magde: Biopolymers **13**, 1–27 (1974)

3.16 D.W. Koppel: Phys. Rev. **10**, 1938–1945 (1974)

3.17 M. Ehrenberg and R. Rigler: Chem. Phys. **4**, 390–401 (1974)

3.18 C. Eggeling, J.R. Fries, L. Brand, R. Günther, and C.A.M. Seidel: Proc. Natl. Acad. Sci. USA **95**, 1556–1561 (1998)

3.19 G. Bonnet, O. Krichevsky, and A. Libchaber: Proc. Natl. Acad. Sci. USA
      95, 8602–8606 (1998)
3.20 H. Qian and E.L. Elson: Biophys. J. **76**, 1598–1605 (1999)
3.21 E.M. Southern: J. Mol. Biol. **98**, 503–517 (1975)
3.22 J.C. Alvine, D.J. Kemp, and G.R. Stark: Proc. Natl. Acad. Sci. USA **74**
      5350–5354 (1977)
3.23 T.J. Kelly and H.O. Smith: J. Mol. Biol. **51**, 393–409 (1970)
3.24 K. Danna and D. Nathans: Proc. Natl. Acad. Sci. USA **68**, 2913–2917 (1971)
3.25 S.M. Koepf: Micro-array technologies for high through-put genetic analysis.
      In: Proc. IBC's International Conference on Massively Parallel DNA Anal-
      ysis – Application of Alternative DNA Diagnostics, San Francisco (1998)
3.26 R.F. Service: Science **283**, 1669 (1999)
3.27 E. Zubritsky: Anal. Chem. **71** (5), 191A–195A (1999)
3.28 W. Sanger: Principles of Nucleic Acid Structure. (Springer, New York 1988)
3.29 K.J. Breslauer, R. Frank, H. Blöcker, and L.A. Marky: Proc. Natl. Acad.
      USA **83**, 3746–3750 (1986)
3.30 K.J. Breslauer: Methods for obtaining thermodynamic data on oligonu-
      cleotide transitions. In: Thermodynamic data for Biochemistry and Biotech-
      nology. H.-J. Hinz, ed. (Springer, New York 1986) pp.402–427
3.31 B. Rauer, E. Neumann, J. Widengren, and R. Rigler: Biophys. Chem. **58**,
      3–12 (1996)
3.32 S. Björling, M. Kinjo, Z. Földes-Papp, E. Hagman, P. Thyberg, and R.
      Rigler: Biochemistry **37**, 12971–12978 (1998)
3.33 M. Sauer, K.–T. Han, R. Muller, S. Nord, A. Schultz, S. Seeger, J. Wolfrum,
      J. Arden–Jacobi, G. Deltau, N.J. Marx, C. Zander, and K.H. Drexhage: J.
      Fluoresc. **5**, 247–261 (1995)
3.34 J. Widengren, J. Dapprich, and R. Rigler: Chem. Phys. **216**, 417–428 (1997)
3.35 M. Kinjo and R. Rigler: Nucleic. Acids. Res. **23**, 1795–1799 (1995)
3.36 E. Uhlmann, A. Peyman, G. Breipohp, and W.D. Will: Angew. Chem. Int.
      Ed. Engl. **37**, 2796–2823 (1998)
3.37 P. Aich, P. Nielsen, and R. Rigler: Nucleosides & Nucleotides **16**, 609–615
      (1997)
3.38 P. Schwille, F. Oehlenschlager, and N.G. Walter: Biochemistry **35**, 10182–
      10193 (1996)
3.39 E. Uhlmann and A. Peyman: Chem. Rev. **90**, 543–584 (1990)
3.40 C.A. Stein and Y.C. Cheng: Science **261**, 1004–1012 (1993)
3.41 S.T. Crooke and C.F. Bennett: Annu. Rev. Pharmacol. Toxicol. **36**, 107–130
      (1996)
3.42 G. Schwab, C. Chavany, I. Duroux, G. Goubin, J. Lebeaur, C. Hélène, and
      T. Saison–Behmoaras: Proc. Natl. Acad. Sci. USA **91**, 10460–10464 (1994)
3.43 V.V. Vlassov, I.E. Vlassova, and L.V. Pautova: Oliogonucleotides and
      polynucleotides as biologically active compounds. In: Progress in Nucleic
      Acid Research and Molecular Biology; vol.57. K. Moldave, ed. (Academic
      Press, San Diego 1997) pp.95–143
3.44 P.S. Miller: Development of antisense and antigene oligonucleotide analogs.
      In: Progress in Nucleic Acid Research and Molecular Biology; vol.52. W.E.
      Cohn, K. Moldave, eds. (Academic Press, San Diego 1996) pp.261–291
3.45 A.V. Orden and R.A. Keller: Anal. Chem. **70**, 4463–4471 (1998)
3.46 T. Winkler, U. Kettling, A. Koltermann, and M. Eigen: Proc. Natl. Acad.
      Sci. USA **96**, 1375–1378 (1999)

3.47 J. Politz, E. Browne, D. Wolf, amd T. Pederson: Proc Natl Acad Sci USA **95**, 6043–6048 (1998)

3.48 P. Wentworth and K.D. Janda: Current Opinion in Biotechnology **9**, 109–115 (1998)

3.49 Z. Földes-Papp, G. Baumann, E. Birch-Hirchfeld, H. Eickhoff, K.O. Greulich, A.K. Kleinschmidt, and H. Seliger: Biopolymers **45**, 361–379 (1998)

3.50 R.J. Lipshutz, S.P.A. Fodor, T.R. Gingeras, and D.J. Lockhart: Nature Genetics Microarray Suppl. **21**, 20–24 (1999)

3.51 D.D.L. Bowtell: Nature Genetics Microarray. Suppl. **21**, 25–32 (1999)

3.52 M.A. Burns, B.N Johnson, S.N. Brahmasandra, K. Handique, J.R. Webster, M. Krishnan, T.S. Sammarco, P.M. Man, D. Jones, D. Heldsinger, C.H. Mastrangelo, and D.T. Burke: Science **282**, 484–487 (1998)

3.53 H.–P. Chou, C. Spence, A. Scherer, and S. Quarke: Proc. Natl. Acad. Sci. USA **96**, 11–13 (1999)

3.54 E. Southern, K. Mir, M. Shchepinow: Nature Genetics, Supplement **21**, (Microarray) 5–9 (1999)

3.55 P. Schwille, F.–J. Meyer–Almes, and R. Rigler: Biohys. J. **72**, 1878–1886 (1997)

3.56 R. Rigler, Z. Földes–Papp, F.–J. Meyer–Almes, C. Sammet, M. Völcker, and A. Schnetz: J. Biotechnol. **63**, 97–109 (1998)

3.57 Z. Földes–Papp and R. Rigler: Quantitative two-color fluorescence cross-correlation spectroscopy in the analysis of polymerase chain reaction. Biol. Chem. (2001) in press

3.58 U. Kettling, A. Koltermann, P. Schwille, and M. Eigen: Proc. Natl. Acad. Sci. USA **95**, 1416–1420 (1998)

3.59 A. Koltermann, U. Kettling, J. Bieschke, T. Winkler, and M. Eigen: Proc. Natl. Acad. Sci. USA **95**, 1421–1426 (1998)

3.60 T. Winkler, U. Kettling, A. Koltermann, and M. Eigen: Proc. Natl. Acad. Sci. USA **96**, 1375–1378 (1999)

3.61 J. Holm, H. Elderstig, O. Kristensen, and R. Rigler: μTAS '96, 85–87 (1996)

3.62 J. Holm, H. Ederstig, O. Kristensen, and R. Rigler: Nucleosides & Nucleotides **16**, 557–562 (1997)

3.63 J. Holm, Z. Földes–Papp, P. Thyberg, and R. Rigler: Single bead analysis and enzymatic DNA degradations in quartz glass microstructures. Personal communication (1998)

3.64 N.J. Dovichi, J.C. Martin, J.H. Jett, and R.A. Keller: Science **219**, 845–847 (1983)

3.65 D.C. Nguyen, R.A. Keller, J.H. Jett, and J.C. Martin: Anal. Chem. **59**, 2158–2161 (1987)

3.66 D.C. Nguyen, R.A. Keller, and M. Trikula: J. Opt.Soc. Am. B **4**, 138–143 (1987)

3.67 E.B. Shera, N.K. Seitzinger, L.M. Davis, R.A. Keller, and S.A. Soper: Chem. Phys. Lett. **174**, 553–557 (1990)

3.68 S.A. Soper, L.M. Davis, and E.B. Shera: J. Opt. Soc. Am. B **9**, 1761–1769 (1992)

3.69 J.H. Jett, R.A. Keller, J.C. Martin, B.L. Marrone, R.K. Moyzis, R.L. Ratliff, N.K. Seitzinger, E.B. Shera, and C.C. Stewart: J. Biomol. Struct. Dyn. **7**, 301–309 (1989)

3.70 L.M. Davis, F.R. Fairfield, C.A. Harger, J.H. Jett, R.A. Keller, J.H. Hahn, L.A. Krakowski, B.L. Marrone, J.C. Martin, H.L. Nutter, R.L. Ratliff, E.B. Shera, D.J. Simpson, and S.A. Soper: Genetic Analysis: Techniques and Applications **8**, 1–7 (1991)

3.71 W.P. Ambrose, P.M. Goodwin, J.H. Jett, M.E. Johnson, J.C. Martin, B.L. Marrone, J.A. Schecker, C.W. Wilkerson, R.A. Keller, A. Haces, P.–O. Shih, and J.D. Harding: Ber. Bunsenges. Phys. Chem. **97**, 1535–1542 (1993)

3.72 P.M. Goodwin, H. Cai, J.H. Jett, S.L. Whang, N.P. Machura, D.J. Semin, A.V. Orden, and R.A. Keller: Nucleosides & Nucleotides **16**, 543–550 (1997)

3.73 K. Dörre, S. Brakmann, M. Brinkmeier, K.–T. Han, K. Riebesel, P. Schwille, J. Stephan, T. Wetzel, M. Lapczyna, M. Stuke, R. Bader, M. Hinz, H. Seliger, J. Holm, M. Eigen, and R. Rigler: Bioimaging **5**, 139–152 (1997)

3.74 J. Stephan, K. Dörre, S. Brakmann, T. Winkler, T. Wetzel, M. Lapczyna, M. Stuke, B. Angerer, W. Ankenbauer, Z. Földes-Papp, R. Rigler, and M. Eigen: J. Biotechnol. **86** (3), (2001) in press

3.75 Z. Larin, M.D. Fricker, E. Maher, Y. Ishikawa-Brush, and E.M. Southern: Nucleic Acids Res. **22**, 3689–3692 (1994)

3.76 K. Anamthawat–Jónsson and S.M. Reader: Genome **38** (4), 814–816 (1995)

3.77 D. Anderson, T.–U. Yu, and M.A. Browne: Mutation Res. **390**, 69–77 (1997)

3.78 H. Zitzelsberger, L. Lehmann, M. Werner, and M. Bauchinger: Histochem. Cell. Biol. **108**, 403–417 (1997)

3.79 Z. Földes-Papp, B. Angerer, W. Ankenbauer, G. Baumann, E. Birch-Hirschfeld, S. Björling, S. Conrad, M. Hinz, R. Rigler, H. Seliger, P. Thyberg, and A.K. Kleinschmidt: Modeling the dynamics of nonenzymatic and enzymatic nucleotide processes by fractal dimension. In: Fractals in Biology and Medicine, Vol.II. G.A. Losa, D. Merlini, T.F. Nonnenmacher, E.R. Weibel, eds. (Birkhäuser, New York 1998) pp.238–254

3.80 M. Augustin, W. Ankenbauer, and B. Angerer: J. Biotechnol. **86** (3), (2001) in press

3.81 Z. Földes-Papp, B. Angerer, W. Ankenbauer, and R. Rigler: J. Biotechnol. (2001) **86** (3), in press

3.82 Z. Földes-Papp, B. Angerer, and R. Rigler: Filling-up reactions with mesophilic and thermostable polymerases and reverse transcriptases for high-density labeling of DNA by complete substitutions of fluorescent dye-dNTPs for their respective natural dNTPs. Personal communication (1997)

3.83 Z. Földes-Papp, B. Angerer, P. Thyberg, M. Hinz, S. Wennmalm, W. Ankenbauer, H. Seliger, A. Holmgren, smf R. Rigler: J. Biotechnol. **86** (3), (2001) in press

3.84 M. Sauer, B. Angerer, W. Ankenbauer, K.H. Dexhage, Z. Földes-Papp, F. Göbel, K.–T. Han, R. Rigler, J. Wolfrum, and C. Zander: J. Biotechnol. (2001) **86** (3), in press

3.85 G. Nishimura, R. Rigler, and M. Kinjo: Bioimaging **5**, 129–133 (1997)

3.86 Z. Földes-Papp, P. Thyberg, S. Björling, A. Holmgren, and R. Rigler: Nucleotides & Nucleosides **16**, 781–787 (1997)

3.87 K. Hori, D.F. Mark, und C.C. Richardson: J. Biol. Chem. **254**, 11598–11604 (1997)

3.88 I. Slaby and A. Holmgren: Protein Expression and Purification **2**, 270–277 (1991)

3.89 S. Forsblom, R. Rigler, M. Ehrenberg, U. Petterson, and L. Philipson: Nucleic Acids Res. **3**, 3255–3269 (1976)

3.90 P.R. Billings, C.L. Smith, and C.R. Cantor: FASEB J. **5**, 28–43 (1991)

3.91 M. Kinjo, G. Nishimura, T. Koyama, Ü. Mets, and R. Rigler: Anal. Biochem. **260**, 166–172 (1998)

3.92 M. Kinjo and G. Nishimura: Bioimaging **5**, 134–138 (1997)

3.93 M. Kinjo: Biotechniques **25**, 706–715 (1998)

3.94 N.G. Walter, P. Schwille, and M. Eigen: Proc. Natl. Acad. Sci. USA **93**,12805–12810 (1996)

3.95 A. Kornberg and T. Baker: DNA Replication. 2nd edn. (Freeman, New York 1992)

3.96 P. Tijssen: Hybridization with nucleic acid probes. In: Laboratory Techniques in Biochemistry and Molecular Biology. (Elsevier, Amsterdam 1999) pp.207–211

3.97 F. Oehlenschlager, P. Schwille, and M. Eigen: Proc. Natl. Acad. Sci. USA **93**, 12811–12816 (1996)

3.98 C. Dang and D. Jayasena: J. Mol. Biol. **264**, 268–278 (1996)

3.99 Y.S. Lie and C.J. Petropoulos: Current Opinion in Biotechnology **9**, 43–48 (1998)

3.100 K.-P. Hopfner, A. Eichinger, R.A. Engh, F. Laue, W. Ankenbauer, R. Huber, and B. Angerer: Proc. Natl. Acad. Sci. USA **96**, 3600–3605 (1999)

3.101 M. Kinjo: Anal. Chim. Acta **365**, 43–48 (1998)

3.102 R. Higuchi, C. Fockler, G. Dollinger, and R. Watson: Bio/Technology **11**, 1026–1030 (1993)

3.103 J. Ju, A.N. Glazer, and R.A. Mathies: Nature Medicine **2**, 236–249 (1996)

3.104 R. Brock, G. Vàmosi, G. Vereb, and T.M. Jovin: Proc. Natl. Acad. Sci. USA **96**, 10123–101128 (1999)

3.105 J. Korlach, P. Schwille, W.W Webb, and G.W. Feigenson: Proc. Natl. Acad. Sci. USA **96**, 8461–8466 (1999)

4 Strain-Dependent Fluorescence Correlation Spectroscopy: Proposing a New Measurement for Conformational Fluctuations of Biological Macromolecules

Hong Qian and Elliot L. Elson

4.1 Introduction

Biological macromolecules are often highly structured and perform complex functions. Due to this complexity, individual macromolecules in a population can differ from each other at any moment in behavior and properties. This variation from molecule to molecule within a population can obscure measurements of structural and functional details. Therefore, the ability to manipulate and characterize individual macromolecules could substantially improve our access to detailed information about molecular structure and function, and is a goal sought avidly during the last three decades. However, analysis of the behavior of single molecules differs fundamentally from that of conventional chemical kinetics: processes involving single molecules are inevitably stochastic because of thermal fluctuations.

Similar considerations apply to a kinetic analysis of a complex biochemical process. The goal of such analysis is to determine a molecular mechanism in terms of simple "elementary" steps. The level of simplicity to which these steps can be reduced depends on the resolution and sensitivity of the experimental methods. For many important processes involving biological macromolecules, e.g., protein folding and the formation of ordered nucleic acid structures, the molecular states connected by the elementary steps are closely spaced in free energy. Therefore, traditional perturbation-relaxation methods would sample many steps in a single measurement, complicating the interpretation of the data. For these processes it is desirable that measurements use the smallest possible perturbation to sample the smallest possible number of states. The result is the most detailed mechanism of the kinetic process. The thermal agitation, determined by the ambient temperature, defines the lower limit of the possible perturbation. Fluorescence correlation spectroscopy (FCS), which measures the kinetics of spontaneous fluctuations, operates at this minimal level of perturbation. In these measurements, macroscopic perturbations are replaced by the thermal energy $\sim k_{\mathrm{B}}T$ (k_{B} is Botzmann's constant and T is temperature in Kelvin). Conventional phenomenological rate constants and transport coefficients are determined from a statistical analysis of the time courses of the spontaneous fluctuations [4.1–

4.3]. FCS has been increasingly applied to studying biological macromolecules at the single molecule level in recent years.

Macromolecular processes are often highly cooperative. By cooperative, we mean that a set of elementary steps is compressed into a single transition that is difficult to decompose into its constituents. Kinetic analysis under conditions in which direct control of the macromolecular conformation prevented a subset of conformational changes could alleviate this problem. For example, refolding of an unfolded protein in vitro is thought often to pass through a condensed, "molten" globular state. The collapsing amplified two color 217-bp DNA process and the folding of a specific structure from this globule are highly cooperative processes in which it is difficult to discern individual conformational steps. If the protein molecule could be held in intermediate states between unfolded extended and disordered condensed forms, it would be possible to measure the kinetics of transitions specifically associated with the relatively smaller collection of microscopic substates within the intermediate state. In this chapter we propose an experimental approach in which conformational fluctuations are observed in a single macromolecule held in a specific range of conformations by an externally applied stress or strain. The independent variable could be either the strain, which controls the molecular conformation, or the stress or force applied to the molecule, which causes the strain. Hence, we name the approach SD-FCS, which could stand for either stress-dependent or strain-dependent FCS.

The possibility of carrying out this kind of measurement has been suggested by recent studies of the unfolding of proteins [4.4–4.6], DNA [4.7–4.9], and other molecules [4.10] under an externally applied tension. Using AFM or laser trapping, an individual macromolecule is subjected to tension which can then force the molecule into an unfolded state. In these experiments, the mechanics of the macromolecules are measured in terms of forces and the intramolecular displacements. Kinetic mechanisms have not been readily accessible using this approach. Our objective is to propose an approach in which FCS measures microscopic conformational changes of macromolecules held in specified intermediate states of unfolding by an externally applied tension. For this experiment to be useful the applied force must be adjusted to maintain the macromolecule in an intermediate state of folding. Therefore the force will be comparable to molecular forces due to thermal agitation of the solvent upon the macromolecule, i.e. to Brownian diffusion and polymeric "entropic spring" forces. The main task of the analysis is to take account of these forces. It is important to emphasize that in the proposed experiments, the molecule is held in "constrained equilibrium" in the intermediate folding state and that the kinetics are measured as fluctuations within the equilibrium state. This contrasts with recent molecular extension experiments, e.g., those cited above, which are carried out under non-equilibrium conditions in which the strain is continually increased.

4.2   Theory

Consider a biological macromolecule, such as a protein composed of a number of independently folding domains arranged in tandem, e.g., the multidomain protein titin [4.11]. One end of the protein is attached to a rigid substratum; the other to a probe, e.g., AFM or laser tweezers, which can apply a specified tensile force, F_{ext}. We suppose that the domains unfold individually, although not necessarily in a pre-specified order, in response to the force. We further suppose that the conditions are arranged so that the protein is mostly folded but is poised near a folding transition. Then, only very small forces will be required partially to unfold the protein. The crucial aspect of the experiment is that very small forces are used to bring the protein into a partially folded state and then to maintain it there *in constrained equilibrium*. In this partially folded equilibrium the conformation of the protein, and in particular the conformations of the various domains of the protein will fluctuate among their folded, partially folded intermediate and unfolded states. By varying the applied force, it should then be possible to scan across different regions of the folding transition. If different domains have different stabilities and/or different sensitivities to force, then, as the applied force is varied, the folding/unfolding transitions of different domains should come into prominence and make larger contributions to the overall conformational signal being measured (in our case conformation dependent fluorescence changes).

The task of the theory is to take into account the different forces and motions that contribute to the overall behavior of the molecule. The applied force, which is specified a priori, is balanced by the elastic properties (conformational entropy) of the molecule. These will be determined by both the unfolded domains, that might be modeled by a Gaussian coil or a worm-like chain, and by the folded domains that are expected to have a much higher stiffness. Since the applied force is small, thermal motion (diffusion) is likely to play a significant role. It is necessary to account for the motions of the domains due to diffusion and the balance between the applied and the intra-molecular nonlinear elastic forces and also for the conformational transitions of the molecular domains among their various conformational states [4.12,4.13].

Let the x_j be the coordinates of the boundary between the j-th and $(j+1)$-th domains. Then, the degree of extension of the j-th domain is measured by $x_j - x_{j-1}$. Suppose that each domain can be in any of ν conformational states. Let the conformational state of the molecule overall be labeled with the index α, which characterizes the distribution of conformational states over the domains of the molecule. For example, if each domain has ν conformational states and there are M domains, then there are ν^M possible values of α. The set of coordinates of the domains $\{x_j\}$ and the conformational state, α, completely determine the overall state of the molecule. Let $P_\alpha(\{x_j\}, t)$ denote the probability that the molecule is in conformational state α with the coordinates of the domain boundaries given by $\{x_j\}$ at time t (Fig. 4.1).

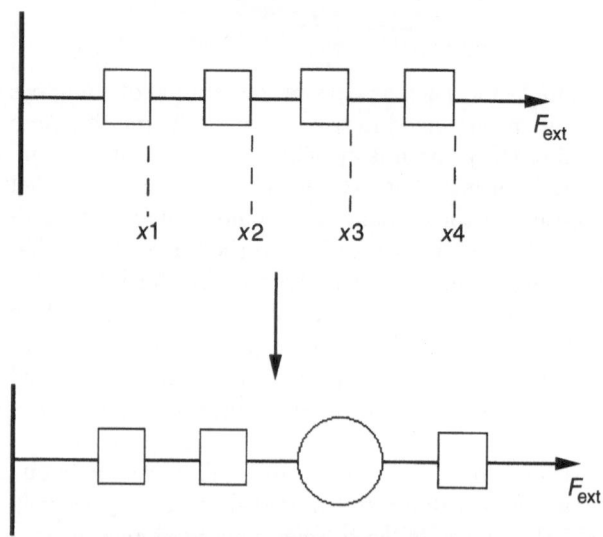

Fig. 4.1. Schematic of model. In this example, a macromolecule with four domains under the influence of an external force, F_{ext}, undergoes a transition of the third domain from the ordered to a disordered state while the other domains remain ordered. The domain boundaries are denoted by x_1 through x_4. The state of the molecule is determined by the values of the domain boundaries, x_i, and the conformations of each of the domains

The equilibrium properties of the molecule, including the probability of occurrence of a particular conformational state and its mean square fluctuation among conformations, can be described with sufficient detail for present purposes using a simple statistical thermodynamic model. Suppose that $\phi_k^{(j)}(x_j)$ is the relative probability (statistical weight) that the j-th domain is in its k-th conformational state and with its boundary at x_j. Then, $q_k^{(j)} = \int \phi_k^{(j)}(x_j)\,dx_j$ is the statistical weight of the j-th domain in its k-th conformational state for all boundary positions. Therefore, $Q^{(j)} = \sum q_k^{(j)}$ is the partition function for the j-th domain and $\Omega = \Pi Q^{(j)}$ is the partition function for the entire molecule. The probability that the j-th domain is in its k-th conformational state is $q_k^{(j)}/Q^{(j)}$. As a simple illustration let us consider a molecule composed of M identical domains, each with partition function $Q = [\sum q_k]$ and $\Omega = Q^M$. Then the average number of domains in the k-th state is $\langle n_k \rangle = Mq_k/Q$. The fluctuation of n_k, i.e. its instantaneous deviation from the mean, is $\delta n_k = n_k - \langle n_k \rangle$. The mean square fluctuation of the number of domains in the k-th state is $\langle (\delta n_k)^2 \rangle = Mq_k(Q - q_k)/Q^2$. Similarly the mean square fluctuation cross-correlation is $\langle (\delta n_j\, \delta n_k) \rangle = -Mq_jq_k/Q^2$.

The value of $P_\alpha(\{x_j\}, t)$ can change due either to motion of the domain boundaries $\{x_j\}$ or to transitions of domains among conformational states. To maintain the equilibrium of the molecule within a conformational transition, the external force must be weak, comparable to thermal forces that drive Brownian motion. Therefore, the domain boundaries move due to both diffusion and as a result of the forces exerted on a domain boundary by the domain to which it is connected. The external force, F_{ext}, acts directly on the final domain of the sequence. The Smoluchowski equation provides a useful framework with which to take account both of the motion of the domain boundaries in space and of the conformational transformations of the domains [4.14,4.15]. The current vector, $J_\alpha(\{x^{(j)}\})$, that describes the "flow" of probability into and out of a specific state, $[\alpha, \{x_j\}]$, accounts for diffusion and the effects of the forces applied at the boundaries of the domains. The diffusion coefficient and the frictional coefficient of the j-th domain both can depend on the conformation of the domain and so are denoted $D_\alpha^{(j)}$ and $\xi_\alpha^{(j)}$, respectively. Then,

$$J_\alpha = \sum_j \left[-D_\alpha^{(j)} \nabla_j P_\alpha(\{x_j\}, t) + P_\alpha(\{x_j\}, t) F_j / \xi_\alpha^{(j)} \right] . \qquad (4.1)$$

Here F_j is the net force exerted at the boundary between the j-th and $(j + 1)$-th domains and ∇_j is the gradient operating on the coordinates of that boundary.

The net rate of change of $P_\alpha(\{x_j\}, t)$ results from the motion of the domain boundaries, $\nabla \cdot J_\alpha(\{x_j\}, t)$, and from the conformational transformations of the domains, $\sum M_{\alpha\beta} P_\alpha(\{x_j\}, t)$, where $M_{\alpha\beta}$ is the rate constant that the molecule will undergo a transition from conformational state β to state α. Thus,

$$
\begin{aligned}
\partial P_\alpha(\{x_j\}, t) / \partial t &= -\nabla_j \cdot J_\alpha + \sum M_{\alpha\beta} P_\beta(\{x_j\}, t) \\
&= \sum_j \left[D_\alpha^{(j)} \nabla_j^2 P_\alpha(\{x_j\}, t) - \left(F_j / \xi_\alpha^{(j)} \right) \cdot \nabla_j P_\alpha(\{x_j\}, t) \right. \quad (4.2) \\
&= \left. -P_\alpha(\{x_j\}, t) \nabla_j \cdot \left(F_j / \xi_\alpha^{(j)} \right) \right] + \sum M_{\alpha\beta} P_\beta(\{x_j\}, t) .
\end{aligned}
$$

Here we have assumed that $D_\alpha^{(j)}$ is independent of $\{x_j\}$.

The dynamics of conformational fluctuations change in correspondence with the force-induced population shifts. The dynamics are characterized in terms of a fluorescence fluctuation autocorrelation function,

$$g(\tau) = \langle \delta i(t + \tau) \delta i(t) \rangle = \langle \delta i(\tau) \delta i(0) \rangle = \lim_{T \to \infty} (1/T) \int_0^T \delta i(t) \delta i(t + \tau) \mathrm{d}t .$$

Where $\delta i(t) = i(t) - \langle i \rangle$ is the fluorescence intensity fluctuation at time t. Thus $\delta i_\alpha(t) = f_\alpha \delta P(t)$, where f_α takes account of the absorption and emission properties of the α'th conformation and instrumental factors such as the excitation intensity and the efficiency of the optical system and $P_\alpha(t) = \int P_\alpha(\{x_j\}, t) \mathrm{d}\{x_j\}$ is the probability of the α'th conformational state irrespective of the position of the domain boundaries. As a consequence of the principle of microscopic reversibility it can be shown that $\Phi_{\alpha\beta}(\tau) = \langle \delta P_\alpha(0)\delta P_\beta(\tau)\rangle = \Phi_{\beta\alpha}(\tau)$. Hence, there are only $m(m+1)/2$ distinct terms in $g(\tau)$ [4.1]:

$$g(\tau) = \sum_\alpha \sum_{\alpha\le\beta}(2 - \delta_{\alpha\beta})f_\alpha f_\beta \Phi_{\alpha\beta}(\tau).$$

4.3 A Simple Example

As a simple illustration we consider a one-domain protein that can be in three conformational states, denoted as f, i, and u, representing folded, intermediate, and unfolded conformations. Motion is restricted to the x axis. To simulate a primitive protein folding landscape, we suppose that the free energy of the j-th conformation can be described as, $\Delta G_j(x) = \gamma_j(x - x_j)^2/2 + \delta G_j$, $(j = f, i, u)$ where the force constants and equilibrium coordinates for the simple harmonic potentials are $\gamma_f \gg \gamma_i > \gamma_u$, $x_f < x_i < x_u$. In the remainder of this chapter, energy will be expressed in term of $k_B T$ units, where k_B is Boltzmann's constant and T is temperature. The δG_j determine the dependence of the free energies of the different conformational states on temperature and/or solvent conditions. Then, the time and space dependence of $P(x, t)$ is given by three coupled equations for $P_f(x, t)$, $P_i(x, t)$, and $P_u(x, t)$. For example, for a constant F_{ext},

$$\partial P_i(x, t)/\partial t = D_i \partial^2 P_i(x, t)/\partial x^2 + (\gamma_i/\xi_i)P_i(x, t)$$
$$+ \left[(\gamma_i(x - x_i) - F_{ext})/\xi_i\right]\partial P_i(x, t)/\partial x$$
$$+ k_{fi}(x)P_i(x, t) - [k_{if}(x) + k_{iu}(x)]P_i(x, t) + k_{ui}P_u(x, t).$$

We defer a detailed discussion of these equations to a later occasion.

For the present purposes it is useful to consider a simpler system to illustrate the effect of the external force on the dynamics of the conformational transitions. We will ignore the effects of diffusion and focus entirely on the effect of the external force on the chemical reaction system. As above, we consider a three state system in which each state is characterized by a harmonic energy profile. The point at which the energy profiles of the f and i conformations intersect, x_{fc}, represents the peak of the activation energy barrier separating the two conformations, $E_{1c} = \Delta G_f(x_{fc}) = \Delta G_i(x_{fc})$. Similarly, the point at which the i and u profiles intersect, x_{uc}, is the peak of the activation energy barrier separating these two conformations, $E_{2c} = \Delta G_i(x_{uc}) = \Delta G_u(x_{uc})$. We suppose that for $x < x_{if}$ or for $x > x_{iu}$ the molecule is in the folded or the unfolded conformation, respectively. For $x_{iu} \ge x \ge x_{fi}$ the molecule is in the i conformation. Upon application of the external force, F_{ext}, the energies of the three states, $j = f, i,$ and u, are

$$\Delta G_j(x) = \gamma_j(x - x_j)^2/2 + \delta G_j + F_{ext}x.$$

The statistical weight of the j-th conformational state is readily calculated as above:

$$q_f = \int_{-\infty}^{\text{xfc}} \exp\left[-\Delta G_{\text{f}}(x)\right] dx$$

$$q_i = \int_{\text{xfc}}^{\text{xuc}} \exp\left[-\Delta G_{\text{i}}(x)\right] dx$$

$$q_u = \int_{xuc}^{\infty} \exp\left[-\Delta G_{\text{u}}(x)\right] dx$$

$$Q = q_{\text{f}} + q_{\text{i}} + q_{\text{u}}.$$

The free energies, again in $k_{\text{B}}T$ units, associated with the specific conformations are $g_k = -\ln(q_k)$. The average fraction of the molecule in the k-th conformation is $\langle P_k \rangle = q_k/Q$ and the autocorrelation amplitudes are $\langle (\delta P_k)^2 \rangle = q_k(Q - q_k)/Q^2$. The cross-correlation amplitudes are $\langle \delta P_j \delta P_k \rangle = -q_j q_k/Q^2$. These provide the initial conditions for solution of (4.4).

The rate constants are proportional to the exponentials of the energy barriers, $\Delta G_{ij}(x)$, that must be crossed from one conformation to the next

$$k_{ij} = A \exp\left[-\Delta G_{ij}\right].$$

For simplicity we choose $A = 1$. Therefore, the k_{ij} are determined by the following:

$$\begin{aligned}
\Delta G_{\text{ui}}(x) &= E_{2\text{c}} - g_{\text{u}} \\
\Delta G_{\text{iu}}(x) &= E_{2\text{c}} - g_{\text{i}} \\
\Delta G_{\text{if}}(x) &= E_{1\text{c}} - g_{\text{i}} \\
\Delta G_{\text{fi}}(x) &= E_{1\text{c}} - g_{\text{f}}.
\end{aligned} \tag{4.3}$$

Defining the rate constants in this way yields the correct equilibrium constants, K_{jk}. For example, $K_{\text{fi}} = k_{\text{fi}}/k_{\text{if}} = q_{\text{f}}/q_{\text{i}}$. This approach is consistent with the Kramers result on the Brownian dynamics of chemical bond disruption [4.16]. Although the rate constants are functions of x because of the effect of the external force on the conformational energies, the probability functions P_j are now functions only of time because we include all of the molecules over the spatial ranges defined above as either u, i, or f.

The kinetic equations for the simple two step conformational system are as follows:

$$\begin{aligned}
dP_{\text{f}}(t)/dt &= -k_{\text{fi}}(x)P_{\text{f}}(t) + k_{\text{if}}(x)P_{\text{i}}(t) \\
dP_{\text{i}}(t)/dt &= k_{\text{fi}}(x)P_{\text{f}}(t) - \left[k_{\text{if}}(x) + k_{\text{iu}}(x)\right]P_{\text{i}}(t) + k_{\text{ui}}(x)P_{\text{u}}(t) \\
dP_{\text{u}}(t)/dt &= k_{\text{iu}}(x)P_{\text{i}}(t) - k_{\text{ui}}(x)P_{\text{u}}(t)
\end{aligned} \tag{4.4}$$

In these equations, x is present only as a parameter determining the effects of the external force on the values of the ks. These equations are solved with initial conditions determined by the appropriate conformational fluctuation amplitudes. Consider the correlation function $\Phi_{jk}(\tau) = \langle \delta P_j(0)\delta P_k(\tau)\rangle$, where $\langle \ldots \rangle$ denotes an ensemble (or time) average. It is readily shown that

$$\Phi_{jk}(\tau) = \langle \delta P_j(0)\delta P_k(\tau)\rangle$$
$$= \sum_s X_k^{(s)} \exp(\lambda^{(s)}\tau) \sum_m (X^{-1})_m^{(s)} \langle \delta P_j(0)\delta P_m(0)\rangle .$$

Here $\lambda^{(s)}$ is the eigenvalue and $X_k^{(s)}$ and $(X^{-1})_k^{(s)}$ are the right and left eigenvectors of the matrix of rate coefficients [4.1]. $\delta P_k(0)$ is the (initial) deviation of the k-th conformation from its equilibrium value. Similarly, for a simple macroscopic conformational change

$$\delta P_j(\tau) = \sum_s X_j^{(s)} \exp(\lambda^{(s)}\tau) \sum_m (X^{-1})_m^{(s)} \delta P_m(0) .$$

Here $\delta P_k(0)$ is the initial (macroscopic) deviation of P_k from its equilibrium value. Hence, $\Phi_{jk}(\tau)$ and $\delta P_k(\tau)$ have the same form, differing only in the interchange of $\Phi_{jk}(0)$ and $\delta P_k(0)$. Therefore, we can obtain $\Phi_{jk}(\tau)$ by solving the system of equations (4.4) using $\Phi_{jk}(0)$ for the initial conditions.

One purpose of this simple example is to compare the effects on fluctuation dynamics of an external force applied to a molecule with the effects of a conventional perturbant such as temperature or a denaturant. We therefore suppose that the conformational free energies depend on temperature, T, as follows: $\Delta G_j(T) = \Delta H_j/(T/T_0) - \Delta S_j$, where T_0 is an arbitrary reference temperature and we have taken the unfolded state with $x = x_u$ and $F_{ext} = 0$, to define the zero level of conformational free energy independent of temperature, $\Delta H_u = 0$ and $\Delta S_u = 0$.

As an informative illustration, suppose that the intermediate state is relatively highly ordered and so has an enthalpy change not much less in magnitude than that of the folded state: $\Delta H_f = -30\,000$ and $\Delta S_f = -75$ and $\Delta H_i = -25\,000$ and $\Delta S_i = -66$, in units of $k_B T$ and k_B, respectively. (We have taken the reference temperature, $T_0 = 1$ K.) Figure 4.2 shows simulated thermal melting curves for each of the conformations for $F_{ext} = 0$ and $F_{ext} = 0.12$. As expected, the conformational transition is shifted to lower temperatures at the higher value of F_{ext}. When $F_{ext} = 0$, the fraction of the molecules present in the intermediate conformation is never greater than 5%. Hence, when $F_{ext} = 0$, the conformational transition behaves effectively as an all-or-none two state process between u and f. In contrast, when $F_{ext} = 0.12$, at $T = 342.2$ K, the temperature at which $P_u = P_i \approx 0.47$, $P_f = 0.05$. Hence, when $F_{ext} = 0.12$, the f state is barely present and the transition behaves as a two state process between the i and u conformations. The "energy landscapes" that determine these different population profiles are shown at their respective transition midpoints in Fig. 4.3. When $F_{ext} = 0$ ($T = 392$ K), it is clear that the intermediate has a higher energy than, and therefore is relatively unstable with respect to, the folded and unfolded conformations.

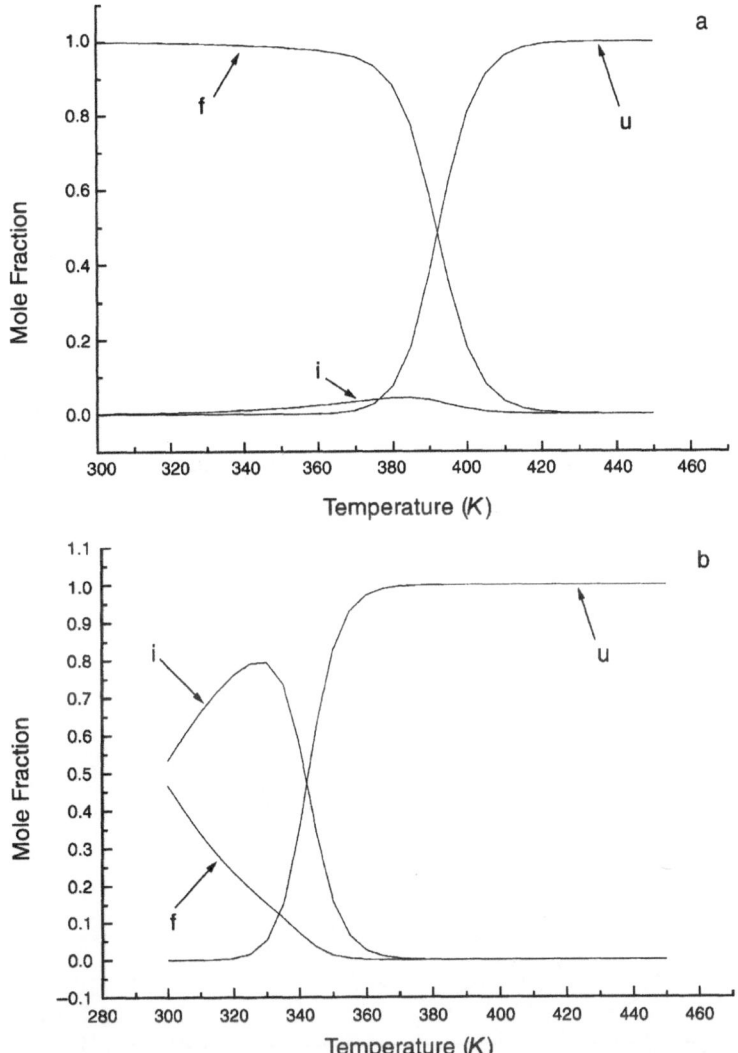

Fig. 4.2 a,b. Thermal melting curves in the presence and absence of tension. The parameters used in these calculations are as follows (in dimensionless units): $\gamma_f = 0.1$, $\gamma_i = 0.01$, $\gamma_u = 0.005$, $\Delta H_f = -30\,000$, $\Delta S_f = -75$, $\Delta H_i = -25\,000$, $\Delta S_i = -66$. $(T_0 = 0)$
The application of force lowers the transition temperature and enhances the presence of the intermediate state. **a** $F_{\text{ext}} = 0$ **b** $F_{\text{ext}} = 0.12$.

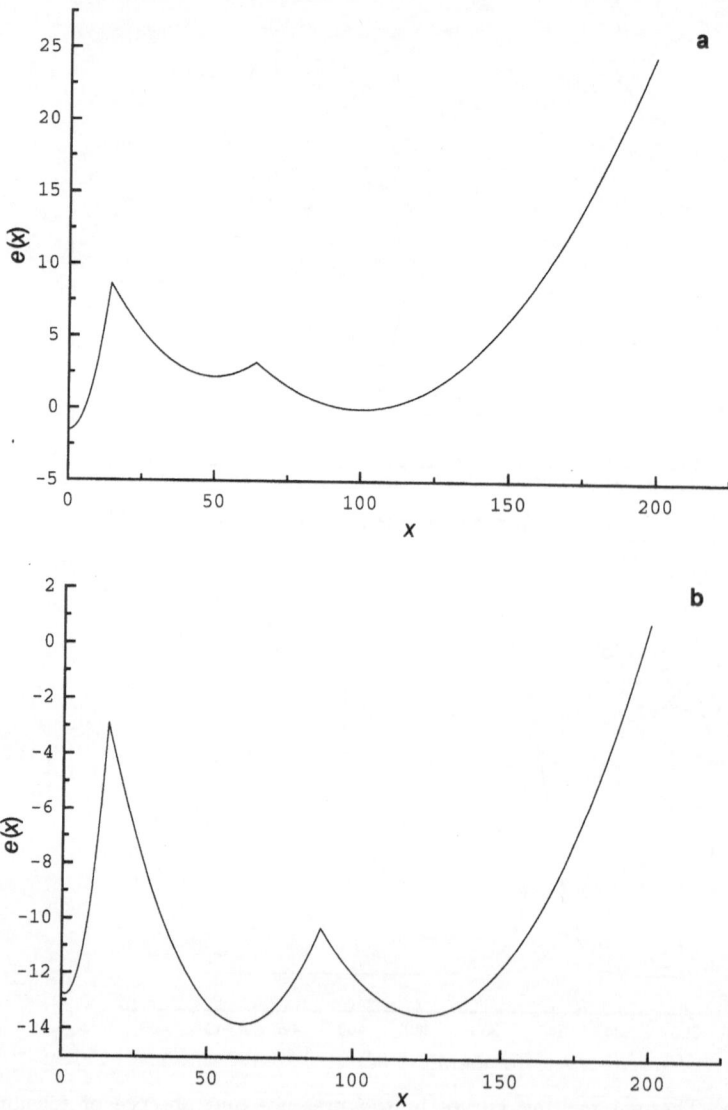

Fig. 4.3 a,b. Energy profiles. Using the parameters listed in the caption to Fig. 4.2, the energy profiles were computed at the center of the thermal transitions for molecules in the absence of force (**a** $F_{ext} = 0$, $T = 392\,\mathrm{K}$) and in the presence of force (**b** $F_{ext} = 0.12$, $T = 342\,\mathrm{K}$). The force has lowered the energy of the intermediate state relative to the folded state, thereby stabilizing the intermediate relative to the folded conformation

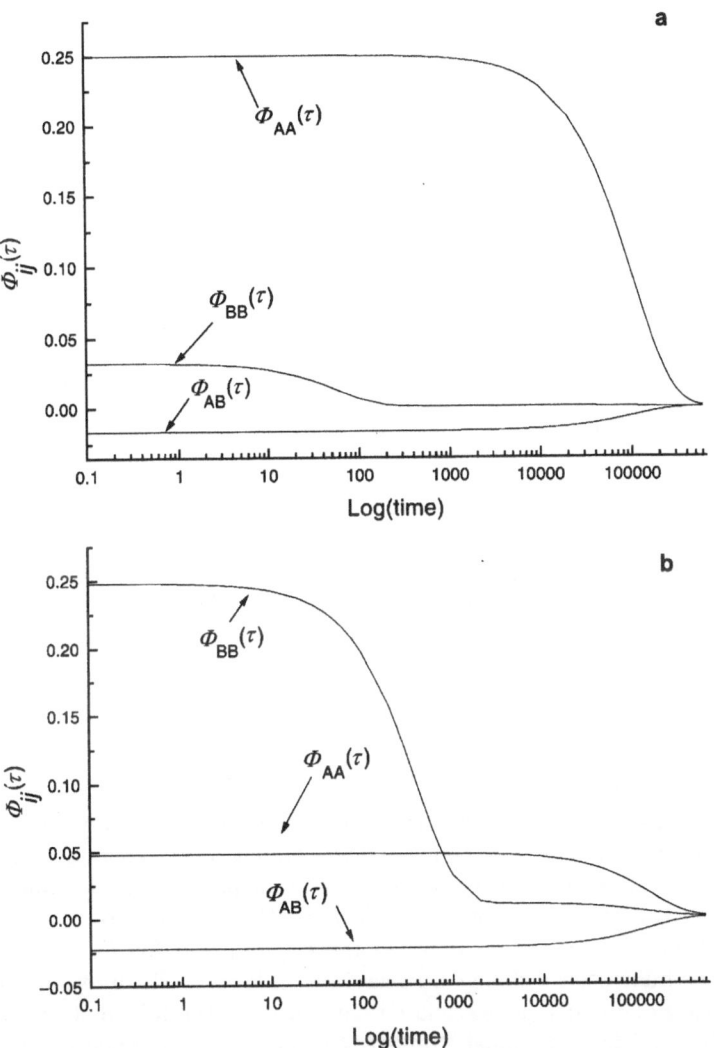

Fig. 4.4 a,b. Conformational correlation functions. Using the parameters listed in the caption to Fig. 4.2, the conformational correlation functions were computed at the center of the thermal transition for molecules in the absence of force (**a** $F_{ext} = 0$, $T = 392\,K$) and in the presence of force (**b** $F_{ext} = 0.12$, $T = 342\,K$). In the absence of force, $\Phi_{AA}(\tau)$, characterized by the longer correlation time, is dominant. Under the influence of the force, $\Phi_{BB}(\tau)$, characterized by the shorter correlation time, is dominant

In contrast, when $F_{ext} = 0.12$ ($T = 342\,\mathrm{K}$), the intermediate and unfolded states are the relatively more stable.

For our three-state example there are only six distinct contributions to $G(\tau)$. To simplify further, we take the fluorescence of the u state to be zero, $F_u = 0$. This reduces the number of contributions to $G(\tau)$ to three:

$$G(\tau) = F_f^2 \, \Phi_{ff}(\tau) + F_i^2 \, \Phi_{ii}(\tau) + 2 F_f F_i \, \Phi_{fi}(\tau)\,.$$

As for any two step chemical reaction system the time course of concentration changes is governed by two relaxation times, $\tau^{(+)}$ and $\tau^{(-)}$. The choice of parameters for our example determines that these two times are widely separated. Approximately, the longer and shorter times, $\tau^{(+)}$ and $\tau^{(-)}$, describe transitions between the f and i conformations and the i and u conformations, respectively. For $F_{ext} = 0$ and $T = 392$, $\tau^{(+)} = 1.03 \times 10^5$ and $\tau^{(-)} = 58.1$. For $F_{ext} = 0.12$ and $T = 342$, $\tau^{(+)} = 1.48 \times 10^5$ and $\tau^{(-)} = 415$. Figure 4.4 presents the individual correlation functions, $\Phi_{jk}(\tau)$ for $F_{ext} = 0$ ($T = 392\,\mathrm{K}$) and for $F_{ext} = 0.12$ ($T = 342\,\mathrm{K}$). Both temperatures were selected to be at the center of the major conformational transition, as shown in Fig. 4.2. When $F_{ext} = 0$, the major conformational transition, between f and i, occurs at the slower rate. The situation is reversed when $F_{ext} = 0.12$; the major transition is between i and u and occurs at the faster rate. Hence, by exerting a prescribed force on the molecule it is possible to reveal the dynamics of processes that can not be observed in the absence of the force. In this instance, the slow all-or-none transition between folded and unfolded forms is observed at zero force; the fast transition between intermediate and unfolded forms dominates in the presence of force.

For the example we have chosen, the intermediate state has a conformational enthalpy nearly as large as that for the folded state. Therefore, it is likely that the fluorescence of the folded and intermediate states are similar, then $F_f \approx F_i$. However, it is also possible that $F_i < F_f$. Figure 4.5 compares the correlation functions computed for $F_i = F_f$ and for $F_i = 0.5\,F_f$. Even when the fluorescence change associated with forming the intermediate is only half that of the change associated with forming the folded state, the dominant observed process under applied tension is the fast transition between i and u. For comparison, a temperature jump measurement simulated for $F_{ext} = 0$ is shown in Fig. 4.6.

4.4   Discussion

4.4.1   SD-FCS

To reach its ordered native state, a randomly disordered biological macromolecule such as a nucleic acid or protein must undergo a myriad of conformational adjustments. Each of its chain elements must find its proper spatial relationship with all other chain elements and form stabilizing interactions.

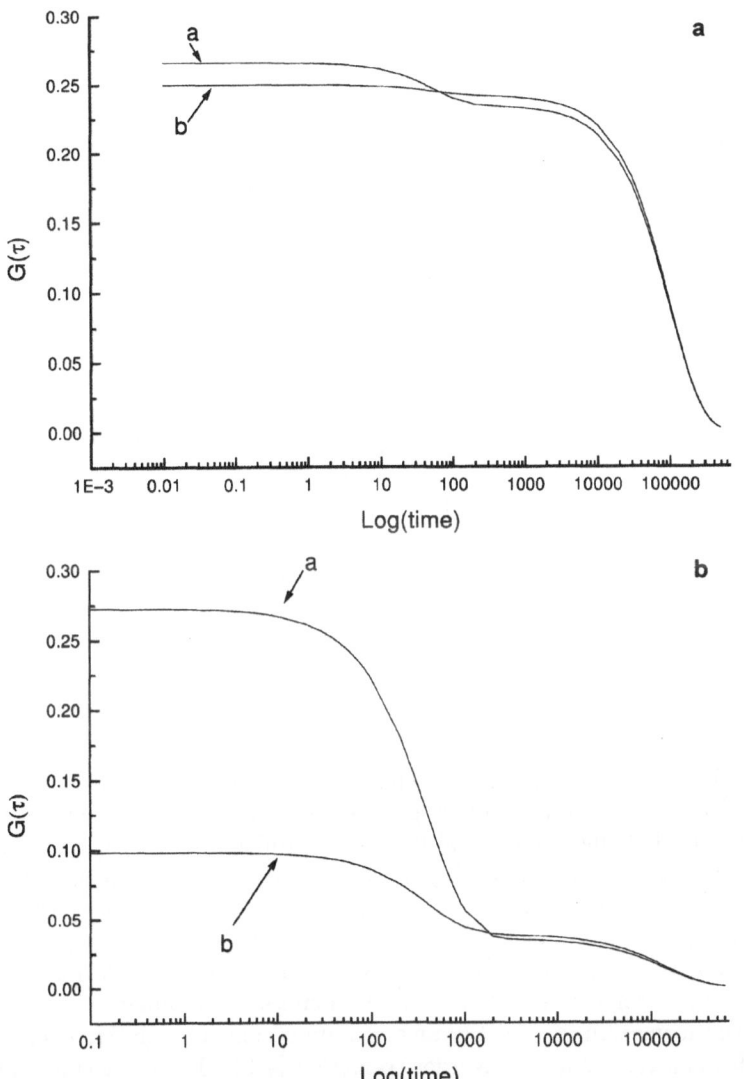

Fig. 4.5 a,b. Fluorescence fluctuation correlation functions. Using the parameters listed in the caption to Fig. 4.2, the fluorescence fluctuation correlation functions were calculated at the center of the thermal transition for two sets of fluorescence weighting parameters: *Curve a* $F_f = F_i = 1$; *Curve b* $F_f = 1$, $F_i = 0.5$ **a** $F_{ext} = 0$, ($T = 392\,\text{K}$) **b** $F_{ext} = 0.12$, ($T = 342\,\text{K}$) the contribution of processes involving the intermediate state is small when $F_{ext} = 0$, therefore changing the weighting factor F_i has little effect on $G(\tau)$, which is dominated by the slow transitions between f and u. When $F_{ext} = 0.12$, decreasing F_i from 1.0 to 0.5 substantially decreases the amplitude of the faster component in $G(\tau)$, but even with this decreased weight the faster component, which characterizes transitions between the i and u states, dominates $G(\tau)$

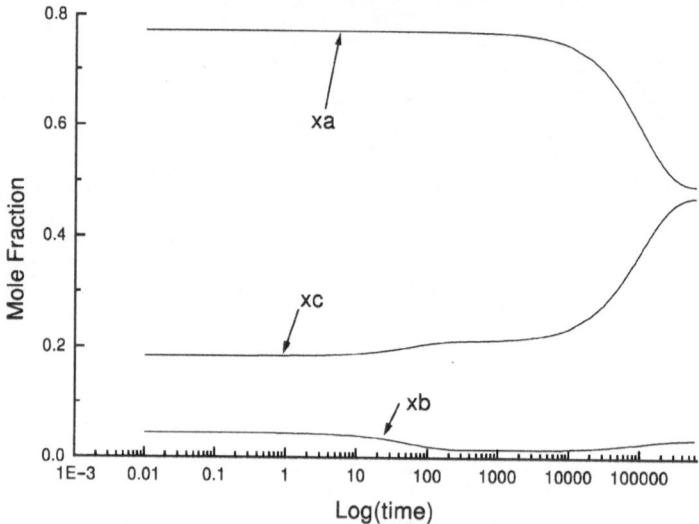

Fig. 4.6. Simulated temperature-jump experiment. Using the parameters listed in the caption to Fig. 4.2, the mole fractions of molecules in each of the conformations were calculated for $F_{ext} = 0$. The temperature was jumped from $T_1 = 385\,\mathrm{K}$ to $T_2 = 392\,\mathrm{K}$. As for the analogous FCS measurement (Fig. 4.5), the transient response to a temperature perturbation in the absence of a tensile force is dominated by the slow quasi-all-or-none transition from the f to the u conformation

It is impractical to regard each of these chain motions and contact formations as an identifiable chemical reaction step. Instead, these motions involve continuous structural changes and small free energy differences giving rise to a continuous free energy landscape. Rapid folding of a protein can occur if the landscape resembles a multi-dimensional funnel through which many parallel folding pathways can be funneled into the native state [4.17]. As the protein passes through the funnel from less to more ordered states, its decrease in conformational entropy, S, is balanced by its increase in stabilizing enthalpy H due to the formation of contacts among amino acids in a native conformation. This is a typical order-disorder transition [4.13]. The early stages of folding are dominated by the entropy; the later stages, by the enthalpy. Thus, the free energy $G(= H - TS)$ first increases and then decreases, giving rise to at least one energy barrier. Specific reaction steps along this pathway may be identifiable if there are additional free energy barriers. The roughness of the landscape determines the rates and number of the distinct steps.

Measurement of the rates of conformational changes in a specific region of the funnel would help to characterize the folding processes in that region and their relationship to the processes that occur in other regions. In principle, FCS provides the possibility to characterize multi-state processes such as protein folding in which the states are closely spaced in free energy because the measured dynamics are driven by the minimal possible perturbation, thermal energy. Therefore, it should be possible to focus on transitions among states within a narrow range of conformational free energy. However, it is frequently observed that protein folding and also other biomacromolecular conformational transitions are highly cooperative. Even when subjected to minimal perturbation these molecules traverse large ranges of conformational states in quasi-all-or-none transitions. Often, a slow, rate-determining step will dominate the overall transition; faster steps appear with relatively small amplitudes [4.18]. In these instances both FCS and perturbation-relaxation chemical kinetics measurements, e.g., temperature jump, have a limited ability to resolve transitions among closely spaced intermediates that have only a small fractional representation compared to the native and unfolded states.

As we have shown using our simplified, illustrative model, it may be possible in favorable instances to obtain information about the dynamics of intermediate conformational changes that are obscured by cooperativity in conventional kinetics or FCS measurements. This might be accomplished by combining conventional FCS with newly developed methods for applying tensile forces to individual protein or nucleic acid molecules. In the absence of the force, the free energies of the intermediate states are high relative to those of the native and completely unfolded states (Fig. 4.3). Therefore, the fraction of molecules present in the intermediate conformations is always small and the transitions among these states will appear in measurements with small amplitudes. However, if the mechanical properties of the intermediate states are such that application of a tensile force to the molecule destabilizes the folded or unfolded form more strongly than it does the intermediates, then the relative proportion of molecules in intermediate conformations will rise and will be more readily observable in measurements. This concept is illustrated simply in the model calculations presented in Sect 4.3.

The exposure by tension of intermediate states that are difficult to see under stress-free conditions could, in principle, be used in conventional perturbation kinetics if there were some way to carry out these measurements on molecules subjected to a stretching force. The advantage of FCS for these measurements is that it can be applied on one or a few molecules, which may be subjected to force more readily than a macroscopic population.

In addition to revealing the kinetics of processes which are difficult to measure using conventional methods, this approach provides information about the dependence of rates on tension. Moreover, the amplitudes of the different fluctuation modes observable in FCS measurements provide information about the effects of tension on the equilibrium characteristics of the forces which stabilize the protein. As the fractional representation of specific intermediate states increases in response to tension, the contributions to the correlation function amplitudes of the transitions among these states are amplified.

For simplicity we have presented the model in terms of a specified constant external force, F_{ext}, applied to the molecule. It may be simpler experimentally to specify the strain. The model can be reformulated to account explicitly for this isometric experiment. When an AFM or laser tweezers are used, the molecule is under neither purely isometric nor isotonic conditions; both force and strain can fluctuate because of the flexibility of the AFM cantilever or the softness of the potential defining the laser trap. The AFM cantilever spring or laser trap potential by which the force is applied to the molecule should be included in the analysis of the dynamic system [4.12]. Alternatively, a feedback system can be used to provide a truly constant F_{ext}

4.4.2   Comparison of SD-FCS with Conventional FCS

Although SD-FCS and conventional FCS are similar formally, they differ substantially in objectives, design, and analysis. Conventional FCS was originally formulated to measure transport and chemical reaction occurring in a population of molecules within an open region of solution (of dimension in the range of micrometers) [4.1,4.3,4.19]. In addition to the thermal forces that drive diffusion, the only external forces taken into account were those that drive systematic transport [4.20]. The analysis takes into account changes in the positions of the molecules considered as geometric points and in their chemical states. In contrast, SD-FCS is meant to deal with motions on less than 1/100 of the above spatial scale within an individual macromolecule anchored in a defined position and subjected to an external tensile force that acts on the molecular conformation. Spatial coordinates are intra-molecular. The theory accounts not only for the external force but also for internal forces due to intra-molecular interactions.

Although we have formulated SD-FCS in terms of a measurement on single molecules, the approach could also be used on a population of molecules if some way could be found to subject each of the molecules to the same forces and strains. This would require that each member of the population be anchored at the same or nearby residues and that the tensile force be applied at the same or similar chain position in each of the molecules. Although this appears a formidable technical challenge, it would be highly advantageous. More data could be acquired from the population than from a single molecule, which would be destroyed by photobleaching within a brief interval of data collection.

SD-FCS experiments differ from recent measurements of domain unfolding in titin [4.4–4.6,4.11,4.21–4.23] tenascin [4.24], and DNA untwisting [4.7,4.8,4.25] in two important respects. The external forces applied in the recent measurements were large enough to stretch the molecules from a fully ordered native state to a fully disordered unfolded or untwisted state. In SD-FCS experiments a smaller force is used to poise the molecule at equilibrium within an unfolding transition. The use of this small extension force requires taking into account the effects of thermal forces on positions of the domain boundaries and therefore on the degree of folding within the domains. This leads to our characterization of the conformational dynamics under tension using a Smolochowski equation (4.2) in which thermal forces and molecular interaction forces balance the applied forces, and in which all of these forces influence the rates of transitions among molecular conformations. SD-FCS differs from current forced unfolding experiments also in allowing the dynamics of conformational changes to be measured by observation of conformational fluctuations in equilibrium. This capability results from combining fluorescence measurements with molecular stretching methods.

4.4.3   Applications and Feasibility

SD-FCS could prove useful in the study of any system in which the molecular conformation is sensitive to an extension force, and in which fluctuations in molecular conformation cause measurable changes of fluorescence. Our simple model could be taken as an example not only of a rudimentary protein folding system but also of DNA untwisting or even a polysaccharide system [4.26]. For proteins, SD-FCS holds the promise of revealing the dynamics of processes such as the rates of molecular conformational changes within the condensed molten globule state that are difficult to observe by conventional means. The helix-random coil transition of natural DNA occurs through the formation of intermediates in which relatively less stable stretches of AT-rich base pairs are melted out to form interior bubbles surrounded by more helical regions with a higher content of GC base pairs [4.27,4.28]. Recently, a measurement of the formation of a "denaturation bubble" in individual tethered DNA molecules has appeared [4.9].

A number of challenges must be overcome to carry out a SD-FCS experiment. The molecule must be attached both to an immobile substratum and to a stretching tool, either an AFM cantilever or a bead that can be held by a laser tweezers. The experiments are simplest to interpret if both of these attachments occur at defined, specific sites within the molecule. The forces or strains to which the molecule is subjected must be held constant at low values compatible with retention of the molecule in equilibrium in a partially folded state. Observation of conformational fluctuations from partially folded molecules under tension will also be a substantial challenge. Molecules must be labeled with fluorophores that will respond to conformational fluctuations with large changes in fluorescence intensity. Photobleaching will limit the time over which data can be collected from any one molecule. Despite these formidable obstacles, published studies have demonstrated that many aspects of the proposed experimental approach are feasible at the current state of single molecule technology. Single molecule measurements of the unfolding of proteins and untwisting of DNA under an applied force are now plentiful. A recent paper has demonstrated the possibility of measuring the fluorescence emitted from a single molecule attached to a force measurement probe [4.29].

4.5   Summary

We have proposed a measurement in which a single protein or nucleic acid molecule is held by an externally applied tension in a partially folded intermediate state. In this state it could be possible to observe the dynamics of conformational changes that would be obscured by conformational cooperativity in conventional experiments. We have presented a simple model that illustrates how this approach might work to reveal states, which could not readily be observed in the absence of the applied force. This measurement approach is applicable to any macromolecular system and could be especially useful in analyzing the processes that dominate protein folding kinetic studies of which began three decades ago [4.30].

References

4.1 E. Elson and D. Magde: Biopolymers **13**, 1–27 (1974)

4.2 E. Elson and W.W. Webb: Annu. Rev. Biophys. Bioeng **4**, 311–34 (1975)

4.3 D. Magde, E.L. Elson, and W.W. Webb: Phys. Rev. Lett. **29**, 705–708 (1972)

4.4 M.S. Kellermayer, S.B. Smith, H.L. Granzier, and C. Bustamante: [published erratum appears in Science 1997 Aug 22;277 1117:(5329)]. Science **276**(5315), 1112–6 (1997)

4.5 M. Rief, M. Gautel, F. Oesterhelt, J.M. Fernandez, and H.E. Gaub: Science **276**(5315), 1109–12 (1997a)

4.6 L. Tskhovrebova, J. Trinick, J.A. Sleep, and R.M. Simmons: Nature **387**(6630), 308–12 (1997)

4.7 J.F. Allemand, D. Bensimon, R. Lavery, and V. Croquette: Proc. Natl. Acad. Sci. USA **95**(24), 14152–7 (1998)

4.8 S.B. Smith, Y. Cui, and C. Bustamante: Science **271**(5250), 795–9 (1996)

4.9 T.R. Strick, V. Croquette, and D. Bensimon: Proc. Natl. Acad. Sci. USA **95**(18), 10579–83 (1998)

4.10 M. Rief, F. Oesterhelt, B. Heymann, and H.E. Gaub: Science **275**(5304), 1295–7 (1997b)

4.11 H.P. Erickson: Proc. Natl. Acad. Sci. USA **91**(21) 10114–8 (1994)

4.12 H. Qian and B.E. Shapiro: Prot. Struct. Funct. Genet **37**, 576–581 (1999)

4.13 H. Qian and S.I. Chan: J. Mol. Biol. **286**, 607–616 (1999)

4.14 M. Doi and S.F. Edwards: The Theory of Polymer Dynamics. Clarendon Press, Oxford, (1988)

4.15 H. Qian: http://xxx.lanl.gov/abs/physics/0007017 (2000)

4.16 H.A. Kramers: Physica (Utrecht) **7**, 284–304 (1940)

4.17 P.G.Wolynes, J.N. Onuchic, and D. Thirumalai: Science **267**(5204), 1619–20 (1995)

4.18 E.L. Elson: Biopolymers **11**(7), 1499–520 (1972)

4.19 D. Magde, E.L. Elson, and W.W. Webb: Biopolymers **13**(1), 29–61 (1974)

4.20 D. Magde, W.W. Webb, and E.L. Elson: Biopolymers **17**, 361–376 (1978)

4.21 M.S. Kellermayer, S.B. Smith, C. Bustamante, and H.L. Granzier: J. Struct. Biol. **122**(1-2), 197–205 (1998)

4.22 P.E. Marszalek, H. Lu, H. Li, M. Carrion-Vazquez, A.F. Oberhauser, K. Schulten, and J.M. Fernandez: Nature **402**(6757), 100–3 (1999)

4.23 M. Rief, M. Gautel, A. Schemmel, and H.E. Gaub: Biophys. J. **75**(6), 3008–14 (1998)

4.24 A.F. Oberhauser, P.E. Marszalek, H.P. Erickson, and J.M. Fernandez: Nature **393**(6681), 181–5 (1998)

4.25 M. Rief, H. Clausen-Schaumann, and H.E. Gaub: Nat. Struct. Biol. **6**(4), 346–9 (1999)

4.26 P.E. Marszalek, A.F. Oberhauser, Y.P. Pang, and J.M. Fernandez: Nature **396**(6712), 661–4 (1998)

4.27 R.B. Inman: J. Mol. Biol. **18**(3), 464–76 (1966)

4.28 R.B. Inman: J. Mol. Biol. **28**(1), 103–16 (1967)

4.29 K. Kitamura, M. Tokunaga, A.H. Iwane, and T. Yanagida: Nature **397**(6715), 129–34 (1999)

4.30 T.Y. Tsong, R.L. Baldwin, and E.L. Elson: Proc. Natl. Acad. Sci. USA **68**(11), 2712–15 (1971)

5 Applications of FCS to Protein-Ligand Interactions:
Comparison with Fluorescence Polarization

Edmund Matayoshi and Kerry Swift

We will discuss several examples from our laboratory of the application of fluorescence correlation spectroscopy (FCS) to studies on protein-ligand interactions in free solution. To demonstrate the usefulness of FCS in such investigations we present situations which span both ligand size (from small molecules of molecular weight MW $\approx 1\,$kDa, to large protein ligands) and affinity (picomolar to micromolar). Furthermore, since FCS technology within the biological research community is available in only a small number of labs, we compare the results obtained by FCS with those obtained on the same samples by fluorescence polarization (FP). Such a comparison is useful because FP is a general and well-established optical biophysical tool for homogeneously monitoring molecular associations. While there clearly is overlap in the types of problems amenable to FCS and FP, the information they provide is complementary, and we believe it will be desirable to use them in tandem whenever feasible. We begin by briefly reviewing and comparing FCS and FP.

5.1 Fluorescence Polarization versus FCS

Some important features of FCS and FP relevant to protein-ligand studies are listed in Table 5.1. The remarks on FCS in the table and in this chapter pertain to the application of single color autocorrelation analyses in which the dynamics of the intensity fluctuations arise specifically from concentration fluctuations of the fluorescent species. In this approach, detection of protein-ligand association therefore requires the ability to resolve in the autocorrelation function (ACF) the translational diffusion coefficients (D_T) corresponding to bound and free states at equilibrium. (For intensity-based discrimination of components, see Chap. 20.) The information in Table 5.1 points to the likely advantages and disadvantages of FCS and FP for any particular problem. It should be noted that in this discussion "FP" includes both steady state anisotropy measurements (symbolized herein by $\langle r \rangle$), and time-resolved anisotropy measurements (symbolized by $r(t)$, and obtainable by either time or frequency domain methods), in which the rotational dynamics are resolved on the picosecond to nanosecond (and sometimes microsecond) time scale.

The simple dependence of FCS correlation times on the global D_T, in contrast to the complex dependence of FP on both the global rotational diffusion coefficient (D_R) and local dynamics (e.g.; due to segmental or domain mobility), underlies the major differences relevant to their implementation in protein-ligand studies. FCS is insensitive to the fluorophore tether length. In contrast, the sensitivity of FP to conformational dynamics local to the fluorophore can be both a blessing and a curse: on the one hand, it can fortuitously yield enhanced sensitivity (particularly needed when the difference in D_R between bound and free fluorescent species is small). On the other hand, it can lead to diminished or unusable $\langle r \rangle$ span if the local depolarizing sources dominate over D_R. To further complicate the interpretation of $\langle r \rangle$, the dynamic contributions may differ in the bound versus free state. Acquiring $r(t)$ will often provide some clarification, but this is hardly guaranteed. Thus, it can be said that, in general, the design and choice of label linker chemistry is relatively straightforward if employing FCS, but can be a frustrating trial and error process for FP applications, where probe optimization consists of balancing considerations of biological activity with those of $\langle r \rangle$ response.

Another point of difference is the magnitude of the minimal mass change between free and bound states which produces a detectable response in FCS versus FP. For the equivalent hydrodynamic sphere, $D_T \propto 1/r$ in the FCS experiment, whereas $D_R \propto 1/r^3$ in the FP experiment. In this respect, FP has a huge advantage in sensitivity, but this advantage can be diminished or completely lost when the molecular mass of the fluorescent component reaches a size such that its rotational correlation time exceeds the excited state fluorescence lifetime by greater than about 5–10-fold. Beyond that, FP registers little response (see below). Most of the new, long wavelength dyes exhibit extraordinary extinction coefficients and quantum yields, but typically have short fluorescence lifetimes of 1 ns or less, which unfortunately compromises their use as FP probes for monitoring diffusion of macromolecules much larger than 10 kDa. Use of new, longer lived probes with lifetimes into the microsecond range [5.1] extends greatly the upper MW limit, but currently there is a limited selection of these probes available. These probes are sensitive to oxygen quenching, so acquisition of lifetime and $r(t)$ data will likely be required. By contrast, there is no upper MW limit in FCS regardless of the fluorescence lifetime, and in practice this feature makes FCS useful particularly as a diagnostic for aggregation (see below).

The operational concentration range for the fluorescent species in FCS and FP differs substantially. Our experience has been that experiments with an optimized FP instrument can be typically run at concentrations at least ten-fold lower and with less photobleaching than analogous FCS-ACF experiments (one photon excitation) under the same conditions. However, it should be noted that below nominal concentrations of ≈ 100 nM, quantitative control over both dyes and macromolecules is usually difficult for all methods due to adsorptive losses. For protein-ligand studies where knowledge of concentrations is essential, this problem can cause quantitative uncertainties in

Table 5.1. Some comparisons of FCS with fluorescence polarization

	FCS	FP
Parameter monitored	Translational diffusion co-efficient (global)	Rotational dynamics (global and local)
Sensitivity to global mass changes (for sphere)	$D_T \propto 1/r$	$D_R \propto 1/r^3$
Molecular mass range for response to ligand binding to receptor	Unrestricted	Narrow (limited by fluorescence lifetime and contribution of local dynamics)
Monitor conformational changes?	No, except for dramatic global effects	Yes, sensitive to local conformational dynamics
Practical fluorophore concentration range	≈ 1 to $100\,\mathrm{nM}$	$\langle r \rangle :\approx 0.1\,\mathrm{nM}$ to $> 100\,\mathrm{\mu M}$ $r(t) :\approx 1\,\mathrm{nM}$ to $> 100\,\mathrm{\mu M}$
Resolution of multiple components	Routinely 2, but depends on relative differences in D_T	$\langle r \rangle$: fit 2 after prior characterization of system; $r(t)$: fit 2 - 3
Segmental motion of label	Not a problem	Depolarizing, can be problematical
Choice of label linker	Free to optimize to minimize interference with biological activity	Usually cannot use long linkers
Data acquisition speed	Slow for full ACF	Fast for $\langle r \rangle$; slow for $r(t)$

the FP experiment (see below), whereas in the FCS measurement one determines the number of fluorescent molecules in the observation volume, and thus has some degree of control. At the other extreme, i.e. at high ligand concentrations, which may be required in order to drive the binding reaction and/or when protein-ligand affinities are weak, FCS has severe limitations because it monitors particle number fluctuations. At concentrations above $\approx 100\,\mathrm{nM}$ (for typical pinhole settings), signal-to-noise in the ACF begins to suffer badly due to the large particle number. In this respect FP is far less restrictive. In the FP experiment, the maximum operating concentration is dictated by probe characteristics and measurement geometry, and if necessary it is not usually difficult to conduct experiments at fluorophore concentrations exceeding $100\,\mathrm{\mu M}$ if the measurement conditions are set properly. An example is given in this chapter which presents still another approach to handling high concentrations.

Acquisition speed is of concern mainly when dealing with large numbers of samples (e.g., high throughput compound screening applications), or for following the kinetics of systems not at equilibrium. It has been our experience that the acquisition of ACFs, with signal-to-noise sufficient for reliably recovering the diffusion coefficients and amplitudes corresponding to two or more components, typically requires replicate averaging of at least 30–60 s each. (However, for some applications where it is not necessary to obtain the diffusion coefficients, other types of analyses may suffice in which the signal averaging requirements are greatly reduced, e.g., a photon counting histogram (Chap. 20), or moment analysis [5.2]) FP $\langle r \rangle$ measurements can be performed rapidly (seconds or less, depending on instrument format) and large numbers of samples can be handled readily using commercial FP microtiter plate readers which have become available recently. As for the automated acquisition of FP $r(t)$ for large numbers of samples, an instrument has been built with 96 well microtiter plate compatibility [5.3], but we are not aware of the current availability of any commercial instrumentation. (This should not be confused with FP $(\langle r \rangle)$ plate readers which also feature "time-resolved fluorescence" capability: the latter refers to *time gated* integration of intensities on a time scale of microseconds or longer, and furthermore without acquisition of the two polarized decays.)

5.2 Experimental Methods

The FCS studies described here were performed on a Zeiss/EVOTEC ConfoCor (Carl Zeiss Jena, GmbH, Germany), which is based on one photon excitation and detection of the fluorescence in a subfemtomolar volume defined by the confocal optics. The first generation ConfoCor features a single avalanche photobleade (APD) detector and hardware correlator, which captures a record of the fluctuation data as an autocorrelation function (ACF). We excited fluorescein and rhodamine – type probes with the 10 mW air cooled argon ion laser (488/515 nm) supplied with the instrument, whereas for experiments with Cy5.5 (Amersham Life Science) we used the 676 nm output from our large frame, water-cooled krypton ion laser (Coherent Innova 400 series) as described in detail in [5.4]. In all FCS experiments the excitation power is routinely attenuated until the triplet fraction (clearly observable in the ACF for times shorter than $\approx 5\,\mu s$) is reduced as far below $\approx 20\%$ as practical. Depending on the fluorophore, this is typically on the order of 100 µW CW at the sample. ACF analysis includes fitting up to two or three discrete components for diffusion times (τ_D) and amplitudes, plus a triplet component.

FCS theory and analysis have been treated previously (e.g., [5.5–5.8] and elsewhere in this volume), and will not be repeated here. However, it should be mentioned that analysis of ACF data for multiple diffusive components across a protein-ligand binding curve is, in general, best accomplished by linking the analyses, for example, by fixing the τ_D for bound and free states with experimentally determined values. In some cases heterogeneity is severe and cannot be satisfactorily resolved, and in these instances (see below) we use an empirically averaged diffusion time (i.e. obtained from a fit of the ACF with a single component) to track changes in the protein-ligand interactions.

FP measurements were performed on an ISS Koala phase fluorometer with a Pockels cell modulator (ISS, Inc., Champaign, IL), or an SLM 48000 MHF phase fluorometer (Spectronic Instruments, Rochester, NY). $\langle r \rangle$ was collected in T-format, and lifetimes and $r(t)$ measured in either sequential or parallel frequency acquisition mode. The following excitation sources were used for the FP studies reported here: a CW argon ion laser (Innova 400 series, Coherent, Inc., Santa Clara, CA) at 488 nm for fluorescein, or 529 nm for tetramethylrhodamine; a helium neon laser (Melles Griot, Irvine, CA) at 543.5 nm for carboxyrhodamine; and a laser diode (Laser Max, Inc., Rochester, NY), at 690 nm for Cy5.5. Further technical details are provided in previous publications [5.3,5.4]. For a discussion of the additivity of $\langle r \rangle$ in mixtures and other aspects of FP theory, the reader is referred to [5.9].

Dynamic light scattering (DLS) measurements were performed with a Nicomp model C370 instrument (Particle Sizing Systems, Santa Barbara, CA) interfaced with a 4 W argon ion laser (Innova 304, Coherent, Inc.) operating at 488 nm. Samples ($\approx$ 50 µl volume) were placed in a 3 mm square cross section quartz cell, thermostatted at 23 °C, and scattering measured at a fixed 90° angle. With this configuration, molecular sizing down to 1 nm can be achieved routinely, and up to three components resolved. Prior to DLS measurements, all samples were centrifuged for 10 min at 14 000 rpm at 20 °C in a thermostatted microcentrifuge in order to remove interfering dust and microparticles.

Fitting of data for equilibrium dissociation constants was accomplished using generic nonlinear least squares algorithms, as provided in programmable commercial software packages. Analyses of antibody-small ligand binding data are discussed in detail in [5.10]. For both FCS and FP analyses, the τ_D or $\langle r \rangle$ values corresponding to free versus bound states are determined under experimental conditions with a fluorescent component alone, or in the presence of a saturating concentration of the second component, respectively.

5.3   HIV Protease

The discovery of HIV protease (PR) inhibitors and their development in the early 1990s into successful pharmaceutical drugs for AIDS therapy [5.11] required the development of assays which quantify inhibitor potency. Drug binding and inhibition mechanisms can be assessed enzymologically [5.12], or by detection of ligand binding at the PR active site in the absence of turnover. Fluorescence-based homogenous assays are most convenient in either case due to their sensitivity and general ease of adaptability to high throughput screening formats. We prepared a carboxy-rhodamine labeled active site inhibitor of HIV-1 PR (D. Kempf and C. Flentge, Abbott Laboratories, unpublished)

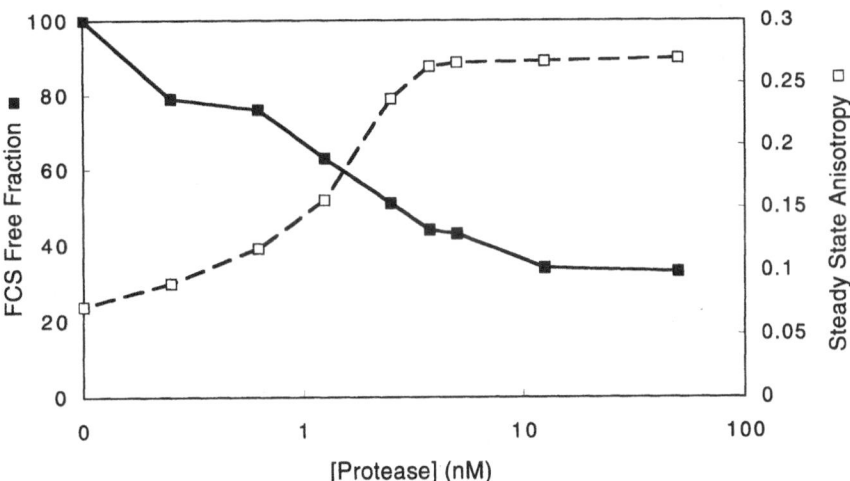

Fig. 5.1. Interaction of the carboxyrhodamine-labeled active site inhibitor RhI (5 nM) with HIV PR. The open symbols represent FP $\langle r \rangle$ values, and the closed symbols are the percentage contribution of the free inhibitor from FCS analysis

for the purpose of studying HIV PR mutants, as well as for developing a competitive binding assay. The inhibitor serving as the core of our PR probe is similar to Abbott's drug Norvir® [5.13]; with the addition of the carboxy-rhodamine moiety, the labeled drug has a MW of ≈ 1.1 kDa. HIV PR is a homodimeric enzyme with a subunit MW ≈ 11 kDa.

The binding of the fluorescent inhibitor ("RhI") to HIV PR, as monitored by FCS and FP, is shown in Fig. 5.1. Measurements of free RhI (in the absence of HIV PR) yielded $\tau_D = 75\,\mu s$ and $\langle r \rangle = 0.065$; fully bound RhI (in the presence of saturating amounts of HIV PR) yielded $\tau_D = 250\,\mu s$ and $\langle r \rangle = 0.265$. An $r(t)$ analysis of the fully bound RhI (lifetime 3.2 ns) yielded two dynamic components: 75% of the decay was attributable to the global rotational correlation time of 20 ns, and the remaining 25% with a correlation time of 400 ps was initially presumed to arise from segmental motion of the label around its linkage. (The fluorescent inhibitor discussed here actually represents the better of two PR active site probes which were originally synthesized: the other probe had a longer and much more flexible linkage between fluorophore and PR inhibitor core, and exhibited $\langle r \rangle = 0.19$ when bound to PR. The reasons for its reduced $\langle r \rangle$ span as compared with the short linker probe RhI are manifest in the $r(t)$ of the bound state: 51%, 450 ps; 49%, 19 ns.)

The FCS ACF at all points of the binding curve fit well to a two-component model using the above values for free and bound τ_D, and the fractional amplitude of the 75 μs component obtained from these analyses is plotted in the figure. Least squares analysis of FP $\langle r \rangle$ data obtained at a probe concentration of 100 pM, using the free and bound values stated above and a simple single site binding model, yielded $K_d = 5$ pM at pH 6.3, and $K_d = 1$ pM at pH 4.8. The higher affinity at lower pH was expected from our previous studies on the pH dependence of HIV PR activity. However, it should be mentioned that fitting of the $\langle r \rangle$ binding data was found to be nonsensical unless the probe concentration was left as an adjustable parameter: the apparent probe concentration obtained in this manner was typically one-half to one-quarter that of the nominal concentration. Such a procedure would be deemed reasonable if either some fraction of the inhibitor preparation is inactive (e.g., if purification of the labeled PR inhibitor failed to remove the unconjugated label completely), and/or is simply unavailable due to adsorptive wall losses sustained at these extremely low (pM) dye concentrations.

In fact, the FCS analyses (Fig. 5.1) show directly that there is indeed an inactive fraction of fluorescent inhibitor present in solution: as much as 30% of the fluorescent probe remains unbound at saturating concentrations of HIV-1 protease. The FCS result also ruled out significant PR dimer dissociation as a factor in the FP binding experiments, which were carried out at subnanomolar concentrations. It should be noted that such a clear cut conclusion cannot be drawn from even the FP $r(t)$ results described above, because the fast 400 ps component cannot be resolved into separate contributions from segmental motion of the bound probe and rotational diffusion of unbound probe. On the other hand, the use of $\langle r \rangle$ was critical in this case for the reliable quantification of the affinity constants, because low intensities prevented FCS measurements from being conducted below 1 nM (which, at 1000-fold above the K_d at pH 4.8, is inappropriate for accurate determination of affinity). In summary, this project represents a good example of the complementary application of FCS and FP, in which each method contributed essential information.

5.4   Death Domain Interactions

Recent research in the area of cytokine regulated apoptosis has revealed a complex sequence of proteolytic reactions, triggered following interactions between Fas and TNF membrane receptors with a host of extra- and intracellular protein ligands [5.14–5.16]. In the Fas model of apoptosis, it has been suggested that trimerization of Fas receptors is brought about by the extra-cellular binding of trimeric Fas ligand (FasL). The resulting trimerized "death domain" of Fas (Fas-DD), located on the cytoplasmic side, will subsequently interact with the death domain of the adaptor protein "FADD" (Fas-associated protein with death domain) as well as other proteins, leading to the activation of a cascade of caspases. The reader is referred to the cited reviews for a detailed discussion. Since most of the hypothesized interactions had yet to be studied as purified components in free solution, we used FCS

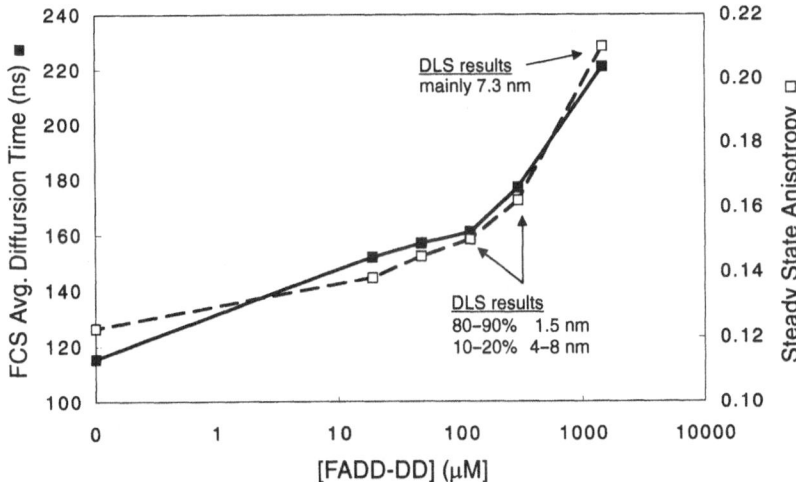

Fig. 5.2. Self-association of FADD-DD. A fixed concentration of 20 nM TMR-labeled FADD-DD was maintained in all samples, with the remainder consisting of unlabeled FADD-DD. *Open symbols* represent FP $\langle r \rangle$ values, *closed symbols* represent FCS τ_{avg} values. At the three indicated points, a DLS analysis was performed

as well as FP and DLS to examine any propensity for self-association of the 14 kDa death domains of FADD ("FADD-DD") and Fas ("Fas-DD"). These studies were done in collaboration with Stephen Fesik and Baohua Huang of the Abbott NMR research group [5.17].

A study of FADD-DD self-association over a protein concentration range of nearly three orders of magnitude is presented in Fig. 5.2. Because of restrictions in the use of high concentrations of fluorophore in FCS or FP, we prepared mixtures of 20 nM tetramethylrhodamine (TMR) labeled FADD-DD with various concentrations of unlabeled FADD-DD (up to 1.5 mM of the latter). Hence, unlabeled protein was generally in huge excess over labeled protein. The results are plotted as the average FCS diffusion time (τ_{avg}) and $\langle r \rangle$ versus FADD-DD concentration. In addition, DLS measurements were performed for the three concentrations.

The FCS analysis of FADD-DD shows a monotonic rise in τ_{avg} from just under 120 µs at 20 nM, to over 200 µs at 1.5 mM, clearly indicating self-association. Similarly, the fluorescence anisotropy increases from ~ 0.12 to ~ 0.21. The DLS results are in qualitative agreement as well: at low total protein concentrations (120 and 300 µM) there is primarily monomer (hydrodynamic diameter 1.5 nm) and a small amount of oligomer (4–8 nm) present. At the highest concentration of FADD-DD the molecular size distribution shifts to mainly a large oligomer (7.3 nm).

The possibility of Fas-DD self-association was also explored (Fig. 5.3). TMR-labeled Fas-DD was held fixed at 15 nM, with mixtures including in-

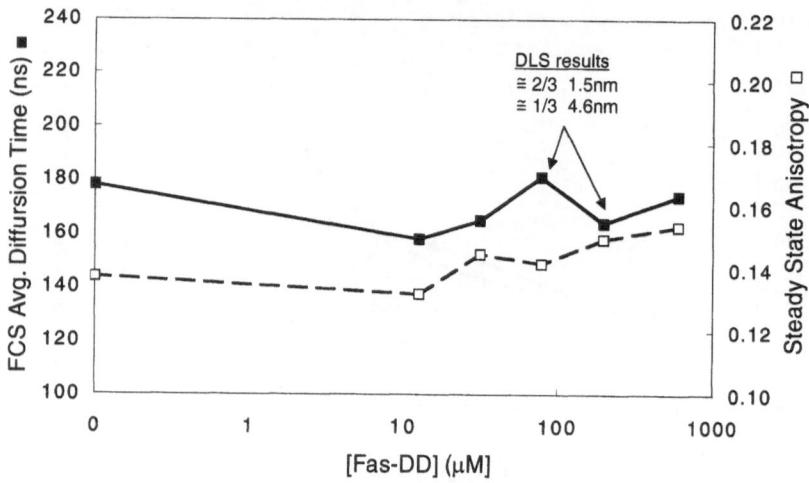

Fig. 5.3. Self-association of Fas-DD. A fixed concentration of 15 nM TMR-labeled Fas-DD was maintained in all samples, with the remainder consisting of unlabeled Fas-DD. DLS analysis was performed at the two sample points indicated

creasing amounts of unlabeled Fas-DD. It is evident that there is very little, if any, dependence of rotational or translational diffusion times on protein concentration over the range of 15 nM to 600 μM, although the absolute diffusion times ($\tau_{avg} \approx 170\,\mu s$) and anisotropies ($\langle r \rangle = 0.145 - 0.155$) are larger than expected for a 14 kDa monomer. DLS was measured at two intermediate points, and the analyses indicated that the samples are indeed *not* monodisperse: approximately two-thirds are monomeric (1.5 nm), while the remainder are oligomeric. Thus, the DLS confirms what is suspected from the FCS and FP experiments, i.e. that while there is no evidence of specific self-association, neither is the preparation homogeneous and monomeric. The two results taken together render slightly ambiguous the conclusion that Fas-DD does not self-associate, because it implies that perhaps a fraction of the preparation of Fas-DD is incorrectly folded, leading to some degree of intrinsic aggregation.

These experiments demonstrate the usefulness of FCS in a somewhat unconventional case, i.e. where associations are being evaluated up to extremely high concentrations of the diffusing species being detected. If one accepts the premise that the fluorescent label can be introduced so as to not perturb interactions significantly between labeled and unlabeled species, then FCS permits investigation of the association reaction over the widest possible concentration range while not placing an upper limit on the largest oligomer which can be detected. The latter capability (see below) is important in order to monitor the potential creation of aggregates at high concentrations, which are likely to be non- specific in nature. In these experiments the FCS data could not be readily resolved into discrete components, and FP provided useful complementary information. The high protein concentrations were ideal for the application of DLS, which greatly enhanced the interpretation of the observations based on fluorescence.

5.5   Antibody-Small Ligand Interactions

The binding of digoxin to an anti-digoxin antibody has importance for clinical and diagnostic assays, and can be studied by FCS or FP [5.18,5.19]. In the experiment shown in Figs. 5.4 and 5.5, a fixed concentration of 1.2 nM fluoresceinated digoxigenin (MW $\approx$ 1.1 kDa) was titrated with bivalent polyclonal antibody (MW $\approx$ 150 kDa).

The τ_D and $\langle r \rangle$ values characterizing bound versus free digoxigenin are obtained by measurement in the presence of saturating antibody versus no antibody, respectively. A two component analysis of the FCS data was carried out by fixing the diffusion times for free fluoresceinated digoxigenin ($\tau_D = 73\,\mu s$) and bound tracer ($\tau_D = 330\,\mu s$) in the analyses and performed at all points in Fig. 5.4, yielding the fractions of free and bound as a function of antibody concentration. These data fit well to a simple single site model [5.10],

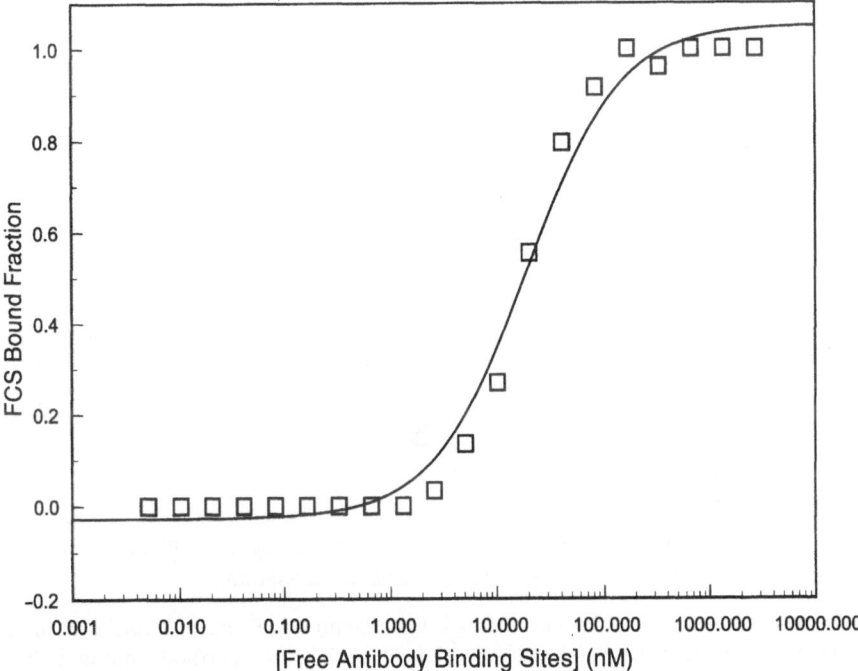

Fig. 5.4. FCS analysis of the interaction of fluoresceinated digoxigenin with anti-digoxin antibody. The fractional contribution of the slow (bound) component is plotted versus free antibody sites, and the *solid line* is the fit

with K_d = 19 nM. It should be mentioned that fluorescein is a somewhat problematical probe for FCS applications: excitation power must be reduced drastically in order to minimize photobleaching and excitation into the triplet state [5.20].

In the FP analysis (Fig. 5.5), the equilibrium $\langle r \rangle$ data were fit using free and bound $\langle r \rangle$ values of 0.017 and 0.235, respectively. The result, K_d = 12 nM, is in good agreement with the FCS analysis. Although the large antibody exceeds the fluorescence lifetime-imposed upper MW restriction for FP experiments by at least five-fold (fluorescein's lifetime is 4 ns), it is unimportant for the purposes of this experiment because one need only have good contrast in the $\langle r \rangle$ values which characterize the free and bound states.

While the binding experiment and analysis in this particular example are more efficiently and accurately carried out by FP, it demonstrates that FCS can in fact be used for quantitative determination of affinity constants. The result serves as a useful benchmark because in other applications (e.g., cellular work) FP may not be appropriate.

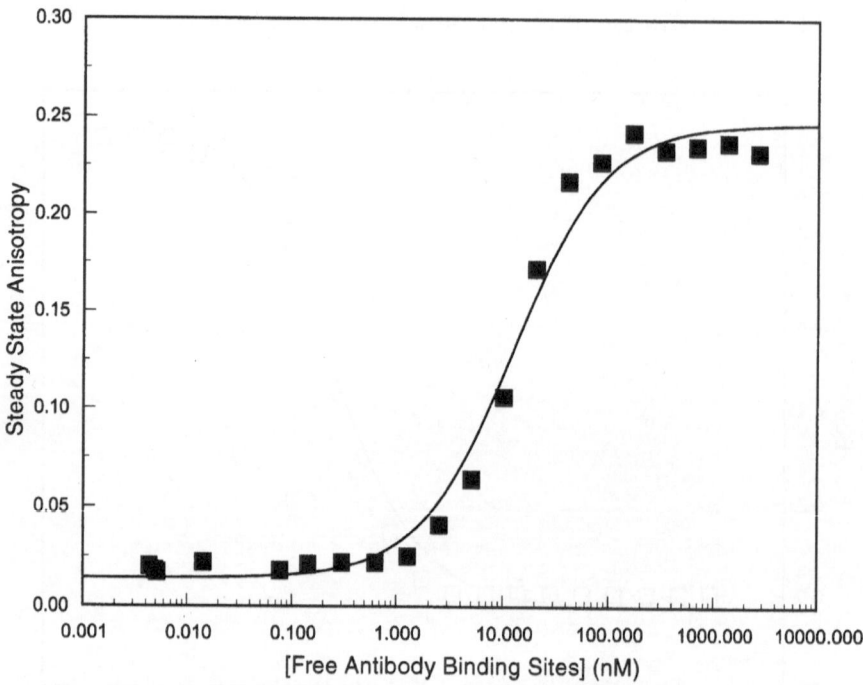

Fig. 5.5. FP analysis of fluoresceinated digoxigenin – anti-digoxin antibody interaction. The observed FP $\langle r \rangle$ value is plotted against free antibody binding sites, and the *solid line* is the fit

5.6   Antibody-Large Ligand Interactions

As a final example, we demonstrate the use of FCS for monitoring protein-ligand interactions where the molecular masses of the components or complexes are well beyond the applicability of FP with traditional nanosecond fluorescent probes. Table 5.2 lists FP and FCS results obtained on three proteins labeled with Cy5.5 in the presence versus absence of excess antibody specific for the respective proteins [5.4].

Examining the results on the left half of the table, which represent $\langle r \rangle$ of the Cy5.5 label, it is obvious that the binding of any of the fluorescent protein ligands to antibody is either completely undetectable, or barely detectable. Aggregated Bovine Serum Albumin (BSA), unintentionally created upon storage, can serve as a demonstration that the $\langle r \rangle$ measurement is unable to distinguish even monomeric from aggregated Cy5.5-BSA (in the absence of antibody). Of course, this result is not surprising given the very short fluorescence lifetime of Cy5.5 (0.8 ns). The result also has implications of general significance because it clearly shows that characterization of preparations by FP alone may well miss the existence of aggregation artifacts. The protein rEGFP, with less than half the mass of monomeric BSA, shows a barely discernible change in the $\langle r \rangle$ of its Cy5.5 label upon binding of antibody, and a somewhat larger change as registered by its intrinsic fluorophore. The latter improvement is not surprising given its slightly longer lifetime ($\approx$ 3 ns for rEGFP).

In contrast with the FP results, the binding of each protein ligand with antibody is easily detected in the FCS measurement by dramatic changes

Table 5.2. Antibody – large ligand interactions probed by FP, FCS

	Fluorescence anisotropy		FCS diffusion time (ms)	
	$(-)$Ab	$(+)$Ab	$(-)$Ab	$(+)$Ab
Cy5.5-BSA (monomeric)	0.24	0.23	0.60	1.3
Cy5.5-BSA (aggregated)[1]	0.25	0.25	1.0	4.4
Cy5.5-CRP[2]	0.29	0.28	0.55	1.5
Cy5.5-rEGFP	0.33	0.35	0.36	1.0
Cy5.5-rEGFP, $\lambda_{ex} = 488$ nm[3]	0.32	0.36	0.17	0.62

[1] Aggregation resulting from long term storage of a labeled BSA preparation.

[2] Human c-reactive protein; a pentamer of 25 kDa subunits.

[3] The intrinsic fluorescence (500–600 nm) of recombinant enhanced green fluorescent protein (MW 27 kDa) was excited at 488 nm. The shorter τ_D thus obtained, compared with that observed by excitation of the attached Cy5.5 label at 676 nm (preceding line in table), is a consequence of the smaller confocal volume attained at the shorter excitation wavelength.

in τ_D. Interestingly, in this experiment the particle number (not shown in the table) can also be used as a diagnostic for binding. Analysis of the ACF shows that upon addition of excess antibody, not only did τ_D increase as indicated, but the particle number in each sample simultaneously dropped by a factor of two. Taken together, these results imply that under the conditions of the experiment, the bivalent antibodies are fully occupied and crosslinking an average of two macromolecular units. The diffusion time is not expected to follow linearly because the antibody-crosslinked complex is not likely to be hydrodynamically symmetric with respect to the uncomplexed molecule. Finally, it should also be noted that the aggregation of Cy5.5-BSA itself (in the absence of antibody) is unambiguously clear from the large increase in the FCS diffusion time.

5.7   Conclusions

FCS is proving to be an extremely valuable and general tool in the bio-physicist's repertoire of methods for studying protein-ligand interactions in homogeneous solution. It offers new capabilities to those provided by classical FP by eliminating the MW ceiling imposed by the fluorescence lifetime, and by overcoming the ambiguity encountered frequently in interpreting changes in $\langle r \rangle$. However, it must be added that FCS will not replace FP in the near future for several practical reasons as discussed here. First, the majority of currently available fluorescent probes is too photolabile for FCS, but satisfactory for FP. Indeed, difficulties are encountered in applying FCS even to that most popular of probes, fluorescein [5.20]. It is likely that many probes which are outstanding for their *biological* compatibility will never be useful FCS probes, yet will remain amenable to FP. Second, the cost of FCS instrumentation at present is prohibitively high for most labs, whereas FP is inexpensive to implement. Finally, we have pointed out that FP responds to a different set of dynamical components than FCS, making it a unique and indispensable molecular "sensor".

The capability of FCS to monitor diffusion of biopolymers spanning an unrestricted size range is unique and perhaps its most powerful feature for protein-ligand studies in general. An important consequence is the ability to detect the aggregation of biopolymers, which is far more ubiquitous than is commonly appreciated. (Prior to the availability of FCS, only DLS had such diagnostic capability in homogeneous solution, but unfortunately DLS is restricted to relatively high concentrations.) We have encountered many examples of protein-ligand systems at "sub-DLS" concentrations in which, without direct evaluation by FCS, we would have had no clear warning of the presence of aggregation at nM to μM concentrations.

In this chapter we have used FP as a reference point for FCS performance, and stressed their *complementary* utility in obtaining a better understanding of the response of any fluorescent probe in studying protein-ligand interactions. The biophysical/spectroscopic strengths and weaknesses of FCS and FP are complementary, and the fundamental information they provide on D_T and D_R of biopolymers is more complete than either alone. While we have dealt only with single color autocorrelation FCS, the current developments in dual color cross-correlation FCS techniques [5.21–5.23], promise to offer still new possibilities for studies on protein-ligand interactions.

Acknowledgements

We are indebted to several of our Abbott colleagues for their contributions to the projects discussed in this chapter. We wish to thank Drs Dale Kempf and Chuck Flentge of Antiviral Research for the synthesis of the carboxyrhodamine labeled HIV PR inhibitor; Drs Baohua Huang and Stephen Fesik of NMR Research for the fluorescent tagging of Fas-DD and FADD-DD molecules (cloned and purified in their lab); and Dr. Sergey Tetin of the Abbott Diagnostics Division for purifying anti-digoxin antibody and collaborating on its biophysical characterization.

References

5.1 X. Guo, F.N. Castellano, L. Li, and J.R. Lakowicz: Anal. Chem. **70**, 632–637 (1998)

5.2 H. Qian and E.L. Elson: Proc. Natl. Acad. Sci. USA **87**, 5479–5483 (1990)

5.3 K.M. Swift and E.D. Matayoshi: In: Advances in Fluorescence Sensing Technology II, SPIE **2388**, 182–189 (1995)

5.4 K.W. Swift and E.D. Matayoshi: In: Advances in Fluorescence Sensing Technology, SPIE **3602**, 68–74 (1999)

5.5 R. Rigler, J, Widengren, and U. Mets: In: Fluorescence Spectroscopy, O.S. Wolfbeis, ed., pp. 13–24 (Springer-Verlag, Heidelberg 1992)

5.6 J. Widengren, U. Mets, and R. Rigler: J. Phys. Chem. **99**, 13368–13379 (1995)

5.7 J. Widengren: "Fluorescence correlation spectroscopy, photophysical aspects and applications". Ph.D. thesis, Dept. of Medical Biochemistry and Biophysics, Karolinska Institute, Stockholm, Sweden (1996)

5.8 U. Meseth, T. Wohland, R. Rigler, and H. Vogel: Biophys. J. **76**, 1619–1631 (1999)

5.9 J.R. Lakowicz: Principles of Fluorescence Spectroscopy, (Plenum Press), New York (1983)

5.10 S.Y. Tetin and T.L. Hazlett: In: Methods **20**, 341–361 (2000)

5.11 A. Wlodawer and J. Vondrasek: Annu. Rev. Biophys. Biomol. Struct. **27**, 249–284 (1998)

5.12 E.D. Matayoshi, G.T. Wang, G.A. Krafft, and J. Erickson: Science **247**, 954–958 (1990)

5.13 D.J. Kempf, K.C. Marsh, J.F. Denissen, E. McDonald, S. Vasavanonda, C.A. Flentge, B.E. Green, L. Fino, C.H. Park, X.-P. Kong, N.E. Wideburg, A, Saldivar, L. Ruiz, W.M. Kati, H.L. Sham, T. Robins, K.D. Stewart, A. Hsu, J.J. Plattner, J.M. Leonard, and D.W. Norbeck: Proc. Natl. Acad. Sci. USA **92**(19), 2484–2488 (1995)

5.14 S. Nagata: Cell **88**, 355–365 (1997)

5.15 D.W. Nicholson and N.A. Thornberry: Tr. Biochem. Sci. **22**, 299–306 (1997)

5.16 P. Villa, S.H. Kaufmann, and W.C. Earnshaw: Tr. Biochem. Sci. **22**, 388–393 (1997)

5.17 B. Huang, M. Eberstadt, E.T. Olejniczak, R.P. Meadows, and S. Fesik: Nature **384**, 638–641 (1996)

5.18 S.Y. Tetin, K.W. Swift, and E.D. Matayoshi: In: Abstracts of 25th FEBS Meeting, 1998, Copenhagen (1998)

5.19 S.Y. Tetin, K.M. Swift, and E.D. Matayoshi: manuscript in preparation (2000)

5.20 E.D. Matayoshi and K.M. Swift: Biophy. J. **78**, 441A (2000)

5.21 M. Eigen and R. Rigler: Proc. Natl. Acad. Sci. USA **91**, 5740–5747 (1994)

5.22 U. Kettling, A. Koltermann, P. Schwille, and M. Eigen: Proc. Natl. Acad. Sci. USA **95**, 1460–1420 (1998)

5.23 K.G. Heinze, A, Koltermann, and P. Schwille: Proc. Natl. Acad. Sci. USA **97**, 10377–10382 (2000)

Part II

FCS at the Cellular Level

6 FCS-Analysis of Ligand-Receptor Interactions in Living Cells

Aladdin Pramanik and Rudolf Rigler

6.1 Introduction

Fluorescence correlation spectroscopy (FCS) in its present form using confocal llumination volumes with single molecule detection sensitivity [6.1,6.2] appears as a foremost tool for the analysis of dynamic processes on cell surfaces as well as in the interior of cells. FCS as used after its introduction in the 1970s [6.3–6.5] involved large excitation volumes and correspondingly long correlation times. In contrast to the single molecule sensitivity of today, FCS was prone to photodestruction, and successful applications at the cellular level were difficult or even impossible. Submicrometer resolution as provided by the confocal setup now allows detailed analysis of cellular processes at defined localization within the cells. FCS has reopened the era of quantitative cell biology which once was pioneered by Torbjörn Caspersson at the Karolinska Institute in his seminal work on UV microspectroscopy [6.6].

In this review we will show how FCS can be used for investigating the interaction of several peptide ligands with their cognate receptors, and how important thermodynamic parameters, including recombination and dissocitation rate constants, as well as binding constants, can be obtained from the analysis of thermal motion. Furthermore, evidence for the existence of allosteric mechanism for signal transduction involving G-proteins will be presented.

6.2 Materials and Methods

6.2.1 Chemicals

The fluorophore rhodamine (tetramethylrhodamine-5-(and-6)-isothiocyanate was purchased from Molecular Probes (Leiden, The Netherlands). Human C-peptide (CP) and IGF-II were purchased from from Eli Lilly (Indianapolis, IA). The amino acid derivatives for the synthesis of galanin (GAL) were bought from Nova Biochem (Läufelfingen, Switzerland). The synthesis and purification of GAL have been described in detail earlier [6.7]. Labeling of CP and GAL with rhodamine (Rh) was performed as described elsewhere [6.7,6.8]. Human insulin was obtained from Novo Nordisk (Denmark), IGF-I from Pharmacia Upjohn (Sweden) and human proinsulin from Sigma Chemical Co. (St. Louis, Mo., USA), while D-C-peptide, randomly scrambled C-peptide and the C-terminal pentapeptide of C-peptide were synthesized by Genosys (Cambridge, UK). Rh-labeled EGF (epidermal growth factor) was bought from Molecular Probes (Leiden, Netherlands). EGF and vascular EGF were purchased from Bachem Feinchemikalien AG (Switzerland). Pertussis toxin was bought from Sigma Chemical Co. All other reagents are of standard grade.

6.2.2   Cell Culture

Binding of CP was conducted in human renal tubular cells, of GAL in the pancreatic insulinoma (Rinm5F) cells 'and of EGF in human diploid fibroblasts. Human renal tubular cells were cultured from the outer cortex of renal tissue obtained from non-diabetic patients undergoing elective nephrectomy due to renal cell carcinomas [6.9] as described earlier [6.8]. The Rinm5F cells were grown to 50% confluency in Gibco RPMI 1640 medium with L-glutamine supplimented with 5% (v/v) fetal calf serum, penicillin (50 units/ml), streptomycin (50 units/ml) in an atmosphere of 5% (v/v) CO_2 enriched air at $37\,^\circ C$. Human diploid fibroblasts were cultured in Gibco minimum essential medium supplimented with Gibco nutrient mixture F-12 M (HAM) and 10% fetal bovine serum.

6.2.3   Fluorescence Correlation Spectroscopy (FCS)

FCS Instrumentation. FCS was performed with confocal illumination of a laser volume element of 0.2 fl in a ConfoCor instrument of Carl Zeiss-Evotec (Jena, Germany, Fig. 6.1), as well in instruments built according to the principles described in [6.1]. As focusing optics a Zeiss Neofluar 40 × NA 1.2 objective for water immersion was used in a epi-illumination setup. For separating exciting from emitted radiation, a dichroic filter (Omega 540 DRL PO_2) and a band-pass filter (Omega 565 DR 50) were used. The Rh-labeled ligands (Rh-EGF, Rh-GAl, Rh-CP) were excited with the 514.5 nm line of an argon laser. The intensity fluctuations were detected by an avalanche photodiode (EG & SPCM 200) and were correlated with a digital correlator (ALV 5000, ALV, Langen, Germany).

Binding Experiment. All binding studies with FCS were carried out on cells cultured in eight-well Nunc chambers (Nalge Nunc Inc., IL, USA) at $20\,^\circ C$. Prior to the experiments cells were washed five times with PBS (NaCl 137 mM, KCl 2.7 mM, KH_2PO_4 1.5 mM, $Na_2HPO_4 \cdot 2\ H_2O$ 8.1 mM, pH = 7.4) and incubated with a binding buffer containing rhodamine-labeled ligand (Rh-L) at room temperature. CP binding was performed in a buffer composed of 20 mM Hepes, pH 7.4, 115 mM NaCl, 24 mM $NaHCO_3$, 4.7 mM KCl,

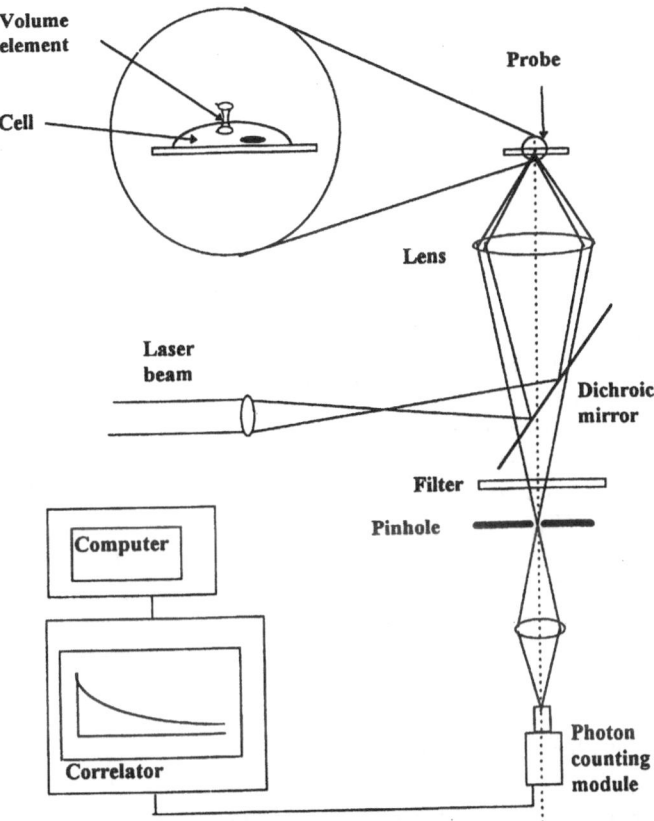

Fig. 6.1. FCS experimental setup. The laser beam from an argon ion laser is sharply focused via a dichroic mirror and a lens to form a tiny cofocal volume element of 0.2 fl. The laser beam is projected from below into an eight-well Nunc chamber containing a monolayer of cultured cells and rhodamine-labeled ligands (e.g., Rh-EGF, Rh-GAL, Rh-CP). After excitation of the Rh-labeled ligand, the emitted light is transmitted via the dichroic mirror, a band-pass filter and a pinhole to a photodetector. The volume element is positioned onto the cell surface using a microscope for detection of ligand binding. The dimensions of the laser beam focus and the pinhole together define the confocal volume element (see magnification) from which fluorescent light is collected. The photodetector operates in photon counting mode, responding with an electrical pulse for each detected photon. The electrical pulses are fed into a digital signal correlator which computes on-line in real time the autocorrelation function of the detected intensity fluctuations. After each measurement, the autocorrelation function is stored on the hard disk of a computer

1.26 mM $CaCl_2$, 1.2 mM KH_2PO_4, 1.2 mM $MgSO_4$, 11.1 mM glucose and 5 mg/ml bovine serum albumin [6.8]. Binding of GAL and EGF was conducted in a buffer composed of Hepes 20 mM, NaCl 140 mM, KCl 5 mM, $MgCl_2$ 1.2 mM, Na_2CO_3 3.3 mM, glucose 5.5 mM, $CaCl_2$ 1.8 mM. pH 7.4 [6.10,6.11]. Binding of ligands (CP, GAL and EGF) was measured after a 60 min incubation of the cells in the presence of 5 nM Rh-ligands. Binding curves for C-peptide, GAL and EGF were obtained by the addition of varying concentrations of Rh-CP, Rh-GAL or Rh-EGF to the incubations, allowing 60 min before determination of binding. Specificity of Rh-ligand binding was demonstrated by the competitive displacement of bound Rh-ligand from the cell surfaces after 3 h incubation of 5 µM of non-labeled ligand added to the cell incubations (post-incubation). Specific binding was also demonstrated by the inability of the Rh-ligand to bind to the cell surfaces when the cells were pre-incubated with non-labeled ligand for 1–3 h.

In a similar fashion, competitive displacement of CP binding was evaluated after pre- and post-incubations with insulin, IGF-I, IGF-II, proinsulin, D-C-peptide, scrambled C-peptide (human C-peptide with the residues assembled in random order) or a segment of the C-terminal part of the C-peptide molecule (EGSLQ). Furthermore, in the same way, competetive displacement of GAL binding was evaluated after pre- and post-incubations with M40, a galanin antagonist and of EGF binding with vascular EGF. Renal tubular cells, Rinm5F cells and human diploid fibroblasts were pretreated with pertussis toxin (1 µg/ml) at 37 °C for 4 h under 95% O_2 and 5% CO_2 before binding studies were carried out.

Measurements of Autocorrelation Functions. The measurments were performed in the solution and on cells cultured in Nunc chambers with Rh-ligand added. The objective used (40 × NA 1.2) enabled illumination of a volume element of 0.2 fl with dimensions of $w = 0.25$ µm and $z = 1.25$ µm, respectively. In order to avoid photobleaching, the exciting intensity was adjusted such that the detected photon count rate did not exceed 3000–4000 counts per molecule and s [6.12].

FCS Data Evaluation

Brownian Motion of Rh-ligands in Solution and on the Cell Surface. The observed fluorescence intensity fluctuations $\delta I(t)$ when correlated with fluorescence intensity fluctuations at time $t + \tau$ yield the normalized intensity autocorrelation function $G(\tau)$:

$$G(\tau) = 1 + \frac{\langle \delta I(t)\, \delta I(t+\tau) \rangle}{\langle I \rangle^2}, \tag{6.1}$$

where the brackets describe the time average and $\langle I \rangle$ the mean fluorescence intensity [6.4,6.5].

The autocorrelation function for Brownian motion of the fluorophore rhodamine molecules in a 3D Gaussian volume element with half axes 'ω' and 'z' is described as

$$G(\tau) = 1 + \frac{1}{N}\left(\frac{1}{1 + \frac{4D\tau}{\omega^2}}\right)\left(\frac{1}{1 + \frac{4D\tau}{z^2}}\right)^{\frac{1}{2}},$$

or

$$G(\tau) = 1 + \frac{1}{N}\left(\frac{1}{1 + \frac{\tau}{\tau_D}}\right)\left(\frac{1}{1 + \left(\frac{\omega}{z}\right)^2 \frac{\tau}{\tau_D}}\right)^{\frac{1}{2}}, \tag{6.2}$$

with $\tau_D = \frac{\omega^2}{4D}$, τ_D being the diffusion time and D the diffusion coefficient.

For the rhodamine-labeled ligand binding to a cell membrane, its diffusion at the membrane takes place only at a 2D surface ($z \gg \omega$) and $G(\tau)$ becomes:

$$G(\tau) = 1 + \frac{1}{N}\left(\frac{1}{1 + \frac{\tau}{\tau_D}}\right). \tag{6.3}$$

When the volume element with half axes $\omega = 0.25\,\mu\text{m}$ and $z = 1.25\,\mu\text{m}$ is projected onto the cell surface, not only bound Rh-ligand diffusing at the cell surface and but also unbound Rh-ligand diffusing above the cell suface will be seen. Then the autocorrelation function for 3D diffusion of the unbound Rh-ligand in solution above the cell surface and 2D diffusion of bound Rh-ligand to membranes on the cell surface is given by [6.8]:

$$G(\tau) = 1 + \frac{1}{N}\left[\left(1 - \sum y_i\right)Q_f^2\left(\frac{1}{1 + \frac{\tau}{\tau_D}}\right)\left(\frac{1}{1 + \left(\frac{\omega}{z}\right)^2 \frac{\tau}{\tau_D^f}}\right)^{\frac{1}{2}}\right.$$

$$\left. + \sum y_i Q^2\left(\frac{1}{1 + \frac{\tau}{\tau_D^{bi}}}\right)\right], \tag{6.4}$$

where y_i is the fraction of membrane bound Rh-ligand diffusing with diffusion time τ_D^{bi} and $(1 - \sum y_i)$ is the fraction of unbound Rh-ligand diffusion with diffusion time τ_D^f. For calculation of parameters of the autocorrelation function $G(\tau)$, nonlinear least square minimization was used [6.13]. Q_i is the quantum yield of the bound ligand represented by y_i. Q_f is the quantum yield of the unbound ligand.

The above mentioned intensity autocorrelation function $G(\tau)$ (6.1–6.4) is valid when the particle size is smaller than the volume element. With a volume element of $1\,\mu\text{m}$ diameter and a 100 nm vesicle size, this condition is fulfilled when FCS has been applied for the studies of molecular interactions of peptides with phospholipid vesicles [6.14]. This condition is also fulfilled with a $1\,\mu\text{m}$ diameter volume element and a membrane size of 4 nm (Fig. 6.2) since the size of the ligand-receptor complex in such a membrane can not be larger than the volume element.

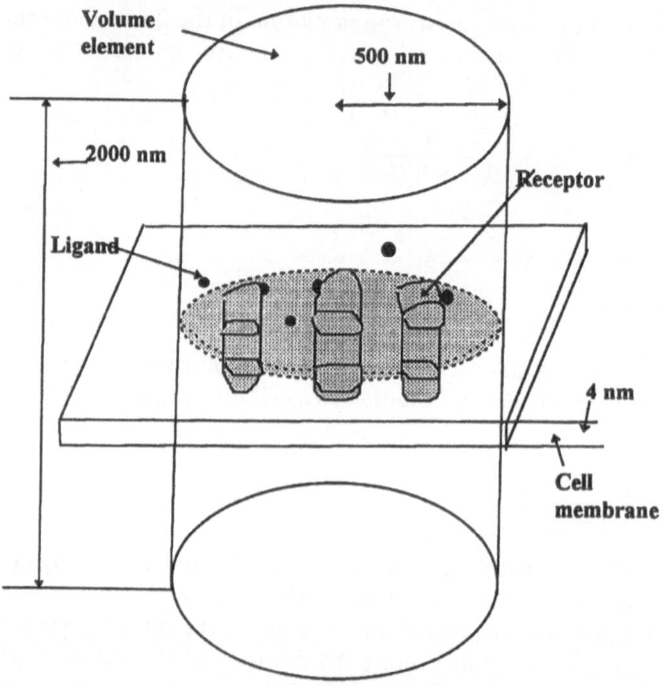

Fig. 6.2. Visualization of the volume element for the ligand-receptor interaction in the membrane of cultured cells

Distribution of the Diffusion Time of Ligand-Receptor Complexes with the CONTIN Algorithm. Instead of evaluating discrete diffusion processes with distinct species (y_i) and their characteristic diffusion times (τ_{Di}), the dynamic motion of receptor complexes in membranes can be distributed with regard to their diffusion times using a distribution function $P(\tau_{Di})$. In order to assess the validity of the presence of discrete ligand-receptor complexes we analyzed $G(\tau)$ representing the distribution of diffusion times according to an expression of multiple components:

$$G(\tau) = 1 + \frac{1}{N} \int_i P(\tau_{Di}) \frac{Q_i^2}{1 + \frac{\tau}{\tau_{Di}}} \left(\frac{1}{1 + \left(\frac{\omega}{z}\right)^2 \frac{\tau}{\tau_{Di}}} \right)^{\frac{1}{2}} d\tau_{Di}. \tag{6.5}$$

To evaluate the distribution function $P(\tau_{Di})$ the CONTIN algorithm [6.15,6.16] was applied.

Analysis of Ligand-Receptor Binding on the Cell Surface. We assume a membrane surface containing receptors with concentration R_o and define

$$L_o = L_f + L_b, \quad \text{where} \quad L_o = \text{total ligand}, L_f = \text{free ligand},$$
$$L_b = \text{bound ligand}$$

$$R_o = R_f + Rb, \quad \text{where} \quad R_o = \text{total receptor}, R_f = \text{free receptor},$$
$$R_b = \text{bound receptor}.$$

A solution with ligands at concentration L_o will be in equlibrium with receptors R_o on the membrane surface. For the situation where the diffusion time τ_D is much faster than the chemical relaxation time $\tau_{\text{chem}} = (k_{\text{ass}} + k_{\text{diss}})^{-1}$, the autocorrelation function $G(\tau)$ is described by [6.17]:

$$G(\tau) = 1 + \frac{1}{N_{L_o}} \left[\left(1 - \sum y_i \right) \left(\frac{1}{1 + \frac{\tau}{\tau_D^f}} \right) \left(\frac{1}{1 + \left(\frac{\omega}{z}\right)^2 \frac{\tau}{\tau_D^f}} \right)^{\frac{1}{2}} \right.$$
$$\left. + \sum y_i \left(\frac{1}{1 + \frac{\tau}{\tau_D^{bi}}} \right) \right],$$
(6.6)

where

$$N_{L_o} = \text{the total number of labeled ligands in the volume element}$$
$$L_b = \sum y_i = \text{the bound ligand and}$$
(6.7)
$$L_f = \left(1 - \sum y_i\right) = \text{the unbound ligand}$$

In order to obtain a binding curve we determine $\sum y_i$ at different ligand concentrations L_o. At saturating concentrations of L_o, all receptors, which are able to bind L_o, are labeled and R_o is given as L_{bs}. To obtain information about the binding constant and the number of ligand binding sites per receptor molecule (n) analysis of the binding isotherm according to Scatchard is performed. Rearrangement of the binding isotherm according to Scatchard provides the relation :

$$\frac{r}{c_f} = K_{\text{ass}}(n - r),$$
(6.8)

with $r = \frac{L_b}{R_o}$, $c_f = L_f$, $K_{\text{ass}} = \frac{k_{\text{ass}}}{k_{\text{diss}}}$, where $K_{\text{ass}} = $ association constant, $k_{\text{ass}} = $ association rate constant, $k_{\text{diss}} = $ dissociation rate constant and 'n' = number of ligand binding sites per receptor molecule. The Scatchard plot is then obtained by plotting $\frac{L_b}{R_o L_f} = K_{\text{ass}}\left(n - \frac{L_b}{R_o}\right)$. In FCS binding experiments, the Scatchard plot is obtained according to (6.9) using relations for L_f and L_b as in (6.7).

$$\frac{\sum y_i}{(1 - \sum y_i) R_o} = K_{\text{ass}}\left(n - \frac{\sum y_i L_o}{R_o}\right),$$

with $R_o = y_s L_o$ at saturation and we plotted

$$\frac{y}{1-y} \cdot \frac{1}{R_o} \quad \text{versus} \quad y\left[L_o^y\right] \cdot \frac{1}{R_o}. \tag{6.9}$$

The fraction of cell surface bound Rh-ligand $(\sum y_i)$ is obtained at the increasing concentration of the Rh-ligand (L_o).

6.3   Results

To highlight receptor binding in the membranes of living cells (cell cultures) using FCS we will review our studies conducted on interactions between the ligands the proinsulin C-peptide (CP) [6.8], the neuropeptide galanin (GAL) [6.10] and the epidermal growth factor (EGF) [6.11,6.18], and their respective receptors. All these ligands were covalently labeled with the fluorophore rhodamine (Rh). The peptide/hormone ligands CP, GAL and EGF and their receptors are of great physiological importance. Since the discovery of the insulin biosynthesis [6.19] it has generally been assumed that C-peptide, the connecting segment of proinsulin, does not possess biological activity of its own. However recent studies have demonstrated beneficial effects of CP in the diabetic state [6.20,6.21]. Galanin is a 29 amino acid long amidated neuroendocrine peptide that displays a variety of important biological actions and is thought to be implicated in several human disorders such as Alzheimer's disease, depression and feeding disorders (see reviews [6.22,6.23]. A tremendous number of studies have focused on EGF and EGF receptor system (see reviews Boonstra [6.24,6.25]. Very recently, the role of EGF in the molecular network communication of physiological processes, in which the EGF receptor operates as a central transducer of heterologous signaling systems, has been demonstrated (see review [6.25].

6.3.1   Background Signal

It is of great importance to take into account background fluorescence signals for the judgement of a specific ligand-receptor interaction when a ligand is tagged with a fluorescent dye. To check the background signals we measured the autofluorescence in cells without the addition of fluorophore-labeled ligands: Rh-CP, Rh-GAL and Rh-EGF. Very little background signal was detected in human diploid fibroblasts (Fig. 6.3a,) pancreatic insulinoma cells (Fig. 6.3b) and renal tubular cells (Fig. 6.3c), whereas huge background signals were observed in endothelial cells from umbilical cord veins (Fig. 6.3d).

6.3.2   Binding of Rh-Ligands to the Cell Membranes

Binding of Rh-CP. Fluorescence intensity fluctuations and autocorrelation functions of Rh-CP free in solution and bound to human renal tubular

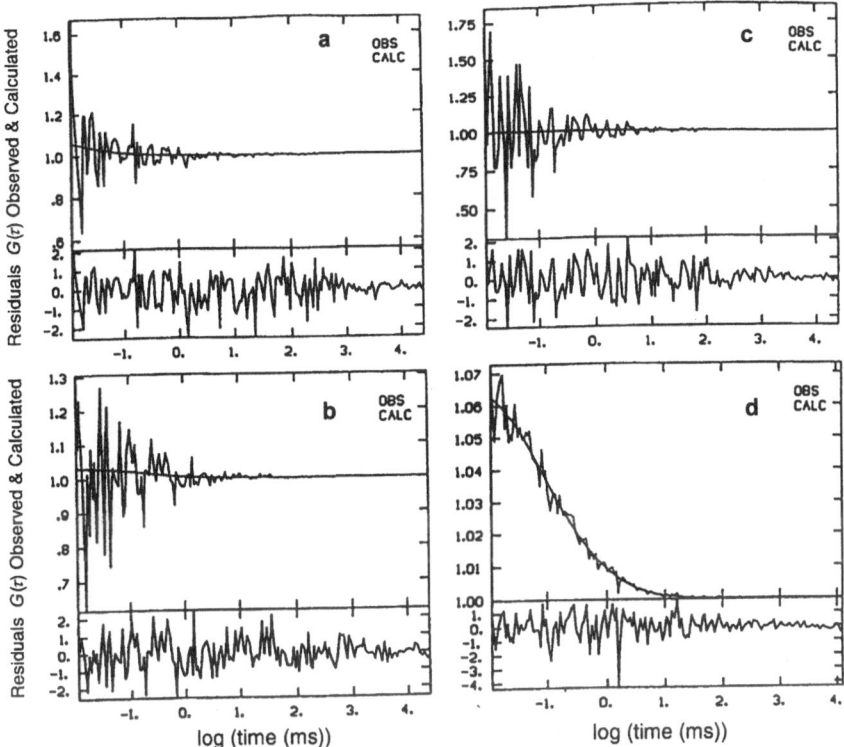

Fig. 6.3 a–d. Background binding in different cells. Autofluorescence in human diploid fibroblasts **a**, pancreatic insulinoma cells **b**, renal tubular cells **c**, and endothelial cells from umbilical cord veins

cell membranes are presented in Fig. 6.4A. Examination of the Brownian motion of unbound Rh-CP in the buffer medium above the cell surface exhibits typical fluctuations (Fig. 6.4A(a)) and a diffusion time (τ_{D1}) of 0.15 ms (Fig. 6.4A(b)). When the volume element is positioned at the level of the cell membrane, an increase and a broadening of the fluctuation peaks are observed (Fig. 6.4A(d)). Correlation analysis of the intensity fluctuations then shows a diffusion process of the cell-bound C-peptide with at least two components characterized by diffusion times of $\tau_{D2} = 1$ ms and $\tau_{D3} = 80$ ms, respectively (Fig. 6.4A(e)). Since the volume element extends into the space above the cell membrane, a small fraction of unbound Rh-CP, ($\tau_{D1} = 0.15$ ms) is also observed.

A

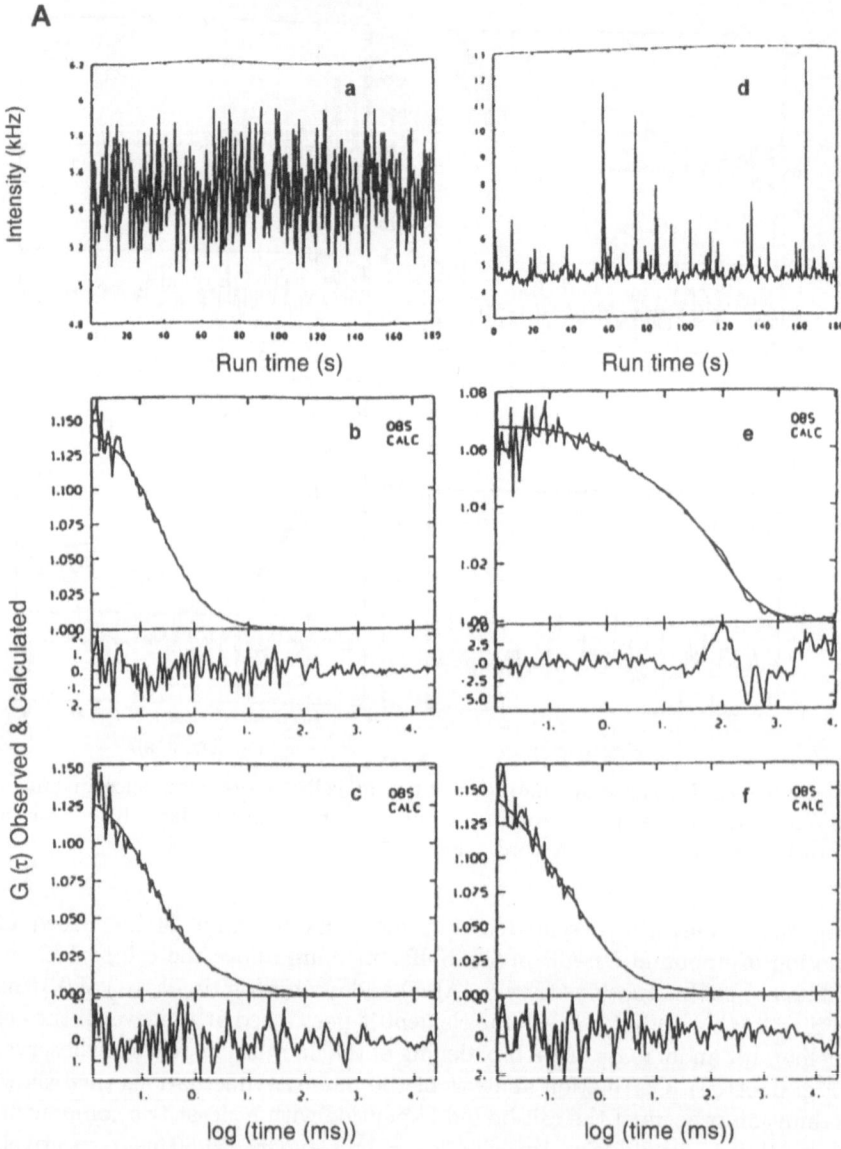

Fig. 6.4 A. C-peptide binding and displacement to the membranes of cultured renal tubular cells. Fluorescence intensity fluctuations **a** and autocorrelation function **b** for Rh-CP (5 nM) free in solution, $\tau_d = 0.15$ ms. Fluorescence intensity fluctuations **d** and autocorrelation function **e** for Rh-CP bound to membranes on the cell surfaces. Diffusion times (τ_D) and corresponding fractions (y) : $\tau_{D1} = 0.15$ ms, $y_1 = 0.1.$; $\tau_{D2} = 1$ ms, $y_2 = 0.15$; $\tau_{D3} = 80$ ms, $y_3 = 0.75$. Autocorrelation functions of displacement of membrane bound Rh-CP by post-incubation of a 1000-fold molar excess of non-labeled C-peptide **f** and non-labeled C-terminal pentapeptide **c**. In post-incubations, renal tubular cells were first incubated with a binding   buffer

B

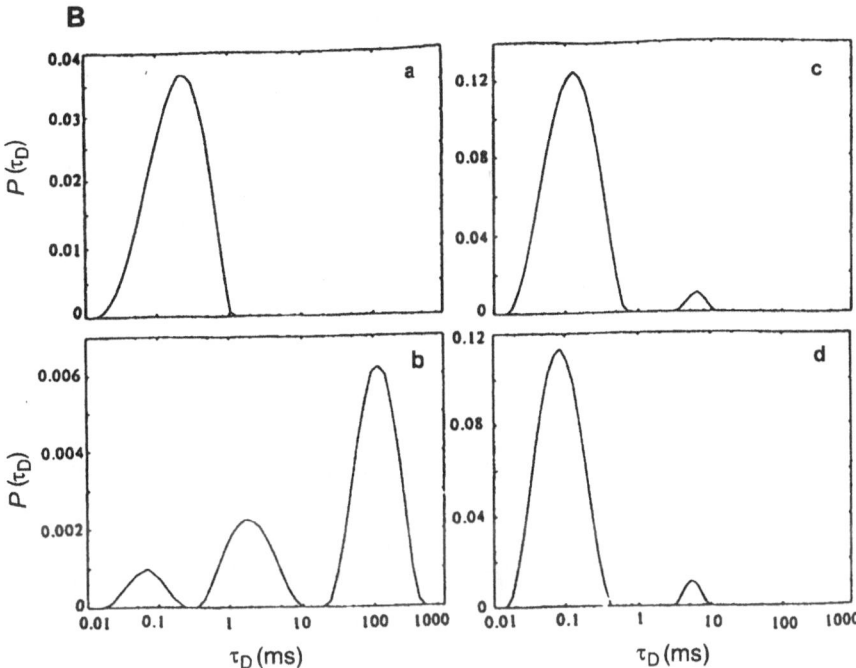

Fig. 6.4 B. CONTIN distributions of diffusion times P(τ_{Di}) of C-peptide binding and displacement to the membranes of cultured renal tubular cells. Rh-CP free in solution **a**, binding Rh-CP to the cell membranes, **b** displacement of membrane bound Rh-CP by post-incubation of a 1000-fold molar excess of non-labeled C-peptide **c** and inhibition of membrane bound Rh-CP by pertusis toxin **d**

Binding of Rh-GAL. Autocorrelation functions of Rh-GAL in solution and bound to the cell membranes are presented in Figs. 6.5A(a) and 6.5A(b), respectively. Figure 6.5A(a) demonstrates that the diffusion time (τ_{D1}) of unbound Rh-GAL is 0.16 ms. As seen in Fig. 6.5A(b), there are two diffusion times $\tau_{D2} = 22$ ms and $\tau_{D3} = 550$ ms. Both the τ_{D2} and τ_{D3} are much longer than τ_{D1}. The total binding of Rh-GAL to the cell membranes is in the range 65–75%.

---

containing 5 nM Rh-CP for 60 min and observed binding was treated as control. Then 5 µM non-labeled CP or non-labeled C-terminal pentapeptide was added to the same cells, incubated for three more hours and checked for displacement. In pre-incubations, cells were first incubated with binding buffer containing 5 µM non-labeled CP or non-labeled C-terminal pentapeptide for 60 min and then 5 nM Rh-CP was added to the same cells and checked for binding for three more hours. The difference between simulated and measured correlation curves (residuals) are indicated in b, c, e and f

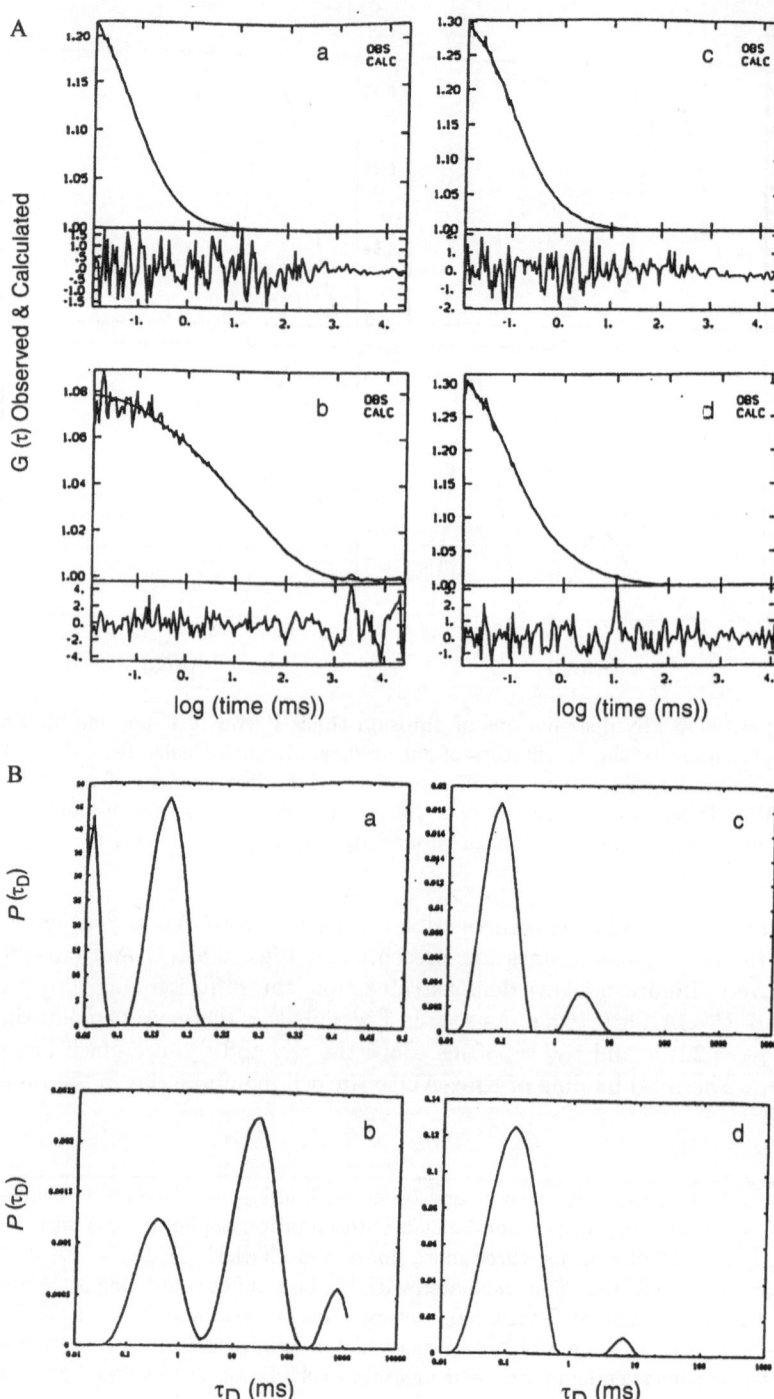

Binding of Rh-EGF. Autocorrelation functions of Rh-EGF in solution and bound to the cell membranes are exhibited in Figs. 6.6A(a) and 6.6A(b), respectively. Figure 6.6A(a) demonstrates that the diffusion time (τ_{D1}) of unbound Rh-EGF is 0.17 ms. As shown in Fig. 6.6A(b), there are two diffusion times $\tau_{D2} = 3$ ms and $\tau_{D3} = 100$ ms. With Rh-EGF, as with Rh-GAL, both τ_{D2} and τ_{D3} are much longer than τ_{D1}. The total binding of Rh-EGF to the cell membranes is in the range of 70–75%. As in the case of Rh-C in Rh-GAL as well as in Rh-EGF a certain fraction of unbound ligands are observed at the cell surface since the volume element extends into the space above the cell membrane.

6.3.3 Presentation of Ligand-Receptor Complexes with Distribution of Diffusion Times

A model-independent analysis of the diffusion processes (6.5) is presented in Figs. 6.4B–6.6B. In each figure (Figs. 6.4B–6.6B) there are three peaks which show distributions of diffusion times of (τ_{Di}) corresponding to unbound Rh-ligands and to receptor-bound Rh-ligand, i.e. these three peaks represent three different species of Rh-ligands: one free (τ_{D1}) and two bound (τ_{D2} and τ_{D3}). Distributions of $\tau_{D1} = 0.15$ ms (Fig. 6.4B(a)), $\tau_{D1} = 0.16$ ms (Fig. 6.5B(a)) and $\tau_{D1} = 0.17$ ms (Fig. 6.6B(a)) correspond to unbound Rh-CP, Rh-GAL and Rh-EGF, respectively. Distributions of $\tau_{D2} = 1$ ms and $\tau_{D3} = 100$ ms (Fig. 6.4B(b), of $\tau_{D2} = 22$ ms and $\tau_{D3} = 700$ ms (Fig. 6.5B(b)),

---

Fig. 6.5. A) Galanin binding and displacement to the membranes of cultured pancreatic insulinoma (Rinm5F) cells. Autocorrelation functions of tetramethyl rhodamine-labeled galanin (Rh-GAL) (5 nM) free in solution **a**, and bound to membranes on the cell surfaces **b**. Diffusion times (τ_D) and corresponding fractions (y) : $\tau_{D1} = 0.16$ ms, $y_1 = 35\%$ **a**, $\tau_{D2} = 22$ ms, $y_2 = 53\%$ and $\tau_{D3} = 700$ ms, $y_3 = 12\%$ **b**. Autocorrelation functions of displacement of cell membrane bound Rh-GAL by 1000-folds of non-labeled GAL **c** and non-labeled M40 **d** after post-incubations. In post-incubations, Rinm5F cells were first incubated with binding buffer containing 5 nM Rh-GAL for 60 min and observed binding was treated as control. Then 5 μM non-labeled GAL or non-labeled M40 was added to the same cells, incubated for three more hours and checked for displacement. In pre-incubations, cells were first incubated with binding buffer containing 5 μM non-labeled GAL or non-labeled M40 for 60 min and then 5 nM Rh-GAL was added to the same cells and checked for binding during three more hours. The difference between simulated and measured correlation curves (residuals) are indicated in a, b, c and d.
B) CONTIN distributions of diffusion times P(τ_{Di}) of GAL binding and displacement to the membranes of cultured Rinm5F cells. Rh-GAL free in solution **a**, binding Rh-GAL to the cell membranes, **b** displacement of membrane bound Rh-GAL by post-incubation of a 1000-fold molar excess of non-labeled GAL **c** and inhibition of membrane bound Rh-GAL by pertusis toxin **d**

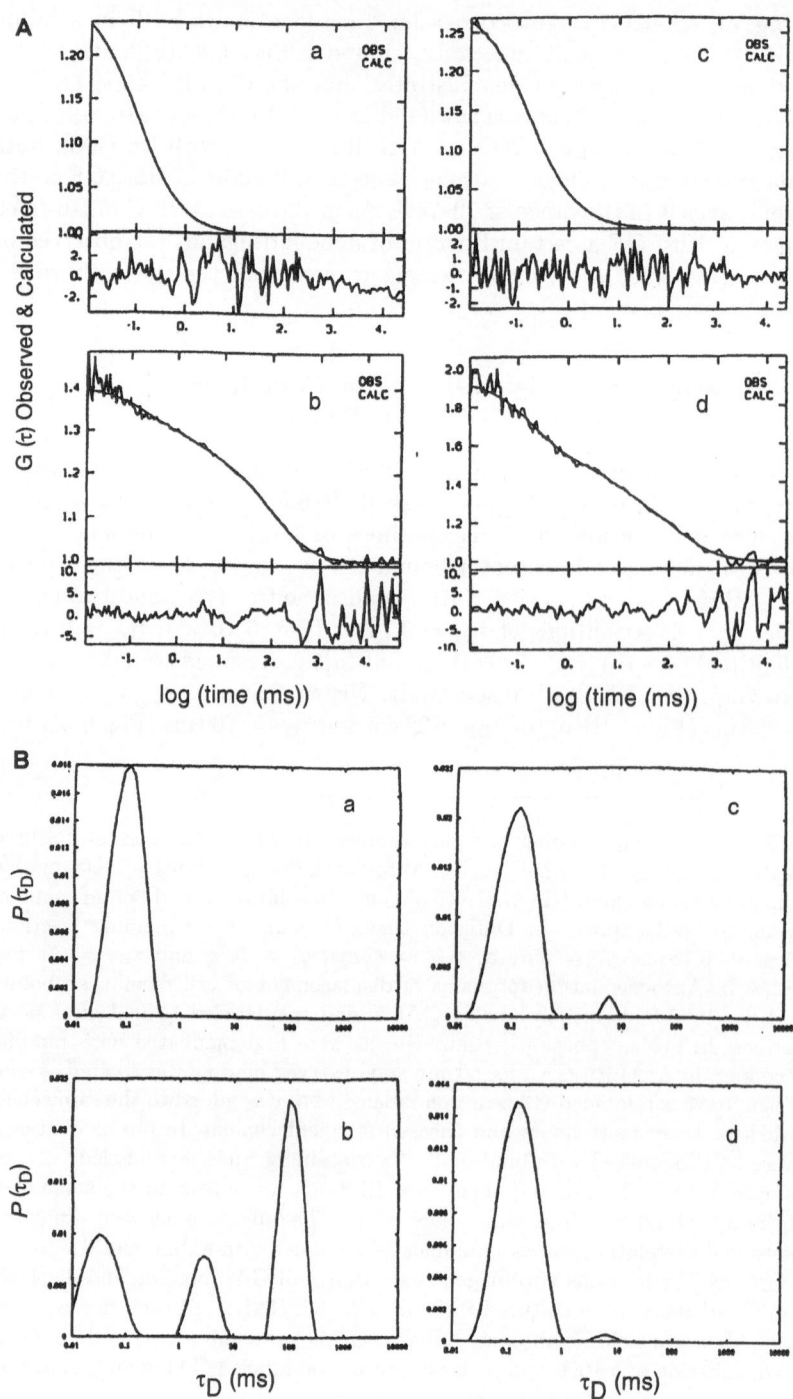

and $\tau_{D2} = 3\,\mathrm{ms}$ and $\tau_{D3} = 100\,\mathrm{ms}$ (Fig. 6.6B(b)), correspond to receptor-bound Rh-CP, Rh-GAL and Rh-EGF, respectively.

As expected, in the presence of a 1000-fold molar excess of non-labeled CP, the two peaks ($\tau_{D2} = 1\,\mathrm{ms}$ and $\tau_{D3} = 100\,\mathrm{ms}$) corresponding to ligand-receptor complexes disappeared due to competitive displacement by non-labeled CP, (Fig. 6.4B(c)). In the presence of a 1000-fold molar excess of non-labeled GAL, the two peaks ($\tau_{D2} = 22\,\mathrm{ms}$ and $\tau_{D3} = 700\,\mathrm{ms}$) corresponding to ligand-receptor complexes, disappeared due to competitive displacement by non-labeled GAL (Fig. 6.5B(c)). Finally, in the presence of a 1000-fold molar excess non-labeled EGF, the two peaks ($\tau_{D2} = 3\,\mathrm{ms}$ and $\tau_{D3} = 100\,\mathrm{ms}$) corresponding to ligand-receptor complexes, disappeared due to competitive displacement by non-labeled EGF (Fig. 6.6B(c)).

6.3.4   Saturation of Binding

Increasing concentrations of Rh-CP in the buffer medium leads to an increased proportion of membrane-bound labeled C-peptide. Saturation of the binding process occured at about 0.8 nM Rh-CP, and 50% binding was found at 0.3 nM Rh-CP (Fig. 6.7a, 6.9). Scatchard analysis reveals a binding process with an equilibrium association constant, K_{ass} of $3.3 \times 10^9\,\mathrm{M}^{-1}$, assuming one receptor binding site per C-peptide ($n = 1$) (Fig. 6.7a, insert). Evidence for the possible existence of a second binding process with a higher affinity and a ligand/receptor ratio of $n < 1$ is also observed.

Fig. 6.6. A) EGF binding and displacement to the membranes of cultured human diploid fibroblasts. Autocorrelation functions of tetramethyl rhodamine-labeled EGF (Rh-EGF) (5 nM) free in solution **a**, and bound to membranes on the cell surfaces **b**. Diffusion times (τ_D) and corresponding fractions (y) : $\tau_{D1} = 0.17\,\mathrm{ms}$, $y_1 = 24\%$ **a**, $\tau_{D2} = 3\,\mathrm{ms}$, $y_2 = 18\%$ and $\tau_{D3} = 100\,\mathrm{ms}$, $y_3 = 58\%$ **b**. Autocorrelation functions of displacement of cell membrane bound Rh-EGF by 1000-folds of non-labeled EGF **c** and non-labeled vascular EGF **d** after post-incubations. In post-incubations, human diploid fibroblasts were first incubated with binding buffer containing 5 nM Rh-EGF for 60 min and observed binding was treated as control. Then 5 µM non-labeled EGF or non-labeled vascular EGF was added to the same cells, incubated for three more hours and checked for displacement. In pre-incubations, cells were first incubated with binding buffer containing 5 µM non-labeled EGF or non-labeled vascular EGF for 60 min and then 5 nM Rh-EGF was added to the same cells and checked for binding for three more hours. The difference between simulated and measured correlation curves (residuals) are indicated in a, b, c and d **B)** CONTIN distributions of diffusion times P(τ_{Di}) of EGF binding and displacement to the membranes of cultured human diploid fibroblasts. Rh-EGF free in solution **a**, binding Rh-EGF to the cell membranes, **b** displacement of membrane bound Rh-EGF by postincubation of a 1000-fold molar excess of non-labeled EGF **c** and inhibition of membrane bound Rh-EGF by pertusis toxin **d**

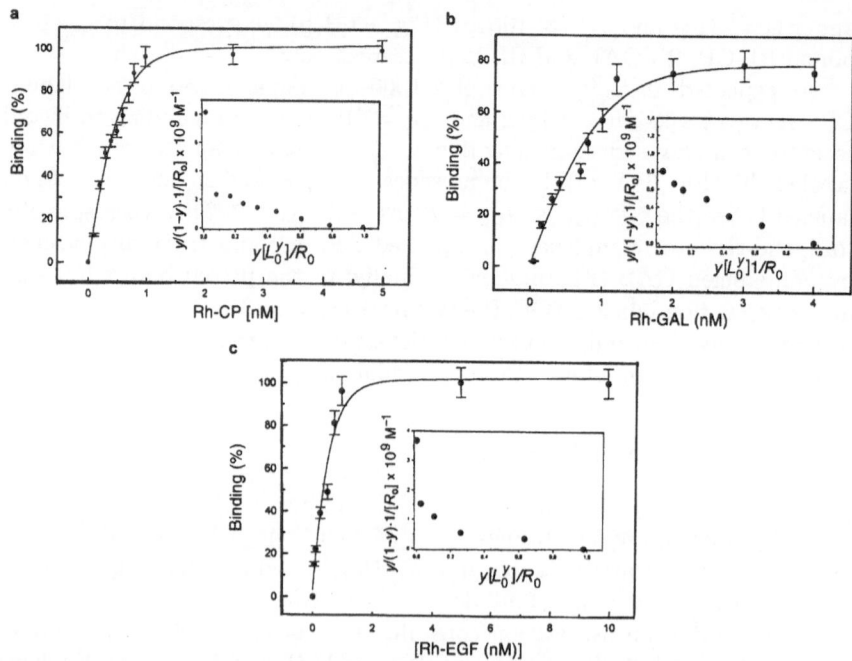

Fig. 6.7. a C-peptide binding curve. Binding of Rh-labeled C-peptide to cell membranes of renal tubular cells. Fractional saturation of the membrane bound Rh-CP (y) as a function of the ligand concentration (L_o) in the binding medium. Cells were incubated with binding buffer containing different concentrations of Rh-CP for 60 min. The Scatchard plot is shown as an insert. Each data point represents the mean of at least six separate measurements. 'y' = bound ligand, '$1 - y$' = free ligand, R_o = total receptor concentartion, L_o = total ligand concentration, **b** Galanin binding curve. Binding of Rh-GAL to cell membranes of Rinm5F cells. Fractional saturation of the membrane bound Rh-GAL (y) as a function of the ligand concentration (L_o) in the binding medium. Cells were incubated with binding buffer containing different concentrations of Rh-GAL for 60 min. The Scatchard plot is shown as an insert. Each data point represents the mean of at least six separate measurements. 'y' = bound ligand, '$1 - y$' = free ligand, R_o = total receptor concentration, L_o = total ligand concentration, **c** EGF binding curve. Binding of Rh-EGF to cell membranes of human diploid fibroblasts. Fractional saturation of the membrane bound Rh-GAL (y) as a function of the ligand concentration (L_o) in the binding medium. Cells were incubated with binding buffer containing different concentrations of Rh-EGF for 60 min. The Scatchard plot is shown as an insert. Each data point represents the mean of at least six separate measurements. 'y' = bound ligand, '$1 - y$' = free ligand, R_o = total receptor concentartion, L_o = total ligand concentration

To test whether the Rh-GAL binding to the specific receptors in the cell membranes is concentration-dependent and saturable, we carried out binding experiments at different concentrations of Rh-GAL. As is seen in Fig. 6.7b,

the increasing concentrations of Rh-GAL clearly led to the elevation of binding and the binding was saturated at nanomolar concentrations of Rh-GAL.
From the Scatchard plot (Fig. 6.7b, insert, (6.9)) of the binding isotherm we
obtain the binding constant $K_{ass} = 0.8 \times 10^9 \, M^{-1}$ and $n = 1$.

We carried out binding experiments at different concentrations of Rh-
EGF in order to check whether the Rh-EGF binding to the specific receptors
in the cell membranes is concentration-dependent and saturable, or not. As
is seen in Fig. 6.7c, the increasing concentrations of Rh-EGF clearly led to an
increased binding and the binding was saturated at nanomolar concentrations
of Rh-EGF. From the Scatchard plot (Fig. 6.7c, insert, (6.9)) of the binding
isotherm we obtain the binding constant $K_{ass} = 1.5 \times 10^9 \, M^{-1}$ and $n = 1$.
Evidence for the possible existence of a second binding process with a higher
affinity and a ligand/receptor ratio of $n < 1$ is also observed.

6.3.5   Specificity and Kinetics of Binding

To determine the specificity of Rh-CP binding, we examined competitive
displacement with non-labeled C-peptide. Cells were incubated with Rh-
CP, and after 60 min a 1000-fold molar excess of non-labeled C-peptide was
added. This results in a reduction of the Rh-CP binding. After 3 h about
85% of the total binding was displaced (Fig. 6.8a), as indicated by the autocorrelation function in Fig. 6.4A(f) and in the model-independent distribution analysis in Fig. 6.4B(c). Moreover, the Rh-CP binding was prevented
when cells are pre-incubated with $5 \, \mu M$ non-labeled C-peptide (data not
shown), demonstrating that all binding sites were already occupied by non-
labeled C-peptide. The dissociation curve plotted on a log scale (Fig. 6.8a,
insert) indicates that displacement of Rh-CP binding occurred in a monoexponential mode. Analysis of the dissociation curve yields a dissociation
time (τ_{diss}) of 2217 s, and a dissociation rate constant (k_{diss}) calculated as
$k_{diss} = \frac{1}{\tau_{diss}} = 4.5 \times 10^{-4} \, s^{-1}$. The association rate constant (k_{ass}) is calculated as: $k_{ass} = K_{ass} \times k_{diss} = 1.5 \times 10^6 \, M^{-1} s^{-1}$. The C-terminal pentapeptide
segment of the human C-peptide molecule (EGSLQ), previously shown to
stimulate Na^+, K^+-ATPase activity [6.26], was tested with regard to its ability to displace bound Rh-CP. A 1000-fold molar excess of the pentapeptide
was found to be as effective as the intact molecule in displacing C-peptide
bound to renal tubular cells (Fig. 6.4A(c)).

To determine whether the binding of Rh-CP to cell membranes is stereo-
specific or not, we incubated cells pre-exposed to Rh-CP with $5 \, \mu M$ of non-
labeled all D-amino acid (enantio) C-peptide for 3 h. The Rh-CP binding was
not displaced by the D-enantio C-peptide. Cells pre-incubated with Rh-CP
were also exposed to $5 \, \mu M$ scrambled C-peptide. Exposure to scrambled C-
peptide did not result in competitive displacement of the Rh-CP binding. In
addition, Rh-CP binding was not displaced by a 1000-fold molar excess
of either insulin, proinsulin, IGF-I or IGF-II. We also carried out binding

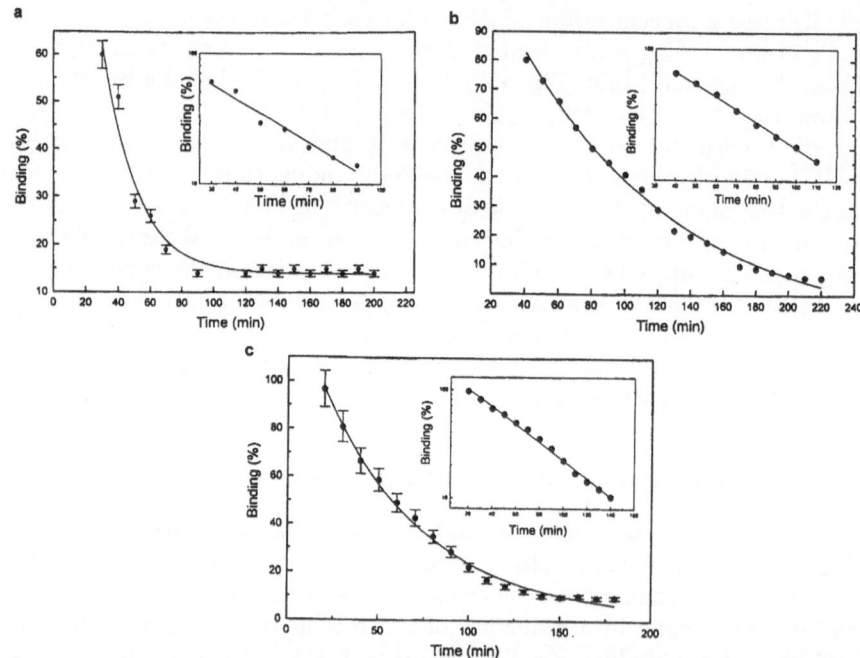

Fig. 6.8. a Time course of displacement of Rh-CP by non-labeled C-peptide. After incubation of cells with 5 nM Rh-CP for 60 min, 5 μM non-labeled C-peptide was added and FCS measurements were carried out at given time intervals. Each data point represents the mean of at least six measurements. A log scale for the binding displacement process is shown as an insert, **b** Time course of displacement of Rh-GAL binding by non-labeled GAL. Cells were incubated with binding buffer containing 5 nM Rh-GAL for 60 min. Then 5 μM non labeled GAL was added to the cells and FCS measurements were carried out at time intervals as shown in the x axis. Each data point represents the mean of at least six separate measurements. A log scale for the displacement of binding process is shown as an insert, **c** Time course of displacement of Rh-EGF binding by non-labeled EGF. Cells were incubated with binding buffer containing 5 nM Rh-EGF for 60 min. Then 5 μM non labeled EGF was added to the cells and FCS measurements were carried out at times (min) as shown in the x axis. Each data point represents the mean of at least six separate measurements. A log scale for the displacement of binding process is shown as an insert

studies with Rh-labeled insulin and renal tubular cells (data not shown). Rh-insulin binding was observed in the nanomolar range and was displaceable after addition of a 1000-fold molar excess of non-labeled insulin.

Similar tests were performed to see whether the Rh-GAL binding to the cell membranes is specifically displaced by non-labeled GAL. Rh-GAL binding in the presence of a 1000-fold molar excess of non-labeled GAL was studied. We incubated the cells with Rh-GAL and after 60 min incubation we measured binding in the control cells. Then we added non-labeled GAL and observed the displacement of Rh-labeled GAL.

As shown in Fig. 6.8b, with time course the presence of non-labeled GAL in the cells resulted in a decrease of the Rh-GAL binding and $\sim 85\%$ of the total binding was specifically displaced after 3 h incubation with non-labelled GAL (5 μM). One of the autocorrelation functions of this displacement and its CONTIN distribution are presented in Figs. 6.5A(c) and 6.5B(c), respectively. Similar displacement was observed when Rh-GAL binding was studied in the presence of a 1000-fold molar excess of non-labeled M40, a non-peptidergic galanin antagonist (Fig. 6.5A(d)) as compared to the galanin antagonist M15 [6.27]. Moreover, the Rh-GAL was unable to bind to cell membranes within 1–3 h when cells were first pre-incubated with non-labeled GAL for 1 h and then Rh-GAL was added to the cells (data not shown), indicating that GAL receptors are already occupied with non-labeled GAL. In this case, only the non-specific binding is observed (data not shown). The plot of the dissociation curve on a log scale (Fig. 6.8b, insert) showes a straight line, indicating that displacement occurred in a monoexponential mode. Furthermore, analysis of the dissociation curve yielded a dissociation time (τ_{diss}) of 2700 s, and a dissociation rate constant (k_{diss}) of $3.7 \times 10^{-4}\,\mathrm{s}^{-1}$. The association rate constant (k_{ass}) was obtained from: $k_{\mathrm{ass}} = K_{\mathrm{ass}} \times k_{\mathrm{diss}} = 3 \times 10^5\,\mathrm{M}^{-1}\,\mathrm{s}^{-1}$.

To check that the Rh-EGF binding to the cell membranes is not only ligand concentration-dependent and saturable but also specifically displaced by non-labeled-EGF, we studied Rh-EGF binding in the presence of a 1000-fold molar excess of non-labeled EGF. We incubated the cells with Rh-EGF and after 60 min incubation we measured binding treated as control. We then added non-labeled EGF to the cells and observed binding with time course. As shown in Fig. 6.8c, with time course the presence of non-labeled EGF in the cells resulted in a drop of Rh-EGF binding, and $\sim 85\%$ of the total binding was specifically displaced after 3 h incubation with non-labeled EGF (5 μM). One of the autocorrelation functions of this displacement and its CONTIN distribution are presented in Figs. 6.6A(c) and 6.6B(c), respectively. Moreover, the Rh-EGF was unable to bind to cell membranes within 1–3 h when cells were first pre-incubated with non-labeled EGF for 1 h and then Rh-EGF was added to the cells (data not shown), indicating that receptors are already occupied with non-labeled EGF. In this case, only the non-specific binding is observed (data not shown). It is noteworthy to mention that a 1000-fold molar excess of non-labeled vascular EGF was unable to displace the Rh-EGF binding to cell membranes (Fig. 6.6A(d)). The plot of the dissociation curve on a log scale (Fig. 6.8c, insert) shows a straight line, indicating that displacement occured in a monoexponential mode. Furthermore, analysis of the dissociation curve yielded a dissociation time (τ_{diss}) of 3460 s, and a dissociation rate constant (k_{diss}) of $2.9 \times 10^{-4}\,\mathrm{s}^{-1}$. The association rate constant (k_{ass}) was obtained from: $k_{\mathrm{ass}} = K_{\mathrm{ass}} \times k_{\mathrm{diss}} = 4.4 \times 10^5\,\mathrm{M}^{-1}\,\mathrm{s}^{-1}$.

6.3.6   Measurement of the Association Rate Constant

In order to measure the association rate constant (k_{ass}) directly we added Rh-EGF to cells and performed a time course of the EGF binding. Given the low concentrations needed to perform an FCS measurement "slow" kinetics can be expected even with diffusion-limited reactions which in classical kinetic studies, like stopped flow or relaxation kinetic measurements, would appear on ms time scales. As is shown in Fig. 6.9, 50% binding was acheived after 30 min when Rh-EGF was added in 1 nM concentrations. With increasing EGF concentrations faster saturation was observed. The time course of the appearance of the ligand-receptor complex LR(t) will follow an exponential time course:

$$LR(t) = LR(0) \exp -(k_{ass}[L] + k_{diss})t$$

From a plot of the inverse relaxation time against the added ligand concentration, the association rate constant can be obtained, as well as from the intercept the dissociation rate constant (Fig. 6.9) The experimentally determined k_{ass} was found to be about $3 \times 10^5 \, \mathrm{M}^{-1} \mathrm{s}^{-1}$ (Fig. 6.9, insert), a value close to $k_{ass} = 4.4 \times 10^5 \, \mathrm{M}^{-1} \mathrm{s}^{-1}$ which is determined indirectly from knowledge of the dissociation rate constant, $k_{diss} = 2.9 \times 10^{-4} \, \mathrm{s}^{-1}$ and

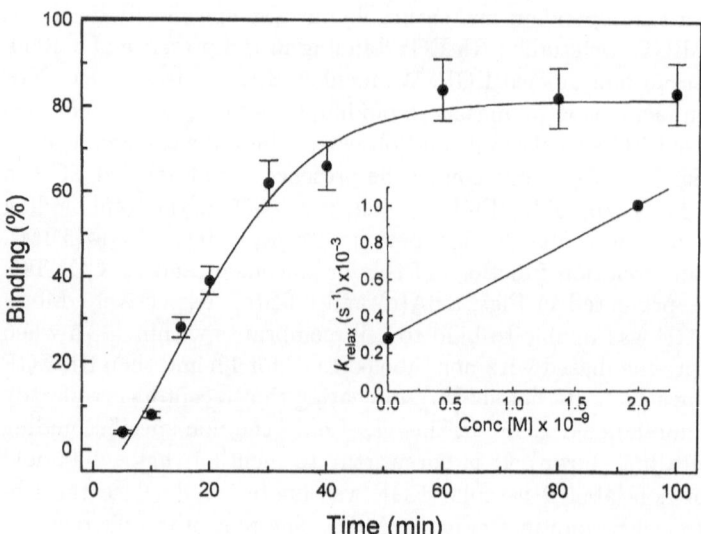

Fig. 6.9. Measurement of the association rate constant for the EGF binding. Experimental conditions are the same as in the Fig. 6.6A. Rh-EGF (5 nM) was added to cells and the Rh-EGF binding to cell membranes were performed at different time intervals. Each data point represents the mean of at least six measurements. Reciprocal time constant (k_{relax}) of the association kinetics of Rh-EGF and its receptor complex as a function of total ligand concentration is shown as an insert. $k_{relax} = k_{on}[EGF] + k_{off}$. The slope yields k_{on} and insert yields k_{off}

$K_{ass} = 1.5 \times 10^9 \, \text{M}^{-1}$ (see above). A close examination reveals that the value for K_{ass} is at least two orders of magnitude to compared to those slow expected for typically diffusion-limited encounters. The observed association kinetics apparently represent a second reaction step following a first diffusion-controlled high affinity encounter [6.10].

6.3.7   Effect of Pertussis Toxin on the Ligand Binding

In order to test whether the C-peptide binding is accompanied by G-protein involvement, we pre-treated cells (renal tubular) with pertussis toxin. As shown in Fig. 6.4B(d) this resulted in the complete loss of the slowly diffusing C-peptide receptor component. A small component of rapidly diffusing (1 ms) complexes (10–15%) remained after the pertussis toxin pre-treatment. This component was found to represent non-specific binding, since it was not displaced by the addition of an excess of unlabeled C-peptide. However, when pertussis toxin pre-treated cells were exposed to Rh-CP concentrations of 50–100 nM, the C-peptide 1 ms binding complex increased to 50%. This component could be displaced very rapidly (within 10 min) by addition of unlabeled C-peptide.

Very similar behavior was found for both GAL and EGF. In order to test whether the GAL binding is also accompanied by G-protein involvement, we pre-treated cells (pancreatic insulinoma) with pertussis toxin. As shown in Fig. 6.5B(d) this resulted in the complete loss of the slowly diffusing GAL receptor component, leaving a small component of rapidly diffusing (1–3 ms) complexes ($\sim$ 10%) which could not be further displaced by the addition of an excess of unlabeled GAL. However, when pertussis toxin pre-treated cells were exposed to Rh-GAL concentrations of 50–100 nM, the rapidly diffusing GAL binding complex increased to 45–50% and this component could be displaced very rapidly (within 10 min) by the addition of unlabeled GAL.

Finally, to check whether the EGF binding is accompanied by G-protein involvement, we pre-treated cells (human diploid fibroblasts) with pertussis toxin. As shown in Fig. 6.6B(d) this also resulted in the complete loss of the slowly diffusing EGF receptor component. As in the previous cases, a rapidly diffusing (1–3 ms) form of the EGF receptor complex could be distinguished after a pertussis toxin pre-treatment. From saturation experiments with Rh-EGF (50–100 nM) followed by rapid displacement by non-labeled EGF within minutes, we conclude that this EGF-receptor form represents a a ligand-receptor complex in a low affinity conformation.

6.3.8   Measurement of IC$_{50}$

In order to establish IC$_{50}$ we performed displacement of the Rh-CP, Rh-GAL and Rh-EGF binding in the presence of increasing concentrations of non-labeled ligands. Figures 6.10 a, 6.10 b and 6.10 c demonstrate the IC$_{50}$ for CP, GAL and EGF, respectively. In the case of GAL we found similar

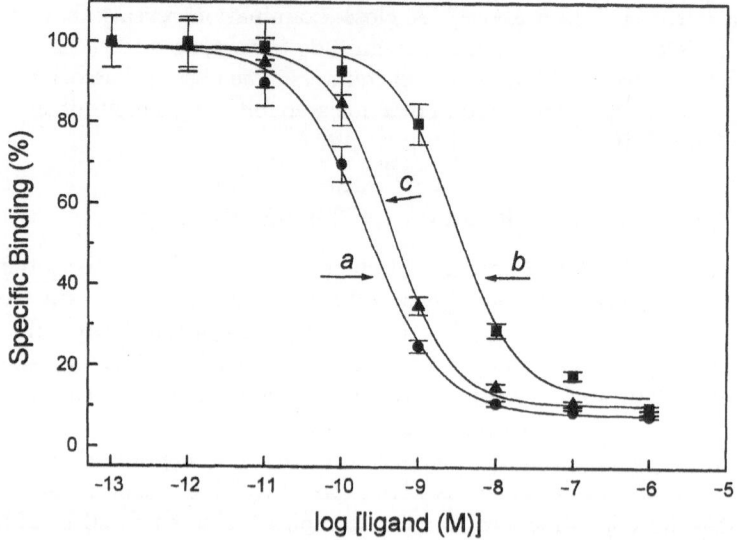

Fig. 6.10. Displacement of binding by incresing concentrations of non-labeled ligands (IC50). Displacement of Rh-CP binding by incresing concentrations of non-labeled CP *a*. Displacement of Rh-GAL binding by incresing concentrations of non-labeled GAL *b*. Displacement of Rh-EGF binding by incresing concentrations of non-labeled EGF *c*

IC_{50} when we also performed displacement of 0.05 nM $[^{125}I\text{-Tyr26}]$-galanin binding by increasing concentrations of Rh-galanin from galanin receptors in the Rinm5F cell line. For all the ligands IC_{50} was in the nanomolar range. For CP IC_{50} was found to be 0.3 ± 0.08 nM, for GAL $1.3.0 \pm 0.30$ nM and for EGF 0.66 ± 0.15 nM. It also is evident that fluorescence labeling of galanin does not affect binding since $[^{125}I\text{-Tyr26}]$-galanin binding is totally displaced by Rh-galanin (data not shown).

6.4 Discussion

6.4.1 Demonstration of Specific Binding

Competitive Displacement. The results demonstrate the presence of specific binding of human proinsulin C-peptide, neuropeptide galanin and EGF to their respective membrane-bound receptors in cultured cells. The specificity of the binding is attested to by the consistent displacement of bound Rh-CP, Rh-GAL and Rh-EGF following the addition of a 1000-fold molar excess of unlabeled C-peptide, GAL and EGF, respectively. In the same way, pre-incubation of cells with excess unlabeled ligands resulted in the failure of Rh-ligands to bind. Further evidence for the binding specificity of C-peptide is obtained from the fact that the addition of excess unlabeled C-terminal pentapeptide was accompanied by the displacement of bound Rh-CP. This is similar to what was observed after addition of intact C-peptide, suggesting that the C-terminal segment is involved in the binding process. The existence of a free C-terminal end of the C-peptide molecule is of importance for the binding, as indicated by the failure of proinsulin to displace bound C-peptide. The binding specificity of GAL was further confirmed when its binding was displaced by the galanin antagonist M40.

The possible existence of different forms of the ligand-receptor complexes stems from the observation that both fast and slowly diffusing complexes are found on the cell surface, both of which can be competitively displaced by unlabeled ligands. Addition of excess unlabeled ligands completely abolishes these complexes but only part of the rapidly diffusing complexes. This indicates that in addition to the specific binding there is also a component of non-specific interaction within the cell membrane.

Cross Reaction. The question may be raised whether the peptide ligands (CP, GAL, EGF) may bind to the receptors of other peptides or peptide hormones. However, neither insulin, IGF-I, IGF-II, nor proinsulin, when added in excess, elicited displacement of bound Rh-CP. Likewise, when rhodamine-labeled insulin was bound to cell membranes, addition of unlabeled C-peptide did not result in displacement of the labeled insulin. Cross reactions of the C-peptide with receptors of insulin or the other tested peptide hormones thus appear unlikely. Similarly, vascular EGF was not able to displace Rh-EGF binding, demonstrating no cross reaction as well.

6.4.2   Nature of Ligand-Receptor Interaction

In the Scatchard representation of the binding processes of CP, GAL and EGF (Fig. 6.7), we have assumed the existence of one binding site per receptor molecule. The reason for this assumption is the receptor concentration cannot be determined directly in FCS measurement but only by the interaction with the labeled ligands. However, a direct determination of the receptor concentration is possible by receptor-GFP fusion, as discussed in Chap. 7. The Scatchard plot shows a binding process with K_{ass} of $3 \times 10^9 \, M^{-1}$ for CP, K_{ass} of $0.8 \times 10^9 \, M^{-1}$ for GAL and K_{ass} of $1.5 \times 10^9 \, M^{-1}$ for EGF (for $n = 1$) (Table 6.1). However, there is also evidence for an additional higher binding affinity extrapolating to $n < 1$ for all these ligands. A possible interpretation of the latter is that one ligand molecule may interact with two or more receptor molecules. In our studies, the Scatchard analysis assumes the existence of independent binding sites. However, binding sites interfering with each other directly or indirectly (negative cooperativity) can result in a non-linear Scatchard plot, as is well established in the case of the insulin/receptor interaction [6.28].

Table 6.1. Association constant (K_{ass}), dissociation rate constant (k_{diss}) and recombination rate constant (k_{rec}).

Receptors	K_{ass} (M^{-1})	k_{diss} (s^{-1})	k_{rec} (M^{-1}s^{-1})
C-peptide	3.3×10^9	4.5×10^{-4}	1.5×10^6
GAL	0.8×10^9	3.7×10^{-4}	3.0×10^5
EGF	1.5×10^9	2.9×10^{-4}	4.4×10^5

6.4.3   Binding Kinetics

The binding of C-peptide, GAL and EGF to their specific binding sites was tested by the time-dependent displacement of Rh-ligands after the addition of excess unlabeled ligands. In the time interval tested (20–120 min), the dissociation kinetics follow a single exponential function characteristic for a relatively slow dissociation process with $k_{diss} = 4.5 \times 10^{-4}\,\mathrm{s}^{-1}$ for CP, $k_{diss} = 3.7 \times 10^{-4}\,\mathrm{s}^{-1}$ for GAL and $k_{diss} = 2.9 \times 10^{-4}\,\mathrm{s}^{-1}$ for EGF. This value is in the same range as observed for insulin [6.27] dissociations from its receptors, or for α-bungarotoxin from the acetylcholin receptor [6.29]. This value is also in the same range as observed for serotonin receptors when the specific interaction between fluorescently labeled ligands and the purified serotonin receptor was measured by the FCS technique in solution [6.30,6.31].

The experiments involving pertussis toxin indicate the existence of ligand-receptor complexes with faster dissociation rates which apparently represent ligand-receptor conformations with lower affinity. More experiments are required to determine the exact values of these rates.

We have demonstrated the possibility of determining directly the association rate constants for ligand-receptor interactions on the cell surface, i.e. in the natural situation. The independent determination of both association and dissociation rates together with the analysis of the Brownian motion of ligand-receptor complexes will provide important details for the mechanism of the ligand-receptor interaction and its coupling to the signal transduction.

6.4.4   Different Ligand-Receptor Complexes
and Binding Sites/Receptor Subtypes

In FCS measurements the fluorescence intensity fluctuations are recorded from only those molecules that diffuse through the confocal laser volume element. The time required for the passage of fluorescent molecules through the volume element is determined by the diffusion coefficient, which is related to the size and shape of a molecule. Thus, diffusion times obtained from the analysis of fluorescence intensity fluctuations with autocorrelation functions allow

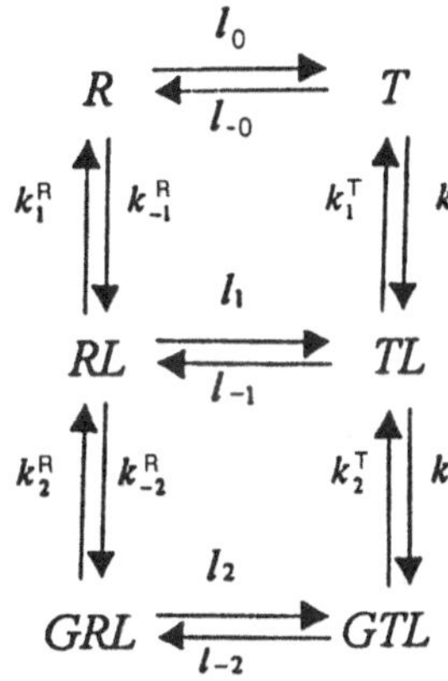

$$R \xrightleftharpoons[l_{-0}]{l_0} T \qquad L_0 = \frac{l_0}{l_{-0}}, \quad L_1 = \frac{l_1}{l_{-1}}, \quad L_2 = \frac{l_2}{l_{-2}},$$

$$K_1^R = \frac{k_1^R}{k_{-1}^R} \qquad K_1^T = \frac{k_1^T}{k_{-1}^T}$$

$$RL \xrightleftharpoons[l_{-1}]{} TL \qquad K_2^R = \frac{k_2^R}{k_{-2}^R}, \qquad K_2^T = \frac{k_2^T}{k_{-2}^T}$$

$$GRL \xrightleftharpoons[l_{-2}]{} GTL \qquad L_1 = L_0 \cdot \frac{K_1^T}{K_1^R}, \quad L_2 = L_0 \cdot \frac{K_1^T}{K_1^R} \cdot \frac{K_2^T}{K_2^R}$$

Fig. 6.11. Diagram for allosteric system in the ligand-receptor interaction

differentiation of faster diffusing and slower diffusing molecules as an analogy for unbound and bound states of ligand molecules, i.e. presentation of free ligand and ligand-receptor complexes. Moreover, in binding studies using the FCS technique, a mixture of several ligand-receptor complexes (components) with different molecular weights and corresponding different diffusion times can be analyzed without any need for separating unbound components from bound ones [6.18]. Compared to conventional methods, FCS allows analysis of the affinity and specificity of the ligand and subtypes of receptors from this mixture directly, without including any separation steps.

Another important point is that since in the FCS diffusion times of a ligand molecule and ligand-receptor complexes are used as characteristic parameters for binding analysis, the distribution of diffusion times with a special algorithm presents a clear cut appearance of binding complexes. Thus, the appearance of two binding complexes (Figs. 6.4B(b), 6.5B(b), 6.6B(b)) through distribution of diffusion times may suggest that these are representations of two different forms or subtypes of C-peptide, galanin or EGF receptors (see the model, Fig. 6.12). In addition, the appearance of two different ligand-receptor complexes most probably suggests that Rh-GAL does not bind to a single homogenous population but to a heterogenous population of receptors, as has been indicated by the galanin antagonist M35 binding to galanin receptors in Rinm5F cells [6.32].

6.4.5   Allosteric Nature of Signal Transduction and Receptor Aggregation

The observation that exposure of the cells to pertussis toxin interferes with the binding of C-peptide, GAL and EGF to their respective receptors (Figs. 6.4B(d), 6.5B(d), 6.6B(d)) provides evidence for the involvement of

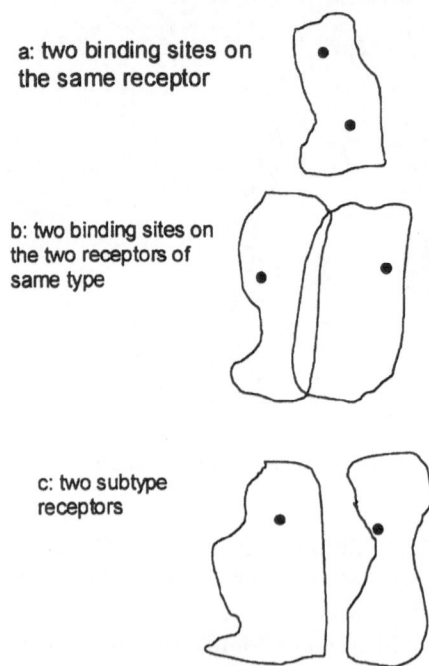

a: two binding sites on
the same receptor

b: two binding sites on
the two receptors of
same type

c: two subtype
receptors

Fig. 6.12 a–c. Model for multiple binding sites. Two binding sites on the same receptor **a**, two binding sites on the two receptors of the same type **b** and binding sites on two subtype receptors **c**

G-proteins in the signal transduction pathway. With the accumulation of detailed structural [6.33] and functional [6.34] data on G-proteins, the hypothesis may be considered that a G-protein which interacts with a ligand-activated receptor could constitute an allosteric system (see the diagram, Fig. 6.11), involving at least two or possibly several conformational states of both the receptor and the G-protein. Following interaction with the specific ligand, certain conformations may become stabilized which may affect the interaction with the G-protein.

Pertussis toxin is known to affect a cystein residue in the C-terminal chain of the α-subunit of the G-protein which interacts with loop regions of the membrane-spanning receptor [6.33]. Pre-incubation of cells with pertussis toxin in our studies was found to abolish the binding of C-peptide, GAL and EGF to the membranes at physiological conditions. These findings may be explained as the result of allosteric actions according to the principle of detailed balance [6.35,6.36] and as shown in the diagram of Fig. 6.11. Addition of higher concentrations of Rh-CP, Rh-GAL and Rh-EGF (50 nM and higher) to pertussis toxin pre-treated cells reveals a rapidly diffusing ligand-receptor complex ($\tau_D = 1$ ms), which in part can be displaced with an excess (50 μM) of unlabeled ligands within a few minutes (data not shown). The observations are compatible with the existence of at least two different ligand-receptor complexes, one with low affinity and high mobility and another with high affinity and low mobility.

However, one should not exclude the possibility that the receptor can aggegate under the influence of G-protein. It has been reported that both association of EGF to cytoskeletal elements of epidermoid carcinoma cells [6.37] and rotational diffusion of EGF receptor on the cell surface [6.38,6.39] can contribute to aggregation. We suggest that the ligand-receptor complex probably can aggregate forming dimers, tetramers and oligomers of receptors. This is in line with the reports where high order autocorrelations and fluorescence correlation spectroscopy are applied to characterize the aggregation process in membranes [6.40,6.41], in cerebrospinal fluid [6.42] and in solution [6.43].

6.4.6   Problems, Limitations, and Precautions

Photobleaching. Intensity fluctuations from dyes or dye-labeled substances (ligand, peptide, protein, oligonucleotides) are measured in FCS. For FCS measurements in cells, determination of the proper laser intensity requires great attention both for the dye and the cell survival. It is very likely that huge laser intensity may induce damage to cells. It has been well-documented that photobleaching is a critical issue in FCS measurements [6.44–6.46]. One has to pay attention to the photobleaching process when dealing with ligand-receptor interaction studies in living cells (cell cultures). It has been shown that if the laser excitiation produces a fluorescence signal intensity in the range of 4000–5000 counts per molecule and s, the survival time of single rhodamine dye molecules is around 4 s [6.11]. In our studies, the diffusion time (τ_{Dmax}) observed for a ligand-receptor complex is of the order of 100 ms. In our FCS analysis of ligand-receptor interactions, photobleaching was avoided by adjusting the laser excitation density such that the detected photon count rate did not exceed 3000–4000 counts per molecule and s to ensure that the characteristic observation times are far below (faster) than the average survival time of the dye tag used.

Labeling and Purity of Dye-Labeled Ligands. In addition to photo-bleaching, the incorporation of fluorophore into ligand is an important factor in receptor binding studies with fluororescently labeled ligands. Ideally the dye-labeled ligands should not contain unbound dye molecules since they may contribute to non-specific interactions. Since dye-ligand complexes usually have a certain shelf life, the purity of dye-labeled ligand complexes should be checked from time to time. It is observed that even 100% dye-labeled ligand loses dye molecules (5–10%) after 2–3 months. A problem concerning ligand tagging is the labeling of low molecular weight ligands. This relates to the limitation of FCS measurements in distinguishing molecular sizes by translational diffusion. Since the increase in molecular weight of a labeled ligand by a factor of eight only doubles its diffusion time, it is sometimes difficult to distinguish the diffusion time of the dye-labeled ligand from that of the dye, particularly if the molecular weight of the ligand is less than that of the dye molecule. However, Häberlein's group has been successful in labeling low molecular weight ligand with rhodamine when the ratio of the molecular weights of the rhodamine-labeled kavain ligand and rhodamine is two [6.47].

State of Cells. FCS analysis of ligand-receptor interactions in cell cultures requires the cells to be in a healthy state. Non-living cells are unsuitable for receptor binding studies with FCS. Furthermore, the lifetime of cells is a limiting factor for FCS measurements with certain cells surviving relatively longer than others. Membranes of dead cells lack the fluidity of a phospholipid environment, which leads to immobile ligand-receptor complexes. One of the ways to reduce background fluorecsence (Fig. 6.3) is to wash the cells four to five times with PBS. Nevertheless, this washing can also induce cell damage. Low ionic strength and temperature of the binding buffer can destroy cells. The pH of the binding buffer must be closer to 7.4 and the use of cold buffers should be avoided.

Focus on the Membrane Surface. If the ligand-receptor interactions are followed on membranes of cultured cells, it is very critical that the focus of the laser beam is properly positioned on the membrane surface. It is difficult to determine if the focus is exactly on the membrane. Our experiences show that if receptor binding is observed, altering the focus 1 μm upward should lead to a reduction of the binding. One should also note that if the receptor is membrane-bound, the binding observed in the cytoplasm mostly is nonspecific. In certain cases, if the focus is in the cytoplasm, one detects the internalization of receptors. If cells are grown in Nunc chambers, it is best to allow them to be attached to the bottom of the chamber. During the FCS measurement, cells should not move, since a change of cell position during the measurement will produce errors.

6.5   Future Perspectives and Cross-Correlation

In FCS, measurements of ligand-membrane interactions at single molecule detection sensitivity in 0.2 fl confocal volume elements [6.2,6.8,6.48] allows the interaction of binding sites with ligands on the molecular level in their native environment on cell surfaces, making it possible to identify a receptor which was not possible before to detect by conventional methods. The beauty of the FCS technique is that there is no need for separating unbound from bound ligand. Thus, with FCS it will be possible to discover new types of receptors as well as new target molecules. Consequently, FCS is a unique technique to perform large scale drug screening at the cellular level, the place of drug "action."

We have discussed that (Sect. 6.4.4. and the model (Fig. 6.12) FCS data analyzed with the CONTIN algorithm (Figs. 6.4B(b), 6.5B(b), 6.6B(b)) represent different receptor conformations and/or just receptor subtypes. Cross-correlation FCS can solve not only problems related to molecular weights of fluorophore-labeled ligands, but also it could be a suitable tool to identify and distinguish binding sites on the same receptor for different ligands. When it is claimed that the binding found in autocorrelation FCS measurement with one fluorophore-labeled ligand is displaced by the same (unlabeled) but not by a second unlabeled ligand, this does not exclude that the second ligand lacks a binding site on this receptor. The latter cannot share the site but may possess a site of its own. In cross-correlation FCS, the diffusion of the interacting molecules labeled with spectrally distinct fluorophores through the detection volume element is measured [6.49,6.50]. If two ligands (e.g., C-peptide and insulin) are labeled with two different fluorophores (e.g., Rhodamine Green and cy5), the cross-related signal will certify to what extent both C-peptide and insulin may share the same receptor with non-overlapping binding sites. Likewise, the possibility of specific complexes involving the interaction of cognate receptors can be elucidated.

The analysis of rotational diffusion and its theoretical background was established when the FCS technique was first developed [6.5]. The important result was that rotational correlations can be obtained independently of the lifetime of the fluorophore, and this has opened the way for the analysis of the slow rotational motions expected to occur at the cellular level. With the FCS analysis in its present form, the analysis of rotational motion is easy to perform [6.17]. With the detailed knowledge of the excited state physics, including triplet kinetics [6.44,6.45], as well as the conditions for single molecule detection [6.48,6.51], the full power of FCS has emerged.

We predict FCS will provide a plethora of new information on processes related to membranes and other parts in the cell.

Acknowledgements

We acknowledge many stimulating discussions with Jerker Widengren, John Wahren, Tamas Bartfai and Anders Zetterberg, all at the Karolinska Institute, on different topics regarding cellular interactions and cellular behavior.

References

6.1 R. Rigler, Ü. Mets, J. Widengren, and P. Kask: Eur. Biophys. J. **22**, 169–175 (1993)

6.2 M. Eigen and R. Rigler: Proc. Natl. Acad. Sci. USA **91**, 5740–5747 (1994)

6.3 D. Magde, E.L. Elson, and W.W. Webb: Phys. Rev. Lett. **29**, 705–711 (1972)

6.4 E.L. Elson and D. Magde: Biopolymers **13**, 1–27 (1974)

6.5 M. Ehrenberg and R. Rigler: Chem. Phys. **4**, 390–401 (1974)

6.6 T. Caspersson and G. Lomaka: In: Introductive to quantitative cytochemistry-II, G.L. Wied and G.F. Bahr, eds., Academic Press, New York and London, 27–56 (1970)

6.7 T. Land, Ü. Langel, G. Fisone, K. Bedecs, and T. Bartfai: Methods Neurosci. **5**, 225–234 (1991)

6.8 A. Pramanik, P. Thyberg, and R. Rigler: Spectroscopy. Chem. Phys. Lipids **104**, 35–47 (2000)

6.9 R. Rigler, A. Pramanik, P. Jonasson, G. Kratz, O.T. Jansson, P.-Å. Nygren, S. Ståhl, K. Ekberg, B.-L. Johansson, S. Uhlén, M. Uhlén, H.Jörnvall, and J. Wahren: Proc. Natl. Acad. Sci. U.S.A. **96**, 13318–13323 (1999)

6.10 M. Söderhäll, U.S. Bergerheim, S.H. Jacobson, J. Lundahl, R. Mollby, S. Normark, and J. Winberg: J. Urology **157**, 346–350 (1997)

6.11 R. Rigler, J. Widengren, P. Thyberg, A. Zetterberg, and A. Pramanik: Submitted (2000)

6.12 S. Wennmalm and R. Rigler: J. Physiol. Chem. **103**, 2516–2519 (1999)

6.13 D.W. Marquardt: J. Soc. Indust. Appl. Math. **11**, 431–441 (1963)

6.14 S.W. Provencher: Computer Physics Commun. **27**, 213–227 (1982a)

6.15 S.W. Provencher: Computer Physics Commun. **27**, 229–242 (1982b)

6.16 R. Rigler and Ü. Mets: Soc. Photo-Opt.Instrum.Eng. **1921**, 239–248 (1992)

6.17 A. Pramanik, A. Juréus, Ü. Langel, T. Bartfai, and R. Rigler: Biomed. Chromatogr. **13**, 119–121 (1999)

6.18 R. Rigler: J. Biotech. **41**, 177–186 (1995)

6.19 D.F. Steiner, D. Cunningham, L. Spigelman, and B. Aten: Science **157**, 697–700 (1967)

6.20 J. Wahren, B.-L. Johansson, H. Wallberg-Henriksson: Diabetologia **37** (*suppl.* 2), 99–107 (1994)

6.21 J. Wahren K. Ekberg, J. Johansson, M. Henriksson, A. Pramanik, B.-L. Johansson, R. Rigler, and H. Jörnvall: A. J. Physiol., Endoezinol Metab **278**, E759–E768 (2000)

6.22 T. Bartfai, T. Hökfelt, and Ü. Langel: Crit. Rev. Neurobiol. **7**, 229–274 (1993)

6.23 K. Bedecs, M. Berthold, and T. Bartfai: Int. J. Biochem. Cell. Biol. **27**, 337–349 (1995)

6.24 J. Boonstra, P. Rijken, B. Humbel, F. Cremers, A. Verkleij, and P.v.B.e. Henegouwen: Cell Biol. Int. **19**, 413–430 (1995)

6.25 E. Zwick, P.O. Hackel, N. Prenzel, and A. Ullrich: Trends Pharmacol. Sci. **20**, 408–412 (1999)

6.26 Y. Ohtomo, T. Bergman, B.-L. Johansson, H. Jörnvall, and J. Wahren: Diabetologia **41**, 287–291 (1998)

6.27 A. Pramanik and S.O. Ögren: Brain Res. **574**, 317–319 (1992)

6.28 P. DeMeyts: Diabetologia **37** (suppl 2), 125–148 (1994)

6.29 B. Rauer, B. Neumann, E. Widengren, J. Rigler, and R. Rigler: Biophys. Chem. **58**, 3–12 (1996)

6.30 A.-P.Tairi, R. Hovius, H. Pick, H. Blasey, A. Bernard, A. Suprenant, K. Lundström, and H. Vogel: Biochem. **37**, 15850–15864 (1998)

6.31 T. Wohland, K. Friedrich, R. Hovius, and H. Vogel: Biochem. **38**, 8671–8681 (1999)

6.32 K. Kask, Ü. Langel, and T. Bartfai: Cell. Mol. Neurobiol. **15**, 653–673 (1995)

6.33 H.E. Hamm: J. Biol. Chem. **279**, 669–672 (1999)

6.34 Z. Farfel, H.R. Bourne, and T. Iiri: N. Engl. J. Med. **340**, 1012–1020 (1999)

6.35 J. Monod, J. Wyman, and J.P. Changeux: J. Mol. Biol. **12**, 88–118 (1965)

6.36 D. Colquhoun: B. J. Pharmacol. **125**, 923–947 (1998)

6.37 F.A.C. Wiegant, F.J. Blok, L.H.K. Defize, W.A.M. Linnemans, A.J. Verkley, and J. Boonstra: J. Cell Biol. **103**, 87–94 (1986)

6.38 R. Zidovetski, Y. Yarden, J. Schlessinger, and T.M. Jovin: Proc. Natl. Acad. Sci. U.S.A. **78**, 6981–6985 (1981)

6.39 R. Zidovetski, D.A. Johnson, D.J. Arnt-Jovin, and T.M. Jovin: Biochem. **30**, 6162–6166 (1991)

6.40 A.G.I. Palmer and N.L. Thomson: Biophys. J. **52**, 257–270 (1987)

6.41 A.G.I. Palmer and N.L. Thomson: Chem. Phys. Liquids **50**, 253–270 (1989)

6.42 M. Pitschke, R. Prior, M. Haupt, and D. Riesner: Nature Med. **4**, 832–834 (1998)

6.43 L. Tjernberg, A. Pramanik, S. Björling, P. Thyberg, J. Thyberg, C. Nordstedt, L. Terenius, and R. Rigler: Chem. Biol. **6**, 53–62 (1999)

6.44 J. Widengren, R. Rigler, and Ü. Mets: J. Fluorescence **4**, 255–258 (1994)

6.45 J. Widengren, Ü. Mets, and R. Rigler: J. Phys. Chem. **99**, 13368–13379 (1995)

6.46 J. Widengren and R. Rigler: Bioimaging **4**, 149–157 (1996)

6.47 G. Boonen, A. Pramanik, R. Rigler, and H. Häberlein: Planta Med. **66**, 7–10 (2000)

6.48 P. Schwille, F.J. Meyer-Almes, and R. Rigler: Biophys. J. **72**, 1878–1886 (1997)

6.49 R. Rigler, Z. Földes-Papp, F.J. Mayer-Almes, C. Sammet, M. Völcker, and A. Schnetz: J. Biotech. **63**, 97–109 (1998)

6.50 R. Rigler, J. Widengren, and Ü. Mets: In Wolfbeis, O.J. (ed.), Fluorescence Spectroscopy (Springer Verlag, Berlin, pp. 3–21, 1992)

6.51 Ü. Mets and R. Rigler: J. Fluorescence **4**, 259–264 (1994)

7 Fluorescence Correlation Microscopy (FCM): Fluorescence Correlation Spectroscopy (FCS) in Cell Biology

Roland Brock and Thomas M. Jovin

7.1 Introduction

Fluorescence correlation spectroscopy (FCS) has traditionally been applied to well-defined in vitro systems, in which only the molecules of interest emit fluorescence. However, the ability to determine absolute molecule numbers, the inherent high sensitivity, and the possibility of measuring molecular diffusion and reactions in minute detection volumes with a dynamic range over nine orders of magnitude in the temporal domain, render FCS a highly attractive technique for acquiring quantitative molecular information in cellular studies. Cellular applications of FCS include two photon excitation [7.1] measurements of fluorescent probes [7.2] and microspheres [7.3], the detection of the intranuclear hybridization of fluorescently labeled oligonucleotides [7.4], the characterization of intracellular autofluorescence [7.5], the comparison of the intranuclear and cytoplasmic diffusion constants of green fluorescent protein (GFP; see below; [7.6]), the binding of proinsulin C-peptide to cell membranes [7.7] and the measurements of epidermal growth factor receptor (EGFR; [7.8])-GFP fusion proteins [7.9].

The analysis of living cells with FCS has to account for the complex chemical nature of the intracellular environment generating an intrinsic autofluorescence background. For this reason, this chapter will first introduce the formalism of FCS in cellular studies (Sect. 7.2). For the analysis of molecular association, a new procedure was devised that does not require the determination of the diffusion constant [7.5]. In cellular systems the structural heterogeneity on the subcellular level is expected to introduce uncertainties in the determination of this parameter. Fluorescent labeling of proteins by chemical modification of reactive amino acid side chains results in an ensemble of products carrying different numbers of fluorophores, compromising the determination of the correct number of molecules N from the autocorrelation amplitude. The same is true for heterogeneous populations of molecular aggregates. As both are likely to be encountered in cellular FCS, the dependence of the relative error in the determination of N on the average labeling ratio is illustrated.

The compartmentalization into organelles with distinct chemical environments imposes specific requirements on instrument design. High positioning accuracy and recording of measurement position are mandatory. Section 7.3

describes fluorescence correlation microscopy (FCM), an extension of FCS optimized for live cell studies. The salient features of FCM include high sensitivity digital imaging, fast switching between point and whole-field fluorescence illumination, high precision positioning along all three directions, automated measurement control, micro-injection, and live cell incubation. A series of applications of FCM in cell biological research illustrating the potential of the technique in the study of metabolic turnover, analysis of GFP fusion proteins, characterization of proteins in different subcellular locations and cell biological screening tasks are presented in Sect. 7.4. The experimental protocols for FCM are exemplified in detail for a fusion protein of GFP with the EGF receptor. The chapter concludes with a discussion of current problems and directions for future research (Sect. 7.5).

7.2 Theory of Cellular FCS

7.2.1 FCS in Multi-component Systems

Cellular FCS has to account for contributions from multiple endogeneous fluorescent species constituting intracellular autofluorescence [7.10–7.13], partitioning of fluorophores between multiple subcellular environments, and uncorrelated background fluorescence from light scatter and other sources. In contrast to conventional and confocal laser scanning fluorescence microscopy, the signal in FCS from intracellular autofluorescence and other background sources alone does not define the lower detection limit for externally introduced fluorophores [7.14] but rather the relative numbers of photons per molecule.

The global autocorrelation function $G_{tot}(\tau)$ is given as a weighted mean of a number L and M individual functions $Diff_i$ and $Diff_e$ for the intrinsic and extrinsic fluorescence contributions, respectively (7.1).

$$G_{tot}(\tau) = 1 + \frac{1}{N}\left(\sum_{i=1}^{L} \phi_i \, Diff_i + \sum_{e=1}^{M} \phi_e \, Diff_e\right). \tag{7.1}$$

For each species i or e, $\phi_{i,e}$ is the fractional weighting factor for the i-th or e-th contribution to the autocorrelation function with the corresponding relative molecular fraction $Y_{i,e} = N_{i,e}/N_{tot}$ and the fluorescence per molecule $fpm_{i,e}$.

$$\phi_{i,e} = \frac{Y_{i,e} fpm_{i,e}^2}{\left(\sum_{i=1}^{L} Y_i \, fpm_i + \sum_{e=1}^{M} Y_e \, fpm_e\right)^2}. \tag{7.2}$$

The fluorescence per molecule fpm is defined as

$$fpm = I/N. \tag{7.3}$$

where I and N are the fluorescence intensity and number of molecules, respectively, of species i or e. This quantity has an enormous analytical potential. Aggregation states can be analyzed by comparison of the *fpm* for the molecule in question with the value for a reference molecule that is known to be in a monomeric state (Sect. 7.2.6). Based on this parameter, the association of fluorescent molecules to intracellular structures has been detected (Sect. 7.2.2; [7.5]), changes in autofluorescence intensity analyzed (Sect. 7.4.2), and the photophysical characteristics of green fluorescent protein in the cytoplasm and nucleus compared [7.6].

Equation 7.4 is the analytical solution for a system of j components, making no explicit distinction between intrinsic and extrinsic fluorophores. The formalism provides a correction for the time-dependent (de)population of the triplet state, assuming equal triplet state characteristics for all components, the fractional population and lifetime of which are given by T and τ_T, respectively [7.15].

$$
G_{tot}(\tau) = 1 + \frac{\left(1 + \frac{T e^{-\tau/\tau_T}}{1-T}\right)\left(1 - \frac{I_B}{I_{tot}}\right)^2}{N_{tot}} \cdot
$$
$$
\sum_j \phi_j \left(1 + \frac{\tau}{\tau_j}\right)^{-1}\left(1 + \left(\frac{\omega_0}{z_0}\right)^2\left(\frac{\tau}{\tau_j}\right)\right)^{-\frac{1}{2}},
\tag{7.4}
$$

where the autocorrelation amplitude ($\tau = 0$) is given by

$$
G_{tot}(0) = 1 + \frac{\left(\frac{1}{1-T}\right)\left(1 - \frac{I_B}{I_{tot}}\right)^2}{N_{tot}} \cdot \sum_j \phi_j
\tag{7.5}
$$

and

$$
I_{tot} = \sum_j i_j + I_B
$$

is the total signal, including the uncorrelated background I_B. The diffusional autocorrelation times τ_j are related to the diffusion constants ($D_j = \omega_0^2/4\tau_j$); ω_0 is the radius of the Gaussian detection function in the optical focal plane, and z_0 is the corresponding axial parameter. The latter leads to a reduction of the fluctuation amplitude, and thereby to the incorrect inference of a higher apparent number of molecules in the detection volume [7.16].

7.2.2 Detection of Molecular Association Without Explicit Analysis of the Diffusion Constant D

In FCS, two independent methods exist for the determination of the number of fluorophores in the detection volume: (1) analysis of the amplitude of the autocorrelation function, leading to N_{ac}; and (2) division of the fluorescence measured in the probe volume by the *fpm* obtained with the same optical configuration in an independent reference measurement, leading to N_c. In the absence of photobleaching, N_c represents a measure of the total number of molecules while N_{ac} identifies freely diffusive molecules. A ratio N_{ac}/N_c close to unity signifies that all the molecules are freely diffusive, whereas a ratio < 1 indicates that the molecules are either partially bound to intracellular structures or confined in their movement within the detection volume. The validity of this concept was demonstrated for microinjected Cy3-labeled dextrans [7.5]. Only for $N_{ac}/N_c \approx 1$ did the diffusion constant derived from the autocorrelation functions correspond to reported values obtained by fluorescence recovery after photobleaching (FRAP; [7.17,7.18]. For microinjected probe molecules, the reference measurement can be carried out in vitro with the same molecules. In the case of the cellular expression of GFP fusion proteins, purified GFP can serve as the standard.

If fluorescently labeled molecules are microinjected, the autofluorescence background can be determined for each cell, enabling a calculation of N_c from the specific increase of fluorescence. For cells expressing GFP fusion proteins the autofluorescence background must either be negligible or the same for transfected and untransfected reference cells.

7.2.3   Determination of N for Distributions of Molecules Carrying Different Numbers of Fluorophores per Molecule

In the analysis of reaction kinetics or binding equilibria by FCS, most experimentalists to date have used singly labeled oligonucleotide probes, free fluorophores or micromolecular ligands carrying only one fluorophore per molecule. If fluorescent labeling is carried out via chemical derivatization of protein side chains, as is usually the case for recombinant proteins and antibodies, an ensemble of products with different numbers of fluorophores will be obtained, generating a multi-component mixture of labeled molecules. For low average labeling ratios it is reasonable to assume a Poisson distribution

Table 7.1. Total fraction of labeled molecules, sum of the fractional weighting factors and the apparent number of molecules derived from the autocorrelation amplitude $G(0)$ as a function of the mean labeling ratio (see Fig. 7.1 and Equations 7.2 and 7.5.). The actual number of molecules in all cases was 1

Labels$_{av}$	Labeled fraction	$\sum \phi_j$	N_{app}
0.5	0.39	0.85	0.33
1.0	0.63	0.79	0.50
1.5	0.78	0.77	0.60
2.0	0.86	0.77	0.67
2.5	0.92	0.78	0.71
3.0	0.95	0.79	0.75

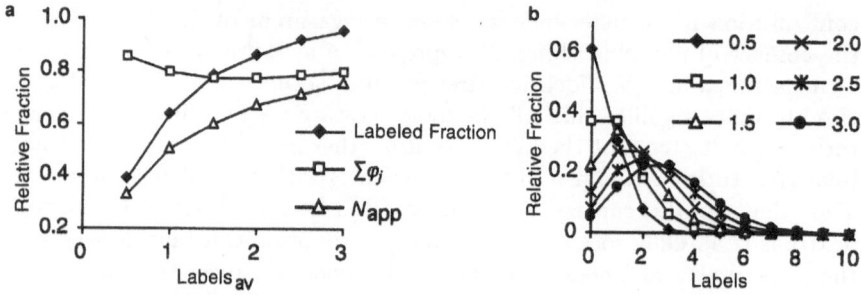

Fig. 7.1. Dependence of the apparent number of molecules on the average labeling ratio. **a** Plots of the total fraction of molecules carrying at least one label, the multicomponent correction factor $\sum \phi_j$ (7.2), and the apparent number of molecules derived from the autocorrelation amplitude $G(0)$ assuming one component. The proper $G(0)$ is related via 7.5 to the product of the inverse of the fraction of labeled molecules and $\sum \phi_j$. **b** Poisson distributions for average labeling ratios up to three labels per molecules. At a ratio of 2, 14% of all molecules are unlabeled while 14% carry ≥ 4 labels, increasing the risk of inactivation. The corresponding values are listed in Table 7.1

for the number of fluorophores per molecule. If the molecular mass of the fluorophore is small in comparison to that of the labeled molecule, the diffusion properties of the ensemble can still be represented by a one-component autocorrelation function. However, the determination of the number of molecules N will be affected (Table 7.1).

For average labeling ratios of 0.5, the actual number of molecules will be underestimated by 65%. In principle it should be possible to correct for this effect via an independent determination of the average labeling ratio, e.g., by UV/VIS spectroscopy and fitting with a multi-component autocorrelation function with relative fractions Y_j derived from the Poisson distribution. However, in reality, the *fpm* is not expected to scale linearly with the number of fluorophores per molecule due to homoquenching [7.19,7.20], causing some deviation from the treatment in Fig. 7.1. Fusion proteins of GFP eliminate this complication as each molecule carries exactly one label.

7.2.4 Intracellular FCS – Approximation of Local Equilibria

The theory of FCS applies to systems at thermodynamic equilibrium [7.21], a condition that is not met by living cells. For cells at rest the concept of a steady-state equilibrium can be applied. The analysis of nucleic acid hybridization [7.22–7.25] and the binding of fluorescent ligands to receptors [7.26] have demonstrated that reactions occurring in the minute range can be approximated adequately for FCS by local temporal equilibria. Intracellular processes such as changes in the association states of proteins in signal transduction should be accessible to FCS in the same way.

7.2.5   FCS in Small Volumes –
The Problem of Fluorophore Depletion

The measurement volume in confocal FCS is usually on the order of a femtoliter [7.27]. In classical cuvette type systems with microliter volumes, even if photodestruction occurs at the focus, the fluorophore concentration can be considered constant [7.27,7.28]. This assumption has been employed for diffusion measurements by continuous micro-photolysis [7.29,7.30]. The volume of a typical cell is only a few picoliters [7.31,7.32] implying that the reservoir of dye molecules is 10^{-6} that of the cuvette and that photodestruction of molecules will have an immediate impact on the signals, leading to a bias in the FCS measurement. First, the average number of molecules in the observation volume no longer remains constant. Second, photodepletion may contribute to the autocorrelation function. Depletion of fluorescent molecules has been observed in FRAP experiments on neurons [7.33]. In general, one seeks in FCS and FCM measurements combinations of dyes and levels of laser irradiance that minimize photodestruction.

7.2.6   FCS-Derived Parameters in Cell Biology

The number of fluorescent molecules in the detection volume and the autocorrelation times for an ensemble of distinct fluorescent molecules are obtained as fitted variables to the measured autocorrelation function (7.4). The following discussion will consider the diffusion constant D instead of the autocorrelation time τ, as both parameters are directly related to each other. Association denotes the tethering of the probe molecules at intracellular structures. For a freely diffusive molecule the number of molecules derived from the autocorrelation analysis N_{ac} equals the number of molecules obtained by dividing the

Table 7.2. Molecular parameters accessible in cellular FCS. The diffusion constant D and the number of molecules N are obtained directly from the fits to the autocorrelation functions. Mobility refers to a quantitative measure of molecular motion based on the diffusion constant, while association aims at distinguishing a freely mobile state from a state in which the molecule is immobilized by binding to intracellular structures

Molecular parameter	Measured parameter(s)
Mobility	D $(\mathrm{cm^2\,s^{-1}})$
Molecule number	N
Association	$N_{\mathrm{ac}}/N_{\mathrm{c}}$
Aggregation	$rfpm$, D, N
Reactions	$rfpm$, D, N

total fluorescence by the *fpm*, yielding N_c. If all molecules are labeled with fluorophores having the same fluorescence efficiency the *fpm* is the ratio of the average fluorescence count rate to the number of molecules. Given a determination of *fpm* for a monomeric reference state of the fluorophore, the number of monomers in an aggregate is obtained directly by calculating the relative *fpm* (*rfpm*) representing the ratio of $fpm_{\text{aggregated state}}$ to $fpm_{\text{reference state}}$. In a similar way reactions can be monitored, if products and reactants differ in their *fpm*, as is the case for redox reactions of flavin and adenine nucleotides (Table 7.2) [7.5].

7.3   Instrumental Requirements for Intracellular FCS

7.3.1   Design of the Fluorescence Correlation Microscope

Our design strategy was aimed at integrating the sensitivity of FCS in a microscope system with the functionalities required for the demands of cell biological investigations. These include live cell incubation on the microscope stage, micro-positioning with recording of the measurement position in all three directions, micro-manipulation and microinjection, digital video microscopy at low light levels to localize cells of interest in real time and conduct time lapse experiments with a minimum of photodamage of the fluorophores, rapid switching between point and whole-field illumination, and measurement of several spectral modalities in parallel (e.g., fluorescence autocorrelations and fluorescence spectra (Fig. 7.2). The latter is central for the analysis of molecular processes inside the cell, e.g., in the response to hormones and growth factors. If the measurement time for an autocorrelation is short in comparison to the reaction time, FCS can be used to follow reaction kinetics [7.23]. However, the temporal progression precludes sequential measurements of the different spectral modalities. To guarantee maximum experimental reproducibility with respect to the temporal and spatial complexity of a cell, a computerized control system was devised that is capable of executing measurement protocols with accurate timing, and recording of the measurement position (Table 7.3).

High Precision Micro-Positioning. The minuteness of the detection volume (about $0.6\,\mu m$ in diameter, $4\,\mu m$ in length) enables measurements in subcellular compartments such as the nucleus, endoplasmic reticulum, Golgi, and mitochondria. To benefit from this spatial resolution in subcellular microspectroscopic measurements, high precision positioning along all three axes with a spatial resolution exceeding $1\,\mu m$ is an indispensable prerequisite. If several positions are to be analyzed sequentially and repeatedly in time lapse experiments, positioning reproducibility also has to fulfill these specifications. In addition, for confocal analysis of flat specimens like surfaces, adherent cells and membranes, precise positioning along the optical axis is mandatory. A

Table 7.3. Overview of FCM-specific functionalities for fully exploiting the minuteness of the confocal sample volume in FCS for the analysis of dynamic processes at the subcellular level

FCM-Specific Functionalities
Live cell incubation
Micro-positioning with recording of the measurement position
Micro-manipulation and microinjection
High sensitivity digital microscopy with high spatial resolution
Measurement of several spectral modalities in parallel
Accurate timing and automated system control
Fast adaptation of the system control to experimental protocols
Rapid switching between point and whole-field illumination

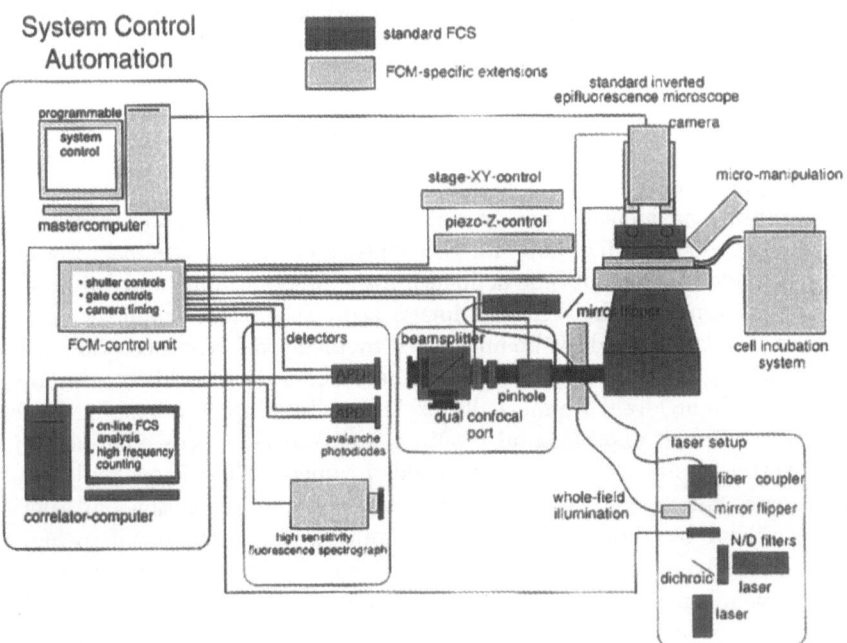

Fig. 7.2. Schematic representation of the FCM setup. System control and automation, confocal detection optics, detectors, and laser sources for excitation are grouped in functional units. Components of a standard FCS and FCM-specific extensions are distinguished

piezo-mechanical focusing device (Pifoc; Physik Instrumente, Waldenbronn, Germany) was selected and introduced between the lens turret and the objective lens. The positioning resolution is 10 nm, and the maximum travel 100 μm. Closed loop operation ensures maximum positioning reproducibility. After manual pre-focusing with the mechanical focusing drive, the axis of the drive is locked and focusing occurs exclusively by software using either buttons for incremental changes or a scanning functionality that records the fluorescence from the photodetector while focusing through the sample. In this way, fluorescence profiles of the sample are generated. Measurement positions are selected with a crosshair cursor. The combination of fluorescence and transmission images with count profiles localizes the measurement volume in all three dimensions.

This method for determining the focal position along the optical axis is superior to visual focusing with respect to sensitivity, speed, and precision. Visual control usually limits the choice of the focus to the focal plane, where fluorescence is the brightest. However, in the case of membrane labeling, for example, one may also wish to obtain measurements within the interior of the cell. Moreover, the image planes at the ocular or camera and the pinhole may deviate from each other i.e. may not fulfill the parfocal condition.

Image Acquisition. FCM relates spectroscopic measurements at the subcellular level to cellular morphology. Image acquisition for the analysis of cellular morphology requires high spatial resolution and low noise. For the localization and identification of cells, a real-time video mode is desirable. Time lapse digital microscopy for the analysis of dynamic cellular processes over extended periods of time depends on the precise timing and automation of the illumination shutters and image acquisition. Photodamage of fluorophores and cells needs to be minimized by high sensitivity detection at low excitation light levels. However, rapid image acquisition and low noise imaging at low light levels are mutually exclusive. To adapt the imaging system to these different experimental requirements, the FCM was equipped with an inverter photocathode intensified video camera (HL5; Proxitronic Funk, Bensheim, Germany). This technology is superior in resolution but inferior in sensitivity by about one order of magnitude compared to microchannel plate (MCP) image intensifiers. However, since cellular signal transduction and morphological changes occur on a minute rather than a subsecond time scale, the sensitivity gap can largely be compensated by image integration on the CCD chip of the camera. At low light levels cells are localized by integration and digitization in continuous mode. At a quarter of the video rate, near real-time operation is still possible. Table 7.4 lists the modes for video operation and image acquisition of the FCM.

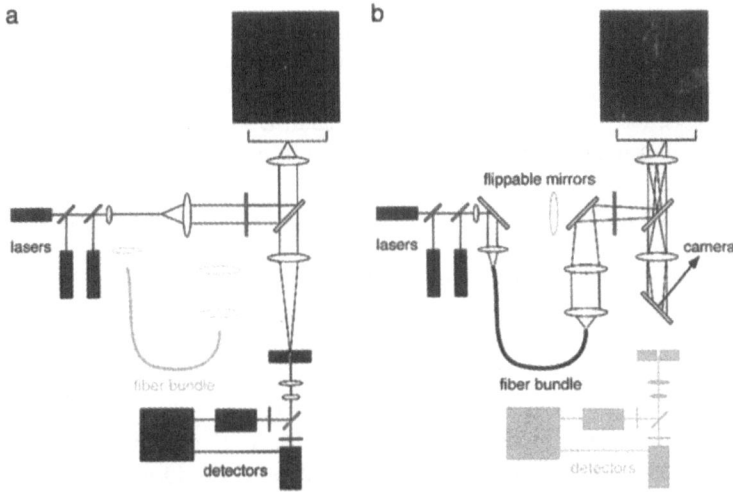

Fig. 7.3 a,b. Changing of illumination modes. **a** Point illumination: For fluorescence correlation measurements the laser light is directed through an optical single-mode fiber, expanded and collimated at the side of the microscope. The optical path is free of moving parts to assure maximum stability of the adjustment. **b** Whole-field illumination: for epifluorescence whole-field illumination the light is coupled into a fiber bundle perpendicular to the single-mode fiber. The light exits the fiber bundle perpendicular to the optical axis for point illumination, is expanded, focused into the back focal plane of the microscope objective, and reflected into the side port of the microscope by a second mirror flipper. Both flippers are motorized and synchronized. The fluorescence micrographs exemplify point and whole-field illumination for a cell, expressing a fusion protein of the human epidermal growth factor receptor (EGFR) and the green fluorescent protein (GFP)

Table 7.4. Operational modes for image acquisition

Experimental Protocol	Mode of Image Acquisition in FCM
Localization/identification of cells	Bright field microscopy on the video screen
	Epifluorescence microscopy on the video screen
Localization/identification with increased sensitivity	Epifluorescence microscopy with continuous frame integration on the CCD chip and display of the digitized image on the computer screen
Time lapse experiments	Automatic timing, shutter control and image acquisition with/without frame integration

Rapid Switching Between Point and Whole-Field Illumination. As image acquisition is an integral part of measurements in cellular FCM, optics were developed for rapid switching between focal measurements and whole-field epifluorescence illumination within less than a second (Fig. 7.3).

Major considerations in the design were switching speed and avoidance of directing the beam for focal illumination over any movable parts for maximum stability of the confocal optical alignment. For whole field illumination, the light is coupled into a fiber bundle by a motorized mirror flipper. At the side of the microscope the light exits the bundle in a direction perpendicular to the optical axis for point illumination, is expanded and focused into the back focal plane of the microscope lens via a second synchronized flippable mirror.

In FCM measurements, the position of the laser focus is recorded by image acquisition and located manually on the video screen, enabling the selection of a subcellular position without exposure of the fluorophores to the focused laser light. Following the measurements, the fluorescence image of the focus and the whole-field transmission and fluorescence images are merged for documentation.

7.4   Applications of Intracellular FCM

7.4.1   FCM in the Analysis of Receptor Diffusion – Measurement Protocols for Intracellular FCM

In this section, the general protocols of cellular FCM are presented for measurements of a fusion protein of the human epidermal growth factor receptor (EGFR, erbB1) with the green fluorescent protein (GFP; EGFR-GFP; [7.34]). The fusion protein has a total molecular mass of $\sim 200\,$kDa including the GFP and the receptor comprising a glycosylated extracellular ligand binding segment, a single plasma-membrane spanning region, and the intracellular tyrosine kinase domain [7.8]. Signal transduction mediated by the EGFR is prototypic for the family of receptor tyrosine kinases (RTK; [7.35]), for which protein-protein interactions constitute the basis for propagating signals inside the cell [7.8].

Binding of epidermal growth factor (EGF; [7.36]), one of the ligands of the receptor [7.37] leads to phosphorylation of tyrosine residues in the C-terminal domain of the receptor. The molecular details of this process are still under dispute. The ligand interacts stoichiometrically with the receptor [7.38]. Ligand-induced receptor dimerization is the favored model for activation [7.35,7.39–7.41]. Analysis of receptor dimerization by fluorescence resonance energy transfer, however, suggested that a subpopulation of receptors with high EGF binding affinity [7.42,7.43] exists as preformed dimers [7.44]. In this case, the ligand is presumed to activate the receptors by altering the mutual disposition of the cytoplasmic domains.

These questions are well-suited for FCM due to its ability to measure diffusion as well as aggregation of fluorescent molecules. A characteristic of the high affinity receptors, complicating FCM analyses, is a close association with the actin cytoskeleton, restricting or possibly abrogating lateral diffusion in the plasma membrane [7.45,7.46]. Because of this circumstance, the detection of immobile molecules needs to be an integral part of the strategy for FCM experiments of cellular EGF receptors. The analysis of photobleaching provides one convenient approach.

For FCM experiments, CHO cells were transfected with an EGFR-GFP construct [7.34]. The GFP was intracellular, fused to the C-terminus of the receptor. The cells were mounted in Hepes-buffered saline (HBS; 10 mM Na-Hepes, 135 mM NaCl, 5 mM KCl, 1 mM $MgCl_2$, 1.8 mM $CaCl_2$, 1% glucose and 0.1% BSA) at room temperature on the microscope stage. The experimental protocol proceeded through the following steps: (1) identification of cells positive for EGFR-GFP fluorescence; (2) adjustment of the laser power to minimize photobleaching, yet achieve good signal-to-noise ratios; (3) characterization of autocorrelations due to cellular autofluorescence using the laser powers chosen in step two; (4) autocorrelation measurements of the EGFR-GFP fusion protein; and (5) data analysis.

In whole-field epifluorescence images, cells expressing little or no fusion protein were easily distinguished from a population of transfected cells differing in their level of EGFR-GFP expression (see Figs. 7.6 and 7.8 below). Cellular integrity was confirmed by morphology based on transmission and fluorescence images.

Before carrying out the autocorrelation measurements of receptor diffusion, different laser powers were tested for their potential to bleach the fluorescence. In autocorrelation measurements the signal-to-noise ratio depends on the number of photons detected per molecule per residence time in the detection volume and is therefore a function of laser excitation power as well as of the diffusion constant [7.47]. Photobleaching of the molecules of interest caused by excessive laser powers needs to be avoided as this will lead to an apparent reduction of diffusion constants [7.28]. The recording of count traces was started and the laser excitation switched on (Fig. 7.4). At a laser power density of 0.4 kW/cm^2, some inital photobleaching occurred (Fig. 7.4). In a second measurement after a four minute interval over which the laser had been switched off, the initial photobleaching was reduced (Fig. 7.4), indicative of bleaching of immobile fluorescent molecules during the first measurement. The intermittent rise in the count-rate may have represented an aggregate of molecules slowly diffusing through the detection volume. However, at a laser power density of 13 kW/cm^2 strong photobleaching was evident. After again reducing the laser power, the fluorescence slowly recovered, indicating that mobile molecules had been locally depleted. A laser power density of 1.3 kW/cm^2 was selected as a compromise between reducing photobleaching and achieving a good signal-to-noise ratio. Measurements at 13 kW/cm^2 were restricted to a rapid discrimination of slowly diffusing molecules from small, freely diffusing molecules relevant for screening applications (Fig. 7.8 [7.9]).

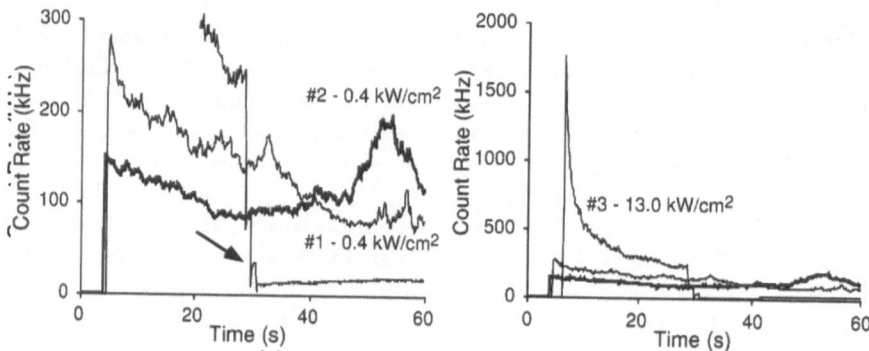

Fig. 7.4. Initial photobleaching at various laser powers. **Left panel:** first, the detection volume was placed inside the cell, as described in the legend to Fig. 7.6, then the recording of count traces was started; after some seconds delay the laser was switched on. All measurements were from one location of a single cell. Measurement #2 was acquired after a 4 min delay with the laser switched off. In measurement #3 the laser was changed back to the low laser power during the measurement (**arrow**). **Right panel:** same as left, with a different ordinate scale to clarify the recovery of fluorescence after the reduction of laser power. Laser powers were calculated according to [7.48]. In this and all following measurements the fluorescence was excited with the 488 nm line of an argon-ion laser (2313-150MLYV, Uniphase, Eching, Germany) and the emission passed through a dichroic mirror (500DRLP, Omega Optical, Brattleboro, VT, USA) and a 515–545 nm detection filter (Delta Light & Optics, Lyngby, Denmark)

Next, we investigated if at the laser powers of $1.3\,\mathrm{kW/cm^2}$ and $13\,\mathrm{kW/cm^2}$ the cellular autofluorescence itself gave rise to autocorrelations. Such auto-correlations arising from autofluorescence had been observed in human epidermoid carcinoma cells as well as in murine fibroblasts using 532 nm laser excitation [7.5]. Figure 7.5 presents a pair of autocorrelation measurements from two different cells. In neither case was an autocorrelation present at the low laser power. At $13\,\mathrm{kW/cm^2}$ the autofluorescence gave rise to a strongly autocorrelated signal in only one instance (Fig. 7.5b). In contrast, autocorrelation functions could be generated for GFP fusion proteins at $1.3\,\mathrm{kW/cm^2}$. This result indicates that GFP has a much higher fluorescence efficiency than the endogeneous fluorophores. Thus, the autofluorescence should be treated as an uncorrelated background signal [7.4] rather than as a component in the autocorrelation function.

Following these controls, autocorrelation functions were acquired for cells expressing the EGFR-GFP fusion protein. Figure 7.6 presents a compilation of all data recorded from a single cell. First, the detection volume was positioned inside the cell, based on the whole-field epifluorescence and transmission images and @z-profiles@ along the optical axis. Autocorrelation measurements were then carried out; the fluorescence count traces as well as autocorrelation functions were recorded. Because the measurement times ex-

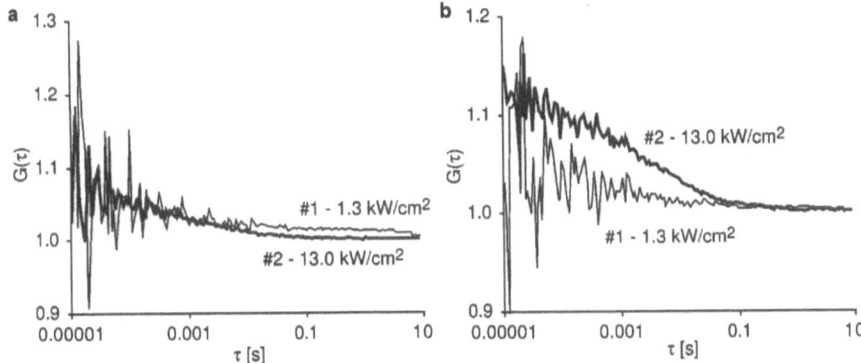

Fig. 7.5. Characterization of intracellular autofluorescence at high and low laser powers in **b** in two different cells **a**, **b**. The laser powers were those used for measurements of the EGFR-GFP fusion protein in other experiments

tended over minutes for several autocorrelation functions, the stability of the focus position along the optical axis was repeatedly controlled. Finally, another set of images and a profile were acquired in order to document the integrity of the cell during the measurement. Autocorrelation measurements were performed at laser power densities of 1.3 and $13\,\mathrm{kW/cm^2}$. The average count rates for the measurements at laser powers one order of magnitude apart were about the same. As these laser powers were far below saturation levels of the GFP, photobleaching, as shown in Fig. 7.4, must have occurred when the excitation was started at $13\,\mathrm{kW/cm^2}$. The amplitude of the autocorrelation function was higher – consistent with a photodepletion of molecules – and shifted towards shorter autocorrelation times. In addition, the relative contribution of a fast component increased with a relaxation time in the millisecond range. A further quantitative analysis of the impact of high laser powers on the derived relaxation constants is given below (Fig. 7.7).

Measurement #1 demonstrated the effect of photodepletion on the autocorrelation functions in general, as well as a problem of the ALV-5000/E autocorrelator (ALV-Laservertriebsgesellschaft, Langen, Germany) board in particular. Photodepletion constitutes a correlated process itself, giving rise to a contribution to the autocorrelation function. If the amplitude of the fluorescence fluctuations due to diffusion is low compared to the extent of photodepletion, i.e. at high molecule numbers, this process may dominate the autocorrelation function. Furthermore, in these cases the internal logic of the autocorrelator board may produce staircase-like artifacts. To ascertain that the long autocorrelation times observed were indeed due to molecular diffusion, the count traces were fitted with monoexponential functions (Fig. 7.6b). The time constants for photodepletion, derived from these fits, differed from the diffusion autocorrelation times by more than one order of magnitude (Fig. 7.7b, arrow).

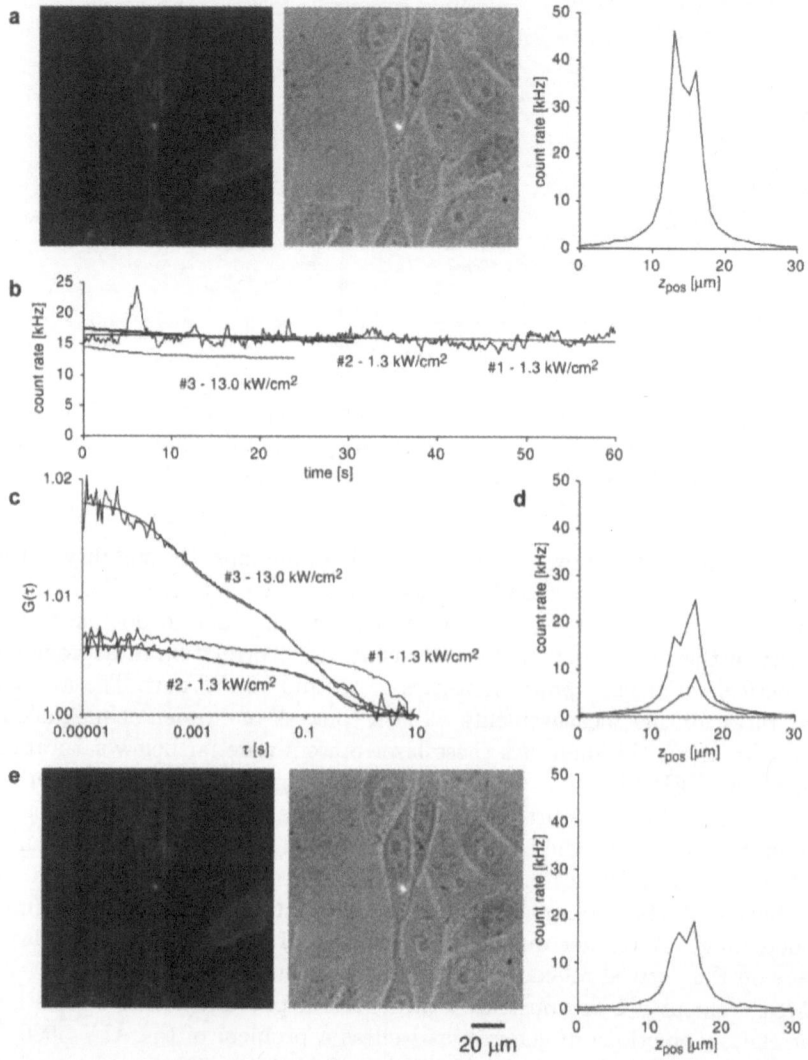

Fig. 7.6 a–e. Compilation of data recorded during autocorrelation measurements for a single cell, expressing the EGFR-GFP fusion protein in the plasma membrane. **a** Localization of the detection volume in the cell in all three dimensions, based on whole-field epifluorescence and transmission images and z-profiles along the optical axis. The z-profiles were acquired at $1.3\,kW/cm^2$. **b** Photodetector count traces recorded with the ALV-5000/E autocorrelator board . The count traces may serve to determine photodepletion rate constants and qualitative features of the observed cellular processes, e.g., frequency of occurrence of high amplitude fluorescence fluctuations. Here, the count traces were fitted with monoexponentials. The derived photodepletion time constants are included in Fig. 7.7b. **c** Autocorrelation functions and fits recorded at 1.3 and $13\,kW/cm^2$. **d** Z-profiles acquired between autocorrelation measurements to verify positioning stability. **e** Confirmation of the cellular integrity by fluorescence and transmission images and count profile

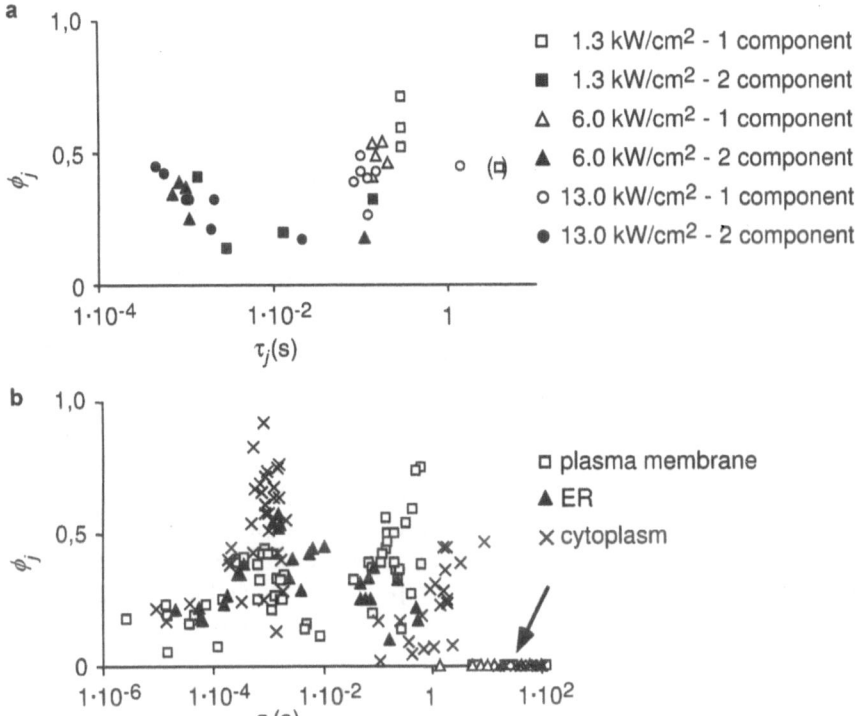

Fig. 7.7 a,b. Statistical analysis of relaxation constants measured for cellular EGFR-GFP fusion proteins. Autocorrelation functions were fitted with two to three diffusion components and an offset term to account for the drift of the autocorrelation function from the baseline. The relative fraction ϕ_j of each component was plotted versus the autocorrelation relaxation time τ_j. **a** Dependence on laser power. With increasing laser power, the autocorrelation time and the relative fraction of the slow component decreases. The bracketed data point was excluded from averaging. **b** FCM in the discrimination of GFP fusion proteins [7.9]; data obtained for GFP fusion proteins localized in the plasma membrane, the endoplasmic reticulum (ER), and the cytoplasm. The photodepletion time constants derived from monoexponential fits to the count traces are included on the abcissa (**arrow**)

The fitting strategy was intended to be model independent. As many components as necessary were included to obtain fits without systematic errors. The evaluation of the results was based on the knowledge of theoretical diffusion limits and typical time constants for photophysical processes characterized by in vitro experiments. Two to three diffusion components yielded fits without systematic errors in most cases. An offset term was included to account for the drift of the autocorrelation function of the baseline.

For the determination of the diffusion constant of the EGFR-GFP fusion protein a series of measurements on a number of cells was carried out at various laser power densities over time intervals of up to 60 s. For each auto-correlation function a major component with a relaxation constant in the upper millisecond range was derived (Fig. 7.7). The mean of the measurements at $1.3\,\mathrm{kW/cm^2}$ was 300 ms, corresponding to a diffusion constant of $7.5 \times 10^{-10}\,\mathrm{cm^2/s}$. For autocorrelation measurements carried out at $13\,\mathrm{kW/cm^2}$ the relaxation constant was shifted to shorter times and the fractional amplitude was diminished (Fig. 7.7a), consistent with the interpretation of the fluorescence recovery present in Fig. 7.4. The mean of all values shown in Fig. 7.7b was $1.9 \times 10^{-9}\,\mathrm{cm^2/s}$, a value greater than the one obtained at low laser powers by about a factor of three.

The diffusion constant of about $1 \times 10^{-9}\,\mathrm{cm^2/s}$ was greater by about one order of magnitude than the values obtained in previous measurements by FRAP [7.49] and single particle tracking [7.50]. We attribute this result at least in part to the apparent reduction of diffusion constants in the presence of photobleaching [7.28], although experiments conducted at even lower laser powers were confirmatory [7.9]. It is conceivable that the transfected fusion protein interacts differently than the wild type receptor with components in the plasma membrane or the underlying cytoskeleton. We also note that a fast diffusing component has not yet been reported for an endoplasmic protein or a transmembrane receptor. In contrast to FRAP, which only detects recovery of fluorescence by diffusion into the detection volume, FCS is also sensitive to small movements occurring within the detection volume. Limited proteolytic degradation and release of free GFP into the cytoplasm may also occur. Control experiments with untransfected CHO cells excluded autofluorescence as the source of the autocorrelations.

In terms of measurement accuracy for cellular measurements of slow diffusing molecules, it is important to note that for a 1 s measurement of rhodamine with a typical autocorrelation time of about $100\,\mu\mathrm{s}$ the statistical sampling of molecules passing through the detection volume still exceeds that for a 60 s measurement of receptors with a 100 ms autocorrelation time by a factor of 17. Subcellular heterogeneity is expected to further broaden the range of detected autocorrelation times.

Up to this point, the experimental protocols required for using FCM in the analysis of a particular molecule of interest have been presented. The integration of high sensitivity imaging and FCS in FCM lends itself to powerful screening applications addressing the molecular and cellular level simultaneously [7.9]. A compilation of the information obtained for GFP fusion proteins at different cellular locations is presented in Fig. 7.8.

7.4.2 FCM in the Analysis of Metabolic Conversions

FCS has the unique ability to directly measure the number of fluorescent molecules without the need of intensity-based calibrations. If a molecule undergoes a structural conversion with a concomittant change of the *fpm*, the number of molecules in each state can be obtained from the autocorrelation

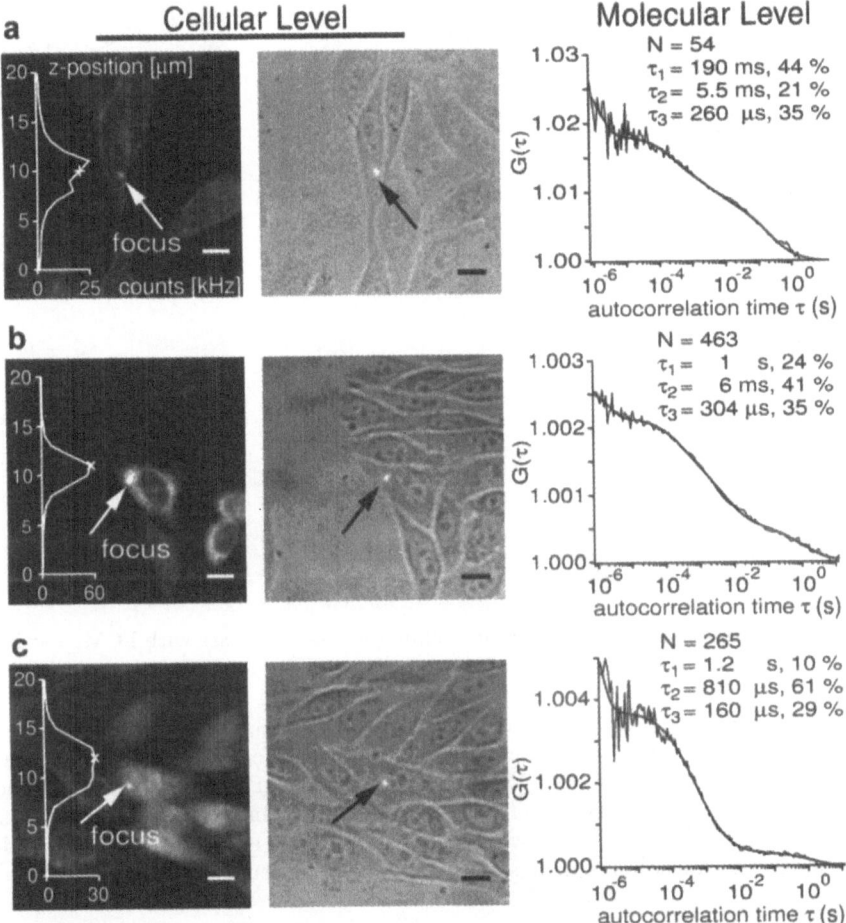

Fig. 7.8 a–c. FCM-based characterization of GFP fusion proteins localized in **a** the plasma membrane, **b** the ER, and **c** the cytoplasm. The cells were chosen from one heterogeneous population of cells transfected with the EGFR-GFP fusion construct. *Left panels*: subcellular distribution in three dimensions. Profiles along the optical axis are superimposed on the fluorescence micrographs; *center panels*: bright-field images; *right panels*: autocorrelation functions fitted with three diffusional components, triplet term, and offset. Bars denote 10 μm

function by formulation of the equation of a two-component system, in which both species differ in their relative *fpm* and the total number of molecules remains constant.

The flavins and nicotinamide adenine dinucleotides constitute an endogeneous reporter system for the metabolic state of a cell [7.12,7.13]; the corresponding *fpm* values are inversely correlated. Oxidative stress leads to an increase of flavin [7.51] and a decrease in nicotinamide adenine dinucleotide

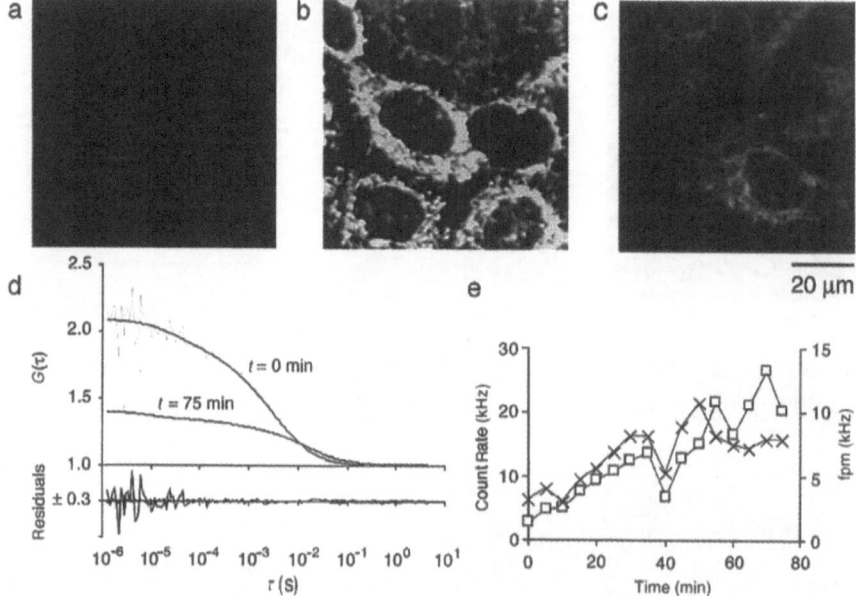

Fig. 7.9 a–e. Characterization of intracellular autofluorescence with FCM. **a–c** Increase of mitochondrial autofluorescence in human epidermoid carcinoma A431 cells; **a** $t = 0$ min, **b** $t = 90$ min, **c** mitochondrial staining with MitoTracker Green FM (Molecular Probes). **d** autocorrelation functions and fits with two diffusive components and a triplet fraction assuming equal *fpm* for all components. **e** Fluorescence count-rate (G) and *fpm* (×) as a function of incubation time on the microscope stage. The number of molecules N was obtained as a fitted parameter to the autocorrelation function without consideration of uncorrelated background. The parallel acquisition of fluorescence spectra was presented in [7.5]

fluorescence [7.13]. In A431 human epidermoid carcinoma cells [7.52] and murine fibroblast-derived HER14 cells [7.53] a time dependent increase in mitochondrial autofluorescence was observed during incubation on the microscope stage (Fig. 7.9a, b, c). This fluorescence increase was accompanied by a decrease in the amplitude of the autocorrelation function (Fig. 7.9d). However, calculation of the *fpm* revealed that the increase of fluorescence intensity and the decrease of amplitude was due in part to a change in the fluorescence characteristics of the cellular autofluorescence (Fig. 7.9e). The fluorescence spectra and reduction of fluorescence upon treatment of the cells with the reducing agent sodium sulfite were in agreement with an oxidation-dependent increase of flavin fluorescence as the source of the cellular autofluorescence.

The application demonstrated here was based on endogeneous fluorophores as reporter molecules. It will be interesting to apply the sensitivity and spatial resolution to processes in signal transduction and metabolism that are accompanied by changes in fluorescent properties of exogeneous reporter molecules,

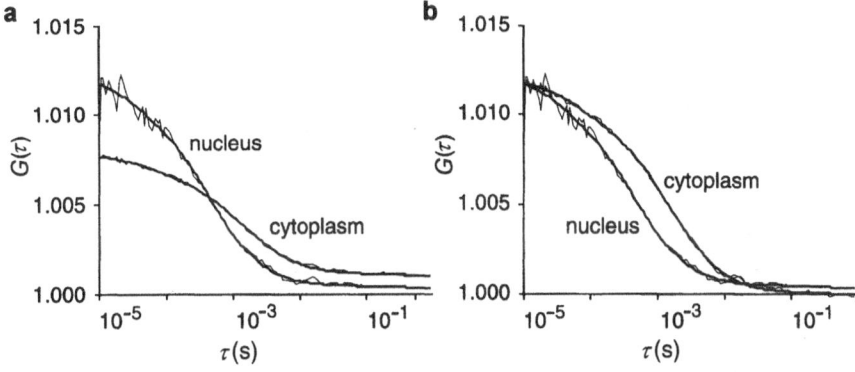

Fig. 7.10 a,b. Autocorrelation functions for transiently expressed nuclear and cytoplasmic GFP. The curves were fitted with one diffusion and one photophysical reaction term before **a** and after **b** normalization of the fluctuation amplitudes by that of nuclear GFP. The focal laser excitation power density was $16\,\mathrm{kW/cm^2}$, and the measurement time was $60\,\mathrm{s}$

e.g., ion-specific dyes. Molecular reporter systems compatible with such an approach have been reviewed in [7.54].

7.4.3   Comparison of Cytoplasmic and Nuclear GFP

To exemplify the application of FCM in the analysis of a protein in distinct subcellular locations, CHO cells were transiently transfected with free GFP. Autocorrelation functions were measured in the nucleus and the cytoplasm for a number of cells. The GFP was the F64L, S65T mutant with humanized codon usage (EGFP; Clontech, Heidelberg, Germany; [7.55]).

The autocorrelation functions were fitted with one diffusion and one photophysical reaction term (Fig. 7.10, Table 7.5; [7.15]). Contributions to the autocorrelation function by triplet state transitions as well as protonation equilibria have been reported for EGFP [7.56]. The time constant for the protonation reaction of EGFP depends on pH as well as on buffer concentration. To delineate potential differences in the diffusion autocorrelation times

Table 7.5. Derived parameters for the autocorrelation functions shown in Fig. 7.10. The fractional contribution of the reaction term is expressed in %

	N	τ_{Diff} (s)	τ_{react} (s)	%
Nucleus	97	4.3×10^{-3}	6.8×10^{-5}	14
Cytoplasm	177	1.4×10^{-3}	6.8×10^{-5}	16

152   R. Brock and T.M. Jovin

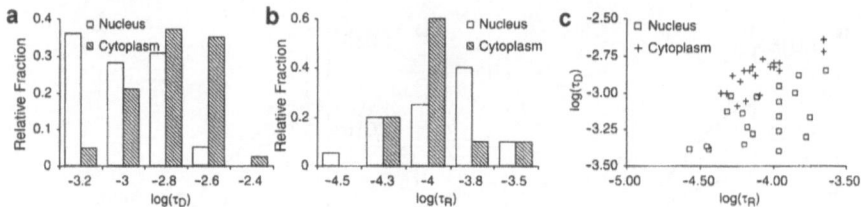

Fig. 7.11 a–c. Comparison of diffusion and reaction relaxation times for cytoplasmic and nuclear GFP. Frequency histograms of diffusion relaxation times **a** and reaction relaxation times **b** in the nucleus and in the cytoplasm. The bins are based on a logarithmic scale. **c** Plots of diffusion relaxation times versus reaction relaxation times. Time constants close to each other may not converge independently. The total number of values are **a** nucleus 39, cytoplasm 40; nucleus 20, cytoplasm 20. In **a** data from two independent experiments, one with varying laser power, one with constant laser power are included for all measurements. **b**, **c** are based on one experiment with a constant laser power density of $16\,kW/cm^2$

of cytoplasmic and nuclear GFP, frequency histograms for the distribution of autocorrelation times were generated (Fig. 7.11). For cytoplasmic GFP the distribution of the diffusion relaxation times was distributed around slightly longer values than the one for nuclear GFP. The means of both populations corresponded to diffusion constants of $3.1 \pm 1.4 \times 10^{-7}\,cm^2/s$ for nuclear and $1.8 \pm 0.7 \times 10^{-7}\,cm^2/s$ for cytoplasmic GFP. The faster diffusion in the nucleus compared to the cytoplasm is consistent with measurements of a 62 kDa dextran [7.57].

The distributions of the reaction relaxation times hardly differed from one another and were centered around 100 to 160 µs. These values are in accordance with the range of values in [7.56]. A plot of diffusion relaxation versus reaction relaxation times was included as a control to determine if both times converged independently from one another in the regression analyses. A slightly positive correlation was apparent for the regression analysis of cytoplasmic GFP.

7.4.4 FCM in Cellular High Throughput Screening

Section 7.4.1 has presented the potential of FCM for obtaining complementary information about molecular distribution and function on the cellular and molecular level. Autocorrelations for free intracellular GFP and a fusion protein of the epidermal growth factor receptor (EGFR) with GFP [7.34] were acquired within 5 to 10 s, measurement times short enough to conduct a few thousand measurements per day. In evaluating the usefulness of FCM for serial cellular screens in a particular application, the diffusion constant, photostability and concentration of the molecule of interest are important determinants.

The slower the diffusion the longer is the required measurement time. For free intracellular GFP a few seconds were sufficient to obtain high quality autocorrelation functions; at low laser powers for EGFR-GFP, data were recorded over 60 s or longer. If both the receptor and ligand are labeled, the need for such long sampling times can be circumvented by the recently introduced method for confocal fluorescence coincidence analysis [7.58]. The sample is rapidly scanned through the detection volume and a coincidence analysis is performed on the fluorescence originating from both fluorophores.

It is expected that with further automation in the positioning of the detection volume at the subcellular level, the versatility of FCS in high throughput screening demonstrated up to now in vitro will gradually become available for FCM in vivo.

7.5   Limitations and Perspectives of Cellular FCM

In the preceding section, the application of FCM to the characterization of plasma membrane receptors, analysis of metabolic conversions inside the cell, and intracellular GFP proteins has been demonstrated. Here, FCM-specific problems that became apparent in the course of this and other work will be discussed, after which the perspectives for future developments will be sketched out. Finally, the present and future potential of FCM will be compared to that of other techniques yielding similar data.

7.5.1   FCM-Specific Problems in Intracellular Research

Requirement of Low Molecular Concentrations. For fluorescence autocorrelation measurements, fewer than one thousand molecules should be present in the detection volume for a number of reasons. At concentrations that are too high the fluctuation amplitude arising from molecular motion and photophysical processes will be too small compared to the fluctuations from other sources (e.g., laser intensity) and noise. Additionally, one may saturate the photon-counting devices. At high fluorophore concentrations, photodepletion has a larger impact on the autocorrelation function than at low concentrations. The higher the fluorophore concentration the larger the amplitude of photodepletion relative to the fluctuations from the fluorophores. Furthermore, the introduction or expression of excess molecules carries the risk of disturbing the physiological equilibria and reactions.

Consequently, the number of fluorescent molecules should be limited to a minimum. However, in our experience, for concentrations below ∼ 50000 molecules per cell, it is increasingly difficult to identify cells based on the whole-field fluorescence images. Although not yet analyzed in detail, whole-field imaging and FCS measurements at the single molecule limit may be incompatible. Generation of clones stably expressing the fusion protein at different levels offers an obvious solution to this problem and has been successfully applied to the EGFR-GFP fusion protein [7.9]. Weakly fluorescent cells could be selected with confidence that fusion proteins were present. For conveniently tuning the biological system to the requirements of FCM, it will be advisable to use inducible promoters with controllable expression levels, e.g., the tetracycline-repressible/inducible system [7.59].

Measurements of Immobile/Slowly Diffusing Molecules. For the EGFR-GFP fusion protein an initial decrease in fluorescence occurred down to $0.4\,kW/cm^2$ – the laser power density required for obtaining autocorrelation curves for slowly diffusing EGFR molecules – due to photobleaching of immobile or less mobile molecules. Immobile molecules are 'invisible' to FCS. At laser power densities below the threshold for photobleaching, they will generate a fluorescence signal from which the number of molecules can be calculated by extrapolating the *fpm* at the low laser power from the value determined in FCS measurements at higher laser powers. The number of molecules derived from the autocorrelation function will reflect the fraction of mobile molecules. Bleaching of mobile molecules in the focus will distort the determination of molecule numbers as well as of diffusion constants. Photobleaching also limits the range over which different autocorrelation times can be measured simultaneously without introducing these artifacts. For free intracellular GFP, a laser power density of $2.5\,kW/cm^2$ was adequate to acquire autocorrelation functions. However, for the EGFR-GFP fusion protein there was already an increased risk of photobleaching at this higher intensity.

To fully define the relative fractions of a molecule within a cell in different diffusional states it will be necessary to devise protocols quantitating the immobile fraction based on the fluorescence recorded at steady state level for very low laser powers, the initial fluorescence decrease at laser powers at which autocorrelation functions are acquired, and the steady state level of mobile molecules during the autocorrelation measurement. As slowly diffusing molecules enter the focus and subsequently bleach, the concept of continuous fluorescence micro-photolysis [7.29] may contribute to the determination of diffusion constants. Autocorrelation measurements performed at different laser powers can help to assess if the autocorrelation times are distorted by photobleaching. Fluorescence recovery experiments can also assist in distinguishing between immobile molecules and molecules diffusing too slowly to escape photobleaching at a given irradiance.

Analysis of Molecules in Confined Subcellular Structures. Autocorrelation functions represent a statistical analysis of a multitude of single events. For this reason, events that occur seldom or with time constants scattering too much to generate an autocorrelation function with a defined relaxation time are difficult to measure with FCS. The EGF-stimulated EGFR-GFP fusion protein represents such a system. Receptor recruitment to coated pits [7.60,7.61] and internalization occur too rapidly for defining the molecular characteristics of the stimulated receptors at any point in time. Furthermore, the high affinity signaling active receptors have been identified as the immobile, cytoskeleton-associated fraction [7.45,7.46]. For this reason, measuring the decrease of the concentration of a mobile, cytoplasmic GFP fusion protein due to conversion into a less mobile or immobile, membrane-bound form seems to be a more suitable task for analysis by FCM in EGFR-dependent signal transduction.

If molecules concentrate in organelles or vesicles, the fluctuations caused by motions of molecules may be negligible compared to the fluctuations caused by the passage of the organelle through the detection volume. FRAP with spot photobleaching was employed to determine the diffusion of GFP fusion proteins in the mitochondrial matrix [7.62]. Measurement times in FRAP are shorter than for FCS. The diffusion constant is derived from only one recovery process, while in FCS statistical averaging from molecules in diffusional equilibrium is required. Moreover, arbitrarily large areas can be bleached in FRAP measurements and the recovery process can be followed in the whole-field image [7.63,7.64].

For the determination of average characteristics from extended subcellular regions with higher signal-to-noise ratios and lower standard deviations, a solution arises from excitation and detection with fringe patterns. So far, cell biological applications with this approach have only been presented with FRAP [7.65]. In vitro FCS measurements of free rhodamine 6G have been published [7.66]. The risk of artifacts from motions of the plasma membrane should be reduced by total internal reflection (TIR)-FCS [7.67] of molecules in the plasma membrane in contact with the coverslip. TIR-based FRAP measurements of the dissociation kinetics of fluorescently labeled EGF on A431 cells have been presented [7.68].

7.5.2   Perspectives in Cellular FCM

Software Autocorrelations. The stimulation of EGFR-GFP fusion proteins with EGF completely changed the appearance of the fluorescence count traces (not shown). Instead of rapid high frequency fluctuations occasional low frequency fluctuations were observed. In the corresponding autocorrelation functions evidence for an additional slow component was present. However, it was impossible to obtain satisfactory three-component fits. In this case, recording of count traces and off-line software correlation would have greatly assisted data analysis. Calculation of autocorrelation functions over time intervals in which low frequency increases of fluorescence are observed would demonstrate if such increases give rise to additional slow components. Autocorrelations could also be calculated separately over intervals with high and low fluorescence. Intervals with high signal levels potentially represent molecular aggregates with distinct diffusion characteristics.

Automated autocorrelation analysis of overlapping time intervals at different temporal resolutions could delineate how frequently an event with a certain relaxation time occurred, i.e. in which fraction of the generated autocorrelation functions the particular relaxation time was present. This would lead to a way for assigning relative fluorescence efficiencies to the distinct contributions. Control experiments are needed to confirm that the fluctuations are due to molecular and not cellular motions. In the analysis of kinetics of non-equilibrium systems (reviewed in [7.24]), overlapping instead of sequential time intervals would yield more data points at a higher temporal resolution.

The count traces also enable the analysis of molecular aggregation, either by high order autocorrelations [7.69,7.70] or by the analysis of the amplitude distribution [7.47]. At the time of the introduction of these methodologies, no applications relevant to cell biology were presented. The ability of both techniques to discriminate between aggregates of different sizes was limited. The advent of better microscopes, larger computer memory for the recording of count traces, and faster computers make it worthwhile to re-evaluate these approaches.

Analysis of Multicomponent Systems. In our experiments intracellular autocorrelation functions were fitted with as many components as necessary to obtain residual-free fits. However, it would be preferable to avoid such a priori assumptions on the composition of the measured system to fully account for the multi-component nature of the intracellular environment. Describing the measured data by a continuous distribution of diffusion coefficients has been applied successfully to the analysis of FCS and FRAP data [7.7,7.72].

Fluorescence Correlation Anisotropy Measurements. The major problem encountered in the analysis of EGFR-GFP diffusion with FCM after stimulation with EGF was cellular motion along the optical axis (data not shown), giving rise to fluorescence fluctuations with relaxation times very similar to slow molecular motions. TIR-FCS has been mentioned as one possible solution for measuring translational diffusion in the plasma membrane. Alternatively, anisotropy decay can detect changes in rotational diffusion due to receptor aggregation [7.73–7.75]. In FCS, anisotropy measurements are independent of long lifetime fluorescence/phosphorescence. The experiment requires the addition of two optical components in the light path: a polarizer and an analyzer before and after the sample, respectively [7.76,7.77].

Cellular Cross-Correlation and FRET-FCS. Both cross-correlation FCS and FRET-FCS probe molecular interactions. In contrast to FRET [7.78], cross-correlation measurements do not require a close proximity of the fluorophores. False negative results arising from labels too distant for energy transfer to occur are avoided. The diffusion through the detection volume of interacting molecules labeled with spectrally distinct fluorophores gives rise to a cross-correlated signal [7.25]. FRET-FCS can yield information on the kinetics of a molecular interaction in thermodynamic equilibrium [7.79]. Close proximity of the fluorophores leads to quenching of the fluorescence in the donor channel and a stimulation of acceptor emission. For cross-correlation FCS fluorophores with minimal spectral overlap are required (e.g., fluorescein and Cy5), to avoid FRET as well as channel cross talk leading to cross-correlation artifacts due to the presence of signal from the short wavelength fluorophore in the long wavelength channel.

In FRET-FCS the choice of fluorophores will determine the molecular distance over which interactions can be probed. Due to the second order dependence of the contribution to the autocorrelation function on the relative fluorescence efficiency of a component (Equation 7.2), the response function decreases with the 12th power of the molecular separation. This fact facilitates the autocorrelation analysis since one can assume that the signal fluctuates between an on-state with 100% efficiency and an off-state with zero efficiency.

These dual channel techniques are a prerequisite for efficiently probing molecular interactions between soluble components in cells. An increase in the molecular weight of a labeled complex by a factor of eight only doubles its diffusion coefficient. For this reason, even binding of a 50 kDa protein to a 300 kDa partner would be difficult to resolve if only the 50 kDa protein were labeled.

Most presently available GFP mutants are spectrally too similar for cellular cross-correlation measurements of GFP fusion proteins devoid of significant channel cross talk. FRET-FCS measurements with these proteins will depend on the availability of laser lines for exciting the short wavelength varieties of GFP, such as Blue-GFP (excitation maximum at 384 nm) or Cyan-GFP (excitation maximum at 435 or 452 nm, depending on the mutant; [7.80]). It therefore seems advisable to combine a GFP fusion protein with a protein labeled in vitro. Rhodamine-like dyes and GFP fusion proteins are appropriate for FRET-FCS, while Cy5-labeled EGF and the EGFR-GFP fusion protein represent a good pair for cross-correlation FCS.

7.5.3 Comparison of FCM with Other Techniques

FCS has only recently been introduced into cell biological research. In contrast to this, single particle tracking (SPT; [7.81]), single molecule imaging, FRAP, and continuous microphotolysis have been employed for the measurement of molecular motion and aggregation for a number of years. It was one of the objectives of this chapter to define the potential of FCM in cellular research and identify primary applications of FCM.

Table 7.6 lists applications that are either inaccessible to other techniques due to technical limitations, such as single molecule detection in FRAP, or

Table 7.6. Primary applications of FCM in cellular research. While most of these aspects can be addressed with other techniques as well, FCM derives its strength from their accessibility with one instrument, the relative ease of use, and its large dynamic range in the temporal domain as well as in the sample concentrations

---

Determination of absolute molecule numbers (concentrations)

Measurements at low molecule numbers, down to the single molecule level

Detection of homoaggregates

Investigation of molecular association at low molecule numbers

Detection of heterologous molecular interactions (cross-correlation FCS)

Measurement of intracellular kinetics of binding reactions and molecular conversions

Analysis of fast photophysical processes

---

require inordinate effort. For example, the determination of molecular concentration from fluorescence intensities requires calibration standards, has to consider the cellular geometry and imaging function, and generally yields little information about the number of components.

FCM derives its strength from the analysis of a variety of molecular parameters that otherwise require dedicated setups, and from the accessibility of diffusion and reaction times extending over more than nine orders of magnitude, i.e. from the nanosecond to the second time range. Molecular processes inside living cells can be investigated simultaneously on the cellular and molecular level. In contrast to FCM, FRAP and continuous micro-photolysis are inherently destructive in nature and are limited to the determination of molecular association, diffusion coefficients, and intracellular transport processes. Single molecule imaging and single particle tracking require very low densities of the label [7.81–7.83]. Three-dimensional motions are very difficult to follow and the low density may compromise the exact determination of localization.

Ligand-receptor colocalization at the single molecule level has demonstrated that molecular interactions can also be detected in a FRET-independent way [7.84]. Unless applied to high affinity interactions sequestering all molecules in the complexes, fluorescence background will compromise these measurements.

High fluorophore concentrations constitute a limit inherent to FCM. However, for many of the problems discussed above, solutions are at hand that will extend the applicability of this technique even further.

References

7.1 W. Denk, J.H. Strickler, and W.W. Webb: Science. **248**, 73–76 (1990)

7.2 P. Schwille, U. Haupts, S. Maiti, and W.W. Webb: Biophys. J. **77**, 2251–2265 (1999)

7.3 K.M. Berland, P.T.C. So, and E. Gratton: Biophys. J. **68**, 694–701 (1995)

7.4 J.C. Politz, E.S. Browne, D.E. Wolf, and T. Pederson: Proc. Natl. Acad. Sci. USA **95**, 6043–6048 (1998)

7.5 R. Brock, M.A. Hink, and T.M. Jovin: Biophys. J. **75**, 2547–2557 (1998)

7.6 R. Brock: Fluorescence correlation microscopy and quantitative microsphere recruitment assay - New methods for quantitative analyses with subcellular resolution. Ph. D. Thesis, Institute of Organic Chemistry. Eberhard-Karls-Universität Tübingen, Tübingen. p. 175 (1999)

7.7 R. Rigler, A. Pramanik, P. Jonasson, G. Kratz, O.T. Jansson, P.A. Nygren, S. Stahl, K. Ekberg, B.L. Johansson, S. Uhlen, M. Uhlen, H. Jornvall, and J. Wahren: Proc. Natl. Acad. Sci. USA **96**, 13318–13323 (1999)

7.8 J. Boonstra, P. Rijken, B. Humbel, F. Cremers, A. Verkleij, and P.v.B.e. Henegouwen: Cell Biol. Int. **19**, 413–430 1995

7.9 R. Brock, G. Vàmosi, G. Vereb, and T.M. Jovin: Proc. Natl. Acad. Sci. USA **96**, 10123–10128 (1999)

7.10 J.E. Aubin: J. Histochem. Cytochem. **27**, 36–43 (1979)

7.11 E.M. Nuutinen: Basic Res. Cardiol. **79**, 49–58 (1984)

7.12 J. Eng, R.M. Lynch, and R.S. Balaban: Biophys. J. **55**, 621–630 (1989)

7.13 H. Schneckenburger, P. Gessler, and I. Pavenstädt-Grupp: J. Histochem. Cytochem. **40**, 1573–1578 (1992)

7.14 K.D. Niswender, S.M. Blackman, L. Rohde, M.A. Magnuson, and D.W. Piston: J. Microsc. **180**, 109–116 (1995)

7.15 J. Widengren, Ü. Mets, and R. Rigler: J. Phys. Chem. **99**, 13368–13379 (1995)

7.16 D.E. Koppel: Phys. Rev. A. **10**, 1938–1945 (1974)

7.17 J. Schlessinger, D. Axelrod, D.E. Koppel, W.W. Webb, and E.L. Elson: Science. **195**, 307–309 (1977)

7.18 K. Luby-Phelps, D.L. Taylor, and F. Lanni: J. Cell Biol. **102**, 2015–2022 (1986)

7.19 F. Perrin: Comptes rendus. **184**, 1097–1100 (1929)

7.20 T. Förster: Fluoreszenz organischer Verbindungen (Vandenhoeck & Ruprecht, Göttingen, 1951)

7.21 E.L. Elson and D. Madge: Biopolymers **13**, 1–27 (1974)

7.22 F. Oehlenschläger, P. Schwille, and M. Eigen: Proc. Natl. Acad. Sci. USA **93**, 12811–12816 (1996)

7.23 N.G. Walter, P. Schwille, and M. Eigen: Proc. Natl. Acad. Sci. USA **93**, 12805–12810 (1996)

7.24 P. Schwille, J. Bieschke, and F. Oehlenschläger: Biophys. Chem. **66**, 211–228 (1997)

7.25 P. Schwille, F.-J. Meyer-Almes, and R. Rigler: Biophys. J. **72**, 1878–1886 (1997)

7.26 B. Rauer, E. Neumann, J. Widengren, and R. Rigler: Biophys. Chem. **58**, 3–12 (1996)

7.27 R. Rigler, U. Mets, J. Widengren, and P. Kask: Eur. Biophys. J. **22**, 169–175 (1993)

7.28 J. Widengren and R. Rigler: Bioimaging. **4**, 149–157 (1996)
7.29 R. Peters: Naturwissenschaften. **70**, 294–302 (1983)
7.30 U. Kubitschek, O. Heinrich, and R. Peters: Bioimaging. **4**, 158–167 (1996)
7.31 D. Cassel, B. Whiteley, Y.X. Zhuang, and L. Glaser: J. Cell. Phys. **122**, 178–186 (1985)
7.32 F. Guilak: J. Microsc. **173**, 245–256 (1994)
7.33 K.J. Angelides, L.W. Elmer, D. Loftus, and E. Elson: J. Cell. Biol. **106**, 1911–1925 (1988)
7.34 R. Brock, I.H.L. Hamelers, and T.M. Jovin: Cytometry. **35**; 353–362 (1999)
7.35 A. Ullrich and J. Schlessinger: Cell **61**, 203–212 (1990)
7.36 S. Cohen: J. Biol. Chem. **237**, 1555–1562 (1962)
7.37 I. Alroy and Y. Yarden: FEBS Lett. **410**, 83–86 (1997)
7.38 M.A. Lemmon, Z. Bu, J.E. Ladbury, M. Zhou, D. Pinchasi, I. Lax, D.M. Engelman, and J. Schlessinger: EMBO J. **16**, 281–294 (1997)
7.39 J. Schlessinger: Trends Biochem. **13**, 443–447 (1988)
7.40 J.M. Sherrill and J. Kyte: Biochemistry **35**, 5705–5718 (1996)
7.41 A. Weiss and J. Schlessinger: Cell **94**, 277–280 (1998)
7.42 M. Gregoriou and A.R. Rees: EMBO J. **3**, 929–937 (1984)
7.43 L.H.K. Defize, D.J. Arndt-Jovin, T.M. Jovin, J. Boonstra, J. Meisenhelder, T. Hunter, H.T.d. Hey, and S.W.d. Laat: J. Cell Biol. **107**, 939–949 (1988)
7.44 T.W.J. Gadella, Jr. and T.M. Jovin: J. Cell Biol. **129**, 1543–1558 (1995)
7.45 A.M. Gronowski and P.J. Bertics: Endocrinology. **133**, 2838–2846 (1993)
7.46 F.A.C. Wiegant, F.J. Blok, L.H.K. Defize, W.A.M. Linnemans, A.J. Verkley, and J. Boonstra: J. Cell Biol. **103**, 87–94 (1986)
7.47 H. Qian: Biophys. Chem. **38**, 49–57 (1990)
7.48 C. Eggeling, J. Widengren, R. Rigler, and C.A.M. Seidel: Anal. Chem. **70**, 2651–2659 (1998)
7.49 E. Livneh, M. Benveniste, R. Prywes, S. Felder, Z. Kam, and J. Schlessinger: J. Cell Biol. **103**, 327–331 (1986)
7.50 A. Kusumi, Y. Sako and M. Yamamoto: Biophys. J. **65**, 2021–2040 (1993)
7.51 M. Nokubo, I. Zs.-Nagy, K. Kitani, and M. Ohta: Biochim. Biophys. Acta. **939**, 441–448 (1988)
7.52 R.N. Fabricant, J.E.D. Larco, and G.J. Todaro: Proc. Natl. Acad. Sci. USA **74**, 565–569 (1977)
7.53 A. Honegger, T.J. Dull, F. Bellot, E.V. Obberghen, D. Szapary, A. Schmidt, A. Ullrich, and J. Schlessinger: EMBO J. **7**, 3045–3052 (1988)
7.54 K.A. Giuliano and D.L. Taylor: Trends Biotech. **16**, 135–140 (1998)
7.55 T.T. Yang, L.Z. Cheng, and S.R. Kain: Nucleic Acids Res. **24**, 4592–4593 (1996)
7.56 U. Haupts, S. Maiti, P. Schwille, and W.W. Webb: Proc. Natl. Acad. Sci. USA **95**, 13573–13578 (1998)
7.57 R. Peters: EMBO J. **3**, 1831–1836 (1984)
7.58 T. Winkler, U. Kettling, A. Koltermann, and M. Eigen: Proc. Natl. Acad. Sci. USA **96**, 1375–1378 (1999)
7.59 M. Gossen and H. Bujard: Proc. Natl. Acad. Sci. USA **89**, 5547–5551 (1992)
7.60 K.D. Brown, Y.-C. Yeh, and R.W. Holley: J. Cell. Physiol. **100**, 227–238 (1979)
7.61 C.E. Futter, A. Pearse, L.J. Hewlett, and C.R. Hopkins: J. Cell Biol. **132**, 1011–1023 (1996)

7.62 A. Partikian, B. Ölveczky, R. Swaminathan, Y. Li, and A.S. Verkman: J. Cell Biol. **140**, 821–829 (1998)

7.63 N.B. Cole, C.L. Smith, N. Sciaky, M. Terasaki, M. Edidin, and J. Lippincott-Schwartz: Science. **273**, 797–801 (1996)

7.64 J. Lippincott-Schwartz, N. Cole, and J. Presley: Trends Cell Biol. **8**, 16–20 (1998)

7.65 H.M. Munnelly, D.A. Roess, W.F. Wade, and B.G. Barisas: Biophys. J. **75**, 1131–1138 (1998)

7.66 R.L. Hansen, X.R. Zhu, and J.M. Harris: Anal. Chem. **70**, 1281–1287 (1998)

7.67 N.L. Thompson and B.C. Lagerholm: Current Opinion Biotech. **8**, 58–64 (1997)

7.68 E.H. Hellen and D. Axelrod: J. Fluoresc. **1**, 113–128 (1991)

7.69 A.G.I. Palmer and N.L. Thompson: Biophys. J. **52**, 257–270 (1987)

7.70 A.G.I. Palmer and N.L. Thompson: Chem. Phys. Liquids. **50**, 253–270 (1989)

7.71 H. Qian and E.L. Elson: Proc. Natl. Acad. Sci. USA **87**, 5479–5483 (1990)

7.72 N. Periasamy and A.S. Verkman: Biophys. J. **75**, 557–567 (1998)

7.73 R. Zidovetzki, Y. Yarden, J. Schlessinger, and T.M. Jovin: Proc. Natl. Acad. Sci. USA. **78**, 6981–6985 (1981)

7.74 R. Zidovetzki, D.A. Johnson, D.J. Arndt-Jovin, and T.M. Jovin: Biochemistry. **30**, 6162–6166 (1991)

7.75 K.L.d. Carraway and R.A. Cerione: Biochemistry. **32**, 12039–12045 (1993)

7.76 S.R. Aragón and R. Pecora: Biopolymers. **14**, 119–138 (1975)

7.77 P. Kask, P. Piksarv, Ü. Mets, M. Pooga, and E. Lippmaa: Eur. Biophys. J. **14**, 257–261 (1987)

7.78 R.M. Clegg: Current Opinion Biotech. **6**, 103–110 (1995)

7.79 G. Bonnet, O. Krichevsky, and A. Libchaber: Proc. Natl. Acad. Sci. USA **95**, 8602–8606 (1998)

7.80 R.Y. Tsien: Annu. Rev. Biochem. **67**, 509–544 (1998)

7.81 R. Simson, B. Yang, S.E. Moore, P. Doherty, F.S. Walsh, and K.A. Jacobson: Biophys. J. **74**, 297–308 (1998)

7.82 L.S. Barak and W.W. Webb: J. Cell. Biol. **90**, 595–604 (1981)

7.83 I. Sase, H. Miyata, J.E.T. Corrie, J.S. Craik, and J.K. Kinosita: Biophys. J. **69**, 323–328 (1995)

7.84 G.J. Schütz, W. Trabesinger, and T. Schmidt: Biophys. J. **74**, 2223–2226 (1998)

8 FCS and Spatial Correlations on Biological Surfaces

Nils O. Petersen

Fluorescence correlation spectroscopy (FCS) provides information about the occupation number and the dynamics of molecules in a small volume. Much of the work presented in this volume is focused on extracting information about the dynamics. Indeed, FCS represents one of the most versatile tools for studying dynamics because of the sensitivity and selectivity afforded by fluorescence and the large range of time scales accessible through the fluctuation analysis. The technique is limited at short times by the speed of electronic data collection (nanoseconds) and at long times by the length of time the molecules spend in the observation volume (milliseconds to seconds).

This chapter will focus on extracting information about occupation numbers with particular emphasis on interpretations in terms of inter-molecular interactions, such as formation of clusters of like molecules or aggregates of different molecules. With this focus in mind, the dynamic information is secondary and even irrelevant. In fact, it now becomes possible to examine systems in which temporal fluctuations are either very slow or completely absent. The key is to recognize that information about occupation numbers can be obtained as easily by the analysis of **spatial** fluctuations as by the analysis of temporal fluctuations.

8.1 The Problem

The fluid mosaic model of cell membranes suggests that, for the most part, the heterogeneous mixture of proteins and lipids is completely mixed and distributed homogeneously across the surface. Modern models of cell membranes recognize that this picture is too simple [8.1,8.2]. Protein enriched domains, such as clathrin coated pits, and glycolipid enriched domains, such as caveolae, have been identified by electron microscopy and through biochemical techniques [8.3–8.6].

Clathrin coated pits function as part of the receptor mediated endocytosis pathway [8.7,8.8]. Specifically, when certain types of membrane receptors are bound by their ligand, they associate with a protein known as adaptor protein 2 (AP-2). This protein, in turn, binds to clathrin triskelia to form a large lattice of proteins on the cytoplasmic surface of the cell membrane. With time, the complex invaginates and forms a coated vesicle that pinches off from the membrane and is internalized within the cell. This represents one interesting example of multiple inter-molecular interactions: ligand-receptor, receptor-adaptor, adaptor-clathrin.

Triton extraction of cell membranes in the cold has revealed insoluble membrane fractions rich in cholesterol, sphingomyelin, glycolipids (including gangliosides) and lipid anchored proteins [8.4,8.5]. These "detergent insoluble glycolipid rich domains"' (DIGs) or "detergent resistant membrane fractions" (DRMs) are proposed to exist as functionally important microscopic domains (70–500 nm) in the intact membrane. It is possible that they are precursors for the structurally distinct, highly invaginated regions known as caveolae. These contain many of the same membrane components, in addition to the structural protein, caveolin. This represents an intriguing example of a phase separation, presumably of lipid-like molecules that prefer to be in a more ordered state.

The organizational heterogeneity seen in the cell membrane, suggests that there are inter-molecular interactions that are functionally important. Thus we must develop tools that allow us to measure these inter-molecular interactions in the membrane of the living cell. A first objective is *to establish the distribution of a molecule in a membrane* – for example, whether a receptor is monomeric or oligomeric. A second objective is *to determine how the distribution may change with time or cellular function* – for example, whether the receptors aggregate following the addition of a ligand. A third objective is *to measure the extent of interaction of one molecule with other molecules* – for example, measure how many receptors interact with adaptor proteins in a coated pit. A fourth objective is *to determine how inter-molecular interactions change* – for example, see how rapidly a receptor moves into a coated pit.

The following sections will demonstrate the following. Auto-correlation analysis of spatial fluctuations in fluorescence can address the first objective. Cross-correlation analysis of spatial fluctuations as a function of time can address the second objective. Cross-correlation analysis of spatial fluctuations attributable to different molecules can address the third objective. In principle, a cross-correlation analysis of spatial fluctuations from different molecules as a function of time can address the fourth objective.

8.2   The Solution

In their first series of papers, Magde, Elson and Webb [8.9–8.11] provided the theoretical framework for the analysis of fluorescence fluctuations deriving from diffusion, chemical kinetics and uniform flow. They showed that for a uniform flow perpendicular to a focused laser beam with a Gaussian intensity profile, the decay of the autocorrelation function would also be Gaussian.

$$g(\tau) = g(0)e^{(\tau/\tau_f)^2} . \tag{8.1}$$

The characteristic time constant for the decay of the Gaussian function, τ_f, is the ratio of the width of the laser beam (w) and the flow velocity (V). The amplitude of the autocorrelation function at zero-lag time contains the usual information about the average number of molecules in the observation volume at any given time.

It is straightforward to recognize that a uniform flow can be imposed by either moving a fixed sample at uniform speed through the laser beam or scanning the laser beam at a constant speed across a fixed sample. The autocorrelation function would have the same functional shape and the same interpretation. The only difference is that the dynamics is imposed by the experiment, so any inherent dynamic information is lost. However, the amplitude of the autocorrelation function should have the same information content.

This transformation from a time domain experiment (a time average) to a space domain experiment (an ensemble average) is an illustration of the ergodic principle, namely that the occupation number should be determined equally well by a time average or an ensemble average, provided the occupation number is an equilibrium property [8.12]. It is likely that the distribution of molecules in solution or on surfaces is at equilibrium and behaves reversibly, even though the occupation number fluctuates. It is less clear that the distribution of molecules in a biological system is at equilibrium. At best we may hope that it is at a steady state. This is an important theoretical problem and a potentially serious limitation of many biophysical tools, including the approaches described in this chapter. Nevertheless, in the absence of direct evidence to the contrary, we will presume that the systems we study are ergodic, or nearly so.

Scanning fluorescence correlation spectroscopy was introduced as a tool to study the distribution of receptors on the surface of biological cells where temporal fluctuations were either absent or too slow to measure [8.13]. It was recognized early that the occupation number measurement could be used to measure the state of aggregation of molecules [8.11] and this concept was explored initially for cell surface receptors with application to aggregation of virus glycoproteins during budding [8.14,8.15].

The initial implementation of scanning fluorescence correlation spectroscopy relied on a precise, linear translation of the sample through the optical focus of a laser beam in a microscope. The sample translator utilized piezo electric biforms to achieve up to 80 μm translations with a precision of each step of 20 nm. The experiment proved to be slow and tedious and of limited general applicability.

The advent of the confocal scanning laser microscope provided an opportunity to make the laser beam scan across the surface in a rapid and linear fashion. Moreover, the confocal microscope produced a two-dimensional map of intensity fluctuations across the surface. Image correlation spectroscopy (ICS) was introduced to take advantage of the increased information content present in the two-dimensional images as well as the increased speed and convenience associated with confocal laser scanning microscopy [8.16]. This chapter will focus on the measurement, analysis and interpretation of image correlation spectroscopy.

8.3   The Experiment

8.3.1   Generating Images Using a Confocal Microscope

The key to successful image correlation spectroscopy is to generate meaningful confocal microscope images. Several factors are important. From a biological perspective, the images must be representative of the phenomenon being studied and must sample a large number of surfaces to achieve statistically meaningful information. From a technical perspective, the sample must be in focus across the detector and photons should be detected in every pixel. The detection system must respond linearly and photon counting is the best way to ensure this. Images collected using nonlinear amplifiers or digital filters will not provide meaningful information in an ICS analysis. None of the pixels can be saturated (at maximum count). It is best if there is a nonzero count in every pixel since this ensures that there is no truncation of small fluctuations by black level setting being set to too high a value.

Most of our experiments have been performed with a Biorad MRC 600 confocal scanning laser microscope in the photon counting mode, at zoom 10×. The gain and black level settings were adjusted to ensure about 10–15 counts in each pixel in the final image, with no counts exceeding 250 in the 8-bit image. The images were collected and stored as 512 × 512 images and in many cases they were the result of multiple scans (up to ∼ 30) over the same region to obtain a reasonable signal-to-noise ratio.

As indicated below, it is important to collect images of the white noise background so that the proper corrections may be made. This is achieved by collecting the same number of scans with the laser beam blocked. This allows background scattering and other random noise contributions to be collected (typically 10–15 counts per pixel in our experiments). In many cases it is also important to collect data on control samples to account for the non-specific binding of chromophores and autofluorescence in the samples. These samples are measured under identical conditions of laser power, black level, gain and photon counting.

8.3.2   Correlation Calculations

Let the intensity in the pixel located at position x, y in the image collected at time t be given by $i(x, y; t)$. The normalized spatial autocorrelation function, $g_i(\xi, \eta; 0)$ for this image is then given by

$$g_i(\xi, \eta; 0) = \frac{\langle (i(x, y; t) - \langle i(x, y; t) \rangle)\, (i(x + \xi, y + \eta; t) - \langle i(x, y; t) \rangle) \rangle}{\langle i(x, y; t) \rangle^2}$$

$$= \frac{\langle (i(x,y;t))\,(i(x+\xi,y+\eta;t)) \rangle - \langle i(x,y;t) \rangle^2}{\langle i(x,y;t) \rangle^2} \tag{8.2}$$

$$= \frac{G_{\mathrm{i}}(\xi,\eta;0)}{\langle i(x,y;t) \rangle^2} - 1 \, ,$$

where the angular brackets indicate the ensemble average, i.e. the average over all spatial coordinates [8.16]. Since there is no time difference in this spatial autocorrelation function, this lag parameter is zero. $G_i(\xi,\eta;0)$ is the unnormalized intensity autocorrelation function [8.9,8.10]. The autocorrelation function may be calculated directly as a summation over all coordinates, but for two-dimensional arrays with 512×512 points, this calculation is prohibitively slow [8.16]. A better approach is to recognize that, in general, the autocorrelation function is the Fourier transform of the power spectrum of a discrete data set. The power spectrum, in turn, is the product of the Fourier transform of the original data set and its complex conjugate. Thus,

$$g_i(\xi,\eta;0) = \frac{\mathfrak{S}^{-1}\left([\mathfrak{S}\,(i(x,y;t))] * [\mathfrak{S}^*\,(i(x,y;t))] \right)}{\langle i(x,y;t) \rangle^2} - 1 \, . \tag{8.3}$$

Here, $\mathfrak{S}$ indicates the two-dimensional Fourier transform in the space coordinates only. Computationally, these calculations can be performed with standard 2D fast Fourier transform (2D-FFT) system routines. However, it is important that the input data array contains 2^n pixels in each dimension and that the Fourier transform calculation is normalized correctly by the number of pixels.

The concept can be extended to a cross-correlation analysis between two images. Let $i(x,y;t)$ represent the intensities in one image and let $j(x,y;t+\tau)$ represent another image collected with a different chromophore and/or at a different time. In this case, we calculate the normalized, spatial cross-correlation function, $g_{ij}(\xi,\eta;\tau)$ as

$$g_{ij}(\xi,\eta;\tau) = \frac{\langle (i(x,y;t))\,(j(x+\xi,y+\eta;t+\tau)) \rangle}{\langle i(x,y;t) \rangle\,\langle j(x,y;t+\tau) \rangle} - 1$$

$$= \frac{\mathfrak{S}^{-1}\left([\mathfrak{S}\,(i(x,y;t))] * [\mathfrak{S}^*\,(j(x,y;t+\tau))] \right)}{\langle i(x,y;t) \rangle\,\langle j(x,y;t+\tau) \rangle} - 1 \, . \tag{8.4}$$

In general, we expect that $g_{ij}(\xi,\eta;\tau) = g_{ji}(\xi,\eta;\tau)$, so that it makes no difference which of the two images is the real and which is the complex conjugate. To date, all calculations we have performed have confirmed this symmetry.

8.3.3   Correlation Function Analysis

For a uniform flow perpendicular to a Gaussian laser beam at its focal point, the autocorrelation function decays as a Gaussian (8.1). Correspondingly, the spatial autocorrelation function will decay as a two-dimensional Gaussian function with a characteristic decay distance given by the width of the laser beam at its focal point [8.16].

$$g_i(\xi, \eta; 0) = g_i(0, 0; 0) \, e^{-(\xi^2 + \eta^2)/w^2} \,. \tag{8.5}$$

If two images are collected from the same sample at the same time ($\tau = 0$), but using two distinct chromophores to generate the images, then the cross-correlation function will also decay as a two-dimensional Gaussian function. The characteristic decay distance is given by the geometric means of the square of the widths of the laser beam used to excite each of the chromophores [8.17].

$$g_{ij}(\xi, \eta; 0) = g_{ij}(0, 0; 0) \, e^{-2(\xi^2 + \eta^2)/(w_i^2 + w_j^2)} \,. \tag{8.6}$$

If two images are collected from the same area with the same chromophore but at different times, then the cross-correlation function will reflect the dynamic changes in the image. For example, the spatial distribution may be fluctuating because of diffusion of the components in the plane of the image. In that case, the spatial decay remains Gaussian, but it is modulated by a diffusive components [8.17–8.19].

$$g_{ij}(\xi, \eta; \tau) = g_{ij}(0, 0; 0) \exp\left[\frac{-(\xi^2 + \eta^2)/w^2}{1 + \tau/\tau_\mathrm{d}}\right] [1 + \tau/\tau_\mathrm{d}]^{-1} \,. \tag{8.7}$$

Importantly, one can monitor the peak of the spatial cross-correlation function, i.e. when ξ and η vanish, as a function of the time between images. This is an ensemble averaged FCS experiment and the cross-correlation function reflects the dynamic processes in the image only [8.19]. For diffusion,

$$g_{ij}(0, 0; \tau) = g_{ij}(0, 0; 0) \, [1 + \tau/\tau_\mathrm{d}]^{-1} \,. \tag{8.8}$$

This logic can be extended to any other dynamic process, including chemical kinetics or flow.

8.3.4 Extracting the Amplitude Information

Equations (8.5–8.8) describe the auto or cross-correlation functions in the limit where the emitting particles are point sources and the width is determined solely by the laser beam. These relations also assume that there is sufficient information to allow complete averaging of low frequency fluctuations or long distance correlations to zero. Accordingly, the only unknown parameter is the amplitude, $g_i(0, 0; 0)$. In reality, many clusters may be physically large enough to affect the width of the correlation function, and invariably, the data is limited by the size of the image. Therefore, it is best to fit the experimental correlation functions to three parameters: the amplitude, $g_i(0, 0; 0)$, the width, w, and an offset, g_0 [8.16]. For example, the autocorrelation functions would be fit to

$$g_i(\xi, \eta; 0) = g_i(0, 0; 0) \, e^{-(\xi^2 + \eta^2)/w^2} + g_0 \,. \tag{8.5a}$$

The experimentally determined width may be compared with the expected or known width of the focused laser beam as a measure of the quality of the data and the assumption. Routinely, we accept data for which the experimental width is within about 30% of the known beam width (measured or calculated, [8.16,8.20]. The offset parameter is included to account for the general observation that the correlation functions do not always decay to zero. This arises since the images are a finite size, and at long lag distances, there is less information available to calculate the correlation function. For example, for a characteristic correlation distance of 0.35 μm and an image dimension of about 15.5 μm, there are only about 50 characteristic fluctuations per dimension of the image. The uncertainty in the offset will then be on the order of $\sqrt{50}/50 \approx 0.14$. This corresponds well with the standard deviation of data on the best samples studied.

It has been demonstrated that the optimum fit is obtained if the fit includes only the data in the correlation function to about three times the width [8.21]. If fewer data are included, the estimate of the offset is poor. If more data are included, the estimate of the amplitude is poor. In our implementation, we use only data for which $(\xi^2 + \eta^2) \leq (3w)^2$, with $3w$ being ~ 32 pixels (~ 1 μm).

The autocorrelation function will always contain a contribution at zero lag-distances from white noise sources [8.22] and care is taken to avoid including the central data point in the fits. In some cases, there is also white noise correlation in the nearest pixels along the rapid scan axis. We frequently exclude that point in the correlation function in the fits [8.22].

The real autocorrelation function is maximal at the zero lag-distances and the fitting is optimized accordingly. The cross-correlation function need not be maximal at this origin, since there may be systematic shifts in the image collected at two different excitation and emission wavelengths or at different times. Accordingly, we search for a global maximum in the correlation function and fit to the data around the coordinates of the maximum (ξ_0, η_0).

$$g_{ij}(\xi, \eta; 0) = g_{ij}(0, 0; 0)\, e^{-2((\xi+\xi_0)^2 + (\eta+\eta_0)^2)(w_i^2 + w_j^2)} \,. \tag{8.6a}$$

Systematic variations in the coordinates of the maximum fit can be noted and accounted for if necessary. For example, if there is a flow process, the changes in the coordinates of the maximum can be used to determine the flow velocity [8.23]. Random variations in these coordinates may result from small instabilities in the microscope and we routinely accept data as valid if the variation in either coordinate is less than about one correlation width, w.

Whenever the true cross-correlation is very small or absent, the fit will be to the largest apparent correlation in the whole function. This will frequently be at large lag-distances, since the information content is poorer because of the limited image dimensions. We record the fraction of images for which the coordinates of the maximum exceed w and interpret this as the fraction of images that show no cross-correlation or cross-correlation below the random correlation. Only the acceptable data are used for further calculations.

8.3.5   Technical Issues

In general, several sources of fluorescence will contribute to the measured photon counting signal (m). These include the signal of interest (s), that from non-specific fluorescence (ns), that from autofluorescence (a) and that from background radiation and electronic noise (wn). The first three will exhibit fluorescence fluctuations with a characteristic spatial dimension determined by the laser beam, whereas the last will contribute as white noise with zero amplitude except at the central pixel. A detailed analysis of these contributions and how they may be dealt with has been published [8.22].

If the four signals are independent (a reasonable assumption given that the origin of the signals are different sources in the cell), then [8.22,8.24]

$$
\begin{aligned}
g_{\mathrm{m}}(\xi,\eta;\tau)\,\langle i_{\mathrm{m}}(x,y;t)\rangle^2 &= \\
&= g_{\mathrm{s}}(\xi,\eta;\tau)\,\langle i_{s}(x,y;t)\rangle^2 + g_{\mathrm{ns}}(\xi,\eta;\tau)\,\langle i_{\mathrm{ns}}(x,y;t)\rangle^2 \\
&= g_{\mathrm{a}}(\xi,\eta;\tau)\,\langle i_{\mathrm{a}}(x,y;t)\rangle^2 + g_{\mathrm{wn}}(\xi,\eta;\tau)\,\langle i_{\mathrm{wn}}(x,y;t)\rangle^2 .
\end{aligned} \tag{8.9}
$$

For true white noise, $g_{\mathrm{wn}}(\xi,\eta;\tau) = 0$ except at the origin which is excluded from the fit. Measurements on cells labeled in the absence of a specific antibody can provide information about the combined contributions from the non-specific and the autofluorescence signals. If appropriate, measurements on unlabeled cells can provide independent estimates of the contributions from the autofluorescence. Careful use of control experiments therefore permits a detailed and quantitative estimate of the contributions from the signal of interest in the system. In one case, it was possible to determine the state of aggregation of a receptor to be about four [8.22,8.24].

In general, we perform the corrections to the correlation calculation to the amplitude of the correlation function only [8.24].

$$
g_{\mathrm{s}}(0,0;0) = \frac{g_{\mathrm{m}}(0,0;0)\langle i_{\mathrm{m}}\rangle^2 - \left[g_{\mathrm{ns}}(0,0;0)\langle i_{\mathrm{ns}}\rangle^2 + g_{\mathrm{a}}(0,0;0)\langle i_{\mathrm{a}}\rangle^2 \right]}{\left(\langle i_{\mathrm{m}}\rangle - [\langle i_{\mathrm{ns}}\rangle + \langle i_{\mathrm{a}}\rangle] - \langle i_{\mathrm{wn}}\rangle \right)^2} . \tag{8.10}
$$

This is simpler and, we suspect, as accurate as is needed for most biological systems. Note that the contribution from the white noise is a correction for the total average intensity only. We try to develop labeling conditions such that the non-specific fluorescence and the autofluorescence contribute less than about 10% of the intensity in the image. Under these conditions, the corrections in the numerator amount to less than one per cent or so. However, the corrections to the intensity in the denominator remain important.

Measurements of tissue culture cells have shown that the largest source of variation in the data arises for differences in the measurements among cells. Thus the population variation is greater than the uncertainty in the experiment. Accordingly, we favor measuring once on every cell only, but measuring a large population of cells. To obtain statistically meaningful estimates, we routinely collect 40 images for each experimental condition. Frequently, the numbers that are measured are based on the average of hundreds of images.

For the comparison of experiments it is necessary to calculate the mean and standard deviation of the data collected. As will be noted below, the amplitude of the correlation function is inversely proportional to the occupation number. Since the mean of the amplitude values will differ from the mean of the inverse of the amplitude values, it is important to resolve which is the more meaningful mean value to calculate. We have resolved the issue as follows:

- In samples where the greatest source of variation in the data is attributable to the error in the experiment rather than to sample to sample variation, it is prudent to average the experimental observable, i.e. the amplitude of the correlation function.
- In samples where the greatest source of variation in the data is attributable to the differences between samples (here cells) rather than the error in the experiment, it is prudent to average the physical variable, i.e. the occupation number.

In almost all the work performed so far, the standard deviation arising from error in the experiment or fitting estimate is less than 15%. In work with cell surfaces, the standard deviation is typically on the order of 30%. For work on biological material, it is best to average the physical variable, namely the occupation number. This approach is reflected in all of the numbers quoted in this chapter.

8.4   Interpretation of Correlation Function Amplitudes

8.4.1   Cluster Densities

Consider the normalized intensity fluctuation as a function of space and time

$$\delta(x, y; t) = \frac{i(x, y; t) - \langle i(x, y; t) \rangle}{\langle i(x, y; t) \rangle} . \tag{8.11}$$

It is well-established that the limit of the autocorrelation function, $g_i(\xi, \eta; \tau)$, as the lag-variables ξ, η and τ vanish, gives the variance of the normalized intensity fluctuations [8.25]. It is equally well established [8.9,8.26,8.27] that when the intensity is a true reflection of the concentration of the species being observed, then the variance in the normalized intensity is also the variance of the concentration fluctuations, which in turn equals the inverse of the average occupation number, $\overline{N}_p$:

$$g_i(0, 0; 0) = \text{var}(\delta i(x, y; t)) = \text{var}(\delta c(x, y; t)) = \frac{1}{\overline{N}_p} . \tag{8.12}$$

This formulation assumes that the particles are independent and non-interacting, and that they are the same. In this case, the average occupation number measures the number of particles in the effective observation volume (or area, in the case of the two-dimensional images). This area is defined by the laser beam width at the focal plane and is $A = \pi w^2$. In this context, a particle can be any correlated collection of molecules: monomers, dimers, oligomers, or larger aggregates. A generic description is a cluster, which may have a number of molecules within it. Hence we define a cluster density as the average number of particles per unit area.

$$CD_i = \frac{\overline{N}_p}{\pi w^2} = \frac{1}{g_i(0,0;0)\pi w^2}. \tag{8.13}$$

8.4.2   Degree of Aggregation

Magde, Elson and Webb [8.9–8.11] showed originally that for mixtures of species, the amplitude of the correlation function is a complex function of the extinction coefficient, ε_k, the quantum yield of fluorescence, Φ_k, and the average concentration, $\overline{C}_k$ of each species, k.

$$g_i(0,0;0) = \frac{1}{V}\frac{\sum_k(\varepsilon_k\Phi_k)^2\overline{C}_k}{\left(\sum_k\varepsilon_k\Phi_k\overline{C}_k\right)^2}. \tag{8.14}$$

Subsequently, we applied (8.14) to the situation where there is a distribution of clusters with a mean number of molecules per cluster of μ and a variance in the distribution of molecules per cluster of σ^2 [8.14]. Assuming that the extinction coefficient scales linearly with the number of molecules in the cluster and that the quantum yield for each molecule is independent of the number of molecules in the cluster, then

$$g_i(0,0;0) = \frac{1}{\overline{N}_\mu}\frac{(\sigma^2+\mu^2)}{\mu^2}, \tag{8.15}$$

which, in the limit of large aggregates in a narrow distribution ($\mu^2 \gg \sigma^2$), yields the simple interpretation that the amplitude of the correlation function is equal to the average number of clusters, $\overline{N}_\mu$.

The basis for the correlation spectroscopy analysis is that the fluorescence intensity is directly proportional to the number of fluorescent molecules in the observation volume, $\overline{N}_m$,

$$\langle i(x,y;t)\rangle = c\overline{N}_m, \tag{8.16}$$

where the constant of proportionality includes the extinction coefficient, the quantum yield, the intensity of the illuminating laser and a factor accounting for the efficiency of transmission and collection of light in the microscope and detection of photons [8.9–8.11,8.14]. This constant can be determined experimentally [8.22]. It is now convenient to define a new experimental variable, the degree of aggregation, DA as

$$DA_i \equiv g_i(0,0;0)\langle i(x,y;t)\rangle = c\frac{\overline{N}_m}{\overline{N}_\mu}\frac{(\sigma^2+\mu^2)}{\mu^2} \cong c\frac{(\sigma^2+\mu^2)}{\mu} \cong c\frac{\overline{N}_m}{\overline{N}_\mu}. \tag{8.17}$$

This parameter should be interpreted as a measure of the average number of monomers (or fluorescent molecules) per cluster. When it is possible to ascertain the value of the parameter c, then the degree of aggregation is an absolute measure. Even when c cannot be determined independently, it is possible to use the degree of aggregation as a relative measure. For example, changes in DA with time or as a consequence of the treatment of cells, should be accurate reflections of changes in the sizes of clusters. This is a convenient parameter for following aggregation or dispersion processes.

8.4.3   Multiple Populations

If there are multiple populations of receptors, then the amplitude of the correlation function is a weighted average of these distributions (8.14).

$$g_i(0,0;0) = \frac{1}{\overline{N}_m^2} \left(\sum_k \overline{N}_k (\sigma_k^2 + \mu_k^2) \right) . \tag{8.18}$$

The k-th population is characterized by a mean (μ_k), a standard deviation (σ_k) and a mean number of clusters $(\overline{N}_k)$. The mean number of monomers is given by $\overline{N}_m$. Application of this approach is limited since the experimentally determined amplitude of the autocorrelation function depends on three parameters for each population, and while some of these are interdependent, there are too many variables. Analysis of higher order autocorrelation terms can provide additional information [8.28,8.29], but for cells it has not yet been demonstrated that the data are good enough to make the analysis meaningful.

It is instructive to consider briefly two extreme models of aggregation:

1. Synchronous growth of clusters by association of oligomers,

$$n\,\mathrm{A} \to \frac{n}{2}\,\mathrm{A}_2 \to \frac{n}{4}\,\mathrm{A}_4 \to \frac{n}{8}\,\mathrm{A}_8 \to \frac{n}{16}\,\mathrm{A}_{16} \to \mathrm{etc}.$$

   This model corresponds to a single distribution of clusters whose size are continually increasing. This model is compatible with (8.15) and (8.17).
2. Reaction of monomers to form clusters

$$(m \times n)\mathrm{A} \to m\,\mathrm{A}_n$$

   if m is a variable, this model predicts that the average aggregate size is fixed and the number of clusters increases at the expense of monomers. If n is a variable, this model predicts that the number of clusters is fixed and the average aggregate size increases at the expense of monomers. In either case, there is a distribution of clusters in equilibrium with monomers at all times, and (8.18) must be used as a two-population model.

In the two-population model where one species is the monomer, it is simple
to show that

$$g_i(0,0;0) = \frac{1}{\overline{N}_m^2} \left(\overline{N}_1 + \overline{N}_\mu \cdot \mu^2 \right) , \qquad (8.19)$$

where $\overline{N}_m = \left(\overline{N}_1 + \overline{N}_\mu \cdot \mu \right)$ is the average number of monomers present in
the system, $\overline{N}_1$ is the average number of monomers existing as monomeric
species and $\overline{N}_\mu$ is the average number of clusters of average size μ. In some
cases, it is possible to obtain enough independent information to analyze that
data and obtain good estimates of all variables in the system. For example,
the total number of monomers may be known from binding studies, and
the average number of large aggregates can be estimated from other image
analysis approaches [8.24]. However, the analysis is sensitive to the model
chosen and correspondingly conclusions are tenuous.

8.4.4 Dynamics of Aggregation

Consider a set of images, $I_1, I_2, I_3 \ldots I_n$, recorded as a function of time from
the same sample using the same fluorescent probe. The amplitudes of the
autocorrelation function of each image ($g_1, g_2, g_3 \ldots g_n$) can be used to obtain the cluster density and the degree of aggregation as a function of time.
These numbers may be interpreted in terms of kinetics of aggregation, as
appropriate, with the understanding that the interpretation will necessarily
be model dependent. To date, there are no good examples of this application.

A set of cross-correlation functions may be calculated between the first
image and the subsequent images, i.e. $g_{12}, g_{13} \ldots g_{1n}$. The amplitudes will reflect changes in the state of aggregation as well as the dynamics of movement
(diffusion, flow) between images. As indicated earlier, this is an ensemble averaged FCS experiment and all the tools for the analysis of the kinetics of
decay discussed in other chapters of this volume are available. The aggregation process can be treated as a chemical reaction [8.9–8.11].

If there are no changes in the state of aggregation, i.e. the amplitudes
of the autocorrelation function are independent of time, then the cross-
correlation functions measure the diffusion or flow processes only [8.19].

8.4.5 Intermolecular Interactions and Colocalization

Consider a pair of images collected at the same time from the same region
of the same surface using two distinct fluorescent probes. The amplitudes of
the autocorrelation function for each image will provide information about
the cluster density and degree of aggregation of the two molecules on the
surface. The cross-correlation function between the two images will reflect

the extent to which the two molecules interact. Specifically, [8.17,8.28,8.30] the amplitude at zero lag-time will be

$$g_{ij}(0,0;0) = \frac{\overline{N}_{ij}}{(\overline{N}_i + \overline{N}_{ij})(\overline{N}_j + \overline{N}_{ij})} = \overline{N}_{ij}g_i(0,0;0)g_j(0,0;0)\,, \quad (8.20)$$

where $\overline{N}_{ij}$ represents the average number of clusters in which both molecules are present. This interpretation is strictly true only for clusters with equal numbers of the two molecules, but to a first approximation, a more liberal interpretation is used.

It is possible to define an experimental parameter that should represent the density of clusters with both molecules present as

$$CD_{ij} = \frac{g_{ij}(0,0;0)}{g_i(0,0;0)g_j(0,0;0)\pi w^2} = \frac{\overline{N}_{ij}}{\pi w^2}. \quad (8.21)$$

This parameter is a measure of the extent to which two molecules are colocalized on the surface. In the case where the two molecules are always associated with each other, the cluster densities of the two separate molecules should be equal to the density of colocalized clusters ($CD_i = CD_j = CD_{ij}$). On the other hand, if there is no colocalization, the cross-correlation function will be small and CD_{ij} will approach zero.

It is instructive to calculate the ratio of cluster densities as

$$F(i|j) = \frac{CD_{ij}}{CD_i} = \frac{g_{ij}(0,0;0)}{g_j(0,0;0)} \quad \text{and} \quad F(j|i) = \frac{CD_{ij}}{CD_j} = \frac{g_{ij}(0,0;0)}{g_i(0,0;0)}. (8.22)$$

Since the clusters that are measured by CD_{ij} is a subset of those in CD_i, the first ratio may be interpreted as the fraction of clusters that contain both molecules i and j relative to all clusters that contains i, or the fraction of clusters with molecules i that are colocalized with molecules j. The second ratio then is the fraction of clusters with j that are colocalized with i.

If pairs of images of two chromophores are collected as a function of time, then the dynamics of the interaction of two different molecules may be determined. For example, as the association of two molecules progresses $(A + B \rightarrow AB)$, the fractions in (8.22) should increase from zero to one.

8.5   Applications to Cell Surfaces

The four objectives established at the beginning of this chapter were to establish the distribution of molecules and determine how this may change, measure the inter-molecular interactions and how they may change. The previous section shows that careful experimental design coupled with detailed analysis and interpretation of the amplitude of auto- or cross-correlation functions can provide qualitative insight and quantitative estimates. This section will highlight some of the examples where image correlation spectroscopy and image cross-correlation spectroscopy have proven useful in understanding the organization of receptors and determining the molecular interactions on cell surfaces.

8.5.1   Receptor Distributions

Concanavalin A Receptors. In early work by St-Pierre and Petersen [8.31], scanning fluorescence correlation spectroscopy was used to measure the distribution of the lectin, Concanavalin A, on the surface of 3T3 cells. They showed that at low concentrations, the lectin binds to disperse, high affinity receptors but at high concentrations it binds to aggregated receptors. While the binding of Concanvalin A is rather non-selective, the experiment demonstrated the importance of measuring the distribution as a function of concentration of ligand added to the cell surface.

Platelet Derived Growth Factor Receptors. Wiseman and co-workers [8.16,8.22,8.24,8.32] have studied the distribution of platelet derived growth factor receptor (PDGFR) on AG 1523 cells in some detail. They showed that, at $4\,^{\circ}\text{C}$, the PDGF receptors are pre-aggregated in clusters with an average of four receptors per cluster ($\mu = 3.9$) and a density of over two clusters per square micrometer ($CD = 2.3\,\mu\text{m}^{-2}$). On average, the AG 1523 cells have a surface area of just over $1 \times 10^4\,\mu\text{m}^2$, providing a total estimate of about 1×10^5 receptors per cell. This compares favorably to the total estimated by binding studies of about 1.5×10^5 receptors per cell. These results demonstrate that ICS can detect all the receptors under saturating conditions and hence the quantitative information is reliable. The images show clearly that the clusters are not uniform in size, but the ICS data provided no information on the range of cluster sizes, only the mean value. At $37\,^{\circ}\text{C}$, the cluster density decreases about three-fold [8.32], suggesting that at physiological conditions, the receptors are even more associated. Treatment of the cells with an erbstatin analog (an inhibitor of tyrosine kinase activity) appears to disperse the receptor to the monomers (the cluster density increases four-fold over that at $4\,^{\circ}\text{C}$ and 12-fold over that at $37\,^{\circ}\text{C}$). Subsequent treatment with PDGF leads to a nearly two-fold decrease in the cluster density, consistent with formation of dimers by the PDGF.

Epidermal Growth Factor Receptors. Using scanning FCS, St-Pierre and Petersen [8.33] found that the epidermal growth factor receptor (EGFR) on A431 cells exist in clusters with as many as 120 receptors at a density of about 7 clusters per square micrometer. For cells with an average area of $3000\,\mu\text{m}^2$, this yields about 2.5 million receptors per cell. These numbers were found to agree well with the total number of receptors on the surface of about 2–3 million as determined by binding data. More recent, but preliminary, work using ICS shows a slightly larger cluster density ($CD = 10\,\mu\text{m}^{-2}$) and smaller clusters ($\mu = 33$), but it is also clear that not all the receptors were detected in these experiments, even though saturating levels of antibody were employed [8.24]. Importantly, the ICS experiments showed clearly that at higher temperatures, the receptors were distributed in more ($CD = 19\,\mu\mathrm{m}^{-2}$) but smaller ($\mu = 11$) clusters. More work needs to be performed on this system.

Transferrin Receptors. Srivastava and Petersen [8.23] studied the dynamics of redistribution of transferrin receptors on the surface of 3T3 fibroblasts and Hep2 epidermal cells. They found that the receptors were distributed fairly homogeneously ($CD = 17$–$21\,\mu\mathrm{m}^{-2}$) at room temperature or on fixed cells. There was no independent information on the total receptor density on these cells, and there was no quantitative estimate of the degree of aggregation. On 3T3 cells, the number of clusters decreased almost three-fold at $37\,^\circ\mathrm{C}$ ($CD = 11\,\mu\mathrm{m}^{-2}$) when compared to $4\,^\circ\mathrm{C}$ ($CD = 27\,\mu\mathrm{m}^{-2}$) [8.42].

Cross-correlation of images measured at 20 s intervals at room temperature showed that there was a steady decrease in the amplitude with time [8.23]. The autocorrelation function amplitudes did not vary with time, therefore the decay had to arise from a dynamic process. The data were fit to (8.8) to yield a 'diffusion' coefficient for the clusters. The diffusion was found to be about three orders of magnitude different from the diffusion of receptor monomers as measured by fluorescence photobleaching experiments. The discrepancy was accounted for a by slow movement of clusters (detected by ICS) coupled with a rapid exchange of monomers among clusters (detected by photobleaching) [8.23].

Transfected HA Mutants. Over the years, Roth and Henis [8.34,8.35] have used a series of mutants of the hemagglutinin protein (HA) from the influenza virus as a model receptor in CV-1 cells. They have shown that the wild-type (HA-wt) does not interact with coated pits, and is free to diffuse in the membrane of the host cells. Mutants with specific modifications in the cytoplasmic tail to introduce a YXXΦ signal sequence (HA-Y532 and HA+8) were impeded in their mobility and were found to be internalized in coated pits. Image correlation spectroscopy measurements on the distribution of HA proteins in the CV-1 cells confirmed that HA-wt was the most uniformly distributed receptor ($CD \sim 20\,\mu\mathrm{m}^{-2}$) while HA+8 was the most aggregated ($CD \sim 3.2\,\mu\mathrm{m}^{-2}$). Since the transfection levels can vary significantly from cell to cell, these experiments were conducted on cells with comparable expression levels on the surface as judged by the total intensity. Under these circumstances, the absolute values of the degree of aggregation are without meaning, but the relative values are helpful.

Receptor-Receptor Interactions. To date, only a limited number of cross-correlation experiments have been performed on receptor pairs by the simultaneous labeling of two receptors. In experiments to show 'proof of principle,' the interactions of transferrin receptors with either PDGFR or EGFR were studied [8.42].

Cells were first exposed to an unlabeled mouse IgG specific for either the PDGF receptor or the EGF receptor. They were then reacted with biotinylated goat-anti-mouse antibody and visualized with avidin-tricolor (an energy transfer dye that absorbs at 488 nm and emits at 640 nm). Subsequently, the cells were exposed to a mouse IgG specific for human CD71 and labeled with fluorescein (absorbs at 488 nm and emits at 530 nm). Images were obtained with an argon ion laser detecting fluorescein first and tricolor second since the former is more susceptible to photobleaching. Pairs of images were cross-correlated to determine the extent of interaction.

As negative controls, pairs of unrelated images were cross-correlated. These revealed that less than 6% of the cross-correlation functions could be fit to (8.6a) at the origin (i.e. with ξ_0 or η_0 less than 10 pixels). Of those that did, the fraction of colocalization was very low ($F(i|j) \sim 0.01$). As a positive control, images were collected from cells labeled first with mouse anti-human CD71 with fluorescein attached and secondly with the biotinylated goat-anti-mouse antibody and avidin-tricolor. This places both chromophores on the same receptors for 100% colocalization. The data showed that all the cross-correlation functions fit at the origin with F(Green|Red) = 1.13 and F(Red|Green) = 0.92. The deviations from unity would be consistent with an energy transfer process, but there is no independent evidence for this. In these experiments, most of the fluorescein is photobleached prior to the accumulation of the red image, so it is not clear how the energy transfer can occur. The deviations are not due to cross-talk between the detectors, since the fluorescence emissions are quite distinct and the filters separate these emissions effectively.

Cross-correlation functions from images on 3T3 cells with transferrin and PDGF-receptors labeled with fluorescein and tricolor, respectively, were all fit well to (8.6a) at the origin. The fraction of transferrin receptor clusters that also contained PDGF receptors was F(Transferrin|PDGFR) = 0.80 while the fraction of PDGF receptor clusters that also contained transferrin was F(PDGFR|Transferrin) = 0.68. This suggests that on average, about 70% of each of these receptors are present in clusters that contain both receptors and about 30% are present in clusters (or as monomers) without the other receptor. These results do not reveal the organization of these clusters or the functional relevance of these inter-molecular interactions.

Cross-correlation functions from images on A431 cells with transferrin and EGF-receptors labeled with fluorescein and tricolor, respectively, generally fit poorly to (8.6a) at the origin. Among those that did fit, the fraction of colocalization was very small (F(Transferrin|EGFR) = 0.019 while F(EGFR|Transferrin) = 0.006). This suggests that there are no significant interactions between transferrin receptors and EGF receptors. Since the EGF receptor is overexpressed in these cells by at least an order of magnitude, these results may not be representative of normal interactions.

It is interesting to note that both transferrin receptors and PDGF receptors are more disperse at lower temperatures and more clustered at higher temperatures. This correlation would be expected if they interact among themselves or with the same structures at all temperatures. In contrast, the EGF receptors are more disperse at higher temperatures and more clustered at lower temperatures. This is compatible with the observation that EGF receptors and transferrin receptors do not interact at any temperature.

8.5.2 Interactions in Coated Pits

Coated pits have been identified as the structures on the surface of cells through which receptor mediated endocytosis occurs. They are known to contain at least three protein components: membrane receptors, adaptor proteins that bind to the receptors and clathrin that binds to the adaptors. The adaptor protein found at the plasma membrane is designated AP-2 and differs immunologically and structurally from AP-1, which is found at the Golgi and AP-3, which is the least understood member of the family. Clathrin is found at the plasma membrane, at the Golgi and as free components in the cytoplasm. Clathrin coated vesicles are also present within the cytoplasm, but they are quickly uncoated under normal conditions.

Clathrin and Adaptor Protein Distributions. The distribution of AP-2 was studied in CV-1 cells using a pair of antibodies specific for the AP-2 type adaptor molecule [8.37]. When the AP-2 is associated with coated pits, we expect the clusters to be quite large. Indeed, the images show a large number of distinct, bright spots, whose individual size is the approximate size of the laser beam. ICS measurements on 439 cells gave an average cluster density of $1.4 \pm 0.2\,\mu\text{m}^{-2}$. Careful analysis of the number of bright spots in a large number of images yielded a density of these spots of $\sim 0.28\,\mu\text{m}^{-2}$, significantly lower than the ICS measurements. Inspection of the cross-sectional intensity in the images revealed a number of small fluctuations, significantly larger than background, but not visible as distinct spots in the images.

A large body of data was subjected to an interpretation that the AP-2 existed in two populations: large aggregates and small aggregates. The large aggregates were presumed to be coated pits (visible in the images). Application of (8.18) for two populations gives

$$g_i(0,0;0) = \frac{1}{\overline{N}_m^2} \left(\overline{N}_f \cdot \mu_f^2 + \overline{N}_c \cdot \mu_c^2 \right) . \tag{8.23}$$

Assuming that $\overline{N}_c = 0.28\,\mu\text{m}^{-2}$ and $\mu_c^2 = 60$ AP-2 molecules per coated pit (one AP-2 for every two clathrin triskelia), they concluded that, on average, there are 50 AP-2 molecules per square micrometer, distributed in two populations. For every coated pit with 60 molecules of AP-2 in each there are five smaller aggregates with about 20 molecules in each for a total of about 50 AP-2 molecules per square micrometer. Treatment of the cells with either hypotonic or acidic conditions disrupted the coated pits and changed the distributions of AP-2 among these two populations. Subsequent work [8.24,8.43] showed that neither the cluster density nor the number of spots per image changed when the cells were permeabilized with saponin. This suggests that the smaller aggregates of AP-2 are associated with the plasma membrane rather than free in the cytoplasm.

The distribution of clathrin in CV-1 cells was studied using a monoclonal antibody specific for the heavy chain of clathrin [8.38]. As for the AP-2, the images showed a large number of distinct, bright spots superimposed on a fairly uniform fluorescence background. ICS measurements on several hundred cells gave an average cluster density of $1.32 \pm 0.08\,\mu\text{m}^{-2}$. This compared with an average density of clearly visible spots of $0.55 \pm 0.10\,\mu\text{m}^{-2}$ indicating that there are two populations of clathrin. Treatment with Saponin led to a reduction of the cluster density to a value comparable to the spot density (CD = 0.50) indicating that the second population of clathrin is free in the cytoplasm. Analysis of the ICS data in detail using the two-population model (8.23) led to the conclusion that there are 0.5–0.6 coated pits per micrometer, that about two-thirds of the clathrin is associated with the coated pit and that the rest is in the cytoplasm.

Receptor-Coated Pit Interactions. In more recent work, Brown and co-workers [8.30] have studied the interaction between the HA protein mutants and the components of the coated pit. The CV-1 cells were transfected with either HA-wt or HA+8 and labeled for the receptor as well as either AP-2 or clathrin. The HA protein was visualized first with fluorescein and the other protein was labeled with rhodamine red. The images were collected separately using one excitation beam at a time to keep cross-over fluorescence below 1% in either direction. As above, the negative control (cross-correlation of unrelated images) showed that less than 1% of the cross-correlation functions could be fit to (8.6a) at the origin and those that fit gave F(Green|Red) ~ 0.03 and F(Red|Green) ~ 0.13. The positive control (labeling of clathrin simultaneously with two different fluorescent antibodies) showed that all cross-correlation functions fit (8.6a) at the origin and gave F(Green|Red) = 0.90 and F(Red|Green) = 1.07. The deviations from unity are not consistent with an energy transfer process, but may reflect the inherent uncertainty in these calculations.

Cross-correlation analysis of pairs of images from cells transfected with the wild-type HA protein yielded less than 20% of the cross-correlation functions that could be fit to (8.6a) at the origin, and the fraction was less than 0.02 in all cases. This result was independent of whether the cross-correlation was for images with AP-2 or clathin. In either case, there are no interactions between HA-wt and the coated pits. In contrast, all the cross-correlation functions obtained for pairs of images from cells transfected with HA+8 could be fit to (8.6a) at the origin. The fraction of HA+8 found associated with AP-2 (F(HA+8|AP-2) = 0.25 ± 0.02) was identical to that for HA+8 and clathrin (F(HA+8|Clathrin) = 0.25±0.03) suggesting that at any given time, about 25% of the receptor is in a coated pit while the rest is free in the membrane. Similarly, the fraction of AP-2 associated with HA+8 (F(AP-2|HA+8) = 0.69 ± 0.07) and the fraction of clathrin associated with HA+8 (F(clathrin|HA+8) = 0.72 ± 0.06) are very close, suggesting that about 70% of these proteins are associated with coated pits that contain the receptor. In agreement with the concept of two populations of AP-2 and clathrin, the remaining 30% of these proteins are outside the coated pits. Interestingly, as the level of expression of HA+8 increases (as judged by the total intensity of the fluorescence) the fraction of HA+8 in the coated pits decreases while the fraction of AP-2 and clathrin associated with the receptor increases toward unity. This indicates that at higher concentrations of receptor, more AP-2 and clathrin is recruited into coated pits [8.30].

8.5.3   Virus Assembly and Fusion

Membrane enveloped virus particles derive their lipid components from the host membrane as they complete an infection and bud off the surface. The viral proteins are either assembled prior to emerging on the cell surface or are expressed on the surface and then caused to aggregate further into the bud. Early work [8.15] showed that it was possible to use scanning fluorescence correlation spectroscopy to distinguish between these mechanisms. Those experiments confirmed that the mobile glycoproteins of the Vesicular Stomatitis Virus emerged and clustered with time on the surface of BHK cells. The less mobile glycoproteins of the Sindbis Virus were initially much more aggregated and could not be induced to aggregate further with a crosslinking antibody. A temperature sensitive mutant of the Sindbis Virus remained less aggregated but the proteins could be easily crosslinked with a second antibody. These data provided corroborative evidence that the Sindbis Virus proteins are assembled prior to emerging on the surface while Vesicular Stomatitis Virus proteins are assembled on the surface.

In more recent work, ICS has been used to study the fusion of Sendai Virus to a host cell. Rasmusson and co-workers [8.39] incorporated a fluorescent lipid derivative into the membrane of the virus prior to adding the virus to the surface of Hep2 or 3T3 cells. They demonstrated that fifteen minutes of incubation at 37 °C was sufficient to disperse a significant amount of the lipid probe into the cell membrane with a corresponding increase in the cluster density. They further showed that the increase in cluster density could be inhibited in a dose-dependent manner by a lipophosphoglycan isolated from *Leishmania donovani*. This work showed that ICS can be used to study the mechanism of virus interactions with cells at the individual cell level rather than in bulk suspensions.

8.5.4   Other Applications and Future Prospects

Lipid Distributions. The applications to lipid distributions are so far very limited. Measurements of the distribution of gangliosides have been reported [8.24], but clearly the cluster density will depend on the efficiency of incorporation. At this stage, it is clear that the lipids will distribute in a very homogeneous manner (CD $\sim 60\,\mu m^{-2}$ or more). However, there are also a few bright spots noted. So far, the origin of these high concentration regions is unclear. It is possible that they represent specialized lipid domains on the surface. In fact, if glycolipid rich domains are prominent on a surface, it should be possible to detect these with image correlation techniques and cross-correlation measurements.

Non-Biological Problems. Image correlation spectroscopy can be applied successfully to images derived from fluorescence microscopy techniques because both the fluorescence intensity is proportional to the concentration and the geometry of the correlator (the laser) is well understood. Other surface science techniques can provide images in which the intensity in the image is related to the surface concentration in a linear fashion. One example is the technique of secondary ion mass spectrometry (SIMS). Here, the surface is bombarded with an ion beam to release ionized material which travels away from the surface. These secondary ions are captured and mass analyzed as the primary ion beam sweeps across the surface. Images can be generated which reflect the number of secondary ions released per unit time per unit area. The SIMS can also be used to obtain successive images as a function of depth into the material. In recent work, Srivastava and co-workers [8.40] have analyzed several sets of SIMS images. They studied the distribution of 'invisible gold' in arsenopyrite samples and showed that at the level of resolution of the ion beam, gold and arsenic are not colocalized in these samples, in spite of apparent similarity of the patterns of distributions of these elements. The advantage of the ICS analysis for samples such as these is that it allows for a semi-quantitative assessment of the interaction of elements at the sub-micrometer to micrometer level.

8.6   Conclusions

This chapter discussed how spatial correlation analysis of images obtained from confocal microscopy can be used to extract information about the distribution of molecules on a surface. The applications have been limited to measurements on cell surfaces, but the technique is more widely applicable.

Clearly, the interpretation of the amplitudes of the correlation function applies equally to any FCS experiment.

The focus of this chapter has been to develop the tools and to show the applications of ICS. It is straightforward to generate the confocal images and to perform the appropriate calculations. However, careful attention must be paid to the development of proper controls and to ensure that the experiments are performed under conditions where the numbers will be meaningful. For example, too low a concentration of an antibody used to label a receptor may lead to problems with autofluorescence, while too high a concentration may lead to interference from non-specific binding of the antibody.

Interpretations can be difficult since in many cases it is necessary to consider complex distribution patterns, including the coexistence of monomeric and aggregated receptors. In selected cases, these interpretations make sense, but the information available is usually less than the number of parameters needed. These interpretations are very sensitive to the models employed.

One of the disadvantages of ICS is that it averages a lot of information into a single average number. Thus, information about the inherent distribution of cluster sizes is lost. While some of this information can be extracted from higher order correlations, it is not clear that there is sufficient accuracy in the data. A more promising approach is to apply techniques such as the photon counting histogram (PCH) technique to images [8.41]. In principle, this approach will provide additional information about the distribution of cluster sizes directly from the distribution of intensities in the image, and will complement the information obtained by ICS analysis.

Image correlation spectroscopy is a relatively new facet of fluorescence correlation spectroscopy. ICS lacks the advantages of FCS to measure dynamic processes, but in return, the information content in an image is sufficient to provide reasonably accurate estimates of the occupation numbers. For applications to cell surfaces, where the dynamics is inherently slow, ICS or variants thereof, are likely to prove very useful.

Image correlation spectroscopy is inherently an averaging technique. The information available in each individual fluctuation is averaged into a single number for an entire image. This is a strength as well as a weakness for biological investigations. Since the biological information appears to vary significantly from one cell to another, it is more important to obtain a large amount of information from many images on many cells rather than to study a single image from a single cell in great detail. In most of the studies discussed here, cluster densities and degrees of aggregation are based on hundreds of images from hundreds of cells. To some, one picture is worth a thousand words. To us, a good number is worth a thousand pictures.

References

8.1 E. Sheets, R. Simons, and K. Jacobson: Curr Opin. Cell Biol. **7**, 707 (1995)

8.2 A. Kusumi and Y. Sako: Curr. Opin. Cell Biol. **8**, 556 (1996)

8.3 T. Harder and K. Simons: Curr. Opin. Cell Biol. **9**, 534 (1995)

8.4 R.G. Parton and K. Simons: Science **269**, 1398 (1995)

8.5 D.A. Brown and J.K. Bose: Cell **68**, 533 (1992)

8.6 R.G. Parton: J. Histochem. Cytochem **42**, 155 (1994)

8.7 M.S. Robinson: Trends Cell Biol. **7**, 99 (1997)

8.8 T. Kirchhausen: Curr. Opin. Struc. Biol. **3**, 182 (1993)

8.9 E.L. Elson and D. Magde: Biopolymers **13**, 1 (1974)

8.10 D. Magde, E.L. Elson, and W.W. Webb: Biopolymers **13**, 29 (1974)

8.11 D. Magde, W.W. Webb, and E.L. Elson: Biopolymers **17**, 331 (1978)

8.12 N. Davidson: Statistical Mechanics (McGraw-Hill, New York, N.Y. 1962)

8.13 N.O. Petersen: Can. J. Biochem. Cell Biol. **62**, 1158 (1984)

8.14 N.O. Petersen: Biophys. J. **49**, 809 (1986)

8.15 N.O. Petersen, D.C. Johnson, and M.J. Schlesinger: Biophys. J. **49**, 817 (1986)

8.16 N.O. Petersen, P.L. Höddelius, P.W. Wiseman, O. Seger, and K.-E. Magnusson: Biophys. J. **65**, 1135 (1993)

8.17 P. Swille, F.J. Meyer-Almes, and R. Rigler: Biophys. J. **72**, 1878 (1997)

8.18 D.E. Koppel, F. Morgan, A.E. Cowan, and J.H. Carson: Biophys. J. **66**, 502 (1994)

8.19 M. Srivastava and N.O. Petersen: Meth. Cell Sci. **18**, 47 (1996)

8.20 N.O. Petersen, S. Felder, and E.L. Elson: In: Handbook of Experimental Immunology (D.M. Weir, L.A. Herzenberg, C.C. Blackwell, L.A. Herzenberg (eds.), Ch. 20 p. 24.1–23 Blackwell Scientific Publications Ltd, Edinburg, 1985)

8.21 A.G. Benn and R.J. Kulperger: Environmetrics **7**, 167 (1996)

8.22 P.W. Wiseman and N.O. Petersen: Biophys. J. **76**, 963 (1999)

8.23 M. Srivastava and N.O. Petersen: Biophysical Chem. **75**, 201 (1998)

8.24 N.O. Petersen, C. Brown, A. Kaminski, J. Rocheleau, M. Srivastava, and P.W. Wiseman: Faraday Discuss **111**, 289 (1999)

8.25 B.J. Berne and R. Pecora: In: Dynamic Light Scattering with Applications to Chemistry, Biology and Physics (John Wiley, New York, 1975)

8.26 M.B. Weissman: Ann. Rev. Phys. Chem **32**, 205 (1981)

8.27 N.O. Petersen and E.L. Elson: In: Methods in Enzymology (C.H.W. Hirs, S.N. Timasheff (eds.), 454–484 Academic Press Inc, New York, 1986)

8.28 H. Qian and E.L. Elson: Proc. Natl. Acad. Sci. USA **87**, 5479 (1990)

8.29 N. Thompson: In: Topics in Fluorescence Spectroscopy, Vol 1 (J.R. Lakowics (ed.), 337–378 Plenum Press, New York, 1991)

8.30 C.M. Brown, M.G. Roth, Y.I. Henis, and N.O. Petersen: Biochemistry **38**, 15166 (1999)

8.31 P.R. St-Pierre and N.O. Petersen: Biophys. J. **58**, 503 (1990)

8.32 P.W. Wiseman, P. Höddelius, N.O. Petersen, and K.-E. Magnusson: FEBS Lett. **401**, 43 (1997)

8.33 P.R. St-Pierre and N.O. Petersen: Biochemistry **31**, 2459 (1992)

8.34 E. Fire, D.E. Zwart, M.G. Roth, and Y.I. Henis: J.Cell Biol. **115**, 1585 (1991)

8.35 E. Fire, O. Gutman, M.G. Roth, and Y.I. Henis: J. Biol. Chem. **270**, 21075 (1995)

8.36 E. Fire, C.M. Brown, M.G. Roth, Y.I. Henis, and N.O. Petersen: J. Biol. Chem. **272**, 28538 (1997)

8.37 C.M. Brown and N.O. Petersen: J. Cell Sci. **111**, 271 (1998)

8.38 C.M. Brown and N.O. Petersen: Biochem. and Cell Biol. **77**, 439 (1999)

8.39 B.J. Rasmusson, T.D. Flanagan, S.J. Turco, R.M. Epand, and N.O. Petersen: Biochim. Biophys. Acta **1404**, 338 (1998)

8.40 M. Srivastava, N.O. Petersen, G. Mount, D. Kingston, and N.S. McIntyre: Surface and Interface Analysis **26**, 188 (1998)

8.41 Y. Chen, J.D. Müller, P.T.C. So, and E. Gratton: Biophys. J. **77**, 553 (1999)

8.42 M. Srivastava, PhD Thesis, University of Western Ontario, London, Canada (1998)

8.43 C.M. Brown, PhD Thesis, The University of Western Ontario, London, Canada (1998)

Part III

Applications in Biotechnology, Drug Screening, and Diagnostics

Applications in Fluorescence
Gene Screening and Diagnostic

9 Dual-Color Confocal Fluorescence Spectroscopy and its Application in Biotechnology

Andre Koltermann*, Ulrich Kettling, Jens Stephan, Thorsten Winkler, and Manfred Eigen

9.1 Introduction

In recent years, fluorescence correlation spectroscopy (FCS) has become an attractive analytical tool for the investigation of biomolecular processes at the single molecular level. This method was invented in the early 1970s by groups at Cornell University, Ithaca, N.Y. [9.1,9.2], and at the Karolinska Institute, Stockholm [9.3,9.4]. During the 1990s, modern confocal optics, new dyes as efficient fluorescent probes, sensitive photon detectors, and fast data processing tools have been introduced, mainly by Rigler and Eigen [9.5,9.6]. Due to these improvements, FCS permits the observation of the dynamics of single molecules in real time while they pass an open volume element of less than a femtoliter, i.e. the size of a common bacterial cell. Nowadays, FCS has found its way into several laboratories and companies all over the world as a tool for basic research as well as for industrial applications such as drug screening.

Dual-color fluorescence cross-correlation spectroscopy (dual-color FCS) was proposed by Rigler and Eigen in the early 1990s [9.6], and was worked out by Schwille ([9.7], see also this issue). In contrast to conventional single-color FCS, the new method uses two perfectly superimposed laser foci, and detects correlated intensity fluctuations that arise from single molecules carrying two spectrally distinguishable fluorescent labels. It allows precise and highly specific detection in spite of a large unspecific fluorescent background; moreover, the method is fully compatible with biological environments. Two important applications have been addressed by the authors: Real-time measurements of enzyme kinetics were successfully performed, and the suitability of dual-color FCS for high-throughput screening has been demonstrated.

Furthermore, the limitations of dual-color FCS with regard to screening purposes led to the development of RAPID FCS (rapid assay processing by integration of dual-color fluorescence cross-correlation spectroscopy). While conventional FCS identifies molecules by their diffusion properties, requiring a considerable amount of analysis time, dual-color FCS simply *counts* double-labeled molecules; therefore, analysis times for simple yes-or-no decisions are much shorter and data evaluation is faster. Analysis times of one second per sample and less were achieved with sample volumes in the range of submi-

I. Covalent binding

II. Single non-covalent binding

III. Dual non-covalent binding

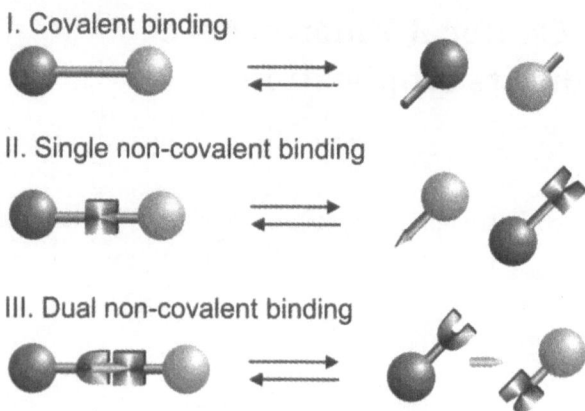

Fig. 9.1. Different assay principles

croliters without decreasing signal intensity. RAPID FCS can process 10^4 to 10^5 samples per day.

The next development in confocal fluorometry in order to enhance the throughput rate which is given by the analysis time per sample is confocal fluorescence coincidence analysis (CFCA). Here, technical improvements have led to a further decrease in analysis times. Besides some modifications concerning the laser source and a controlled external enhancement of concentration fluctuation, a new data evaluation procedure for a specific detection of double-fluorescent molecules was introduced. CFCA has clear advantages over RAPID FCS concerning analysis speed and suitability for screening. Analysis times of 100 ms per sample and below allows for an ultra-high-throughput rate of 10^6 samples per day for evolutionary biotechnology, where selection by screening is needed.

Dual-color FCS, RAPID FCS and CFCA each employ two different fluorescent labels with well-separated emission spectra and generally allow monitoring any reaction in which a chemical bond or physical association between those fluorophores is established or broken. In contrast to single-color FCS, where the result of diffusion analysis is needed, these technologies simply count double-labeled molecules and therefore give a high degree of freedom in biochemical and cellular assay design, as shown in Fig. 9.1.

Depending on the need of the application, the same assay can be used for kinetic studies as well as for high-throughput screening and/or ultra-high-throughput screening.

The striking advantages of the dual-color setup are clearly demonstrated in the cleavage of small substrates, where single-color FCS suffers from only small changes in the diffusion time of substrates and products. We demonstrated for the first time the suitability of dual-color FCS to monitor and screen a specific enzymatic cleavage reaction. Among a huge variety of different enzymatic cleavage and ligation reactions which we have carried out with

the described technologies, we choose the cleavage of the specific endonuclease
EcoRI for a presentation in more detail. The applied methods, from kinetic
studies in real time via throughput rates for screening of catalytic activity to
ultra-high-throughput screening of the cleavage of a double-stranded DNA
substrate by EcoRI, will be described in the following sections.

9.2   Real-Time Monitoring of Enzymatic Activity by Dual-Color FCS

Kinetic studies on enzymes are among the most important tools for understanding biological interactions at the molecular level. In combination with
new approaches in genetic engineering and structure determination, in recent
years major efforts have been made to develop more sensitive and precise techniques for characterizing the kinetics of enzymatic reactions. With the help
of these techniques, it was possible to develop efficient assays for analyzing
catalytic parameters such as turnover rates, substrate specificity, as well as
regio- and stereospecificity; they constitute a major part of the biochemical
and pharmaceutical research. The dual-color cross-correlation method circumvents the necessity of evaluating the diffusion characteristics of the product fractions; therefore, simple mathematical evaluation is adequate [9.7,9.8].
The instrumental setup allows sensitive detection of molecules which bear two
different fluorescent labels. This can be achieved by a proper combination of
excitation and detection schemes utilizing two fluorescent labels that are spectrally well-separated. This topic is described in great detail by Schwille (see
this issue). The experiment is designed in such a way that the reaction step
to be investigated converts double-labeled educts into single-labeled products
or vice versa. In the past few years, different assays have been designed such
as hybridization kinetics [9.7] or prion aggregation [9.9]. Monitoring of enzymatic reactions by dual-color FCS was first applied by Kettling et al. (1998).

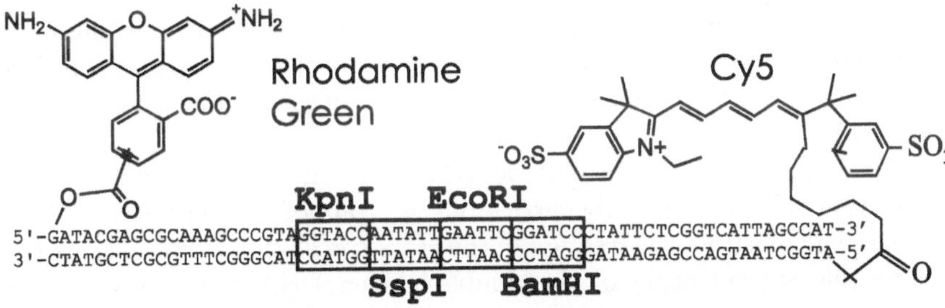

Fig. 9.2. The double-stranded DNA substrate. Sequence of the double-stranded
DNA substrate with fluorophore labels Cy-5 and Rhodamine Green at its 5'-ends,
respectively. The recognition sequences for different specific endonucleases are indicated in *rectangles*

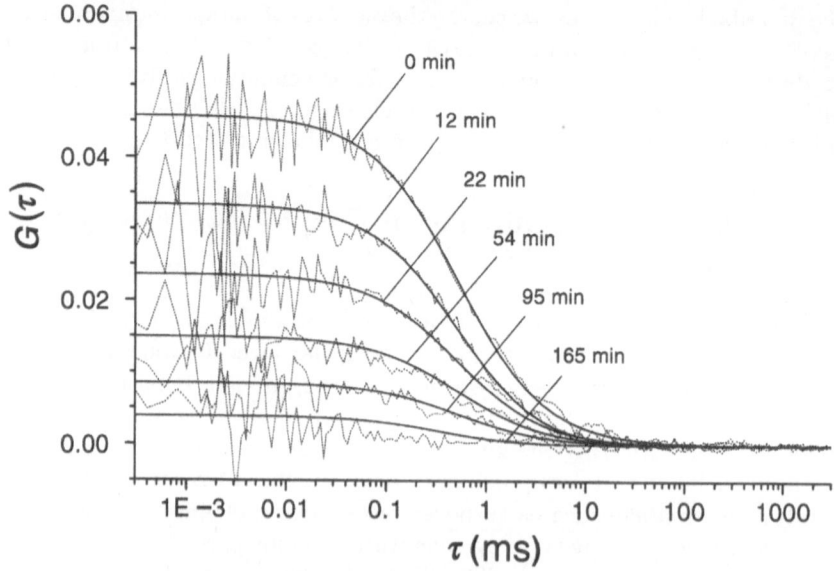

Fig. 9.3. Cross-correlation curves at different time points during an endonucleolytic cleavage reaction. 10 nM labeled DNA, 80 nM unlabeled DNA and 1.6 nM EcoRI were incubated in the reaction buffer at 27 °C. Dotted lines are the original data, the fitted curves are given in solid lines. During the reaction the cross-correlation amplitude $G_{gr}(0)$, which is a measure of the reaction progress, gradually decreases

In this work, the cleavage reaction of a double-stranded DNA molecule catalyzed by the restriction endonuclease EcoRI for analyzing enzyme kinetics was setup. The DNA molecule was labeled with a red and a green dye at opposite ends (Fig. 9.2), and the catalyzed reaction was monitored on-line using a dual-color fluorescence cross-correlation spectrometer prototype, which is described in detail by Schwille (see this issue).

Cleavage of the double-stranded DNA substrate by EcoRI (Fig. 9.2) breaks the chemical linkage between the two different fluorophores resulting in loss of the cross-correlation signal. The time course of this enzymatic reaction can be precisely and homogeneously monitored in solution by dual-color FCS (Fig. 9.3).

The fluorescence signals can be measured continuously, and the cross-correlation analysis is usually carried out at a rate of one per minute. During the cleavage reaction, the fraction of the cross-correlation curves, corresponding to the concentration of the double-labeled DNA substrate, decreased successively; the second important parameter – the average diffusion time of cross-correlating entities – remained constant, indicating a highly specific detection of the double-labeled DNA substrate by this method. FCS measurements usually require pico- to nanomolar concentrations of fluores-

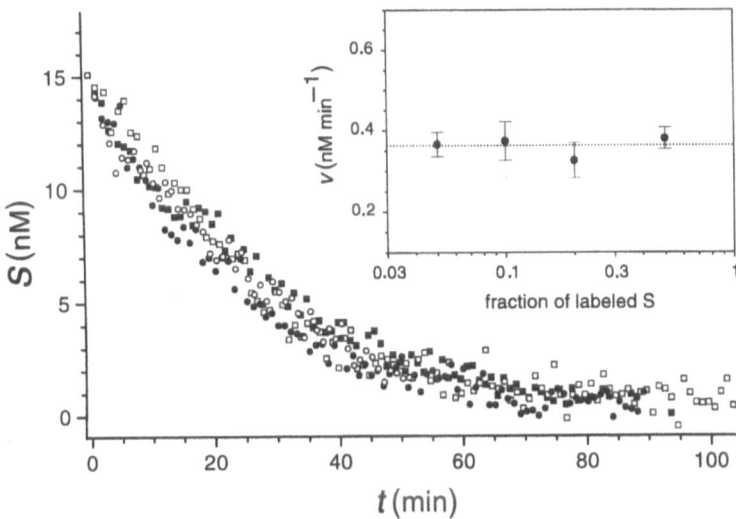

Fig. 9.4. Influence of different indicator fractions. Real time cleavage reactions were monitored at 27 °C applying 160 pM EcoRI and 15 nM substrate concentration containing different amounts of labeled substrate (7.5 nM, *solid squares*; 3.0 nM, *open circles*; 1.5 nM, *open squares*; 0.75 nM, *solid circles*) resulting in fractions ranging from 0.5 to 0.05. Inset: Evaluated initial slopes were plotted versus fraction values

cent molecules for optimal correlation analysis. Here, a kinetic analysis with broader ranges of substrate concentrations was achieved by adding unlabeled substrate to the reaction sample. Cleavage kinetics using different ratios of labeled to unlabeled substrate ranging from 0.05 to 0.5 were obtained by plotting the evaluated substrate concentrations versus time (Fig. 9.4). These measurements revealed identical kinetics, providing evidence that the fluorophores attached at the ends of the DNA molecule did not interfere with the enzyme's catalytic action (Fig. 9.4, inset); therefore, the labeled substrate served as a one-to-one indicator.

The class II restriction endonuclease EcoRI catalyzes the cleavage of the phosphodiester bond between the guanosine and the adenosine monomer of the palindromic recognition sequence GAATTC in each of the two strands. Dual-color FCS detects cleavage reactions by observing the separation of two different fluorescent labels, resulting from scissions in both strands of a single molecule; consequently, the kinetics that is observed refers to an overall reaction rate. Cleavage reactions were monitored over a wide range of substrate concentrations (from 1 to 130 nM) by adding a defined amount of unlabeled to a constant amount of labeled substrate, the latter serving as

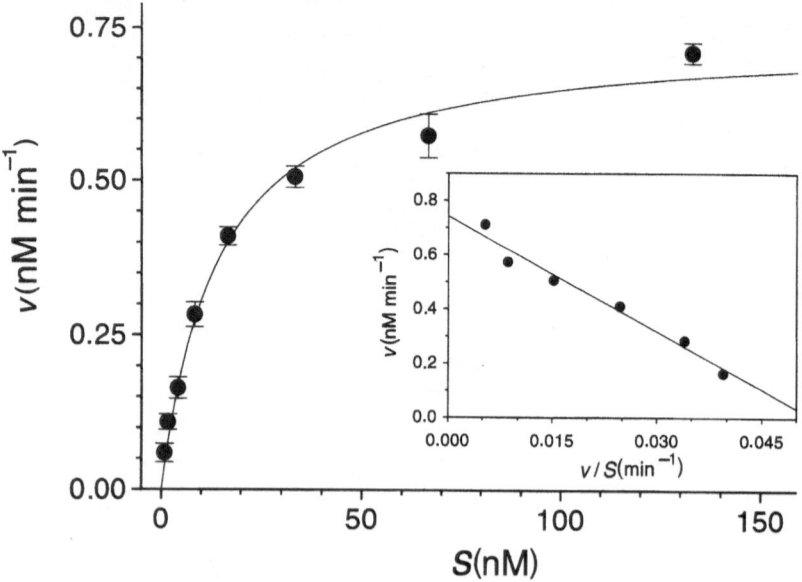

Fig. 9.5. Michaelis-Menten plot. Labeled DNA at a final concentration of 0.8 nM was mixed with different amounts (0–130 nM) of unlabeled DNA and incubated with 160 pM EcoRI in the reaction buffer at 27 °C. The reactions were monitored on-line and the initial rates v were derived by linear regression of data points of the first 5–20 min. *Inset:* Calculations from an Eadie–Hofstee plot lead to a K_M value of (14 ± 1) nM and v_{max} of (0.74 ± 0.03) nM min^{-1}

an indicator. From a linear regression of initial slopes at different substrate concentrations it was confirmed that EcoRI catalysis obeys the Michaelis–Menten equation (Fig. 9.5). From an Eadie–Hofstee plot (Fig. 9.5, inset) a K_M of (14 ± 1) nM and a k_{cat} of (4.6 ± 0.2) min^{-1} was derived, and the suitability of dual-color FCS for measuring kinetics in the nanomolar range was confirmed.

The ability of the cross-correlation analysis to sensitively quantify the enzyme activity is shown in Fig. 9.6. With an initial substrate concentration of 0.8 nM (below the K_M) a linear relation between initial reaction rates and enzyme concentrations was obtained over a concentration range of more than one order of magnitude (Fig. 9.6, inset). Enzyme activity was detected down to 1.6 pM. The extremely low enzyme activities detected by dual-color FCS demonstrate the enormous sensitivity of this technique.

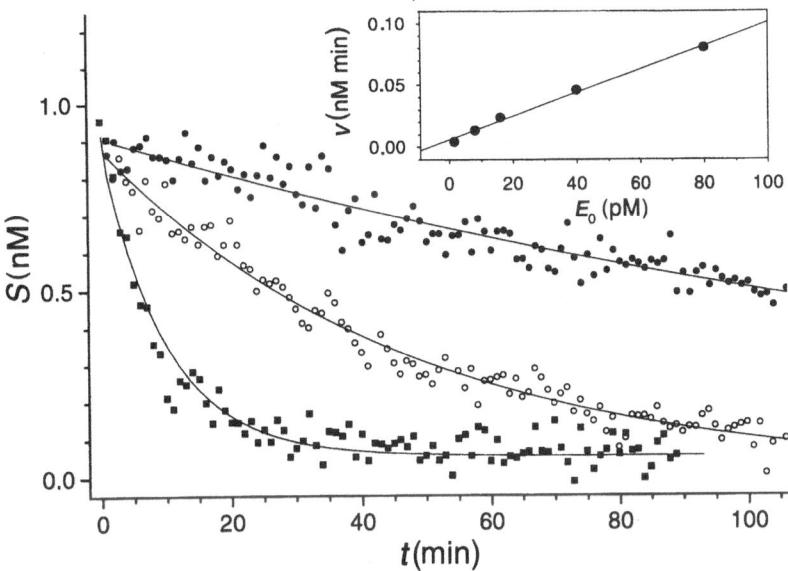

Fig. 9.6. Reaction rates at different enzyme concentrations. Endonucleolytic reactions were carried out in the reaction buffer at 27 °C with 0.8 nM double-labeled DNA substrate and different enzyme concentrations ranging from 1.6 to 80 pM. Time courses are shown for $E_0 = 1.6$ pM (*solid circles*), 8 pM (*open circles*) and 80 pM (*solid squares*). *Inset:* Initial rates v were calculated from linear regression of the data points of the first 5–50 min; the plots of these inital rates versus enzyme concentrations E_0 indicate a clearly linear relationship between v and E_0

9.3   RAPID FCS and CFCA for Screening Applications

High-throughput screening with dual-color FCS was demonstrated by Koltermann *et al.* [9.11] and termed RAPID FCS (rapid assay processing by integration of dual-color fluorescence cross-correlation spectroscopy). Analysis times of one second per sample and less were achieved with endonucleolytic assays with sample volumes of a few microliters. With the reported assay for endonucleolytic cleavage of EcoRI and additional restriction enzymes (BamHI and SspI) which cleave the double-stranded DNA substrate (Fig. 9.2), as well as a restriction enzyme without a recognition site in the substrate (HindIII), a simulated HTS was performed (Fig. 9.7).

Here, endpoint determinations for enzymatic activity were screened with different analysis times per sample. The lower limit of analysis times with RAPID FCS is between one and two seconds, depending on the desired tolerance for false positives or false negatives. This leads to a possible throughput of 10^4 to 10^5 samples per day. Therefore, this technology is an ideal tool for screening applications with moderate throughput rates. The combination of a

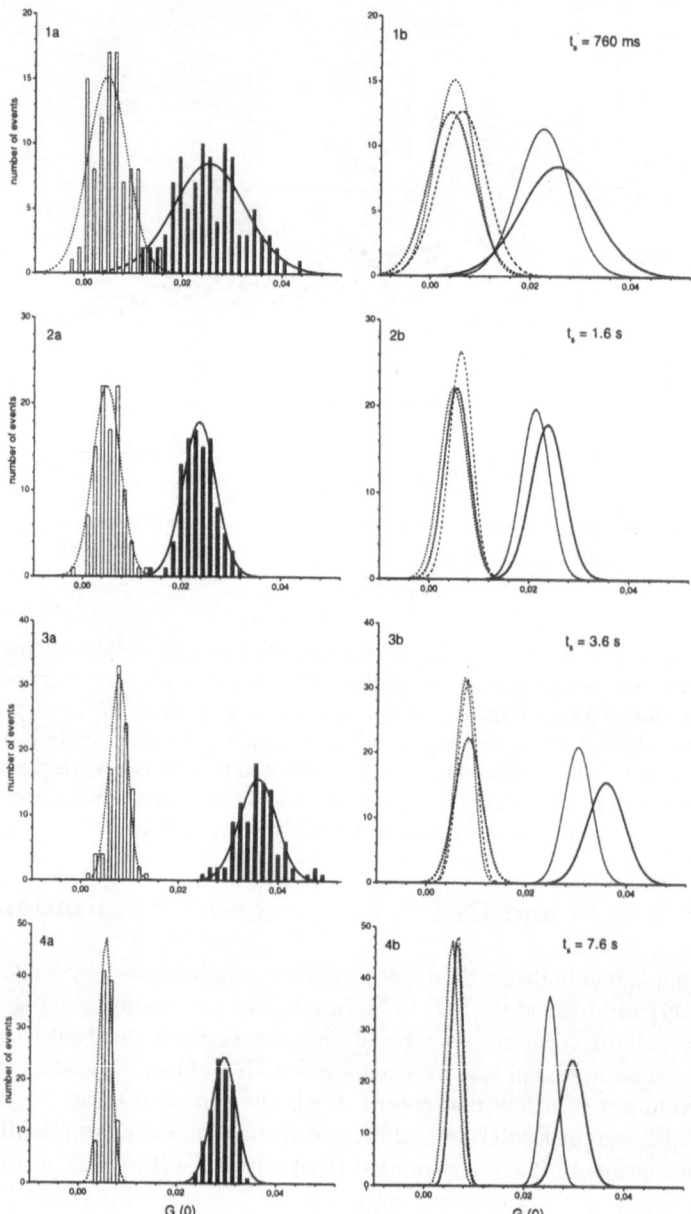

standard dual-color FCS set-up with nanotechnology will gain access to progressive selection strategies in evolutionary biotechnology, in which rare and specific binding or catalytic properties have to be screened in large numbers of samples.

To increase the moderate throughput rates of RAPID FCS for ultra-high-throughput screening applications, CFCA (confocal fluorescence coincidence analysis) was introduced by Winkler et al. [9.12]. The combination of one-beam two-color excitation, relative movement between sample volume and focal element as well as a data evaluation by coincidence analysis lowers the analysis time per sample by more than a tenth. The design of the CFCA setup is shown in Fig. 9.8.

Data evaluation by CFCA identifies fluctuations that arise coincidently in the two different spectral emission ranges, thereby quantifying the number of double-labeled fluorescent molecules. There is a variety of possible algorithms that may solve this task. As a first approach, Winkler et al. [9.12] introduced the normalized product sum algorithm:

$$K(n) = \frac{\sum_m N_1(m) N_2(m)}{\sum_m N_1(m) \sum_m N_2(m)} \times n \,. \tag{9.1}$$

$K(n)$ represents the coincidence value as a measure of the frequency of coincident events in the two detection channels. $N_1(m)$ and $N_2(m)$ are the number of counts in the different emission ranges in time channel m, and n is the total number of time channels in the trace. The analysis times are determined by the width of the time channels and the number of channels n. Data are recorded and evaluated on-line by the reported algorithm and therefore minimizes the readout parameter to one value without any further fitting routine as is required with single-color FCS, dual-color FCS or RAPID FCS.

←———

Fig. 9.7. Application of RAPID FCS for a simulated high-throughput screening for restriction endonucleolytic activities. The samples were screened in a circular manner for a total of 500 measurements. *Left-hand* histogram plots of the obtained cross-correlation amplitudes G(0) (1a–4a) compare the distributions of samples with (BamHI, **open bars**) and without specific endonucleolytic activity (HindIII, **solid bars**) for analysis times of 760 ms, 1.6 s, 3.6 s and 7.6 s. The *right-hand* histogram plots (1b–4b) show the ability to separate the different restriction endonucleases by their extracted Gaussian fittings with pure substrate (**solid**), HindIII (**bold solid**), BamHI (**bold dotted**), EcoRI (**dotted**), SspI (**dashed**). The overlaps were 2.6–5.4% (760 ms), 0.1% (1.6 s), < 0.002% (3.6 s) and $\leq 10^{-5}\%$ (7.6 s). Assays were performed in 5 μl volumes using 10 nM labeled DNA substrate, incubated for 3 h at 37 °C with 0.25 U/μl HindIII, 0.1 U/μl BamHI, 0.25 U/μl EcoRI, 0.08 U/μl SspI and without enzyme addition

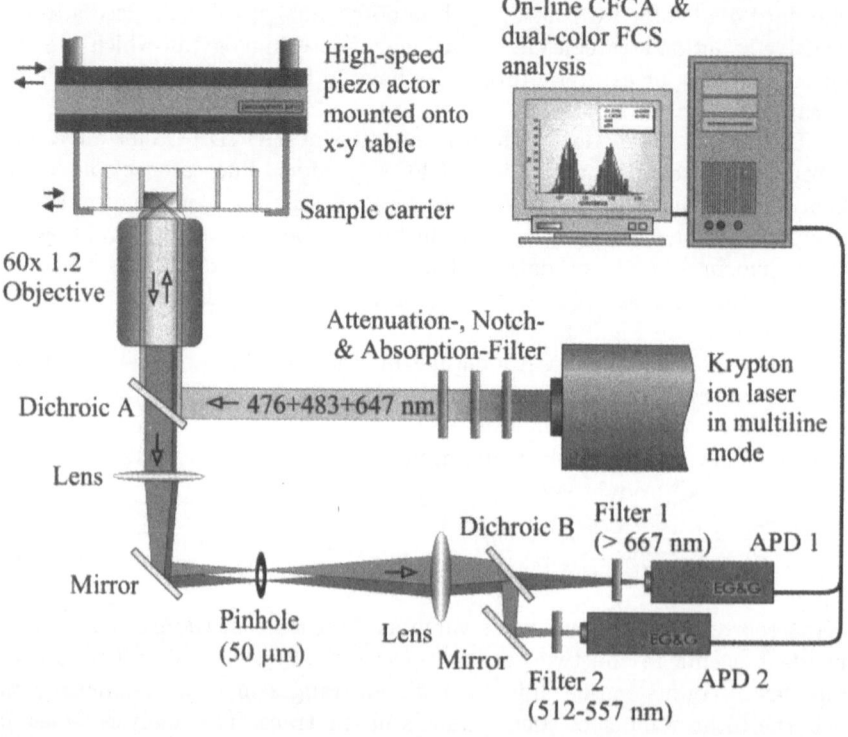

Fig. 9.8. The confocal fluorescence coincidence analysis (CFCA) setup. Fluorescence excitation was achieved by epi-illumination of a water-immersion objective $(60 \times /1.2\,\mathrm{W})$ with radiation of wavelengths $476/483\,\mathrm{nm}$ and $647\,\mathrm{nm}$ from a krypton ion laser operating in multiline mode. The dichroic beamsplitter A reflected at $< 502\,\mathrm{nm}$ and $585\text{--}655\,\mathrm{nm}$ and transmitted at $502\text{--}585$ and $> 655\,\mathrm{nm}$. Additional laser lines at $531\,\mathrm{nm}$ and $568\,\mathrm{nm}$ were blocked with a custom-made excitation/notch filter. Appropriate relative laser excitation power for both wavelengths was obtained by the combination of an absorption glass filter and an attenuation filter (OD 0.6). The sample holder was connected to a high-speed, two-dimensional piezo-actor which in turn was mounted onto a high-precision x-y mechanical positioning table. Fluorescence photons were separated at dichroic beamsplitter B and, after filtering in the red and the green channel, were imaged onto two avalanche photo-diodes (APD). Digital pulses of APD were recorded either by a dual input multiscaler PC card for analysis of time traces at high temporal resolution, or by an on-board processor PC card which was programmed to perform on-line data processing. Alternatively, for comparison with dual-color FCS, the APD pulses were cross-correlated by a PC correlator card

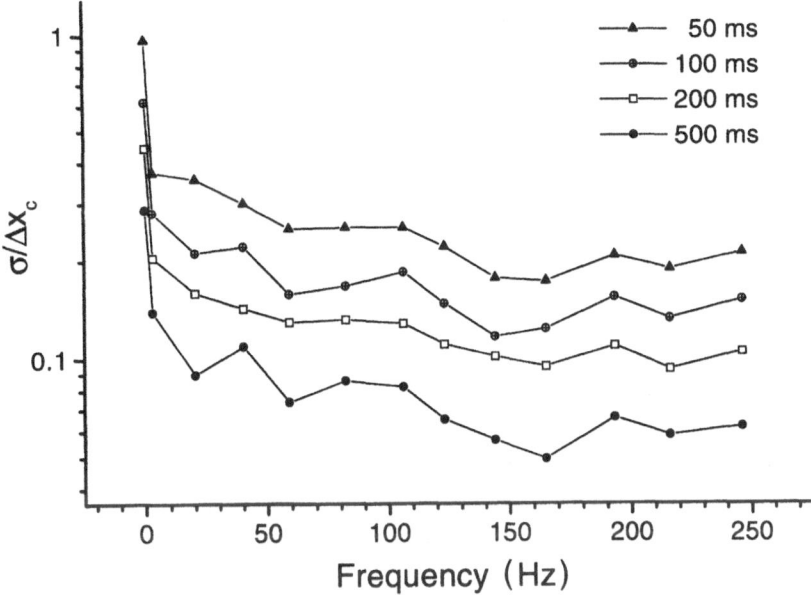

Fig. 9.9. Relative standard deviation $\sigma/\Delta x_c$ of Gaussian curve fits applied to distributions of K versus sample oscillation frequency at analysis times ranging from 50 to 500 ms. The abscissa indicates the frequency in the y-direction which was superimposed by a constant frequency of 3 Hz in the x-direction (except for the oscillation-free case). Coincidence analysis was performed using an on-board processor PC card with a time channel width of 12.5 ms. The curves showed similar courses with a large reduction in relative standard deviation when the oscillation of the sample was started. Towards higher frequencies, a further moderate decrease could be noticed. At shortened analysis times the curves were shifted in parallel towards increasing $\sigma/\Delta x_c$

In order to characterize CFCA, such essential parameters as time channel width, frequency of oscillation, and analysis time per sample have been examined systematically (Fig. 9.9 and Fig. 9.10).

K resulting from the normalized product sum algorithm (see (9.1)) served as a measure of coincidence. For each parameter combination, a set of 300 measurements of a model sample (10 nM double-labeled, double-stranded DNA) was evaluated. To characterize the resulting distributions of K, Gaussian curve fits were applied. The extracted standard deviation s was standardized by a division by the distance Dx_c between the actual center point (x_c) and the center point of an ideal sample without any coincidence $(x_{c,0} = 1.0)$. This relative standard deviation $\sigma/\Delta x_c$ then was considered as a measure of the detection error. Generally, the detection error strongly depended on the oscillation frequency and analysis time. In contrast, changing the time

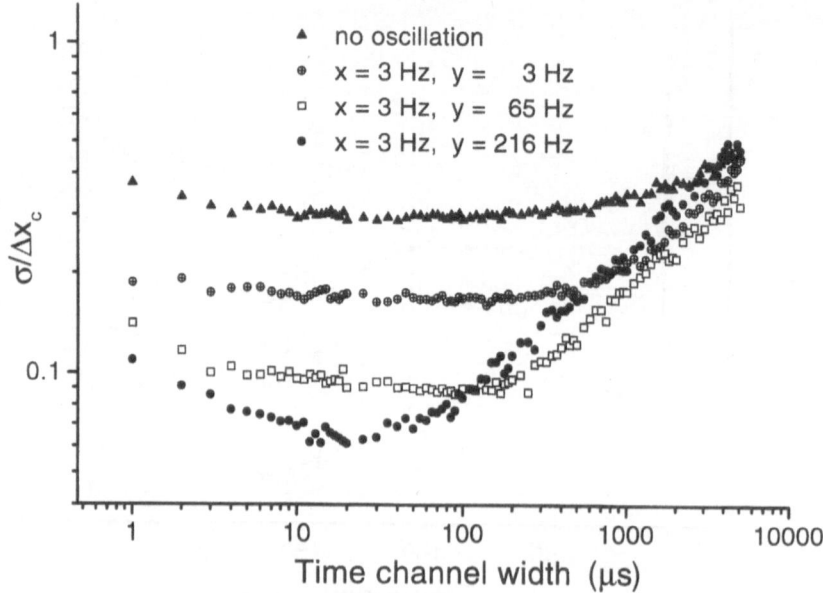

Fig. 9.10. Relative standard deviation $\sigma/\Delta x_c$ of Gaussian curve fits applied to distributions of K as a function of the time channel width at different frequencies of sample oscillations. K was evaluated from multiscaler time traces with analysis times per sample of 500 ms. At every frequency there was a range of minimum $\sigma/\Delta x_c$ in the central part of the covered time channel width. At time channel widths which were of the order of magnitude of the triplet state lifetime ($< 5\,\mu s$) or which were larger than the average residence time t_{res} (which was a function of oscillation frequency), the relative standard deviation increased significantly. The effect of increased frequency was even more pronounced. As the frequency increased, $\sigma \Delta x_c$ diminished drastically. Simultaneously, the plateau of time channel width at minimum relative standard deviation narrowed down to the range of 10–30 μs at 216 Hz

channel width had only a minor influence; only at the highest oscillation frequencies applied, which corresponded to shortest molecular residence times in the observation volume, the range of optimal time channel widths became quite narrow. Particularly noteworthy was the large decrease of the detection error in a sample oscillating at 3 Hz compared to a resting sample. We assume this phenomenon to be mainly caused by the suppression of photobleaching effects.

The properties of CFCA show the great importance of the introduced improvements for fast and reliable yes-or-no decisions in screening applications. The procedure of measuring the number of coincident events in two distinct time traces combined with piezo-driven sample oscillations leads to a considerable reduction of the tolerance of spot checks in investigated biochemical systems. By rapidly assaying a large number of positive and negative samples of uncleaved and cleaved double-stranded DNA substrate with EcoRI (see Fig. 9.2) analysis times of 200 ms with an overlap between substrate and product distributions of less than 0.2% could be achieved. At 100 ms the overlap is less than 2% which still corresponds to a tolerable number of false negative samples. Even at 50 ms, the intersection of the distributions amounted of less than 10%. These results constitute a considerable advance in the effort to shorten the analysis times for screening biomolecular samples. Compared to RAPID FCS, the results presented here lead to a further increase in analysis speed of up to ten times.

9.4 Applications in Evolutionary Biotechnology

Designing biomolecules by evolutionary approaches opens a new area in biotechnology. The idea to transfer natures principles of Darwinian evolution, i.e. variation and selection, to the molecular level and to apply it to the directed evolution of molecules is the underlying concept of *evolutionary biotechnology* [9.13]. Important prerequisites are a comprehensive understanding of the mode of molecular evolution as well as the ability to apply its principles to experimental systems in order to create and optimize molecular functions with scientific or economic value. In recent years, evolutionary approaches turned out to be the most successful techniques for molecular design if one deals with biopolymers of more than a few monomers [9.14,9.15]. Selection of variants with properties more similar to the intended phenotype than the average is one of the most crucial steps. The most common strategy for selection is to couple the desired molecular feature directly to the amplification rate of a replicator unit (in vitro) or the growth rate of a host system (in vivo). Many successful examples are reported in detail, which have shown the power of this nature-like strategy, where the desired molecular feature influences the amplification rate. Unfortunately, this strategy is strongly dependent on the coupling of a molecular feature to the amplification rate and thereby limited to a few exceptional cases.

Several alternative selection strategies have been devised in recent years and technically realized. They can be subdivided into three different categories: (i) natural selection by coupling the molecular feature to amplification, (ii) selection by physical separation of the molecular feature and (iii) selection by screening of the molecular function. Whereas the natural selection and physical separation are strongly limited, selection by screening offers the broadest access to different molecular functions. In principle, every molecular function which can be screened by a corresponding technique can be subjected to evolutionary approaches. The challenge addressed in our laboratory is to set up and integrate RAPID FCS and CFCA in processes for selection by screening to generate novel biomolecules. This strategy using confocal fluorescence spectroscopy was firstly proposed by Eigen and Rigler

200   A. Koltermann et al.

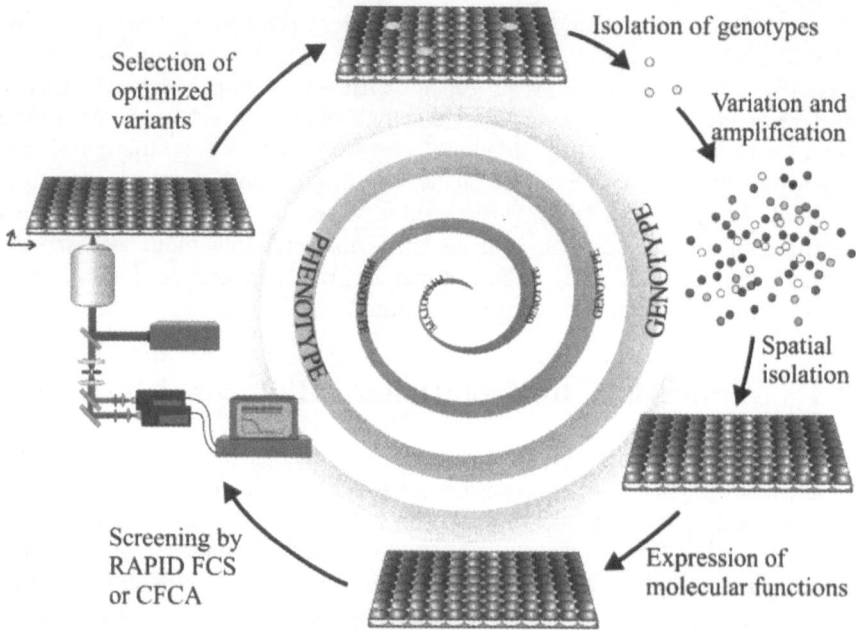

Fig. 9.11. Evolutionary biotechnology is the cyclic interaction of variation, amplification and selection in a defined technical environment. All steps are separated by technical means. Expression of function is only required when phenotype and genotype carriers are separate molecules. A library of genotypes is dispensed into compartments of a sample carrier. After expressing the molecular phenotype, individual variants are screened by RAPID FCS or CFCA. The integration of RAPID FCS or CFCA is needed for the screening of individual molecular variants. Corresponding genotypes are isolated and subjected to the next round of optimization in order to increase the level of intended functions

in the early 1990s [9.6]. The experimental realization is schematically shown in Fig. 9.11.

Besides short analysis time per sample and broad access to different biochemical reactions, which is discussed in the preceding sections, some more major issues must be addressed in the implementation of RAPID FCS and CFCA for selection by screening in evolutionary biotechnology. Additional steps, as pointed out in Fig. 9.11, are spatial isolation of individual clones and expression of molecular functions. Spatial isolation of individual clones can be carried out by simple dilution of a population of different clones into different compartments. Therefore, miniaturized sample carriers were developed. Different designs were realized using volumes ranging from 100 nl to 5 µl per well. Because of the tiny volume element of confocal fluorescence spectroscopy, biochemical reactions can be carried out in these sample carriers without any decrease in signal, but provisions against evaporation and

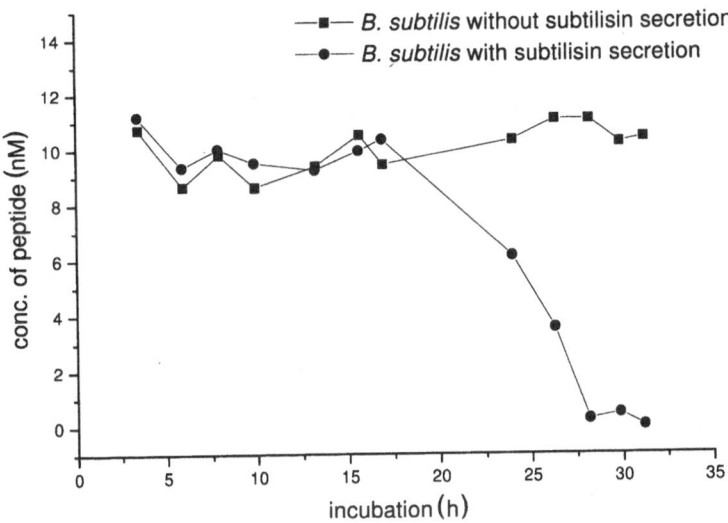

Fig. 9.12. Monitoring the degradation of 10 nM double-labeled peptide in a growing
B. subtilis culture

non-specific adsorption must be taken into account. The standard sample
volume is one μl per sample and below. For dispensing bacterial cultures
or assay buffers in the sub-microliter range, standard piezo-driven pumps or
magnetic valves are successfully applied. The expression of molecular functions is monitored by dual-color confocal fluorescence spectroscopy. In biological environments, even in crude samples where, besides the double-labeled
substrate, bacteria, culture media and metabolic products are present, dual-
color confocal fluorescence spectroscopy shows great sensitivity in monitoring
biochemical reactions, as shown in Fig. 9.12.

The only difference in the experimental setup presented in Fig. 9.12 was
the application of two different clones. One clone carries a plasmid which
encodes the subtilisin E gene for secretion whereas the other clone carries a
reference plasmid lacking the subtilisin E gene. The monitoring of the precise
degradation of low amounts of double-labeled peptide substrate after several
hours shows the enormous sensitivity of this method in biological environments.

9.5   Outlook

The present developments and innovations in the field of dual-color confocal fluorescence spectroscopy show the great potential of this single-molecule
based technology. The broad application to a variety of different biochemical
reactions as well as the short analysis times per sample are the first steps
towards the development of a full-grown detection technology. It is assumed

that analysis times per sample can be reduced by at least one order of magnitude. The intrinsic features of miniaturized dimensions and extreme sensitivity may lead away from the statistic analysis of an ensemble of molecules towards the analysis of a single molecule. Integration of pulsed laser systems for two-photon excitation (see Chap. 14) and time-resolved analysis [9.16] will be the next milestones in impoving dual-color based confocal techniques.

On the other hand, confocal fluorometric techniques have already been successfully applied in life science industries as high-throughput screening tools for drug discovery and design (see Henco et al., Stöckli et al., Moore et al., this issue). Nowadays, dual-color confocal techniques are being, or will be, implemented to high-throughput screening in the near future.

In evolutionary biotechnology, the integration of dual-color confocal techniques such as CFCA for selection by screening of biomolecular functions was recently established with throughput rates of more than 100 000 samples in one day. They have the potential of more than one million samples per day. The application relevance of these techniques can be outlined by the flexibility for precise monitoring different biochemical reactions in short analysis times. Further improvements in different fields such as miniaturization, detection efficiency, parallelization, improved optics, evaluation of collected data and assay developments will push this limit up to several millions of samples per day in the near future. The combination with nanotechnology will further reduce sample volumes and applications to flows in micro- and nanostructures will lead to ultrasensitive analysis tools. Nowadays, the problem is not longer single molecule detection which "easily" can be achieved below nanomolar concentrations. The challange is to search single molecules at dramatically low concentrations of femto- to attomolar (10^{-15} to 10^{-18} M) and to screen for rare events among large numbers of samples.

References

9.1 D. Magde, E.L. Elson, and W.W. Webb: Phys. Rev. Lett. **29**, 705–708 (1972)
9.2 E.L. Elson, D. Magde, and W.W. Webb: Biopolymers **13**, 29–61 (1974)
9.3 M. Ehrenberg and R. Rigler: Chem. Phys. **4**, 390–401 (1974)
9.4 M. Ehrenberg, and R. Rigler: Quart. Rev. Biophys. **9**, 69–81 (1976)
9.5 R. Rigler, U. Mets, J. Widengren, and P. Kask: Eur. Biophys. J. **22**, 169–175 (1993)
9.6 M. Eigen and R. Rigler: Proc. Natl. Acad. Sci. USA **91**, 5740–5747 (1994)
9.7 P. Schwille, F.J. Meyer-Almes, and R. Rigler: Biophys. J. **72**, 1878–1886 (1997)
9.8 P. Schwille: Fluoreszenz-Korrelations-Spektroskopie: Analyse biochemischer Systeme auf Einzelmolekülebene. Dissertation. Technische Universität Braunschweig (1996)
9.9 J. Bieschke and P. Schwille: Fluorescence Microscopy and Fluorescent Probes **2**, 81–86 (1998)
9.10 U. Kettling, A. Koltermann, P. Schwille, and M. Eigen: Proc. Natl. Acad. Sci. USA **95**, 1416–1420 (1998)

9.11 A. Koltermann, U. Kettling, J. Bieschke, T. Winkler, and M. Eigen: Proc. Natl. Acad. Sci. USA **95**, 1421–1426 (1998)
9.12 T. Winkler, U. Kettling, A. Koltermann, and M. Eigen: Proc. Natl. Acad. Sci. USA **96**, 1375–1378 (1999)
9.13 M. Eigen and W. Gardiner: Pure. Appl. Chem. **56**, 967–978 (1984)
9.14 A. Koltermann and U. Kettling: Biophys. Chem. **66**, 159–177 (1997)
9.15 U. Kettling, A. Koltermann, and M. Eigen: Current Topics in Microbiology & Immunology **243**, 173–185 (1999)
9.16 C. Eggeling, J.R. Fries, L. Brand, R. Günther, and C.A.M. Seidel: Proc. Natl. Acad. Sci. USA **95**, 1556–1561 (1998)

10 Nanoparticle Immunoassays: A new Method for Use in Molecular Diagnostics and High Throughput Pharmaceutical Screening based on Fluorescence Correlation Spectroscopy

F.J. Meyer-Almes

10.1 Introduction

Fluorescence correlation spectroscopy (FCS) has been developed into a versatile, specific and sensitive method applicable to the life sciences. DNA-DNA, protein-DNA, ligand-receptor, dye-DNA, and protein-protein interactions, among others, have been successfully analyzed using FCS [10.1–10.11]. Recently, special applications in the field of high throughput pharmaceutical screening have been described [10.12,10.13].

In order to obtain a rapid and precise measurement of the interaction between two molecules using FCS, there must be a substantial difference between the translational diffusion constant of a free molecule and that of the same molecule when it is part of a molecular complex. To achieve this change, typical FCS assays require that the smaller of the two interacting molecules be tagged with a fluorescent dye. If only the larger binding partner can be labeled with a fluorophore, the binding event can not be measured using FCS since the resulting change in translational diffusion upon complex formation will not be sufficient to be detected. Nanoparticle binding assays eliminate this limitation of conventional FCS.

This work describes several types of nanoparticle binding assays, in particular nanoparticle immunoassays (NPIAs) which feature the use of antibodies (see Fig. 10.1). Nanoparticles bearing capture antibodies, protein A, or streptavidin, can be used to perform a variety of assays which capitalize on the advantages of using FCS to study biological interactions. Using this strategy it is possible to use FCS to evaluate the binding of a large fluorescent antibody to a much smaller antigen, thereby eliminating the requirement of conventional FCS that the smaller molecule bear the fluorescent tag, and capitalizing on the broad availability of fluorescent antibodies. NPIAs allow for the design of FCS experiments without regard to the mass differences between antigen and antibody. The nanoparticles which are used in NPIAs must exhibit several features. They must behave as true diffusing molecules in order to satisfy the fit model which is used for evaluation. This condition is fulfilled if the diameter is smaller than 300 nm. The nanoparticles must be

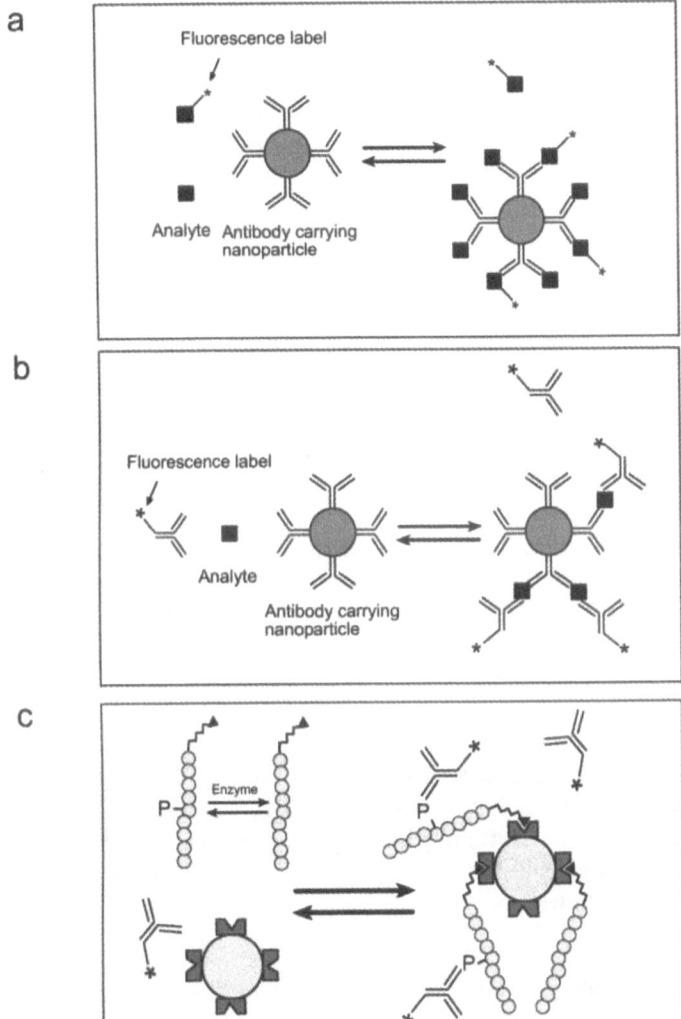

Fig. 10.1 a–c. Schemes of different types of nanoparticle immuno assays (NPIAs): **a** Competitive NPIA: With increasing analyte concentration the binding degree of the analyte conjugate decreases and, concurrently, the particle number per observation volume increases. **b** Sandwich NPIA: With increasing analyte concentration the binding degree of fluorescent reporter antibody rises and the number of particles inside the focus decreases simultaneously. **c** Enzyme activity NPIA: The example of a phosphatase (e.g. CD45) assay is chosen. A phosphorylated (P) and biotinylated (**filled triangle**) peptide is dephosphorylated by a phosphatase. Following the enzyme reaction, streptavidin (■)-coated nanoparticles and fluorescent antiphosphotyrosine antibody which binds exclusively the phosphorylated peptide are added. At high enzyme activity no antibody molecule is bound to the nanoparticles. If the enzyme activity is low or suppressed by an inhibitor, the antibodies bind to the nanoparticles and thus the binding degree increases while the number of particles in the focus is declining

stable as colloids in solution and show no tendency to aggregate. Furthermore, non-specific binding of molecules, in particular the fluorescent ligand, to the particle surface must not occur.

In this work, the theoretical background for FCS-based nanoparticle immunoassays is developed and applications in the fields of diagnostics and pharmaceutical screening are described. In conclusion, the impact of NPIAs on both fields is discussed.

10.2   Theory

NPIAs are characterized by the binding of many ligands to a particle which carries many receptor molecules. In the following sections, the theoretical basis for the FCS analysis of such systems is described.

10.2.1   Competitive NPIA

In the case of a competition NPIA, the system of chemical reactions is described by

$$
\begin{aligned}
L + R &\overset{K_1}{\rightleftharpoons} RL \\
C + R &\overset{K_2}{\rightleftharpoons} RC,
\end{aligned}
\tag{10.1}
$$

with C representing the non-fluorescent ligand which competes with the fluorescent ligand L for binding to the receptor R. RL and RC are the complexes of L and R, and C and R, respectively, with K_1 and K_2 being the corresponding dissociation constants which describe the binding affinity. The concentrations of the molecular species involved in the chemical equilibrium can be calculated as follows:

$$
\begin{aligned}
[L]_{eq} &= [L]_0 - [RL]_{eq} \\
[C]_{eq} &= [C]_0 - [RC]_{eq} \\
[RL]_{eq} &= \frac{[L]_{eq}[R]_0 K_1}{1 + K_1[L]_{eq} + K_2[C]_{eq}} \\
[RC]_{eq} &= \frac{[C]_{eq}[R]_0 K_2}{1 + K_1[L]_{eq} + K_2[C]_{eq}}.
\end{aligned}
\tag{10.2}
$$

The equations can be solved numerically using such computer programs as Mathcad (Mathsoft), Mathematica (Wolfram Research) or Scientist (Micro-Math).

The probability p that a receptor molecule is occupied by ligand L is simply given by:

$$
p = \frac{[RL]_{eq}}{[R]_0}
\tag{10.3}
$$

10.2.2   Sandwich NPIA

In a sandwich NPIA, a ternary complex is formed. This complex consists of the fluorescently labeled reporter molecule A, the analyte H and the receptor molecule R which is linked to the nanoparticles. The reaction system is described by:

$$A + H \overset{K_\alpha}{\Leftrightarrow} AH$$
$$A + R \overset{K_\beta}{\Leftrightarrow} HR \qquad\qquad (10.4)$$
$$AH + R \overset{K_\beta}{\Leftrightarrow} AHR.$$

The equilibrium concentrations of all species involved can be obtained from the numerical solution to the following equations:

$$K_\alpha = \frac{[A]_{eq}[H]_{eq}}{[AH]_{eq}}$$

$$K_\beta = \frac{[H]_{eq}[R]_{eq}}{[HR]_{eq}}$$

$$K_\beta = \frac{[AH]_{eq}[R]_{eq}}{[AHR]_{eq}} \qquad\qquad (10.5)$$

$$[A]_{eq} = [A]_0 - [AH]_{eq} - [AHR]_{eq}$$

$$[H]_{eq} = [H]_0 - [AH]_{eq} - [AHR]_{eq} - [HR]_{eq}$$

$$[R]_{eq} = [R]_0 - [AHR]_{eq} - [HR]_{eq}.$$

K_α and K_β are the dissociation constants of the above-stated system of chemical equations (10.4). A is the fluorescent-detection molecule, H is the analyte and R the receptor molecule which is located onto the surface of the nanoparticle. The analyte has two independent binding sites, one which binds the reporter molecule, and the other which binds the receptor molecule. Furthermore, the formation of the ternary complex AHR must not be prevented or influenced by the preliminary formation of binary complexes AH or HR. In the proposed model it is assumed that the affinities of H to R and AH to R are the same. The equilibrium concentrations are noted by subscript eq and the total species concentration by subscript 0.

The probability p that a receptor molecule R is occupied by the binary complex AH, thus forming the ternary complex AHR, is given by:

$$p = \frac{[AHR]_{eq}}{[R]_0}. \qquad\qquad (10.6)$$

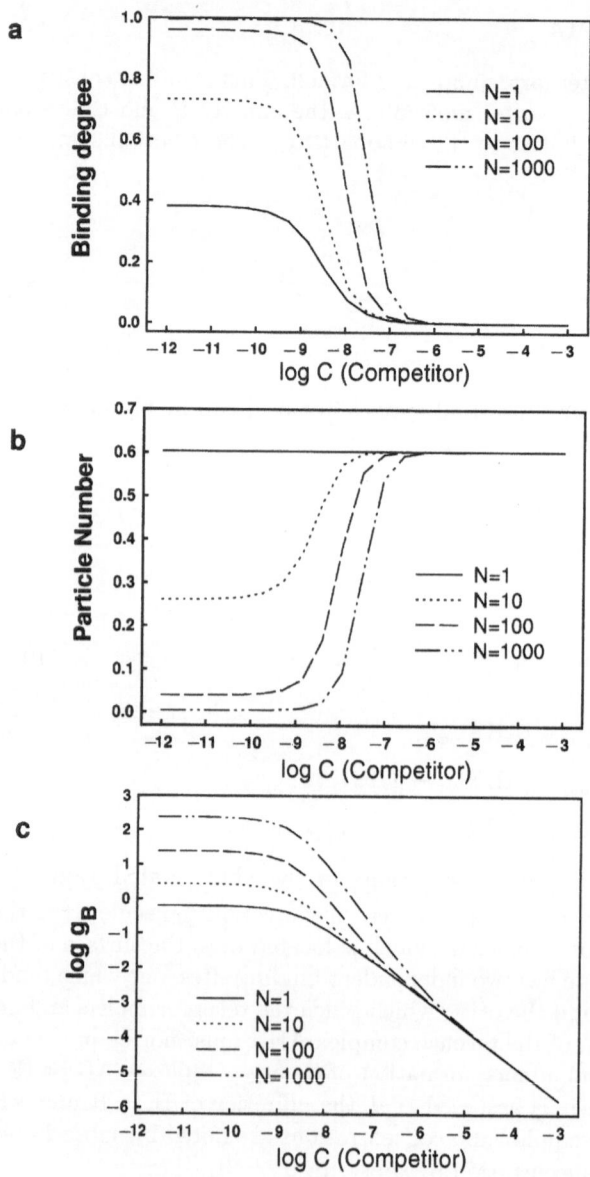

Fig. 10.2 a–c. Simulated competition curves of **a** binding degree, **b** mean particle number per observation volume, **c** correlation amplitude, g_B, of bound ligand were generated using (10.2) and (10.12) with parameters $K_1 = K_2 = 1\,\text{nM}$, $[L]_0 = [R]_0 = 1\,\text{nM}$ and different numbers N of receptor molecules per nanoparticle. The corresponding IC_{50} values are summarized in Table 10.1

Table 10.1. Calculated maximal signal differences and IC_{50} values of competitive NPIAs at different combinations of total ligand, L, and receptor, R, concentrations and different numbers, N, of receptor molecules per particle with respect to read-out parameters binding degree, mean particle number per observation volume and correlation amplitude, g_B, of ligands bound to particles.

$[L]_0/nM$	$[R]_0/nM$	N	Binding Degree		Particle Number		g_B	
			Δ_{max}	IC_{50}/nM	Δ_{max}	IC_{50}/nM	Δ_{max}	IC_{50}/nM
1	1	1	0.38	3	0	-	0.63	3
		10	0.74	4	0.33	2.5	2.8	2
		100	0.97	10	0.56	8	25	1.5
		1000	0.99	35	0.6	30	250	1.5
1	10	1	0.90	20	0	-	1.5	20
		10	0.95	30	0.25	20	2.8	20
		100	0.99	50	0.53	40	15	15
		1000	1.0	100	0.59	100	135	15
10	10	1	0.73	17	0	-	0.1	17
		10	0.96	35	5	30	0.9	8
		100	0.99	100	5.9	90	9	8

10.2.3 Autocorrelation Amplitudes

Each particle carries N receptor molecules. The propability that a particle binds n ligand molecules L is therefore:

$$p(n) = \frac{N!(1-p)^{N-n}p^n}{(N-n)!n!} . \tag{10.7}$$

Since $p(n)$ is a binominal distribution, the mean number and the mean of the square of the number of ligands bound per particle are:

$$\overline{n} = \sum_n np(n) = pN \tag{10.8}$$

$$\overline{n^2} = \sum_n n^2 p(n) = pN(1 - p + Np) . \tag{10.9}$$

The autocorrelation amplitude g_i of a diffusing species i is given by:

$$g_1 = \frac{c_i q_i^2}{\left(\sum_i c_i q_1\right)^2} = \frac{c_i q_i^2}{I^2} . \tag{10.10}$$

c_i and q_i denote the concentration and the relative fluorescence intensity, respectively, detected per molecule of species i. For simplicity q_L is chosen to be 1 and there is no change in fluorescence intensity per molecule when the ligand L binds to its receptor molecule. Then the fluorescence intensity of a particle is proportional to the number of bound ligands. This applies for the majority of fluorescent dyes used for FCS.

The amplitude of the unbound ligand L is then simply:

$$g_L = \frac{[L]}{I^2}.$$

(10.11)

The amplitude of the ligands bound to particles g_B can be calculated accordingly using (10.9) and (10.10):

$$\begin{aligned} g_B &= \frac{1}{I^2} \sum_n [P]_0 p(n) n^2 \\ &= \frac{1}{I^2} [P]_0 N p (1 - p + Np), \end{aligned}$$

(10.12)

with $[P]_0$ being the concentration of nanoparticles.

The apparent degree of binding determined using FCS is defined as $Y = g_B/(g_L + g_B)$ and the mean number of particles per observation volume is defined as $1/(g_L + g_B)$. Consequently, g_B can be deduced from a conventional FCS experiment by dividing the calculated degree of binding Y by the corresponding mean number of particles within the focus.

10.3   Material and Methods

10.3.1   Substances

The 17-β-estradiol conjugates and TAMRA-theophylline (for chemical structures see Table 10.2) were synthesized by EVOTEC BioSystems AG (Hamburg). The anti-estradiol-antibody was kindly provided by Dr. Schumacher (Institute of Fertility Research, Hamburg, Germany). The anti-theophylline antibody was purchased from Europa Research Products (Cambridge, UK). 17-β-estradiol and theophylline were obtained from Sigma. The protein-A-coated nanoparticles used in the experiments were from Immunicon Corporation (USA) and MicroCaps (Rostock, Germany). Monoclonal mouse anti-α-hCG antibody (Biotrend) was coupled to Alexa 568 (Molecular Probes) according to manufacturers protocol. From the ratio of the absorption at 280 and 577 nm a dye loading of 4.7 Alexa molecules per antibody was estimated. Another monoclonal mouse anti-β-hCG antibody (Chemicon) was biotinylated using biotin-LC-sulfo-NHS-ester (Pierce) according to standard protocols. Highly purified hCG (Chemicon) was used as analyte. TAMRA-P1 peptide (TAMRA-aminohexanoyl-EQEDEPEGIpYGVLF-NH$_2$) was synthesized at EVOTEC BioSystems AG, TRITC-antiphosphotyrosine antibody and the CD45 tyrosine phosphatase were obtained from Sigma and Biomol Research Laboratories, respectively.

10.3.2   Equipment

ConfoCor® (Carl Zeiss, Jena/EVOTEC BioSystems AG, Hamburg) equipped with standard filter sets for excitation with a helium-neon laser (either 543 nm or 632 nm line, 40 μW, Uniphase), C-APOCHROMAT 40×/1.2 objective (Carl Zeiss Jena), pinhole 50 μm. The photon count signal was autocorrelated over 20 (theophylline assay), 60 s (estradiol and CD45 assay), or 5 min (hCG assay), respectively. Each data point represents the mean of 10 subsequent measurements.

Data evaluation: Evaluation of the autocorrelation curves was carried out with a Marquardt nonlinear least-square fitting routine using the following two-component model corresponding to free and bound conjugate [10.3]:

$$G(\tau) = \left[1 - T + T \exp\left(\frac{-\tau}{\tau_T}\right)\right] N^{-1}$$

$$\left[\frac{1 - Y}{\left(1 + \frac{\tau}{\tau_{\text{free}}}\right)\sqrt{1 + \frac{r_0^2}{z_0^2}\frac{\tau}{\tau_{\text{free}}}}} + \frac{Y}{\left(1 + \frac{\tau}{\tau_{\text{bound}}}\right)\sqrt{1 + \frac{r_0^2}{z_0^2}\frac{\tau}{\tau_{\text{bound}}}}}\right] \quad (10.13)$$

where T is the average fraction of dye molecules in the triplet state with relaxation time τ; N is the total average number of fluorescent molecules in the observation volume; Y is the relative concentration fraction of bound conjugate; τ_{free} and τ_{bound} define the average diffusion times for free and bound conjugate molecules through the observation volume. The parameters r_0 and z_0 are the lateral and axial distances between the coordinate where the Gaussian emission light distribution reaches its maximum value and the point where the light intensity decreases to $1/e^2$ of the maximum value (observation volume).

10.3.3   Reactions

The concentration of the conjugates (TAMRA-estradiol and TAMRA-theophylline), the corresponding antibody and the nanoparticles were kept constant in a specific series of experiments. The competitor (17-β-estradiol, theophylline) concentration varied from 0.1 to 5000 nM. All incubations and measurements were carried out at room temperature (22 ± 2 °C).

The competitive NPIAs were carried out as follows: The TAMRA conjugates of estradiol and theophylline, respectively, were mixed with the corresponding unlabeled molecules. The mixture was then incubated with the corresponding unlabeled antibody at a final dilution of 1:10 000 at 37 °C for 15 min. In the final step a colloid suspension of nanoparticles in PBS was added and the whole mixture was incubated at 37 °C for an additional 15 min. The assay buffer was PBS (140 mM NaCl, 3 mM KCl, 8 mM Na$_2$HPO$_4$,

1.5 mM KH_2PO_4, pH 7.4, 0.05% Tween 20). Some experiments with JA190-estradiol conjugates were carried out in E-NAP buffer (200 mM sodium acetate, 150 mM NaCl, 0.2% BSA, pH 4). It was shown independently that antibody and protein-A-coated nanoparticles could be premixed yielding similar results.

The assay protocol of the hCG sandwich NPIA was as follows: Buffer, Alexa 568 coated anti-α-hCG antibody, and hCG antigen were incubated at 37 °C for 15 min. Subsequently, biotinylated anti-β-hCG antibody and nanoparticles were added, incubating for 15 min after each addition. This protocol can be simplified by simply premixing Alexa-anti-α-hCG, biotinylated anti-β-hCG antibodies and nanoparticles for 15 min at 37 °C and adding unknown analyte to the mixture. The ternary binding reaction was determined to be completed after less than 5 min (data not shown). In the first step of the CD45-NPIA, the CD45 phosphatase enzyme is allowed to dephosphorylate the N-terminal-biotinylated, phosphorylated peptide TAMRA-P1, which serves as the substrate. After 5 min at 25 °C the enzyme was deactivated and a mixture of streptavidin-coated nanoparticles and TRITC-anti-phosphotyrosine antibody was added to quantify the amount of intact educt.

10.3.4 Simulations and Data Fitting

The Feldman equations (10.2), which describe the competition of conjugate and corresponding unlabeled molecules for binding to a common receptor molecule, were solved numerically using the programs MathCad or Scientist. The experimental competition curves were simultaneously fit to (10.2) and (10.12) with the parameters K_1, K_2, which are the K_D-values for the conjugate and the competitor, respectively, number of receptor molecules per nanoparticle, N, the concentration of nanoparticles, $[P]_0$, and conjugate, $[L]_0$, using the Levenberg–Marquardt fitting routine implemented in the program Scientist. The data of the hCG sandwich NPIA were fit to (10.5) and (10.12) with parameters K_α and K_β, N, $[P]_0$ and $[L]_0$, using the same fit algorithm.

10.4 Results

10.4.1 Simulations

Binding curves for competitive NPIAs were simulated in order to determine which assay parameters are important for sensitivity in terms of IC_{50}. The IC_{50} is defined as being the concentration of the unlabeled ligand C which inhibits 50 percent of the binding of the labeled ligand L to particles. In all simulations of FCS correlation amplitudes, the concentration of labeled and unlabeled ligand in mol/l was multiplied by 6.023×10^8 in order to approximate the number of molecules occupying the FCS observation volume of 10^{-15} liter in a FCS measurement. The K_1 and K_2 values were fixed at

1 nM. The parameters $[L]_0$, $[R]_0$, $[C]_0$ and N were varied and are noted for each simulation.

The simulations show that the degree of binding measured using FCS will decrease with increasing competitor concentration, as expected. An interesting phenomenon noted in the simulations is that the higher the number of receptor molecules per particle, the higher the degree of binding measured with FCS despite the fact that the ratio of free to bound ligand does not change (see Fig. 10.2). This is due to differences in weighting the brightness of free ligand molecules and particles with several ligands bound to them in the FCS analysis algorithm. Put another way, this means that the degree of binding is overestimated when more than one fluorescent ligand is bound per particle. This means that the binding signal could be enhanced by using particles which carry many receptor molecules.

When the stoichiometry of ligand to receptor molecules is 1:1, the absolute number of particles per observation volume does not change upon binding. Using nanoparticles which carry more than one receptor molecule per particle, FCS enables the precise detection of the resulting decrease in the absolute number of particles per observation volume upon binding (see Fig. 10.2, Table 10.1). At the same time, the fluorescence emission per nanoparticle increases. With increasing competitor concentration, the correlation amplitude of the bound ligand fraction, g_B, vanishes since it is the quotient of binding degree and particle number per observation volume. The greater the number of receptor molecules attached per particle, the greater the maximum value of the fractional correlation amplitude. When only binding degree or the number of particles per observation volume are considered, assay sensitivity in terms of IC_{50}-values declines as the number of receptor molecules per nanoparticle increases (see Table 10.1). However, evaluation of the correlation amplitude for particle bound complexes g_B should yield a sensitivity which is at least as high as that for simple binary binding ($N = 1$).

10.4.2   Experiments

A series of binding experiments using TAMRA-estradiol and TAMRA-theophylline were made in order to verify the theoretical expectations. The binding of small fluorescent conjugates to nanoparticles should result in increased diffusion times, greater fluorescence emission per particle and a reduced particle concentration when evaluated using conventional FCS. In particular, the prediction that the IC_{50}-value will be smaller when evaluating the results on the basis of the correlation amplitude of the bound conjugate, g_B, instead of the binding degree, Y, is tested.

Theophylline-Assay: Theophylline is a drug used in the treatment of asthma and apnoea in premature infants [10.14]. Its serum concentration must be closely monitored because even at the therapeutic dosage, some toxic effects can occur.

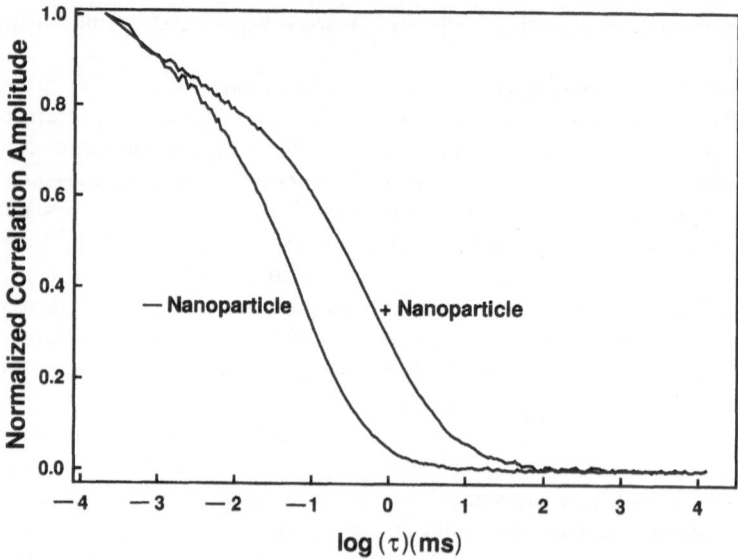

Fig. 10.3. Normalized autocorrelation curves of the fluorescence emission of TAMRA-theophylline and its complex with anti-theophylline antibody and nanoparticles coated with protein A. The corresponding diffusion times are 70 and 860 µs, respectively

In the theophylline assay presented here nanoparticles coated with protein A were used (MicroCaps, Rostock). These particles varied only slightly from one another in size, with a mean diameter of about 90 nm. The small diameter as well as the narrow size distribution account for the high reproducibility of the FCS results (Fig. 10.3). The assay principle is shown in Fig. 10.1a. The diffusion times for unbound TAMRA-theophylline and the complex consisting of TAMRA-theophylline, anti-theophylline and nanoparticles coated by protein A were 70 and 860 µs, respectively. The amount of TAMRA-theophylline used resulted in 3.4 particles per observation volume, corresponding to a concentration of 22.6 nM given a focus size of 0.25×10^{-15} l.

The IC_{50} value calculated from a plot of binding degree, Y, versus theophylline concentration was about 140 nM (Fig. 10.4). When the correlation amplitude, g_B, of the TAMRA-theophylline complex was plotted against theophylline concentration, the IC_{50} value is shifted slightly lower towards concentrations of approximately 65 nM.

Both sets of competition data, of Y as well as g_B, were fit to (10.2) and (10.12) using the Levenberg–Marquard algorithm. For fitting purposes, the concentration of TAMRA-theophylline was fixed at 22.6 nM. The data fit very well to the proposed model.

The mean number of anti-theophylline antibodies immobilized onto the surface of a nanoparticle was calculated to be 1.57. This number is rather

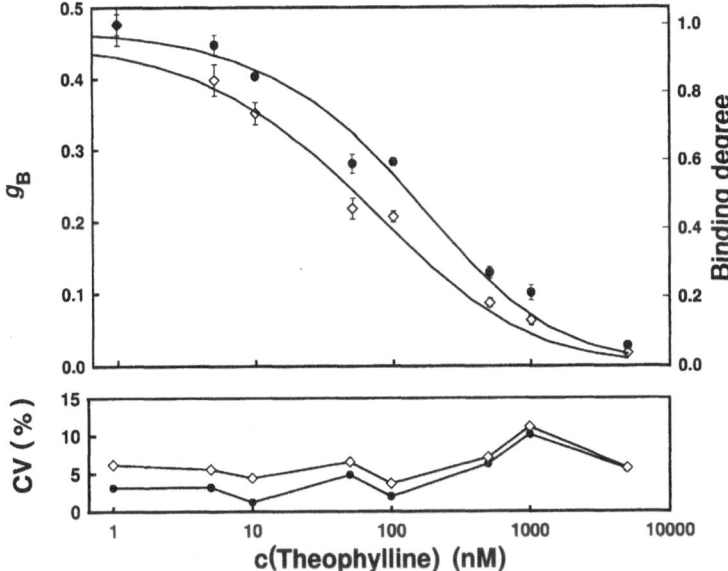

Fig. 10.4. *Top panel*: the correlation amplitude g_B ($\diamond$) and the binding degree ($\bullet$) of a competitive NPIA consisting of TAMRA-theophylline, anti-theophylline antibody, protein A coated nanoparticles ($\varnothing = 80\,\text{nm}$) and different concentrations of competing unlabeled theophylline. The **solid lines** are a simultaneous fit of both data sets to (10.2) and (10.12) yielding the parameters: $K_1 = 3.1 \times 10^{-13}\,\text{M}$, $K_2 = 2.5 \times 10^{-12}\,\text{M}$, $[P]_0 = 14\,\text{nM}$, $N = 1.57$ and $[L]0 = 22.6\,\text{nM}$. *Bottom panel*: the CV values of the corresponding data points from the *top panel* are plotted

small, but agrees with the small change in particle number from 3.4 to 2.1 measured in presence of 1000 and 1 nM theophylline, respectively. The calculated concentration of nanoparticles is 14.2 nM and the K_D values for TAMRA-theophylline and the competitor theophylline are 3×10^{-13} and $2.5 \times 10^{-12}\,\text{M}$, respectively.

These binding affinities appear to be somewhat high compared with typical antigen-antibody K_D values of 10^{-12} to $10^{-9}\,\text{M}$. The fit is strongly dependent on the parameters conjugate concentration, $[L]_0$, number of receptors per particle, N, and particle concentration, $[P]_0$. For example, if the K_1 value is fixed at $10^{-11}\,\text{M}$, the resulting fit is of similar quality to that made by varying K_1, and the resulting K_2 is $8.3 \times 10^{-11}\,\text{M}$. The parameters $[P]_0$ and N did not change significantly compared to the fit made with variable K_1, and were calculated to be 14.1 nM and 1.59, respectively. Taken together, the data agreed with the proposed model and the prediction of greater assay sensitivity when using the correlation amplitude g_B could be proven.

17-β-Estradiol-Assay: Accurate measurement of the circulating level of estradiol is important for assessing ovarian function and monitoring follicular development for assisted reproduction protocols. Estradiol plays an essential role throughout the human menstrual cycle [10.15]. Since it is suspected that coupling a fluorophore to a molecule as small as estradiol could have an impact on its binding specificity and affinity to its antibody, a set of fluorescent estradiol conjugates were synthesized and studied. The results are summarized in Table 10.2. All estradiol conjugates, except for the BODIPY™ derivative, bound to an anti-estradiol antibody. Binding of the BODIPY™ derivative could not be evaluated because of the occurence of severe aggregation, most probably due to the hydrophobicity of the dye. Of the remaining conjugates, only Cy5-estradiol exhibited a significant decrease in affinity for the anti-estradiol antibody. All other conjugate-antibody interactions remained strong. The IC_{50} values for the bound conjugates were in the nanomolar range, except for the Cy5 conjugate, which had an IC_{50} of $> 20\,\mu M$ due to the low affinity of the anti-estradiol antibody to the Cy5 conjugate. TAMRA-estradiol was used for the estradiol NPIA.

In contrast to the theophylline assay, protein-A-coated nanoparticles (Immunicon Corporation) with a mean diameter of 160 nm and a relative broad distribution of diameter were used. This resulted in markedly worse FCS results even though measurement time was extended to 60 s instead of 20 s. On the other hand, the antibody binding capacity of the Immunicon nanoparticles was much higher than those from MicroCaps.

Theoretically, the change in particle number should be more pronounced the greater the number of antibodies per nanoparticle. Furthermore, it is expected that the assay will become much more sensitive when g_B is used as the readout parameter instead of the degree of binding. The mean particle number per FCS observation volume decreases from 1.7 to 0.17 upon 100% binding. The IC_{50} values obtained when using the correlation amplitude and the binding degree are 3 and 20 nM, respectively (see Fig. 10.6). The transition range of the FCS-based binding curve is shifted to higher 17-β-estradiol concentrations with respect to the transition range of the autocorrelation amplitude of the nanoparticle bound complex the more receptor molecules are immobilized per nanoparticle. Consequently, it is possible to extend the dynamic range of a NPIA compared to conventional immunoassays by examining both FCS binding degree and autocorrelation amplitude data. The experimental data could be fit well to the theoretical model yielding the parameters $K_1 = 1.4 \times 10^{-12}\,M$, $K_2 = 5.9 \times 10^{-13}\,M$, $[P]_0 = 0.17\,nM$, $N = 34$ and a concentration of TAMRA-estradiol of 15 nM (see Fig. 10.5). The difference in the IC_{50} values obtained from competition experiments when using correlation amplitude, g_B, versus the binding degree is greater than with the theophylline assay. This is in agreement with the model predicting this difference to be greater the more receptor molecules are presented per nanoparticle.

Table 10.2. Chemical structures, half-saturation points and IC_{50} values for complexes consisting of estradiol conjugates/TAMRA-theophylline and the corresponding antibodies. The half-saturation point is defined as the dilution of antibody solution which binds half of the total estradiol or theophylline conjugate. The competitors were unlabeled, estradiol and theophylline, respectively. The concentrations of all investigated conjugates were in the range of 1–5 nM.

Conjugate	Chemical structure	Half-saturation point/antibody dilution	IC_{50}^7 / nM
TAMRA-diaminodioxaoctan-**estradiol**		< 1:20000 [1,3]	20 [3,5]
Cy5-diaminodioxaoctan-**estradiol**		ca. 1:60 [1,4]	> 20000 [4,5]
Resorufin-diamino-dioxaoctan-**estradiol**		1:3500 [1,3]	10 [3,5]
BODIPY630-diaminodioxaoctan-**estradiol**		No binding [1,3]	n.d. aggregation
JA190-diaminodioxa-octan-**estradiol**		1:10000 [1,4]	1 [4,5]
JA190-diaminoethan-**estradiol**		1:10000 [1,4]	1 [4,5]
JA190-(diaminodioxa-octan)$_2$-**estradiol**		1:10000 [1,4]	2 [4,5]
TAMRA-**theophyllin**		1:50000 [2,3]	100 [3,6]

[1] Anti-estradiol antibody, [2] Anti-theophylline antibody, [3] PBS-buffer, [4] E-NAP-buffer [5] Competitor 17-β-estradiol, [6] Competitor theophylline, [7] On the basis of binding degree as determined using FCS

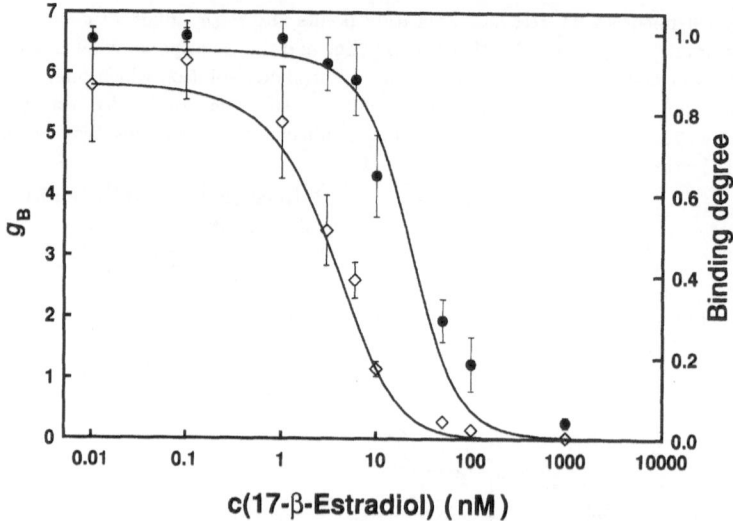

Fig. 10.5. Correlation amplitude, $g_B(\diamond)$, and degree of binding, ($\bullet$), of a competitive NPIA consisting of TAMRA-estradiol, anti-estradiol antibody and protein-A-coated nanoparticles ($\varnothing = 160\,\text{nm}$) in presence of different concentrations of competing unlabeled estradiol. The **solid lines** are a simultaneous fit of both data sets to (10.2) and (10.12) yielding the parameters: $K_1 = 1.4\times10^{-12}\,\text{M}$, $K_2 = 5.9\times10^{-13}\,\text{M}$, $[P]_0 = 0.17\,\text{nM}$, $N = 34$ and $[L]_0 = 15\,\text{nM}$

Human Chorionic Gonadotropin Assay: Human chorionic gonadotropin (hCG) is a glycoprotein with two non-covalently bound subunits. The α-subunit is similar to those of luteinizing hormone (LH), follicle-stimulating hormone (FSH), and thyroid-stimulating hormone (TSH) [10.16,10.17]. The β-subunit of hCG differs from other pituitary glycoprotein hormones, which results in its unique biochemical and immunological properties. HCG is synthesized by the cells of the placenta and is involved in maintaining the corpus luteum during pregnancy. It is detected as early as one week after conception.

In pregnancy, the levels of hCG increase exponentially for about 8 to 10 weeks after the last menstrual cycle. Later in pregnancy, about 12 weeks after conception, the concentration of hCG begins to fall as the placenta begins to produce steroid hormones. Other sources of elevated hCG values are ectopic pregnancy, threatened abortion, micro-abortion and recent termination of pregnancy, or chorionic-, embryonal- or terato-carcinoma.

HCG is a large molecule and thus is well-suited for a sandwich NPIA. For such an assay, a pair of monoclonal mouse antibodies were used one directed against the α-subunit and the other against the β-subunit of hCG. Since the α-subunit of hCG is similar to those of LH, FSH and TSH, this sandwich assay can easily be modified to detect these analytes by simply labeling the corresponding anti-β-hCG-antibody and attaching the anti-α-hCG-antibody to the nanoparticles.

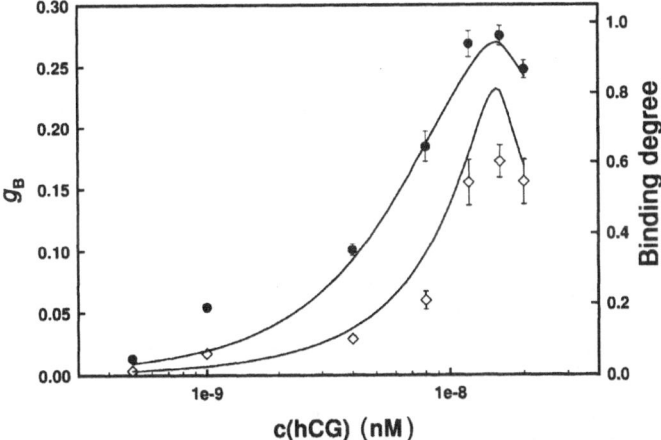

Fig. 10.6. Correlation amplitude, g_B (◇), and degree of binding, (●), of a sandwich NPIA consisting of Alexa-568 labeled anti-α-hCG antibody, anti-β-hCG bearing nanoparticles in presence of different hCG concentrations. The **solid lines** are a simultaneous fit of both data sets to the sandwich assay model (10.5) and (10.12). The determined parameters are: $K_\alpha = 3.4 \times 10^{-11}$ M, $K_\beta = 4.1 \times 10^{-12}$ M, $[P]_0 = 5.4$ nM, $N = 2.6$ and $[L]_0 = 16$ nM

For the hCG-sandwich-assay streptavidine-coated nanoparticles from Miltenyi Biotec with a mean diameter of about 150 nm were used. These nanoparticles showed a very broad size distribution comparable to the Immunicon Prot A nanoparticles. Like these particles, the binding capacity was much greater than that of the nanoparticles from MicroCaps. In contrast to competition experiments the binding degree of nanoparticle bound complexes rose with increasing hCG concentration because the reporter antibody (Alexa-anti-α-hCG) is bound to the nanoparticles via hCG antigen (Fig. 10.1b). However, above a concentration threshold where all nanoparticles are saturated by hCG, increasing concentrations of hCG which is not bound to nanoparticles capture more and more reporter antibodies thus decreasing the percentage of nanoparticle bound complexes. For a diagnostic hormone assay this threshold must be shifted to concentrations which are higher than any realistic hCG concentration. In order to apply the model for the sandwich NPIA (10.5) it is assumed that all biotinylated anti-β-hCG antibody molecules were attached to the nanoparticles. The data could be fit to the model although the quality of the data is not optimal (Fig. 10.6). With increasing hCG concentrations the degree of binding reaches a maximum and decreases again when the binding capacity of the nanoparticles has been exceeded. This behavior can be explained by the sandwich assay model. The K_d-values K_α and K_β are estimated to be 3.4×10^{-11} M and 4.1×10^{-12} M which appear to be very small. As pointed out before, K_α and K_β do not have a great impact on the fit result. If both K_d-values are chosen to be 10^{-10} M, the data could also be well fit without a significant change in the parameters $[P]_0$, $[L]_0$ and N. The estimated particle concentration, $[P]_0$, is 5.4 nM, the conjugate concentration, $[L]_0$, 16 nM and the number of anti-β-hCG antibodies per particle, N, is determined to be 2.6.

CD45 Pharmaceutical Screening Assay: CD45 phosphatase is a transmembrane protein tyrosine phosphatase. It is the most abundant leukocyte cell surface glycoprotein (leukocyte common antigen, LCA) and plays a pivotal role in antigen-stimulated proliferation of T cells. The enzyme dephosphorylates the negative regulatory tyrosine residue of LcK, a Src-family protein tyrosine kinase, thereby serving as a positive regulator. Inhibitors of CD45 tyrosine phosphatse activity would potentially lead to novel drugs for the suppression of immune and inflammatory reactions. The assay principle is schematically drawn in Fig. 10.1c.

An ELISA assay has been previously described [10.13] using recombinant CD45 D1-D2, and a biotinylated phosphotyrosine peptide as substrate. After the enzymatic (inhibition) reaction was performed in homogeneous solution, the N-terminal biotinylated peptide was captured on a streptavidin-coated microtiter plate to enable subsequent detection using a mouse anti-phosphotyrosine/anti-mouse horseradish peroxidase-antibody cascade.

This ELISA assay format was easily adapted to FCS. One approach was to replace the biotin by tetramethylrhodamine and directly measure the binding of the substrate to anti-phosphotyrosine antibody [10.13]. The approach presented here takes advantage of the interaction between the biotinylated phosphotyrosine peptide and the anti-phosphotyrosine antibody. The ELISA assay protocol was directly transferred into a homogeneous nanoparticle immunoassay. Instead of the streptavidin-coated microplate, streptavidin-coated nanoparticles were used and the antibody was replaced by its TRITC conjugate. Since there is essentially no mass difference between the TRITC antibody and its complex with the phosphorylated substrate peptide this interaction cannot be measured using conventional FCS.

Without CD45 a maximum of TRITC-antiphosphotyrosine antibody was bound via unprocessed phosphorylated substrate peptide to the slow diffusing nanoparticles concurrently reducing the mean particle number from by a factor of about 4 as determined using FCS. The diffusion times of the TRITC-labeled antibody and its complex with nanoparticles were 0.42 ms and 2.8 ms, respectively. If increasing concentrations of CD45 are used the binding to nanoparticles and the correlation amplitude, g_B, vanishes.

At CD45 concentrations above 5 nM the data are comparable to control experiments without nanoparticles. If inhibitory substances would be present during the enzyme reaction, binding to nanoparticles increases using the CD45-NPIA.

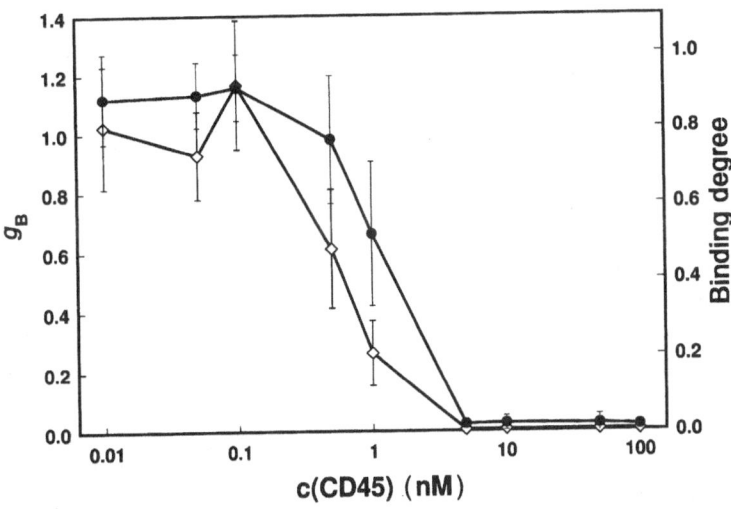

Fig. 10.7. Correlation amplitude, g_B (◊), and degree of binding, (•), of the CD45 tyrosine phosphatase assay consisting of TRITC-anti-phosphotyrosine antibody, biotinylated and phosphorylated peptide substrate and streptavidin coated nanoparticles ($\varnothing = 160\,$nm) after action of different concentrations of CD45 tyrosine phosphatase for 5 min at 25 °C

10.5   Discussion

Nanoparticle immuno assays (NPIAs) extend conventional FCS applications. Using NPIAs enables the measurement of binding events independent of mass differences. This is achieved by immobilizing one binding partner onto the surface of nanoparticles thus generating the slow diffusing binding partner for FCS readout.

Immunoassay protocols can be easily converted into soluble homogeneous NPIAs. This is particularly useful, because a wealth of quantitative assays employed in diagnostics and pharmaceutical screening are immunoassays, e.g. RIAs, EIAs, ELISAs. The versatile colloidal nanoparticles can be used as a solid phase which remains in solution. Nanoparticles can therefore be conveniently handled like a true solution. Since NPIAs are homogeneous assays, time-consuming washing steps are omitted and the overal assay time can be reduced from several hours to half an hour and less. FCS detection of NPIAs allows for reducing reagent consumption and makes existing assays readily adaptable to high throughput screening applications.

The conversion of heterogeneous immunoassays to homogeneous NPIA format comprises only few steps involving synthesis of fluorescent antigens/ antibodies, immobilization of antibody on the surface of nanoparticles and optimization of concentration of antigens, antibodies and nanoparticles.

For a competitive assay, a fluorescent conjugate of an antigen must be prepared. If a small antigen is to be conjugated to a fluorophore, the lack of a functional group for coupling or the drop in affinity of the specific antibody could cause problems. Just as is common in existing forms of non-isotopic immunoassays antigene conjugates with biotin may be used directly. In addition, derivatives of many small compounds such as hormones like estradiol, testosterone, progesterone, are available which contain functional groups for linking a fluorophore. In the case of estradiol, antibody affinity has been investigated for six fluorescent conjugates. In fact, five out of six conjugates could be bound and four conjugates could be shown to bind specifically to the anti-estradiol antibody. Protein antigens can be labeled with fluorophors or biotin according to standard procedures. In particular, the coupling of dyes to antibodies usually does not affect affinity to its antigen and virtually all antibody and protein conjugates worked in our assays. Thus, we conclude that it is generally possible to synthesize fluorescent antigens or antibodies which work in immunoassays and equally well in NPIAs. The capture antibody or antigen, if an antibody is to be detected, will be easily attached to the nanoparticles using biotin/streptavidin, protein A, anti-species secondary antibody, for example. The many existing building blocks of nanoparticles and capture antibodies enhance the flexibility of NPIAs. Since the development of an NPIA always involves similar steps, their establishment can be done very efficiently according to standard procedures. This helps trim a major bottleneck in pharmaceutical screening, namely, the duration of screening assay development.

The applicability of NPIAs is not restricted to classical binding reactions, but can be extended to enzyme reactions, provided educt and product could be discriminated, such as using binding to an antibody. This assay principle has been demonstrated with the CD45-assay. The CD45 assay is an excellent example for the straightforward tranferability of existing ELISA or other immunoassays to NPIAs. All ELISA reagents could be used. The only modification was that the anti-phosphotyrosine antibody had to be fluorescent and a TRITC conjugate of this antibody was even commercially available. Consequently, the assay could be established in an extremely short period of time. The measuring times for NPIAs are not yet sufficient for ultra-high throughput screening in the pharmaceutical lead finding industry. But preliminary results of NPIA measurements using EVOTEC's FIDA-technology (in preparation) appear to be promising in order to achieve measuring times of a few seconds per assay with acceptable coefficients of variation.

In conclusion, NPIAs represent a new general assay method in the area of diagnostics and pharmaceutical screening. NPIAs feature maximal flexibility in assay design because of their quasi mass independence and the many available building blocks, including antibodies, nanoparticles, biotinylated antigens, etc. Existing heterogeneous immunoassays can be simply and quickly transferred to homogeneous miniaturizable NPIAs. The assay principle can not only be applied to binding interactions, but also to enzyme reactions. NPIAs have a similar sensitivity as conventional immunoassays. Their great potential for use in molecular diagnostics is substantiated by the fact that NPIAs are homogeneous, therefore possessing a short overal assay time, are miniaturizable without loss in performance and have a quick read-out. The versatility of NPIAs opens new ways to improve the efficiency of drug development by shortening the time for screening assay setup.

Acknowledgements

The synthesis of the estradiol and theophylline conjugates by Dr. Eloisa Lopez-Calle and the excellent technical assistence of Silke Christoph and Susanne Harzendorf as well as fruitful discussions with Dr. Joachim Krämer are greatfully acknowledged. I also thank Dr. Schumacher for the gift of anti-estradiol antibody and Rodney Turner for critical comments on the manuscript.

References

10.1   J. Widengren and R. Rigler: Cell Mol. Biol. **44**, 857–79 (1998)

10.2   U. Kettling, A. Koltermann, P. Schwille, and M. Eigen: Proc. Natl. Acad. Sci. USA **95**, 1416–1420 (1998)

10.3   P. Schwille, F.-J. Meyer-Almes, and R. Rigler: Biophys. J. **72**, 1878 (1997)

10.4   F.-J. Meyer-Almes, K. Wyzgol, and M.J. Powell: Biophys. Chem. **75**, 151–160 (1998)

10.5   R. Rigler, Z. Földes-Papp, F.-J. Meyer-Almes, C. Sammet, M. Völcker, and A. Schnetz: J. of Biotechnology **63**, 97–109 (1998)

10.6   N.G. Walter, P. Schwille, and M. Eigen: Proc. Natl. Acad. Sci. USA **93**, 12805–12810 (1996)

10.7   M. Pitschke, R. Prior, M. Haupt, and D. Riesner: Nature Medicine **4**, 832–34 (1998)

10.8   P. Schwille, J. Bieschke, and F. Oehlenschlager: **66**, 211–28 (1997)

10.9   S. Maiti, U. Haupts, and W.W. Webb: Proc. Natl. Acad. Sci. USA. **94**, 11753–57 (1997)

10.10   A. Koltermann, U. Kettling, J. Bischke, T. Winkler, and M. Eigen: Proc. Natl. Acad. Sci. USA **95**, 1421–1426 (1998)

10.11   T. Winkler, U. Kettling, A. Koltermann, and M. Eigen: Proc. Natl. Acad. Sci. USA **96**, 1375–8 (1999)

10.12   S. Sterrer and K. Henco: J. Receptor Signal Transduction Research **17**, 511–520 (1997)

10.13   M. Auer, K.J. Moore, F.-J. Meyer-Almes, R. Guenther, A.J. Pope, and K.A. Stoeckli: Drug Discovery Today **3**, 457–465 (1998)

10.14   R. Uauy, D.L. Shapiro, B. Smith, and J.B. Warshaw: Pediatrics **55**, 595–598 (1975)

10.15   L. Speroff, R.H. Glass, N.G. Kase (Eds.): Clinical Gynecologic Endocrinology and Infertility, 4th edn. (Baltimore, Williams and Wilkins, 1989) 91–119

224   F.J. Meyer-Almes

10.16 P.T. Russel: In: L.A. Kaplan and A.J. Pesce (Eds.), Clinical Chemistry: theory, analysis and correlation, (St. Louis, CV Mosby, 1989) p. 572
10.17 L.A. Kaplan: In: L.A. Kaplan,and A.J. Pesce (Eds.), Clinical Chemistry: theory, analysis and correlation, (St. Louis, CV Mosby, 1989) p. 938

11 Protein Aggregation Associated with Alzheimer and Prion Diseases

Detlev Riesner

11.1 Introduction

The brains of patients with Alzheimer's disease (AD), Creutzfeldt–Jakob disease (CJD), Kuru and others are characterized by an abundance of amyloid deposits. As long as the amyloid deposit could be described only by its histopathological morphology, these and other diseases appeared to have common molecular grounds, although the clinical pictures were different. Rudolf Virchow was the first who described the deposits in Alzheimer patients brains as "Eiweiss-Stärke" (protein-starch) and therefore gave the name "amyloid". Due to the work of Beyreuther and colleagues (see [11.1]) and Prusiner and colleagues (see [11.2]) it became known that the Alzheimer deposits consisted predominantly of the β-amyloid-peptide and those of CJD and Kuru of the prion protein, respectively, and that both proteins are encoded by a host nuclear gene. The molecular mechanism, cellular location and disease etiology, however, are quite different. Besides CJD and Kuru more human diseases like Gestmann–Sträussler–Scheinker disease and familial fatal insomnia, but also animal diseases like scrapie in sheep and bovine spongiform encephalopathy (BSE) are closely connected with a pathological form of the prion protein, therefore denoted generally as "prion-diseases". Most strikingly, and in contrast to AD, prion-diseases are transmissible, which led also to the name prion as proteinaceous infectious agent [11.3]. It is not yet known whether formation of insoluble deposits is a prerequisite of infectivity, but it is generally accepted that a pathological isoform of the prion protein, the so called Scrapie-isoform, PrP^{Sc}, is the major component of the brain deposits as well as of the infectious agent. During the infection event PrP^{Sc} is formed by a post-translational process from the cellular isoform PrP^C. The relationship between the aggregate formation, the conformational transition of PrP^C into PrP^{Sc}, and generation of infectivity and pathogenicity will be one of the major subjects of this chapter.

The Alzheimer specific β-amyloid peptide (Aβ) is derived also from a host product, the so-called amyloid precursor protein (APP; 695-770 amino acid residues, see [11.4]). Unlike PrP, however, it is not produced by a conformational change, but by proteolytic cleavage. Depending upon the cerebral tissue predominantly the 40 residues Aβ-(1-40) or the 42 residues Aβ-(1-42) are cut out from the outer membrane part of the trans-membrane protein

APP. Secretases which are specific peptidases, are involved in the cleavage, and predominantly the peptide Aβ-(1-42) plays a key role in promoting AD. An extensive literature exists on mutations in APP and homologous proteins called presenilins and their connections with Aβ amyloid formation and AD [11.5]. Aggregate formation during AD is not restricted to the extra-cellular space; the tau-protein can dissociate from the tubulin spindle and form disease-associated intra-cellular fibrils [11.6]. This chapter, however, is restricted to amyloid formation of the Aβ-peptide and its association with the disease.

11.2   Prion-Protein Multimerization

11.2.1   Conformation and State of Aggregation

PrP^{Sc} is derived from PrP^C by a post-translational process. Limited proteolysis of PrP^{Sc} produces a N-terminal truncated protein, designated PrP27-30, under conditions in which prion infectivity is retained; PrP27-30 resists further proteolysis by proteinase K[11.2]. The transformation of PrP^C into PrP^{Sc} involves a profound conformational change: PrP^C consists of over 40% α-helix and little β-sheet, while PrP^{Sc} exhibits about 30% β-sheet and less than 30% α-helix [11.7,11.8].

As mentioned above, PrP^{Sc} can be isolated from the tissue only as insoluble aggregates. The insolubility of PrP^{Sc} and PrP27-30 complicated conformational studies and has prevented the application of methods like X-ray crystallography or nuclear magnetic resonance (NMR) spectroscopy. While some structural features were predicted by computational methods [11.9], more recently the structure of recombinant PrPs in aqueous buffers have been determined by NMR spectroscopy [11.10–11.13]. The protein is composed of three α-helices and a small two-stranded antiparallel β-sheet.

Attempts to solubilize prion infectivity have been fraught with difficulties [11.14]. Sonication of prion rods in the presence of 0.2% SDS produced soluble, spherical oligomers of PrP27-30, which were no longer infectious. These oligomers exhibited a high α-helical content and could readily be digested with proteinase K, properties that are reminiscent of PrP^C [11.15]. Solubilized PrP27-30 will be designated furtheron as sPrP.

Studies were carried out on PrP27-30 and recombinant PrP (90-231) [11.16] to analyze the physical states of PrP in the presence and absence of SDS [11.17]. Whether changing the conditions from 0.2% SDS to physiological conditions induces structural transitions and multimerization was tested. The concentration of SDS was reduced by mere dilution with aqueous buffer. As assessed by differential centrifugation at 100 000 g for one hour and gelelectrophoresis of the supernatant and pellet fractions (Fig. 11.1), PrP remained soluble at concentrations of SDS above 0.03%. At 0.02% SDS about 50% of the protein remained soluble and 50% readily precipitated. At 0.01% SDS virtually all of the PrP27-30 was found in the pellet fraction.

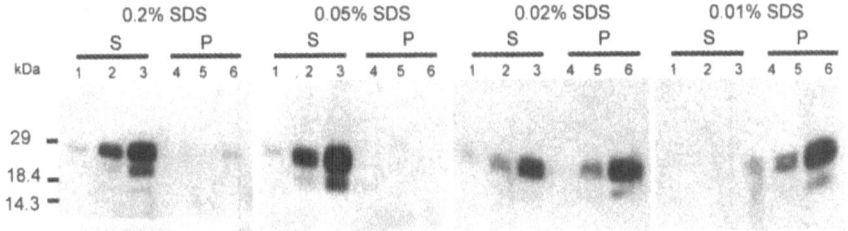

Fig. 11.1. Distribution of sPrP between the supernatant and the pellet of centrifugation at 100 000 g for 1 h. Soluble PrP27-30 at 0.2% SDS and after diluting the SDS to a concentration of 0.05%, 0.02% and 0.01%. S designates the supernatent and P the pellet fraction, respectively. The numbers above the slots indicate the relative concentrations; slot 3 without dilution, slot 2 after 3-fold dilution and slot 1 after 9-fold dilution. Samples were incubated overnight at 37 °C in non-silanized tubes and centrifuged at 100 000 g for 1 h. After separation in supernatant and pellet and resuspension of the pellet, SDS-PAGE and Western blot analysis with chemiluminescence detection was peformed. The positions of marker proteins are depicted at the left margin. Figure with permission from [11.17]

Similar results were obtained with recombinant PrP(90-231), referred to as rPrP (data not shown). Reduction of the SDS concentration from 0.2% to 0.01% induced complete multimerization. As concluded from an analysis by MALDI mass spectroscopy the multimeric state of rPrP after dilution to 0.01% SDS is essentially free of SDS and therefore refered to as "after removing SDS" [11.17].

The influence of the temperature was also analyzed. At intermediate SDS concentration, i.e. 0.02%, PrP remained predominantly soluble at 4 °C, but was completely aggregated at 54 °C. From this temperature dependence it has to be concluded that aggregation of PrP is driven by hydrophobic interactions.

To examine the effect of transfering sPrP27-30 or rPrP from 0.2% SDS into aqueous buffer on its secondary structure, CD was used [11.17]. As shown in Fig. 11.2, sPrP27-30 in 0.2% exhibited an α-helix content above 40%. After dilution of the SDS to 0.01% the α-helical content was reduced and a substantial amount of newly formed β-sheet was found. Only qualitative data are given because of a high variability with the different secondary structure determination programs that were used. Although not quantitatively evaluated, it should be noted that β-sheets formed at least partially on account of random coil and not solely on account of α-helix. An intermediate spectrum was obtained by incubation with 0.02% SDS. With recombinant PrP a similar structural transition was observed. However, the amount of random coil structure was higher in 0.2%, and only in lower SDS concentrations, for example 0.1% SDS was the α-helix content of rPrP close to that of sPrP in 0.2% SDS. Control proteins showed after incubation with 0.2% SDS an enlarged amount of random coil. After dilution of the SDS to 0.01%

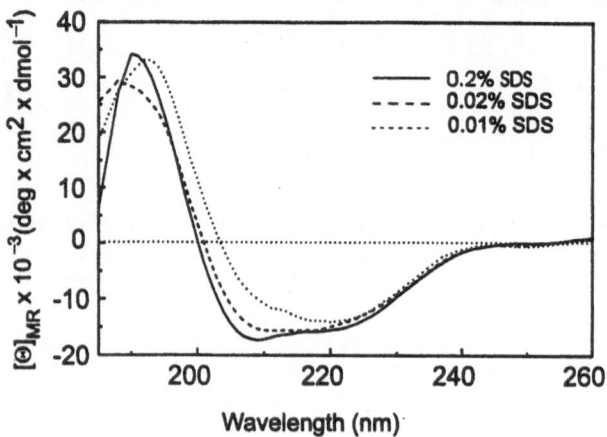

Fig. 11.2. Circular dichroism (CD) spectra of sPrP in 10 mM sodium phosphate, pH 7.2 in the presence of various SDS concentrations. Concentrations of SDS: 0.2% (**solid line**), 0.02% (**dashed line**) and 0.01% (**dotted line**). Figure with permission from [11.17]

all proteins showed their original structure distribution without any further aggregation. This finding demonstrates the effect of 0.2% SDS is reversible in accordance with the complete removing of SDS from multimeric PrP (see above), and secondly the aggregation as a consequence of removing SDS is a specific feature of PrP and not a common feature of proteins.

The size and shape of multimeric PrP27-30 after removing the SDS was studied by confocal fluorescence correlation spectroscopy (FCS) and analytical ultracentrifugation. An analytical ultracentrifuge was equipped with a fluorescence detection system for increased sensitivity [11.18] and with a biohazard safety containment device to prevent contamination of the laboratory with prion infectivity [11.19]. The molecular weight of sPrP27-30 in 0.2% SDS was homogeneous but varied from preparation to preparation, depending on the wattage of sonication, between 20 kDa and 100 kDa [11.19]. It was monomeric in 0.2% SDS in all of the studies reported here. After dilution of the SDS multimers of PrP27-30 were detected by analytical ultracentrifugation; the size of the multimers exceeded, however, the quantitative evaluation limits of the method.

Ultrastructural studies with electron microscopy showed that multimeric rPrP immediately after removing SDS formed amorphous particles of approximately 24 ± 8 nm (Fig. 11.3). After overnight incubation at 37 °C, these multimers assemble into larger structures. In some cases the amorphous particles in panel A are visible within the larger structures in panel B. Increas-

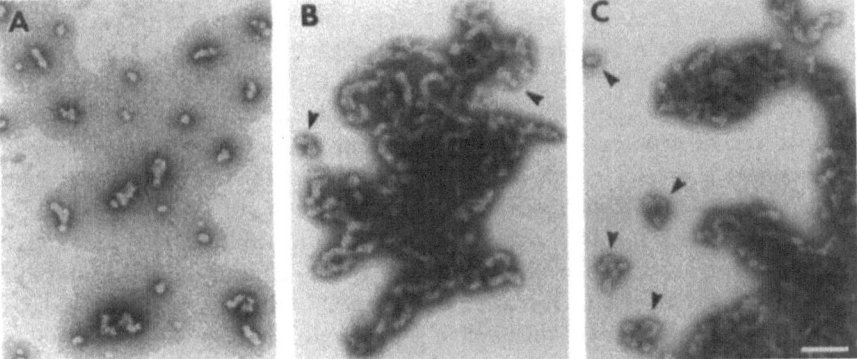

Fig. 11.3. Electron micrographs of negatively stained multimers of rPrP. (A) Multimers of rPrP obtained directly after dilution to 0.01% SDS. Amorphous oligomers are present in large numbers and represent the early stages of higher multimerization. (B) Sample as in A after additional overnight incubation at 37 °C. (C) Sample as in B after additional 200 000 g centrifugation of 2 h and resuspension in a small volume of supernatant. Negative staining with 2% uranyl acetate, bar = 100 nm. Figure with permission from [11.7]

ing the concentration of the rPrP multimers by ultracentrifugation produced particles that are more dense and tightly packed. However, regular fibrillar structures as described for the β-amyloid peptide (see below; [11.20]) were not seen with PrP.

Since PrP27-30 lost its resistance to proteolysis by proteinase K due to solubilization in 0.2% SDS, how much resistance could be re-established by re-aggregation sPrP27-30 in 0.01% SDS was determined. Stochiometric concentrations of proteinase K were used and the kinetics of the digestion determined. While the resistance of multimeric PrP27-30 to digestion with proteinase K was measurable, it was low compared to that found with rPrP (data not shown). The rPrP in 0.2% SDS was degraded within 5 min, whereas multimeric rPrP after removing SDS, incubation overnight and subsequent ultracentrifugation at 250 000 g for two hours protected the protein from proteolysis for more than 15 min. Digestion for one hour degraded the rPrP multimers. These data are in accordance with some recognizable structure but not regular fibrils. Most probably, only highly regular structures like prion rods exhibit the well-known extraordinary resistance to proteolysis.

11.2.2   Analysis of Multimerization by FCS

The method of fluorescence correlation spectroscopy (FCS) is very suitable to analyze the process of multimerization over a very wide range of molecular weights. Therefore, this method was applied to prion-protein multimerization [11.17]. Basic features of the method, the principles of the analysis and the handling of the instrument have been described by Rigler and co-workers

[11.21,11.22] and are outlined in other chapters. As shown below conditions were found whereby labeling of sPrP27-30 with the fluorophor Cy2 did not affect the multimerization properties of PrP. Only minute amounts of the sample are needed. Use of the instrument and handling of the infectious prion samples can be done easily under safe conditions S2 [11.23]. Furthermore, the measurements are so fast that kinetic studies in a time range longer than a minute can be carried out (see below).

The results of the FCS are presented as autocorrelation functions (Fig. 11.4); the average diffusion time is defined by the half-height of the step-like autocorrelation function. The sPrP27-30 in 0.2% SDS (Fig. 11.4A) showed a diffusion time of about 340 μs that corresponds to a molecular weight of about 23 kDa assuming a spherical shape. After removing SDS and incubation for three hours, the diffusion time increased to 750 μs, corresponding to a molecular weight of about 250 kDa. For the multimers a spherical shape was assumed, as was a molecular packing close to one. In contrast to the autocorrelation curves of sPrP27-30 in 0.2% SDS (Fig. 11.4A), the curves of PrP27-30 after removing SDS (Fig. 11.4B) were variable, indicating a broad distribution of molecular weights. Increasing the concentration by centrifugation raised the molecular weights of the multimers (Table 11.1). Calculating the molecular weights from the diffusion times gave an average of about 5 MDa and, in addition, single, even larger multimers with about ten times the diffusion times were observed. Similar results were obtained also with rPrP.

Table 11.1. Determination of molecular parameters for sPrP and rPrP for different incubation conditions.

Samples		Diffusion time (ms)	Molecular mass (kDa)
sPrP	0.2% SDS	0.34	23
	0.01% SDS	0.75	250
	0.01% SDS centrifuged	2.02*	4800*
RPrP	0.2% SDS	0.28	16
	0.01% SDS	0.89	510
	0.01% SDS centrifuged	1.94*	5300*

*Diffusion times were corrected for optical adjustment as described by [11.17]. Centrifugation: 250 000 g for two hours; the asterisk designates a mean value.

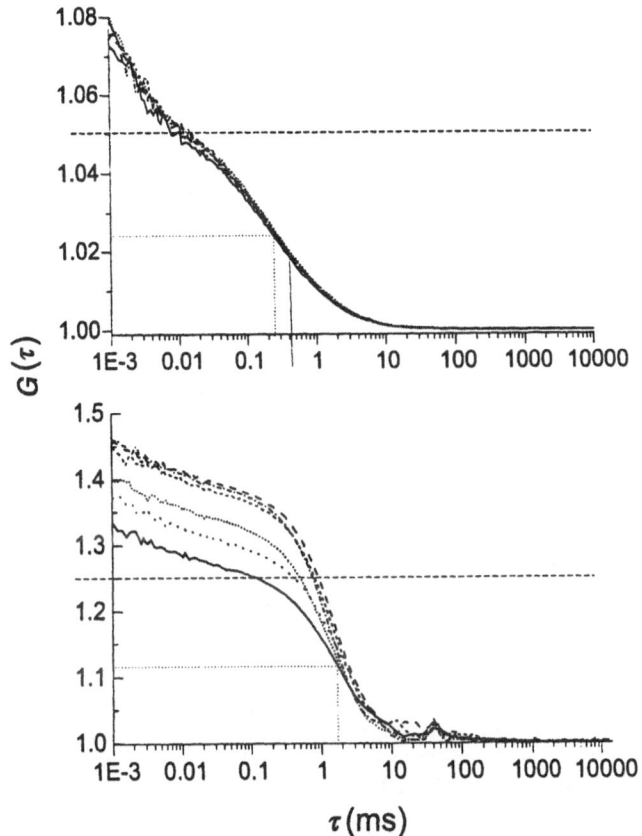

Fig. 11.4. Fluorescence correlation spectroscopy (FCS) of sPrP after 16 h incubation in 10 mM sodium phosphate, pH 7.2 at 0.2% SDS (A) and 0.01% SDS (B). The autocorrelation function of the fluorescence intensity $G(\tau)$ is shown. The dashed line represents the adapted saturation level of $G(\tau)$ at the short diffusion times. For calculation of $G(\tau)$ see the Chaps. 19 and 22. The point of inflexion (**dotted line**) represents the diffusion time of the macromolecule through the volume of fluorescence detection. Six independent scans are shown for each sample. The PrP concentration was 5 ng/µl; PrP was labeled with fluorophor Cy2. Figure with permission from [11.17]

11.2.3   Influence of Fluorescence Labeling on the Multimerization Reaction

FCS yields best results if the fluorophore acts only as a label without affecting the reaction. This might be particularly sensible for a multimerization process and has to be checked specifically for every pair of multimerizing protein and fluorescent dye [11.23]. For example, it was found that some dyes induce

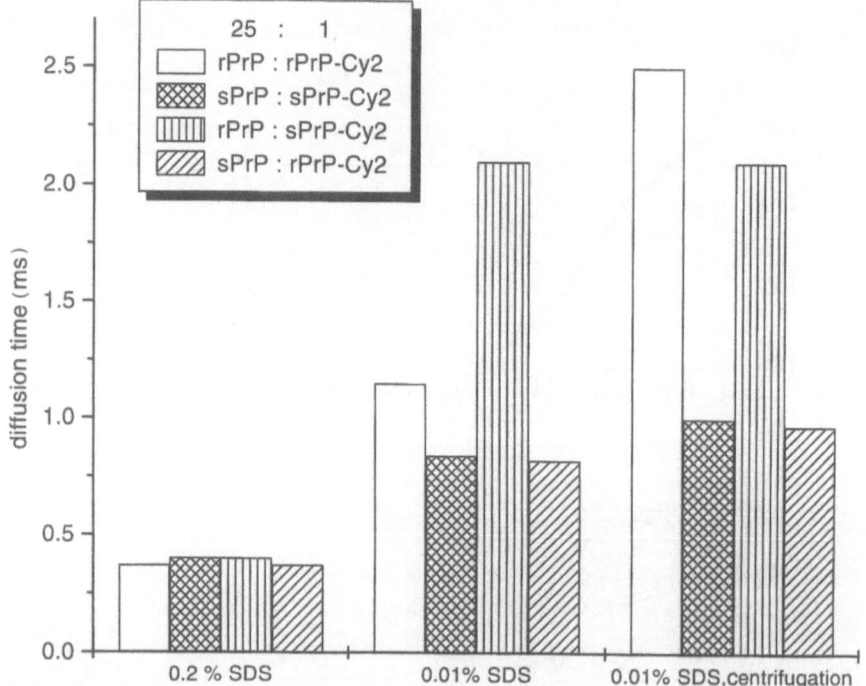

Fig. 11.5. Homo- and heterologous multimerization of labeled and unlabeled PrP. An amount of 5 ng/μl of unlabeled sPrP or rPrP were mixed with 0.2 ng/μl of sPrP · Cy2 or rPrP · Cy2, respectively. Incubation for 16 h, centrifugation for 1 h, 100 000 g. Figure with permission from [11.23]

multimerization at 0.2% SDS, a concentration which keeps PrP completely soluble without a dye.

The fluorophor Cy2 (Amersham, UK) was found to be very suitable for analyzing the multimerization of PrP, if the ratio of labeled to unlabeled PrP was restricted to 1:25. Under physiological conditions Cy2 carries one negative charge which prevented multimerization in 0.01% SDS if every PrP molecule carried a label. If unlabeled PrP was in excess, however, as mentioned above, the labeled PrP is co-incorporated into the aggregate of unlabeled PrP. This was found with aggregates of sPrP as well as those from rPrP. However, the multimers from sPrP and rPrP exhibit significantly different diffusion times under otherwise identical multimerization conditions. This different behavior was analyzed with labeled sPrP as well as labeled rPrP as probe. It is shown in Fig. 11.5, that only the nature of the unlabeled excess PrP and not the nature of the probe determined the specific diffusion time. It was concluded from this result that the probe does not significantly affect the multimerization of PrP.

11.2.4   Kinetics of Spontaneous Multimerization

It has been shown by Lansbury and co-workers [11.24,11.25] in studies on peptide fragments of PrP and on β-amyloid peptides, that the kinetics of association is affected by the presence of multimeric seeds. Therefore, the kinetics of multimerization of rPrP was studied in the absence of pre-existing seeds, so-called spontaneous multimerization, and in the presence of seeds, so-called seeded multimerization [11.23].

Spontaneous multimerization was induced by transfering sPrP or rPrP from 0.2% SDS to aqueous buffer, i.e. 0.01% SDS. The CD-spectrum was changed within 90 s, which was the shortest time for recording a CD spectrum in the instrument used (Jasco J 715). Typically, transitions from a predominantly α-helical conformation to one with a raised β-sheet content were observed. No appreciable change was found over the next 15 min.

The kinetic measurements of sPrP with FCS showed a large increase in the diffusion time (Fig. 11.6) which could be resolved into three processes:

i) faster than one minute, i.e. the time of resolution of FCS (Fig. 11.6A)
ii) in the time range of 20 min (Fig. 11.6A), and
iii) long-term process in the time range of hours to days (Fig. 11.6B).

In the time period up to 20 min, i.e. the second step, a doubling of the diffusion time, that represents an increase in the molecular weight to 200–250 kDa for spherical particles, was detected. The apparent diffusion time of sPrP in 0.2% SDS scattered around 250 μs, which served as a baseline for the kinetic curve. Extrapolation of the measured diffusion times in the absence of SDS to an incubation time of zero, displayed a higher diffusion time than the baseline. The extrapolated diffusion time was interpreted as a dimerization of the sPrP in the first step. The third step resulted in much larger multimers with diffusion times longer than 1 ms corresponding to molecular weights of more than 1 MDa (Fig. 11.6B). In contrast to the oligomerization products detected in step ii) (Fig. 11.6A), the size distribution of the large multimers in step iii) was much more heterogeneous (Fig. 11.6B). In addition, some extremely large multimers with diffusion times longer than 5 ms were observed, but these particles were not included in the data of Fig. 11.6B.

Similar results were obtained with rPrP. Preliminary studies varying the concentration of rPrP over a wide range exhibited a strong dependence of the increase of the molecular weight upon the concentration of PrP. If it was assumed that the first step which is too fast to measure was a dimerization, then the oligomers are formed from association of dimers. For this step a rate constant of about $10^5\,\mathrm{M^{-1}s^{-1}}$ could be estimated [11.26].

11.2.5   Seeded Multimerization of PrP

Spontaneous multimerization, as described above, leads after longer incubation, i.e. longer than one hour, to very large aggregates which are visible

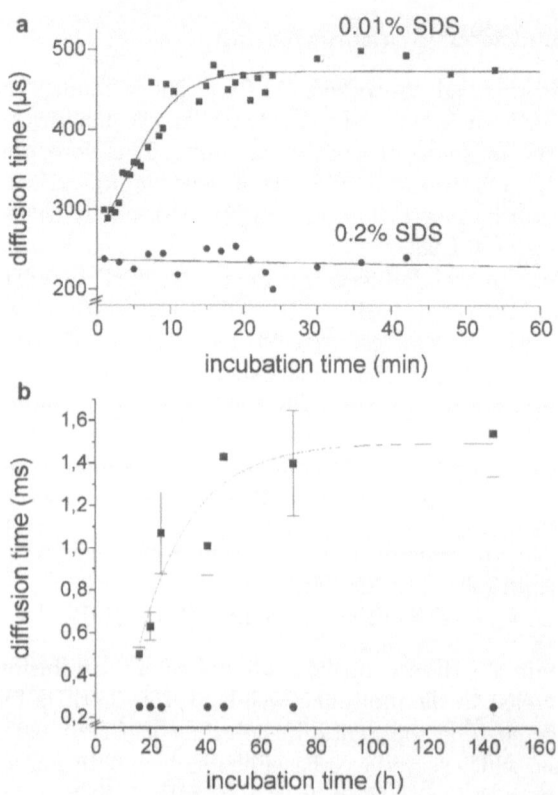

Fig. 11.6 a,b. Kinetics of multimerization as measured by FCS after incubation in the presence of 0.2% SDS (**circles**) and after dilution to 0.01% SDS and subsequent incubation (**squares**). **a** represents the time range of the first 60 min., **b** the subsequent long time range up to 145 h. PrP unlabeled: 5 ng/μl; PrP labeled with Cy2: 0.2 ng/μl. Figure with permission from [11.17]

in FCS as single very high peaks. These peaks exceed the remainder of the fluctuating signal sometimes by an order of magnitude. The autocorrelation function $G(\tau)$ cannot be applied for evaluation, because the number of peaks is too small and therefore the statistics are too low. If fluorescently labeled monomeric PrP is added to those large aggregates which do not carry a fluorophore, the association kinetics of monomers to pre-existing seeds can be studied directly [11.23]. In Fig. 11.7 the fluorescence signal is compared for two experiments: In Fig. 11.7A fluorescent PrP is diluted from 0.2% SDS to 0.01% SDS in the absence of PrP-seeds; a fluctuating signal with increasing τ during repeated scans was detected which is characteristic of spontaneous multimerization. In Fig. 11.7B , however, from the start of the fluorescence recording onwards, i.e. a few seconds after dilution of 0.2% SDS large peaks were visible. Consequently, fluorescent monomeric PrP was attached to the

a   b

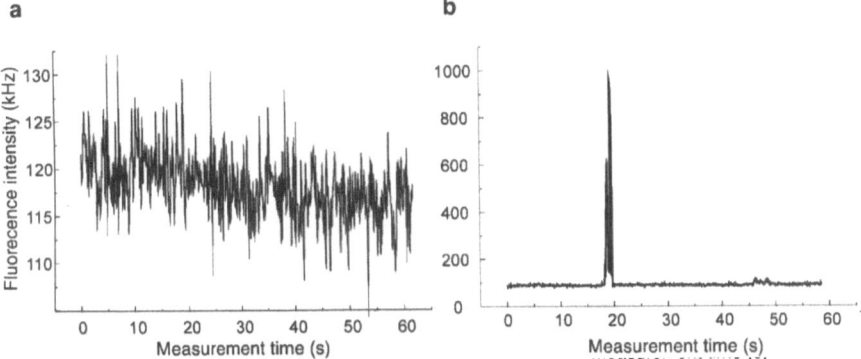

Fig. 11.7 a,b. Time course of the fluorescence intensity of sPrP-Cy2 in FCS. **a:** in the absence of pre-existing PrP-aggregates; **b:** in the presence of pre-existing aggregates of sPrP (unlabeled) acting as seeds for multimerization of further labeled and unlabeled sPrP. Concentration of sPrP was 5 ng/µl, of sPrP-Cy2 0.1 ng/µl. In **a** unlabeled and labeled sPrP were mixed before multimerization was induced by dilution of 0.2% SDS to 0.01%; in **b** unlabeled PrP was incubated after dilution of 0.2% SDS to 0.01% for 48 h before labeled sPrP was added. Figure was modified from [11.23]

multimeric seeds in a fast reaction. The hight of the peaks is characteristic of the number of fluorescent PrP molecules attached to the seed, and the width of the peak indicates the diffusion time of the complex through the laser beam.

In Fig. 11.7B one peak is seen in a scan over one minute. Within statistical variation this number stays constant for 30 min to an hour, i.e. before spontaneous aggregation of PrP generates additional very large aggregates (see above). All pre-existing seeds in the solution, however, are marked with fluorescent PrP before the first fluorescent scan is recorded. Therefore, the number of peaks counted in the first 30 min is proportional to the number of seeds present in solution. A linear relationship could be shown experimentally over several orders of magnitude (see. also below in Fig. 11.13; [11.23]). In a preliminary study it was shown that the number of prion particles in a solution can be determined by counting the number of seeds for multimerization. This effect is the basis for a diagnostic method which was exploited further the Alzheimer's disease (see below).

Seeded multimerization described here shows qualitatively a similar dependence on SDS-concentration as the spontaneous multimerization described above. Quantitatively, however, the dependencies are different; spontaneous multimerization occured only below 0.02% SDS, whereas seeded multimerizaiton was observed up to 0.05% SDS [11.23].

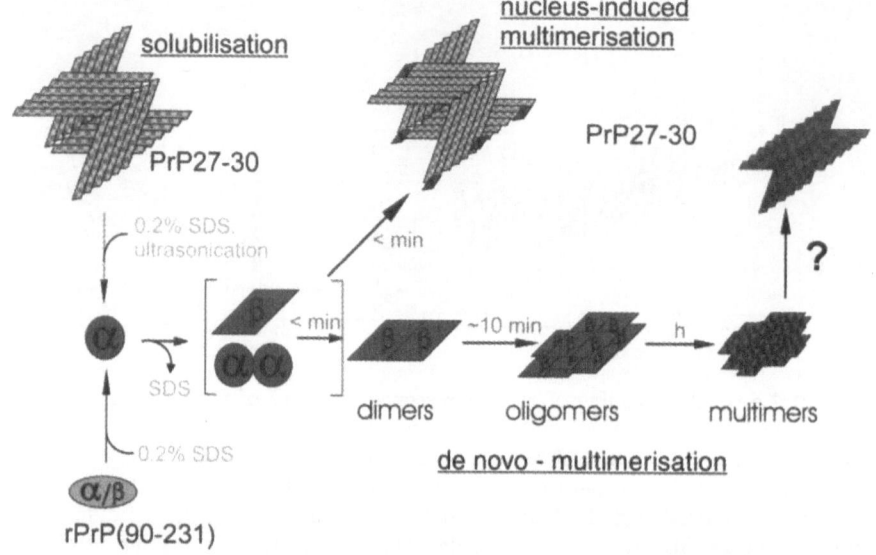

Fig. 11.8. Summary sheme of conformational transitions and steps of multimerization. Figure was modified from [11.23]

11.2.6 Summary of PrP Conformational Transitions

The results on the mechanism of the conformational transitions of PrP are depicted schematically in Fig. 11.8. Soluble, monomeric PrP with high α-helix content was produced either from prion rods by addition of 0.2% SDS and sonication or from aggregates of recombinant PrP by SDS alone, respectively. A series of conformational changes of either sPrP or rPrP occurred upon removing the SDS by dilution. Using FCS for the state of multimerization and CD for the secondary structure, at least three steps were identified. In the time range of less than one minute, a secondary structural transition from predominantly α-helical to one containing increased β-sheet content was observed. It is known that protein folding or refolding can occur in the millisecond range and does not proceed in a single step, but rather through distinct conformational intermediates with non-native, non- denatured structures often referred to as molten globules [11.27,11.28]. The combined results from CD and FCS indicate that the secondary structure transition occurs concomitantly with a dimerization reaction. The secondary structure transition might precede or follow the dimerization; both possible transient states are depicted in Fig. 11.8. The second step, as determined by FCS, is an oligomerization in the time range of minutes. These oligomers are stable for hours, and further multimerization to extended molecules occurs in the time range of hours to days; additional centrifugation accelerates this process.

The multimers show some internal structure but not a regular fibrillar structure. Furthermore, it was shown [11.37] that the multimers did not regain infectivity. Therefore, one has to assume that at least one conformational transition is still missing, designated by a question mark in Fig. 11.8. In that transition, infectivity and possibly fibrillar structure would be acquired. It has been tacitly assumed that, in principle, infectivity could be acquired by PrP alone without any second component.

In addition to the spontaneous multimerization also the seeded multimerization is depicted in Fig. 11.8. Possibly the same transient state is required for binding to pre-existing prions as forming β-sheeted multimers. But the slow process of forming larger aggregates up to a PK-resistent state is circumvented by direct attachment to the multimeric PrP in prion particles.

11.3   Amyloid β-Peptide Multimerization

Since the chemical nature of the AD associated plaques was identified as an aggregate of the Aβ-peptide the multimerization process of the peptide has been studied in vitro by a number of techniques. Light scattering, ultracentrifugation, different spectroscopic techniques, microscopy, calorimetry, chromatography and electrophoresis were among them, but these will not be reviewed here. When FCS was developed it was obvious that this method provided several advantages for studying β-peptide multimerization. The high sensitivity of FCS, allowing analysis in the nanomolar range or to detect single aggregates (see above) opens the new possibility to study details of the seeding process. Furthermore, components with different molecular weights and corresponding different diffusion times can be analyzed without any preceding separation. In one recent study the spontaneous multimerization [11.20] was analyzed, identifying several intermediate states before fibril formation, and in another study [11.38] FCS was utilized as a diagnostic principle in AD patients for detecting AD-associated aggregates.

11.3.1   Spontaneous Multimerization

Spontaneous multimerization was studied with Aβ(1-40). Similarly to the investigation of the prion protein the multimerization of the β-peptide was induced by transferring it into aqueous buffer (50 mM Tris-HCl, 150 mM NaCl, pH 7.4); however, lyophilized samples of the Aβ(1-40) were used as starting material, whereas the PrP-samples (see above) and the Aβ-peptides in the experiments described later were dissolved in 0.2% SDS before transferring them to water. Rhodamine-Lys-Aβ(1-40) was used as the labeled peptide; it carried the label only at the N-terminus and was present in the solution at always less than 1% of the unlabeled Aβ-peptide.

The FCS measurements of free rhodamine and labeled Aβ(1-40) yielded diffusion times of 40 μs and 90 μs, respectively. When compared with the diffusion time of 40 μs for the free rhodamine (443 Da) the labeled Aβ corresponds to a molecular weight of about 5 kDa (theoretical value 4370 Da), assuming diffusion of spherical particles. It was interpreted in accordance with other results from the literature [11.29] that the population of molecules observed in the freshly prepared solution represents a mixture of monomers and dimers. The measurements with FCS after 2 min up to 24 h are shown in Fig. 11.9. Within the first 2 min after dissolution of the Aβ-peptide no change in diffusion time and thus no induction of an aggregation was observed (Fig. 11.9a–c). After 10 min very large aggregates were detected and after 40 min the aggregation reached a maximum (Fig. 11.9d–f). Characteristic large fluorescence intensity peaks were visible (Fig. 11.9d) as known also from the large PrP aggregates. The diffusion time for these aggregates was found to be 3 s (Fig. 11.9e). After 80 min of incubation the diffusion time had changed from 3 s to 0.3 s, and no further changes were detected between 2 h and 24 h.

The results from the FCS measurements were compared with those of an electron microscopic analysis. After 40 min incubation, when the longest diffusion time (3 s) was observed in FCS, a network of amorphous structures and diffuse short fibrils were observed. From 3 s diffusion time a hydrodynamic radius of 60 μm was calculated which, however, cannot be associated with the structures from electron microscopy in a quantitative manner. After the long incubation of 24 h fibrils, often arranged as networks or in bundles with a diameter of about 100 nm, were observed. When the diffusion time in FCS observed after one day (0.3 s) was used to calculate the length of a rigid rod with the diameter of 100 nm, a length of 55 μm was obtained which is somewhat larger than the 5–20 μm observed in electron microscopy (EM). A more quantitative agreement cannot be expected, since isolated rods were evaluated from FCS and a network was actually observed in EM.

To study the influence of the concentration of Aβ on the Aβ multimerization, Aβ(1-40) was dissolved directly in 50 mM Tris-HCl and 150 mM NaCl (TABS) with Aβ concentration ranging from 1 μM to 300 μM. The aggregation was strongly dependent on concentration and the critical concentration for rapid multimerization was found to be 50 μM (Fig. 11.10). Below this concentration only monomers/dimers were detected. Above this concentration the polymerization to large aggregates proceeded rapidly after a phase lag of a few minutes.

In model studies it could also be shown that FCS is suitable to screen drugs for their capability to inhibit the pathogenic multimerization of Aβ [11.20]. When the multimerization process was carried out in the presence of an Aβ binding peptide (Asn-Lys-Leu-Val-Phe-Phe-Ala, 320 μM), multimerization amounted to only 5 to 10% as compared to the amount of multimers in the absence of such a peptide or in the presence of a non-binding peptide (data not shown).

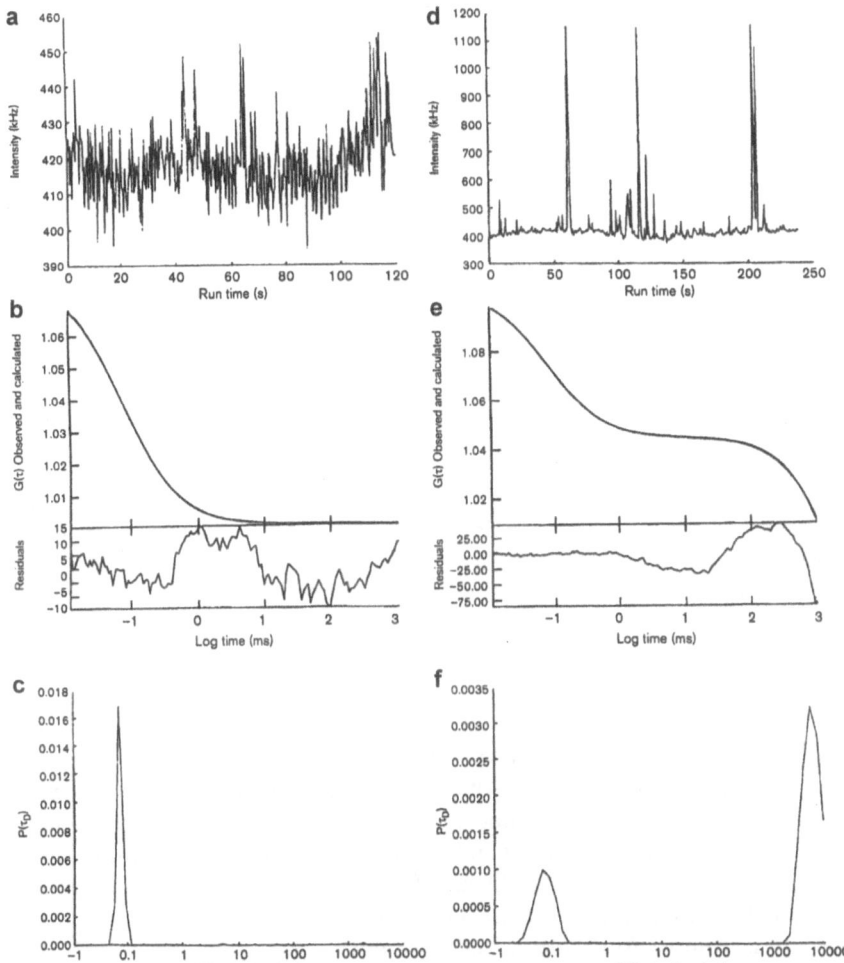

Fig. 11.9 a–f. FCS of the multimerization of Aβ using rhodamine-labeled Aβ(1-40) as probe. Aβ(80 μM) was incubated with Aβ-rhodamine (10 nM). Fluorescence intensity was measured after 2 min **a** and 40 min **d** incubation. The corresponding fluorescence intensity autocorrelation functions $G(\tau)$ were calculated as described in Chaps. 9 and 22. Observed and calculated data points are completely overlapping. Their difference is shown as residuals underneath (Fig. **b,e**). **b** The monophasic form of the curve indicates similar diffusion times (90 μs) of all species present. **e** The biphasic distribution after 40 min incubation indicates the presence of very large aggregates with a diffusion time of 3 s and of the initial monomers/dimers. **c,f** The corresponding distribution of diffusion times $P(\tau_o)$ are shown. For further details cf.[11.20]. Figure with permission from [11.20]

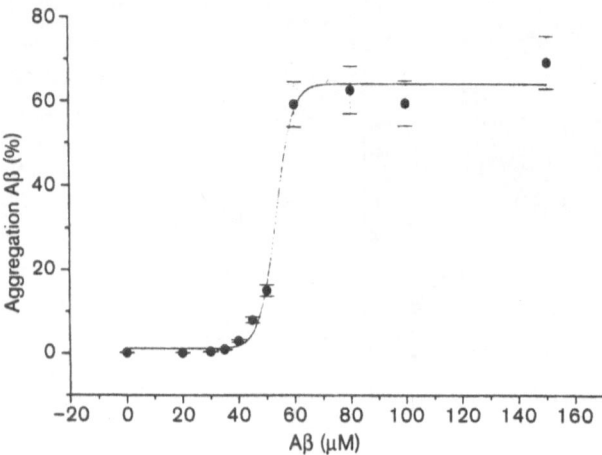

Fig. 11.10. Concentration dependence of Aβ multimerization. Different concentrations of Aβ with 10 nM rhodamine-labeled Aβ were incubated for 40 min. The percentage of Aβ aggregates was calculated from the distribution of diffusion times $P(\tau_0)$ (compare to Fig. 11.9). Figure with permission from [11.20]

11.3.2   Seeded Aggregation as a Diagnostic Tool

As shown above Aβ-peptides can multimerize after longer times (20 min) into larger aggregates that are detected by FCS as large peaks. The first steps in aggregate formation of Aβ are represented by a more or less continuous increase in the diffusion time corresponding to the formation of smaller oligomers. Both phenomena depend critically upon the concentration of Aβ and its molecular size (40 or 42 amino acids), and furthermore, upon incubation time and solvent conditions. Whereas in the preceding paragraph multimerization of Aβ(1-40) in high-salt buffer was examined and strongly cooperative multimerization at concentrations above 50 μM Aβ was found, in another study [11.23] the longer peptide Aβ(1-42) in lower ionic strength was tested for multimerization, and multimerization was detected, in fact, at concentrations as low as 100 nM Aβ-peptide. As described with PrP the process of seeded multimerization was induced by adding labeled Aβ-peptides to pre-existing multimeric particles. Investigation were made into whether Aβ multimers in the cerebrospinal fluid (CSF) of AD patients might act as "seeds" for polymerization [11.38]. Fluorescence-labeled Aβ 1-42, which was kept soluble in aqueous buffer containing 0.2% SDS was added to human CSF samples, thereby diluting it to 0.02% SDS. The final Aβ concentration was less than 100 nM, far below the critical solubility limit. Samples from the non-AD control group produced a fluctuating signal in the first 20 min after addition of the Aβ (Fig. 11.11a); however, those from AD-patients produced high-intensity peaks (Fig. 11.11b). Rarely, fluorescence bursts were also produced by samples from the control group, which might be caused by

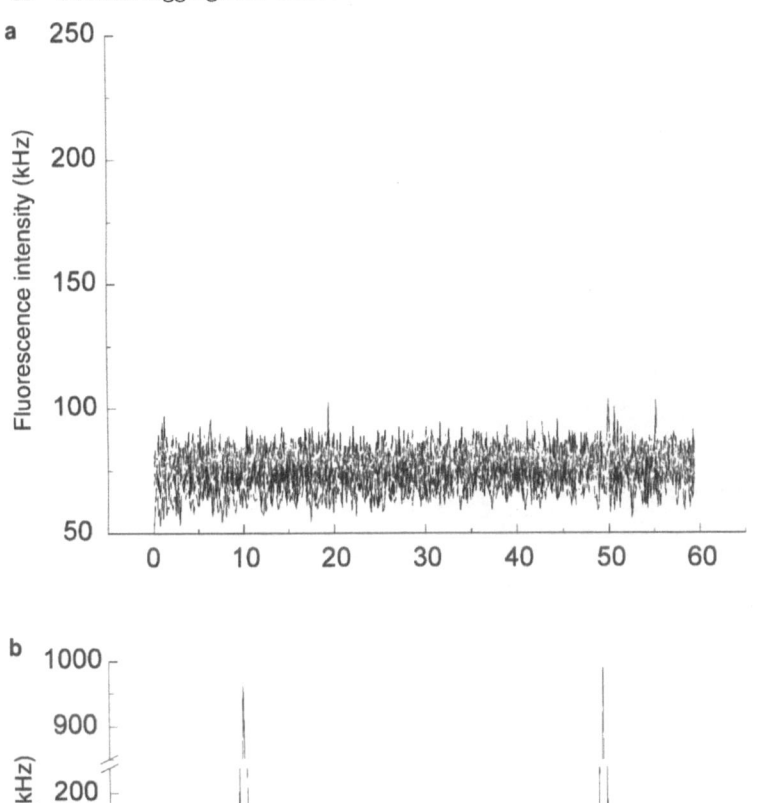

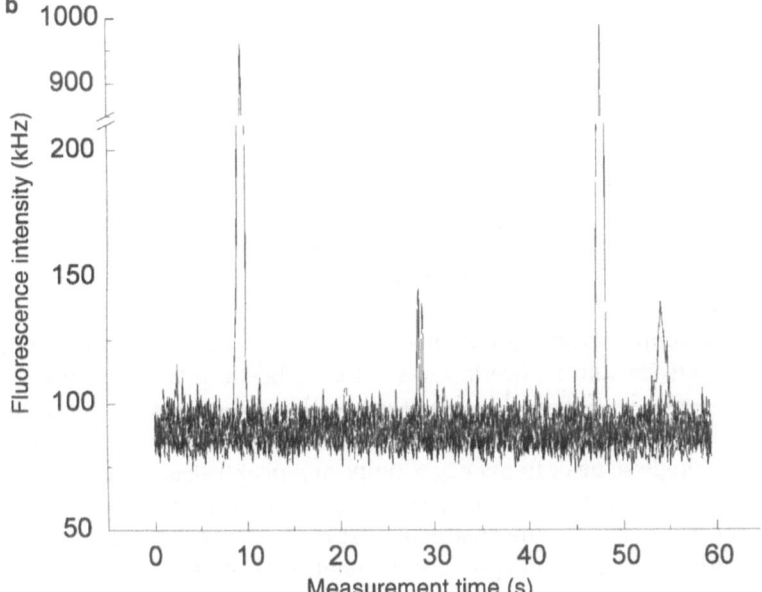

Fig. 11.11 a,b. Time course of the fluorescence intensity in FCS after the addition of fluorescence-labeled Aβ(1-42) to control **a** and AD **b** CSF. Each large peak in **b** represents a single multimeric particle to which labeled Aβ(1-42) is bound. Twenty scans of 60 s observation time each are recorded in series and superimposed. AD-specific peaks are more than three times the average fluctuation of the fluorescence intensity. Figure with permission from [11.38]

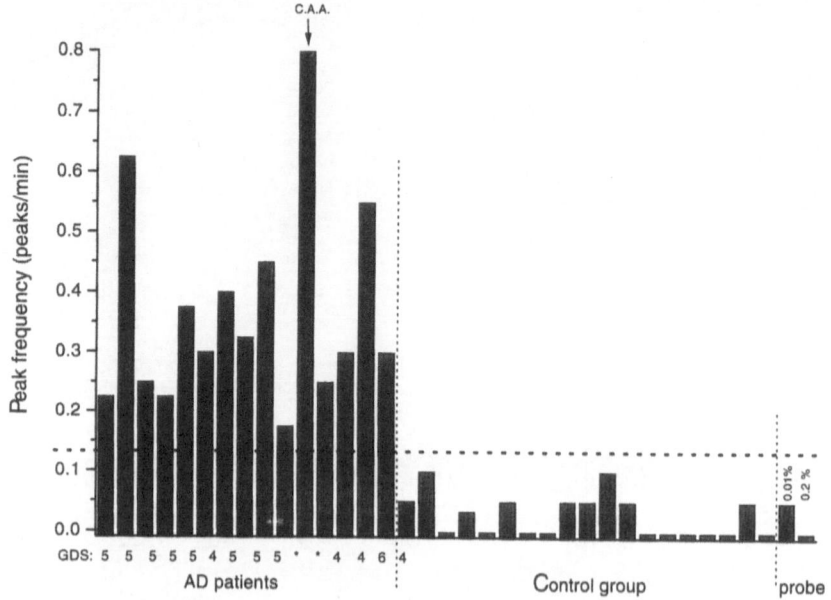

Fig. 11.12. Selective detection of labeled Aβ(1-42) deposited on multimeric particles in the CSF of AD patients and controls. Probe measurements represent the background from labeled Aβ(1-42) in phosphate buffer, either in 0.2% SDS or after dilution of 0.01% SDS, with a few events of spontaneous multimerization. C.A.A., the patient with cerebral amyloid angiopathy; GDS, global deterioration scale; *, GDS not done. Figure with permission from [11.38]

spontaneous multimerization of the fluorescent Aβ during the time span of the experiment (about 30 min) or by deposition of Aβ on other structures (for example, cell fragments) in the CSF. The number of peaks counted in 20–30 min in samples from 15 patients is more than those from the control group of 19 persons, and no overlap was detected between these groups. The results are shown in Fig. 11.12. A patient with cerebral amyloid angiopathy (a condition closely related to AD, with marked vascular Aβ deposition) produced the highest peak frequency. Comparing peak frequencies measured in the control group with those from spontaneous multimerization of the Aβ probe in phosphate buffer indicates that most of the unspecific peaks in these analyses might orginate from spontaneous multimerization of the probe.

Several tests were carried out in order to analyze the nature of the particles detected in the CSF of AD patients. Immunoprecipitation with the monoclonal Aβ antibody 6E10 reduced the peak frequency in AD CSF by an order of magnitude. Addition of Congo Red, which is known to inhibit the fibrillogenesis of Aβ [11.30], also lowered the peak frequency of Aβ-incorporation. Increased metal ion concentrations on the other hand might favor aggregation [11.31,11.32]. In fact, addition of 1 mM $ZnCl_2$ resulted in a higher number of peaks. From these three independent lines of evidence it was concluded that the particles detected in CSF of AD patients are mostly multimeric Aβ-proteins. The Aβ multimers could be detected in the CSF only from AD patients, whereas it is known that the amount of soluble Aβ in CSF of AD patients is either unchanged or decreased [11.33,11.34]; a decrease of soluble Aβ might result from Aβ multimer formation. Therefore, the specific detection of Aβ multimers in the CSF could be developed into a diagnostic test for AD.

To optimize the detection of Aβ multimers in CSF, the peak frequencies after the addition of Aβ 1-42 or Aβ 1-40 to CSF samples were compared and analyzed for their dependence on the SDS concentration after dilution. Aβ 1-42 produced significantly higher peak frequencies when used as a probe for AD-specific particles in the CSF, and a SDS concentration of 0.02% was the best compromise between sensitivity and high specifity, that is suppression of spontaneous multimerization (data not shown). The linearity of the FCS-determined peak frequency with the concentration of AD-specific particles could not be tested with CSF samples from AD patients because the number of those particles could not be determined independently. Therefore,

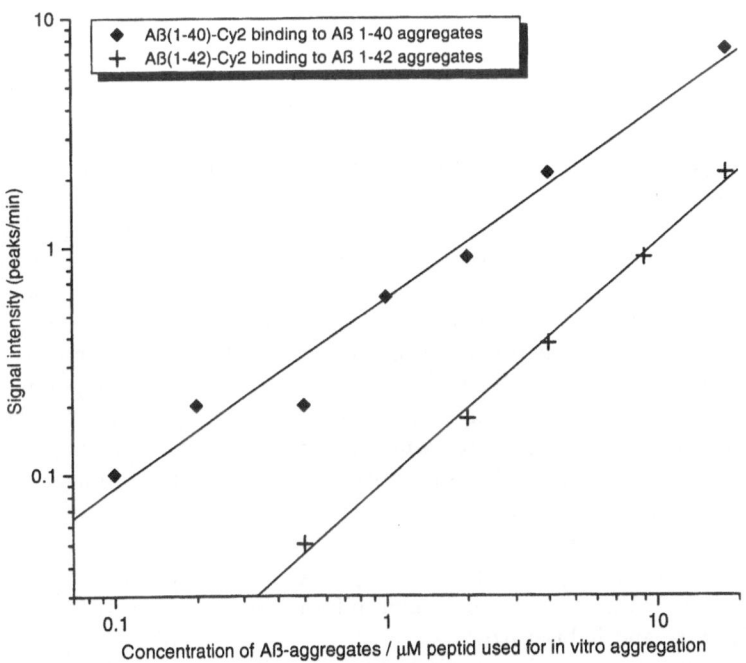

Fig. 11.13. Sensitivity of seeded multimerization of Aβ(1-40)Cy2 on Aβ(1-40) aggregates (**rhombus**) and of Aβ(1-42)Cy2 on Aβ(1-42) aggregates (**cross**). The experiment was carried out as described in Fig. 11.7. Incubation in 0.01% SDS, concentratin of the Aβ-peptide was 0.4 ng/μl. Figure from [11.23]

the linearity was tested with synthetic Aβ multimers and could be verified over almost two orders of magnitude (20 ng to 1000 ng multimer per 20 µl sample volume) with an average deviation of about 15%. In Fig. 11.13 those results are shown for Aβ(1-42)-Cy2 incorporation in Aβ(1-42)-aggregates and for Aβ(1-40)-Cy2 incorporation in Aβ(1-40)-aggregates. In contrast to measurements on CSF samples from AD patients where the Aβ(1-42) probe was more sensitive as compared to the Aβ(1-40)-probe, the synthetic pairs Aβ(1-40)-Cy2/Aβ(1-40)-aggregates produce more peaks in FCS as compared to Aβ(1-42)-Cy2/Aβ(1-42)-aggregates.

11.4 Synopsis

When multimerization of the prion protein or of the Aβ peptide was studied by FCS, it was possible to differentiate between spontaneous and seeded multimerization. For both proteins intermediate states during the multimerization could be identified by FCS. Within minutes after the monomeric or dimeric proteins are transfered into aqueous buffer, a polymorphic network of hundreds of monomers is formed. Whereas PrP has undergone a conformational transition from predominantly α-helical to a partially β-sheeted structure and the β-sheeted structure can be detected in the aggregates, the Aβ-aggregates still show random coil structure when analyzed by CD. Only during the consecutive slow formation of the fibrillar structure does the spectrum change to that of β-sheets; for PrP the formation of a fibrillar structure could be induced so far only with fragments of PrP [11.35] but not with full length PrP or sPrP27-30. One should note that the conformation of PrP and Aβ are different at zero time of the multimerization; PrP in 0.2% SDS is predominantly α-helical with some coiled parts whereas Aβ after dissolving is apparently completely in random coil structure. This may be why β-sheets form more rapidly in PrP. Rigler and colleagues [11.20] argue that the molecular interactions within the polymorphic network are still weak, and strong interactions are formed only during fibrillogenesis which can be detected then by CD or calorimetry [11.29]. They discussed also the fact that the polymorphic network was not detected in dynamic light scattering [11.36] as a lack of change of the index of refraction, which occurs only during figrillogenesis. The nuclei which are responsible for cooperativity (see Fig. 11.10) might form within the network. In summary, different methods allow one to detect different intermediate states; FCS in agreement with electron microscopy and ultracentrifugation is capable of analyzing large diffuse networks, whereas only aggregates with stronger molecular interactions are detected also by CD, calorimetry and light scattering.

Seeded multimerization was exploited in a specific situation: In the cerebrospinal fluid (CSF), very large multimers of Aβ were present, to which monomeric fluorescent Aβ-peptides were added. Additional results indicated that the major component of the particles detected in the CSF of AD patients are Aβ multimers. The same principle was applied to detect prions in scrapie-infected tissue, when a labeled PrP probe was used. Work is in progress to search for PrP-aggregates in the CSF of patients with Creutzfeldt–Jakob disease [11.39] or of animals with scrapie or BSE [11.23]. At present the main emphasis is on the biophysical basics of this new method; more extensive clinical studies will detail its sensitivity and specificity as a diagnostic test and demonstrate the relation between FCS-determined peak frequency and the clinical and pathological stage of AD. The inclusion of preclinical familial AD cases will demonstrate its validity in testing for early AD. Since Aβ-multimerization is a critical step in AD pathogenesis one possibility for therapy development is inhibitors of the multimerization as outlined above, and FCS would be a suitable method for screening those inhibitor compounds.

Acknowledgements

The author is indebted to Drs Th. Appel, M. Pitschke, and K. Post for many stimulating discussions and to Ms. H. Gruber for help in preparing the manuscript.

References

11.1 K. Beyreuther, P. Pollwein, G. Multhaup, U. Monning, G. Konig, T. Dyrks, W. Schubert, and C.L. Masters: Ann. N.Y. Acad. Sci. **695**, 91–102 (1993)

11.2 S.B. Prusiner: Molecular biology of prion diseases. Sience **252**, 1515–1522 (1991)

11.3 S.B. Prusiner: Science **216**, 136–144 (1982)

11.4 D.J. Selkoe: Annu. Rev. Neurosci. **17**, 489–517 (1994)

11.5 J. Hardy: Trends Neurosci. **20**, 154–159 (1997)

11.6 E.-M. Mandelkow, O. Schweers, G. Drewes, D. Bierant, N. Gustke, B. Trinczek, and E. Mandelkow: Ann. N.Y. Acad, Sci. **777**, 96–106 (1996)

11.7 K.-M. Pan, M. Baldwin, J. Nguyen, M. Gasset, A. Serban, D. Groth, I., Mehlhorn, Z. Huang, R.J. Fletterick, F.E. Cohen, and S.B. Prusiner: Proc. Natl. Acad. Sci. USA **90**, 10962–10966 (1993)

11.8 J. Safar, P.P. Roller, D.C. Gaydusek, and C.J. Gibbs, Jr: J. Biol. Chem. **268**, 20276–20284 (1993)

11.9 Z. Huang, J.-M. Gabriel, M.A. Balwin, J.R. Fletterick, S.B. Prusiner, and F.E. Cohen: Proc. Natl. Acad. Sci. USA **91**, 7139–7143 (1994)

11.10 R. Riek, S. Hornemann, G. Wider, M. Billeter, R. Glockshuber, and K, Wüthrich: Nature **382**, 180–182 (1996)

11.11 R. Riek, S. Hornemann, G. Wider, R. Glockshuber, and K. Wüthrich: FEBS-Lett. **413**, 282–288 (1997)

11.12 T.L. James, H. Liu, N.B. Ulyanov, S. Farr-Jones, H. Zhang, D.G. Donne, K. Kaneko, D. Groth, I. Mehlhorn, S.B. Prusiner, and F.E. Cohen: Proc. Natl. Acad. Sci. USA **94**, 10086–10091 (1997)

11.13 D.G. Donne, J.H. Viles, D. Groth, I. Mehlhorn, T. James, F.E. Cohen, S.B. Prusiner, P. Wright, and H.J. Dyson: Proc. Natl. Acad. Sci. USA **94**, 13452–13457 (1997)

11.14 M. Gasset, M.A. Baldwin, R.J. Fletterick, and S.B. Prusiner: Proc. Natl. Acad. Sci. USA. **90**, 1–5 (1993)

11.15 D. Riesner, K. Kellings, K. Post, H. Wille, H. Serban, D. Groth, M.B. Baldwin, and S.B. Prusiner: J. Virol. **70**, 1714–1722 (1996)

11.16 I. Mehlhorn, D. Groth, J. Stöckel, B. Moffat, D. Reilly, D. Yansuro, W.S. Willet, M. Baldwin, R. Fletterick, F.E. Cohen, R. Vandlen, D. Henner, and S.B. Prusiner: Biochemistry **35**, 5528–5537 (1996)

11.17 K. Post, M. Pitschke, O. Schäfer, H. Wille, T.R. Appel, D. Kirsch, I. Mehlhorn, H. Serban, S.B. Prusiner, and D. Riesner: Biol. Chem. **379**, 1307–1317 (1998)

11.18 B. Schmidt, W. Rappold, V. Rosenbaum, R., Fischer, and D. Riesner: Colloid Polym. Sci. **268**, 45–54 (1990)

11.19 M. Pitschke, K. Post and D. Riesner: Colloid Polym. Sci. **107**, 72–76 (1997)

11.20 L.O. Tjernberg, A. Pramanik, S. Björling, P. Thyberg, J. Thyberg, Ch. Nordstedt, K.D. Berndt, L. Terrenius, and R. Rigler: Chem. Biol. **6**, 53–62 (1999)

11.21 R. Rigler, U. Mets, J. Widengren, and P. Kask: Eur. Biophys. J. **22**, 169–175 (1993)

11.22 R. Rigler: J. Biotech. **41**, 177–186 (1995)

11.23 M. Pitschke: In: Mechanismus der Multimerisierung von Prion-Proteinen als Basis eines biophysikalischen Diagnoseverfahrens von Spongiformen Encepahlopathien und der Alzheimerschen Erkrankung (Thesis, Heinrich-Heine-Universität Düsseldorf 1999)

11.24 J.H. Come, P.E. Fraser, and P.T. Lansbury: Proc. Natl. Acad. Sci. USA **90**, 5959–5963 (1993)

11.25 J.T. Jarett and P.T. Lansbury, Jr: Cell **73**, 1055–1058 (1993)

11.26 O. Schäfer: In: Biophysikalische Analyse verschiedener Aggregationszustände von rekombinantem Prion-Protein (Diplomarbeit, Heinrich-Heine-Universität Düsseldorf 1997)

11.27 O.B. Ptitsyn: FEBS Letters **285**, 176–181 (1991)

11.28 G.E. Schulz and R.I.J. Schirmer: In: Principles of Protein Structure (Springer, New York, Berlin, Heidelberg, Tokyo 1979)

11.29 E. Terzi, G. Holzemann, and J. Seelig: J. Mol. Biol. **252**, 633–642 (1995)

11.30 A. Lorenzo and B.A. Yankner: Proc. Natl. Acad. Sci. USA **91**, 12243–12247 (1994)

11.31 A.I. Bush, W.H. Pettingell, G. Multhaup, M. Paradis, J.P. Vonsattel, J.F. Gusella, K. Beyreuther, C.L. Masters, and R.E. Tanzi: Science **265**, 1464–1467 (1994)

11.32 A.M. Brown, D.M. Tummolo, K.J. Rhodes, J.-K. Hofmann, J.S. Jacobson, and Sonnenberg-Reines: J. Neurochem. **69**, 1204–1212 (1997)

11.33 R. Motter, D. Vigo-Pelfrey, D. Kholodenko, R. Barbour, K. Johnson-Wood, D. Galasko, L. Chang, G. Miller, C. Clark, and R. Green: Ann. Neurol. **38**, 643–648 (1995)

11.34 R.M. Nitsch, G.W. Rebeck, M. Deng, U.I. Richardson, M. Tennis, D.B. Schenk, D. Vigo-Pelfrey, I. Lieberburg, R.J. Wurtman, and B.T. Hyman: Ann. Neurol. **37**, 512–518 (1995)

11.35 K.M. Lundberg, C.J. Stenland, F.E. Cohnen, S.B. Prusiner, and G.L. Millhauser: Chem. Biol. **4**, 345–355 (1997)

11.36 A. Lomakin, D.S. Chung, G.B. Benedek, D.A. Kirschner, and D.B. Teplow: Proc. Natl. Acad. Sci. USA **93**, 1125–1129 (1996)

11.37 K. Post, D.R. Brown, M. Groschup, H.A. Kretschmar, and D. Riesner: Arch. Virol. **16**, 265–273 (2000)
11.38 M. Pitschke, R. Prior, M. Haupt, and D. Riesner: Nature Med. **4**, 832–834 (1998)
11.39 J. Bieschke, A. Giese, W. Schulz-Schaeffer, J. Zerr, S. Poser, M. Eigen, and H. Kretschmar: Proc. Natl. Acad. Sci. USA **97**, 5468–5473 (2000)

Part IV

Environmental Analysis and Monitoring

12 Application of FCS to the Study of Environmental Systems

Konstantin Starchev, Kevin J. Wilkinson, and Jacques Buffle

12.1 Introduction

The key role of submicrometer sized colloids and biopolymers in the transport of trace metals and organic pollutants in waters and soils is now well-documented. Through covalent, electrostatic or hydrophobic interactions [12.1,12.2], a large proportion (often 40–90%) of trace compounds may be adsorbed on aquatic colloids [12.3,12.4]. Subsequently, trace compounds can be transported long distances on stable colloids, or quickly sedimented on flocculating colloids and biopolymers. In order to thoroughly understand these processes, information on the exact nature of the colloids and biopolymers is required.

In natural waters, a key question to be resolved is the exact role of natural organic matter (NOM). NOM is composed of large numbers of biopolymers with poorly defined structures (see below). It is generally accepted that NOM will stabilize inorganic colloids in natural waters [12.5,12.6]. On the other hand, for certain specific groups of biopolymers, in particular the high molecular weight polysaccharides and peptidoglycans, the opposite phenomena, i.e. a destabilization, has been shown to occur [12.7,12.8]. In the water treatment process, biopolymers play similar dual roles, either by facilitating or by making more difficult the desired coagulation step. A thorough understanding of the properties and interactions of the major colloids and biopolymers is therefore key to the prediction of the circulation of trace metals or organic pollutants in waters, soil or sediments and to the efficient elimination of suspended matter via floc formation in the water treatment process. These concepts are also essential to our understanding of microstructures such as those occurring in soils and sediments and the transport of chemical compounds through these media.

A large number of sophisticated and sensitive instrumental techniques have been developed to determine the *chemical* composition of inorganic colloids (e.g., atomic absorption, X-ray fluorescence, ICP-MS) and organic biopolymers (e.g., HPLC, mass spectrometry). On the other hand, the number of techniques to study *supramolecular physical* properties such as the size, conformation, or fractal dimensions of the submicrometer colloids, large biopolymers or their aggregates is quite limited. In particular, static light scattering and photon correlation spectroscopy are neither selective nor sensitive enough for the measurement of most environmentally relevant concentrations of the smaller colloids and polymers in natural systems [12.9,12.10]. Microscopic techniques such as transmission electron microscopy (TEM) and atomic force microscopy (AFM) are informative but require significant skill and time in order to obtain representative results [12.9,12.11,12.12]. In that respect, fluorescence correlation spectroscopy (FCS) offers significant potential advantages for environmental systems (see Sect. 12.2.3 for a detailed discussion). Until recently, FCS has been applied mainly to biochemical and pharmaceutical studies, and only rarely to environmentally relevant, especially aquatic, systems. Prior to the routine application of FCS, it will be necessary to take into account the conditions which distinguish environmental systems from most others, i.e. very low concentrations of colloids and biopolymers, a large size polydispersity and chemical heterogeneity, and the tendency to aggregate.

12.2   Nature and Characteristics of Aquatic and Terrestrial Colloids and Biopolymers

In this Chapter, *colloid* is used for any organic or inorganic entity large enough to have supramolecular structure and properties (e.g., the possibility of a modification of the conformation or surface properties) but small enough so as not to sediment quickly (hours-days) in the absence of aggregation. Colloids are thus typically in the size range of 1 nm to 1 μm. Most organic and inorganic colloids are negatively charged [12.1,12.13,12.14]. The major components of aquatic systems are listed in Fig. 12.1.

12.2.1   Nature of the Major Aquatic and Terrestrial Colloids

The major *inorganic colloids* include aluminosilicates (clays), silica and Fe oxyhydroxides. Aluminosilicates are angular sheet-like particles (Fig. 12.2). Electron microscopy reveals that iron oxyhydroxides and silica colloids are often near spherical, although silica can also be found as irregularly shaped diatom debris. Calcium carbonate is also present in aquatic systems but is generally found in the larger size fraction.

Therefore, as a first approximation, the inorganic colloids, despite their variable shapes, can be treated as compact entities with a negative surface charge [12.13]. Two important characteristics of the inorganic colloids will influence both the rate at which they form aggregates as well as aggregate structure:

i) their irregular shape and physico-chemical heterogeneity,
ii) their large size distribution.

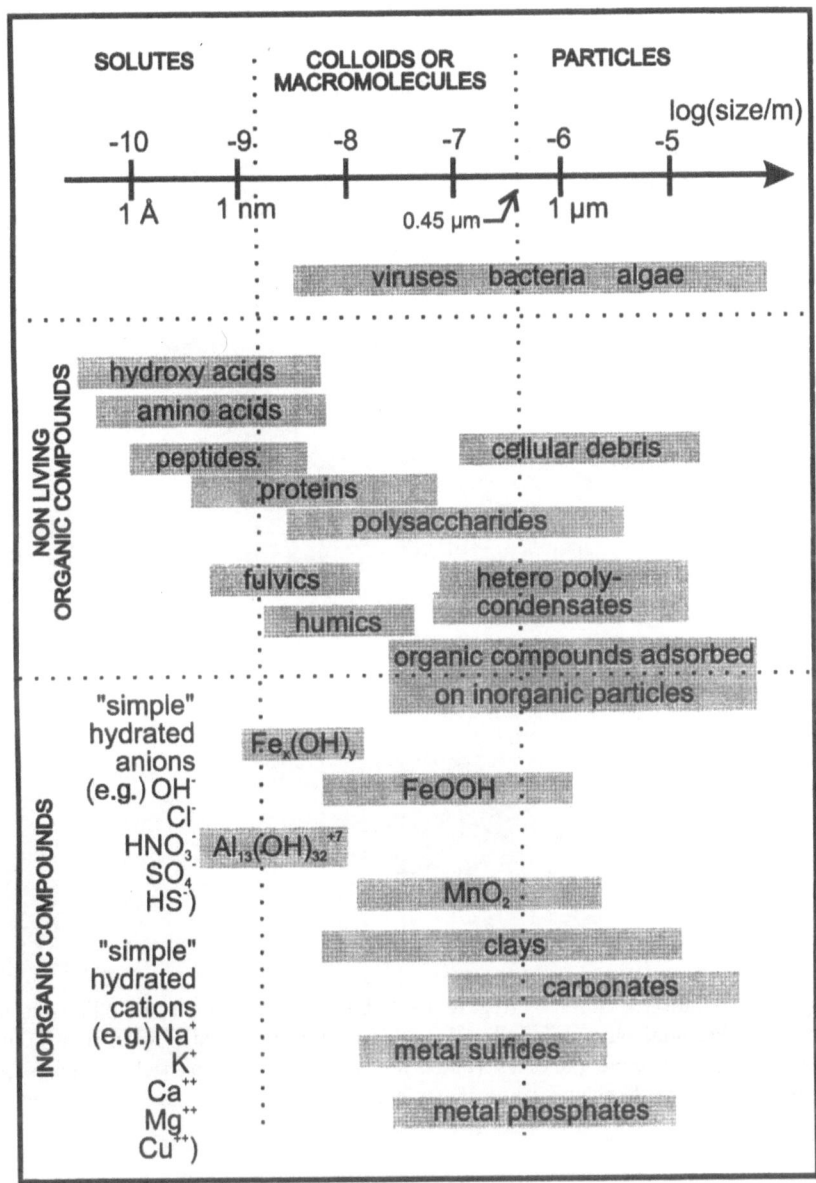

Fig. 12.1. Schematic representation, by size, of some of the important organic and inorganic components of natural waters [12.14]

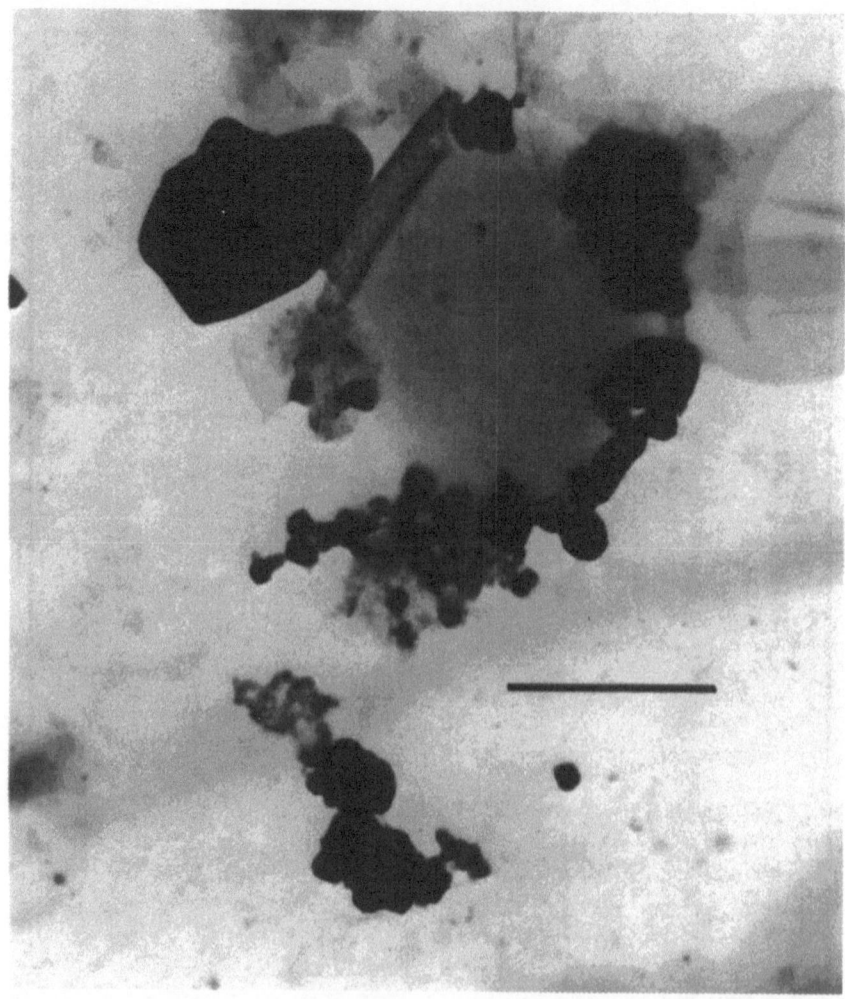

Fig. 12.2. Compact heteroaggregate from Lake Bret, Switzerland. The picture shows a spheroidal particle of silica (gray at center) aggregated with many much smaller iron hydroxide particles (black spheroids), a clay particle (black angular particle at top left) and some biological debris. The scale bar corresponds to 250 nm [12.13]

The overall size distribution of the colloids follows a Pareto power law distribution (Fig. 12.3; [12.15]) over several orders of magnitude in water, sediments and soils [12.14,12.16]. Interestingly, size distribution peaks do exist in specific size ranges [12.15], resulting from specific production and/or elimination processes. Detailed knowledge of the exact causes and nature of such size distributions is required in order to allow the interpretation of the corresponding environmental processes.

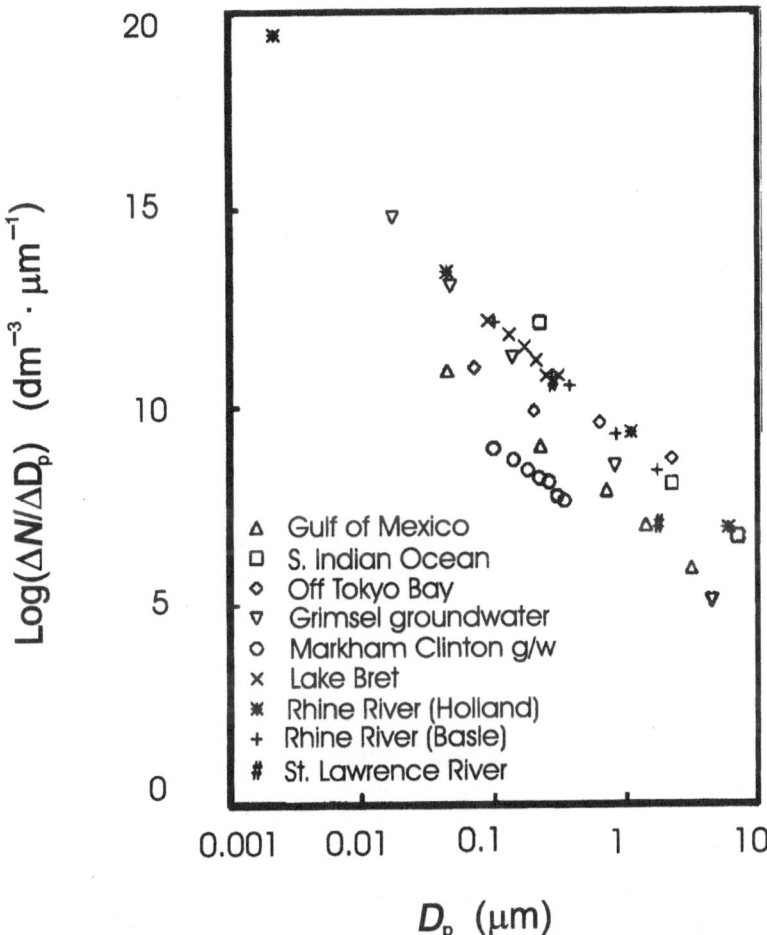

Fig. 12.3. Size distributions based on the particle number for different natural aquatic systems [12.10]

In water, soils and sediments, NOM includes two major groups of refractory biopolymers produced from the transformation of microbial exudates. The *polysaccharides* and *peptidoglycans* are high molecular weight (10^4–10^6 Da) biopolymers that are released from the cell walls of plankton (Fig. 12.4; [12.13,12.14]). They are generally neutral or slightly negatively charged due to carboxylate groups. Due to their association into double or triple helices stabilized by hydrogen or calcium bridges [12.17] and their high degree of hydration (up to 80%), structurally rigid fibrils are often formed. In many cases, the total length of the fibrils may be 1 μm or greater, whereas their thickness is not more than a few nm [12.13]. These rigid biopolymers may represent 10–30% of surface water NOM [12.8]. More flexible large

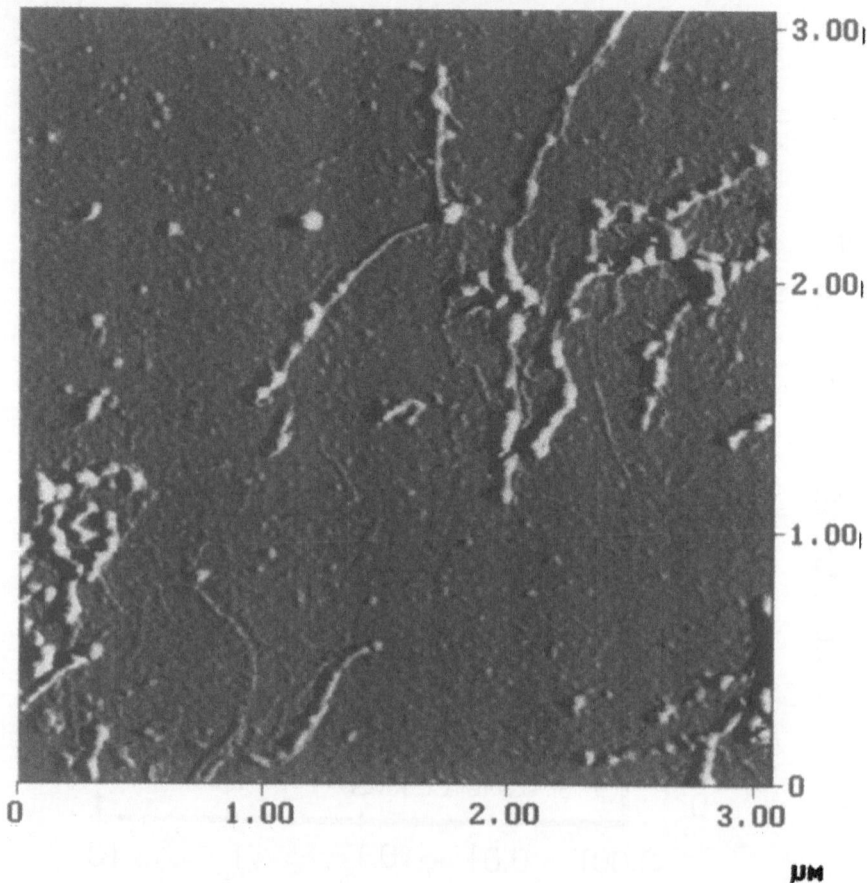

Fig. 12.4. AFM images of fibrils extracted from the pelagic zone of the Gulf of Mexico. Compact colloids appear to be associated with most fibrils [12.18]

biopolymers, such as the alginates in sea water, might also be important; however, their role is presently not well-known.

The other major group of aquatic NOM are the *humic substances* (HS), originating from the degradation of higher plants and micro-organisms into small molecules, followed by their recombination into chemically heterogeneous small polymers (molar mass typically in the 500–5000 range). These compounds may be discriminated into two main groups [12.13,12.14]: (i) those produced by the degradation of micro-organisms in the water column, which are largely aliphatic with a lower charge density (-2 to -5 meq g^{-1}), and (ii) the soil-derived fulvics which are largely aromatic, have a high proportion of carboxylic and phenolic groups, and a higher charge density (-6 to -11 meq g^{-1}). The soil-derived fulvics are ubiquitous in surface freshwaters where they often represent 40–80% of NOM. They are the most well-

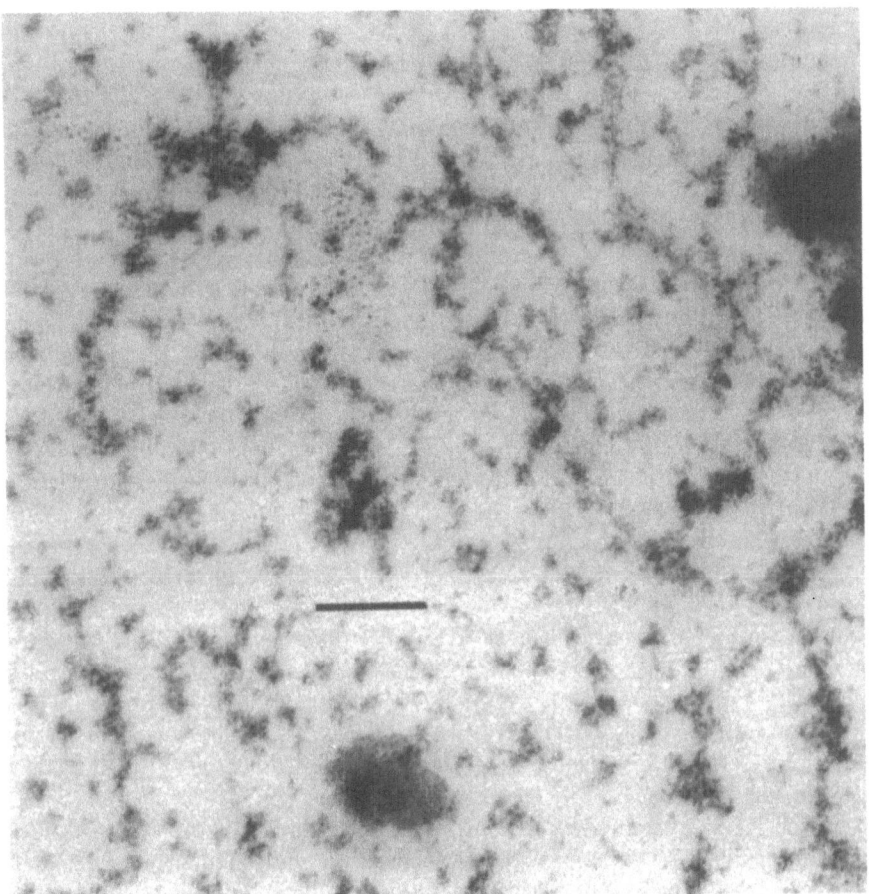

Fig. 12.5. TEM image of humic rich NOM from a Lake Bret sample. This image is interpreted as having individual fulvic macromolecules (the smallest black points), homoaggregates (association of black points) and the association of fulvics with fibrillar compounds. The scale bar corresponds to 100 nm [12.13]

characterized of the aquatic NOM, generally resembling 1–3 nm small spheres (Fig. 12.5; see also Sect. 12.4). Furthermore, they are highly hydrated and can self-aggregate due to the presence of hydrophobic moieties (Fig. 12.5). Due to their aromatic structure, they are naturally fluorescent, which allows for their direct study by FCS without the need for labeling (Sect. 12.4).

12.2.2   Aggregation Processes and Aggregate Structure

A major feature of aquatic and soil colloids and biopolymers is their heterogeneity in any given environmental medium. This heterogeneity leads to the formation of heteroaggregates (aggregates of colloids or polymers of different natures) which constitute the microscopic basis of soil, sediment and floc structures. The heteroaggregates strongly influence the circulation and behavior of chemical compounds in environmental media.

Aggregation between colloids and/or biopolymers results from the competition between repulsive (mostly electrostatic) and attractive (van der Waals, hydrophobic bonding, chemical coordination, hydrogen bonds) forces. These opposing forces define an energy barrier, which, if overcome by the kinetic energy of the particles moving in the water, will lead to aggregation at a rate which depends on the maximum energy of the barrier. In homoaggregation (aggregation between colloids of the same nature), the electrostatic forces between particles are generally repulsive, whereas in heteroaggregation they may be attractive, thus facilitating the aggregation process. Homoaggregation is the only case for which a detailed theory presently exists (e.g., see [12.19–12.21]. Despite the diversity of natural particles, application of the principles valid for the homoaggregation of hard spheres has been relatively successful for aquatic inorganic colloids larger than $0.1\,\mu m$, possibly because they are often coated with a comparatively thin layer of negatively charged fulvics [12.6]. It is likely that the adsorbed layer fixes the surface charge and chemical reactivity of different types of particles at similar values.

On the other hand, the situation is quite different for smaller ($< 0.1\,\mu m$) colloids [12.15,12.22]. With or without a humic coating, they may aggregate with the slightly charged and comparatively larger rigid fibrils (bridging flocculation). Large, loose flocs are thus formed with structure which is very different from the compact aggregates of inorganic colloids (Fig. 12.2). Furthermore, the flocs can quickly grow to sizes large enough to sediment from solution [12.13,12.23], whereas weeks or months should be necessary to form aggregates larger than a micrometer, based on the coagulation of the submicrometer compact colloids alone. The dynamic simulation of homo- and heteroaggregation processes using small inorganic colloids and comparatively large rigid biopolymers produces floc structures similar to those observed by electron microscopy or AFM.

Therefore, as a first approximation, colloidal aquatic systems can be considered to be a mixture of three types of components: inorganic colloids, fulvics and large biopolymers. Their interactions will depend on several factors including:

- the respective ratios of the inorganic colloids, fulvics and large biopolymers, in particular, the rigid fibrils
- the charge density of the fulvics, large biopolymers and colloids
- the polydispersity of the inorganic colloids and the large biopolymers
- the conformation and flexibility of the large bridging biopolymers
- the diffusional properties of the three types of compounds [12.15,12.21].

12.2.3   Potential Advantages and Limitations of FCS for Environmental Applications

Many of the aforementioned properties can be studied by means of FCS, which makes this technique a particularly useful tool for environmental studies. However, there are some potential limitations, several of which are briefly summarized below. Section 12.3 discusses how some of the limitations can be overcome. Section 12.4 discusses a few applications of FCS to natural colloids.

A major advantage of FCS for environmental studies is the high sensitivity of the technique which allows the study of natural colloids in situ at realistic concentrations. This is not the case for other in situ techniques such as static light scattering (SLS) and photon correlation spectroscopy (PCS) for which colloid concentrations must be used which exceed natural ones by orders of magnitude, particularly in the smaller size ranges. For example, the concentrations of inorganic colloids are typically less than $1\,mg\,L^{-1}$; those of the humics $1–10\,mg\,L^{-1}$, whereas the polysaccharides occur naturally at less than $1–2\,mg\,L^{-1}$. The sensitivity limit of PCS is $10\,mg\,L^{-1}$ for $14\,nm$ particles [12.10] and more than $100\,mg\,L^{-1}$ for polysaccharides with molecular weights of $400\,kD$. Low molecular weight fulvics are almost impossible to measure with PCS. In contrast, diffusion coefficients can be determined by FCS for a $0.1\,mg\,L^{-1}$ suspension of $14\,nm$ silica particles or for a $1\,mg\,L^{-1}$ solution of humics.

Another important characteristic of FCS is its selectivity to specifically labeled molecules. The selectivity is a major advantage in chemically heterogeneous media and should allow the study of formation kinetics and the structures of complex heteroaggregates. The selectivity of FCS should also allow the measurement of diffusion in complex media like gels [12.24], concentrated solutions [12.25] or on cell surfaces [12.26], where methods such as PCS are not appropriate due to the strong background signal. It is clear that insight into processes in complex media are key to our understanding of the bio-uptake of pollutants or nutrients in soils, sediments or bacterial flocs.

One important limitation of the FCS technique is that there are only a few naturally fluorescent aquatic or terrestrial colloids and biopolymers, of which the humic substances have been, by far, the most well studied [12.27]. Humics present broad excitation and emission bands (typical width at half height $\sim 100\,nm$) with maxima near 350 and $450\,nm$, respectively (the exact values may vary according to the origin of the humics). Nearly all other aquatic colloids are either non-fluorescent, or even fluorescence quenchers. For example, the iron oxyhydroxides are a ubiquitous colloidal component, which, despite their low concentration, are an important fluorescence quencher. As with the fulvics, they often coat other colloids or polymers. Therefore, it is generally quite difficult to label the natural colloids since: (i) special fluorophores (e.g., with spacers) should be used in order to avoid interactions between the fluorophore and potential quenchers, and (ii) the labeling procedure should not perturb the conformation of the biopolymers. Significant work is still required in order to develop appropriate labeling procedures. This problem does not occur when working with most model colloidal compounds.

Apart from the problems mentioned above, a major limitation of FCS for environmental applications is the large polydispersity of all natural colloidal systems. This problem is even more important in systems where aggregation occurs, since even the aggregation of initially monodisperse colloids will lead to polydisperse systems. In PCS, the measured average particle size and the particle distribution are strongly biased towards larger sizes since the intensity of the scattered light generally depends on R^6, where R is the particle radius. The signal due to particles with a radius which is smaller than that of the major particle fraction is thus negligible. On the other hand, the size weighting factor in FCS depends on the nature of the labeling. In cases where different size fractions are labeled with the same number of fluorophores, a number distribution can be obtained which is not biased towards larger fractions. For example, this is often the case for studies of DNA or proteins which are labeled at one end of the chain. In other cases, particularly in environmental studies, colloids must be labeled by adsorption of the fluorophores to the particle surface. In this case, since the FCS signal is proportional to the square of the fluorescence intensity and the surface is roughly proportional to R^2, the relative weight of the different fractions will depend on R^4. In any case, the bias towards the larger particle sizes should be smaller than that of PCS.

A final aspect to consider is that classical FCS theory is limited to studies of infinitely small particles and/or to particles with only one fluorescent center. Recent studies examining particles with diameters which are comparable to the width of the laser beam has demonstrated that, in such cases, the apparent diffusion time and particle concentrations are larger than the actual values [12.28]. This factor may limit studies in the case of very polydisperse natural samples, or for studies of aggregation processes leading to sedimentation, where the mechanisms leading to the formation of aggregates of several micrometers remains to be understood. Although these limitations are not always easy to overcome, some examples of the ongoing work in this direction are given in Sect. 12.3. In any case, the potential advantages of the FCS technique for environmental samples clearly outweigh the disadvantages.

12.3   Development of FCS for its Application to the Study of Environmental Systems

12.3.1   Colloids With Sizes Comparable to the Beam Width

An analytical solution of the FCS autocorrelation function, $g(t)$, has only been published for infinitely small particles (molecules) diffusing in a three-dimensional Gaussian beam [12.29,12.30]:

$$g(t) = g(0) \left(1 + \frac{t}{\tau_1}\right)^{-1} \left(1 + \frac{t}{p^2 \tau_1}\right)^{-1/2} , \qquad (12.1)$$

where $g(0)$ is the autocorrelation function at time 0, τ_1 is the characteristic diffusion time of the particle through the sample (confocal) volume, t is the delay time, and p is the structure parameter or the ratio between the transversal and longitudinal radius of the sample volume, $p = \omega_1/\omega_2$.

When the particle size becomes comparable to the size of confocal volume, some fluorophores will be illuminated more strongly than others depending on the distribution of the fluorophores on the particle and the intensity profile of the confocal volume. In this case, the autocorrelation function will depend on the particle size and shape. Nevertheless, for disc shaped particles with radii (R) that do not exceed the beam radius (ω_1) (i.e., $R/\omega_1 \leq 1$), it has been shown by 2D Monte Carlo simulations [12.28] that Equation (12.1) remains valid after replacing τ and $G(0)$ by their effective (eff) values.

$$\tau_{\text{eff}} = \tau_{\text{inf}} + \frac{R^2}{4D} = \frac{\omega_1^2 + R^2}{4D} \tag{12.2}$$

$$\tau_{\text{eff}}(0) = G(0)_{\text{inf}} \frac{\omega_1^2}{\omega_1^2 + R^2} = G(0)_{\text{inf}} \left(\frac{1}{1 + \left(\frac{R}{\omega_1}\right)^2} \right), \tag{12.3}$$

where the subscript inf refers to infinitely small particles compared to ω_1, and D is the diffusion coefficient of the particles. Equations (12.2) and (12.3) apply to particles with fluorophores which are uniformly distributed on the surface. The physical meaning of (12.2) and (12.3) is that large particles spend a longer time in the beam because parts of them are illuminated long after the particle center has left the confocal volume. This effect is illustrated in Fig. 12.6 which shows the results for the simulation of the diffusion of disc particles in a 2D Gaussian beam. The ability of (12.2) and (12.3) to describe the simulated results is better for particles with $R/\omega_1 \leq 1.0$. For $R/\omega_1 \geq 3.0$ (insert of Fig. 12.1), the shape of the autocorrelation curve is changed and can no longer be represented by Equations (12.1) to (12.3).

Experimental verification of these results has been performed with standard latex particles with narrow size distributions ($\leq 5\%$ in most cases). Particles were labeled by adsorption of R6G and their size was measured simultaneously by FCS and PCS. The data are shown in Fig. 12.7 along with data obtained by computer simulation. The expected apparent size calculated from τ_{eff} using (12.2) is shown by the dotted line. Excellent agreement was found between sizes measured by PCS and FCS using theory developed for "infinitely" small particles (i.e. for the 30, 60 and 100 nm particles; i.e. $R/\omega_1 \leq 0.2$). As predicted by (12.2), the deviation between sizes measured by PCS and apparent sizes obtained by FCS is significant for larger particles. The small deviations between the experimental results and the theoretical curves based upon (12.2) are most likely due to sample polydispersity.

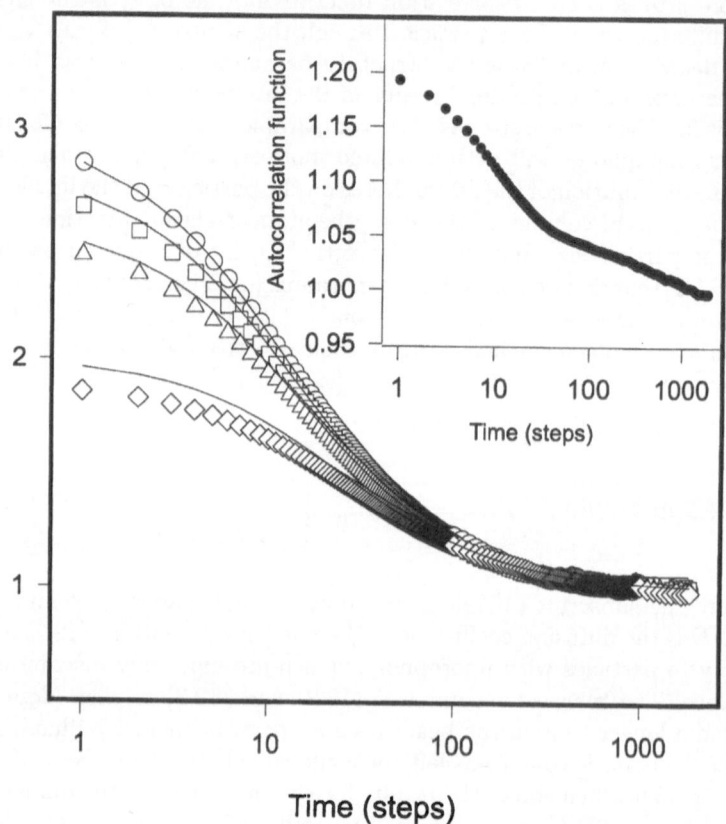

Fig. 12.6. Autocorrelation function of disk particles in a 2D Gaussian beam obtained by Monte Carlo simulations. $R/\omega_1 = 0.1$ **circles**, 0.3 **squares**, 0.5 **triangles**, 1.0 **diamonds**. The lines are obtained using (12.2) and (12.3). *Inset:* $R/\omega_1 = 3.0$ [12.28]

12.3.2 Polydisperse Systems

The various physico-chemical methods used to determine the particle size of polydisperse systems may provide different average values based upon the operating principles of the technique. For example, z-average diffusion coefficients are determined by PCS; weight-average molecular masses by SLS and number average molecular weight by techniques such as membrane osmometry. The combination of these values can be used to determine the sample polydispersity. Thompson [12.31] has pointed out that the weight-average molecular weight can be calculated from the zero-time FCS autocorrelation function. However, this approach is only valid if the fluorescence yield of each molecule in the suspension is proportional to its molecular weight. On the

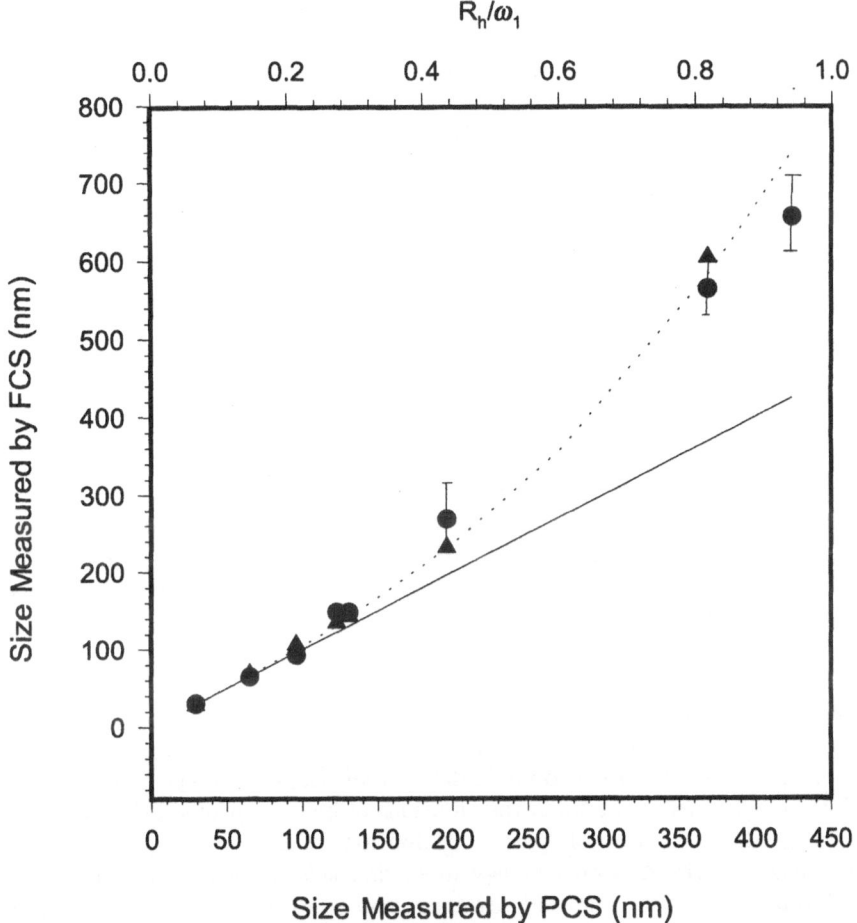

Fig. 12.7. Hydrodynamic sizes of the latex particles as determined by FCS and PCS. The **circles** are the experimental measurements. Computer simulation is represented by the **triangles**. The **dotted line** is the apparent size of the particles calculated using (12.2). The **solid line** gives the | : | correlation with sizes measured by PCS. The error bars represent standard deviations from five replicate measurements is given by **bars**. The top scale gives an estimate of the R/ω_1 ratio [12.28]

other hand, if the fluorescence of each species is constant, i.e. independent of the molecular weight, then a number-average molecular weight is obtained.

Results obtained by fitting the decay of the FCS autocorrelation function can be strongly influenced by the polydispersity of the suspension. In this case, the forced fitting of (12.1) is not very helpful because it yields an ill-defined average for τ_1. Furthermore, the method of cumulants proposed for

PCS [12.32] is not applicable to FCS because of the strong nonlinearity of
the FCS autocorrelation function, even in the case of monodisperse solutions.
On the other hand, the method of histograms has been successfully employed
[12.33] to determine size distributions from the FCS autocorrelation function.
In this case, the FCS diffusion time scale is divided into a finite number (n)
of intervals (i). The corresponding fraction, c_i, of particles in each interval is
then represented by a bar height and the corresponding FCS autocorrelation
function is given by:

$$g(t) = \sum_{i=1}^{n} c_i \left(1 + \frac{t}{\tau_i}\right)^{-1} \left(1 + \frac{t}{p^2 \tau_i}\right)^{-1/2} , \qquad (12.4)$$

with the normalizing condition:

$$\sum_{i=1}^{n} c_i = g(0) . \qquad (12.5)$$

The bar heights are varied in order to minimize the differences between the
calculated and experimental correlation function. However, such an approach
is an ill-posed problem [12.34], as several different distributions, indistinguishable from each other within experimental error, may be found from the same
data. In addition, the calculations are unstable, i.e. small changes in the correlation curves, such as those found for replicate measurements, may give
rise to considerable differences in the obtained polydispersity distribution.
These kinds of problems are common not only for polydispersity measurements, but also for other kinds of determinations, e.g., image reconstruction
[12.35], calculation of affinity distributions of heterogeneous sorbents [12.36],
and the determination of spatial distributions of fluorescent groups in latex
particles [12.37]. A good strategy in solving this problem is the imposition
of additional constraints and regularization conditions which eliminate the
solutions which are not physically possible or are not acceptable based on
prior knowledge of the system. For example, in the case of natural samples,
one constraint is that the concentration of the particles cannot be negative.
Furthermore, a smooth size distribution can be assumed (regularization condition). Formally, it is necessary to minimize the function $\rho(c)$ with respect
to each c_i value simultaneously:

$$\rho(c) = \frac{1}{t_2 - t_1} \int_{t_1}^{t_2} w(t) \left[g_e(t) - g(t)\right]^2 dt + \lambda_m \left[1 - \frac{\sum_i c_1}{m}\right]^2 + \lambda_r \sum_i c_i''^2 , (12.6)$$

where $g_e(t)$ is the experimentally determined correlation function, $g(t)$ is the
correlation function calculated using (12.4) at each step of the minimization,
$w(t)$ is a weighting factor related to the experimental error of each data point
of the correlation function, m is the total mass of the particles, in this case,
to $g_e(0)$, c_i'' is the second derivative of c_i with respect to τ, and λ_m and λ_r

are Lagrange multipliers (λ_r is also called the regularization parameter). The following non-negative constraint is imposed:

$$c_i \geq 0. \tag{12.7}$$

The first term of (12.6) is the chi-squared value which minimizes the differences between the experimental correlation function and the calculated one. The second and third terms weighted by λ_m and λ_r bias the solution towards a function with some expected properties, i.e. constant "mass" (second term) and smooth distribution (third term).

Results based upon computer simulations of Gaussian size distributions are shown in Fig. 12.8. In this case, λ_r was varied from 0 to 10^{-4}. For 0 or small values of λ_r ($< 10^{-5}$), the fit was unstable. For values of λ_r ($\geq 10^{-4}$) the fit is poor and the distribution peaks are flattened too much. It is clear from these results that λ_r must be optimized based on methods developed in the literature [12.36].

Bi-modal distributions are important in FCS studies, as it is common to work with solutions containing some free fluorophores in the presence of the labeled polymers or colloids. It is generally accepted that two monodisperse components can be distinguished if their diffusion times differ by at least a factor of two [12.38]. However, in such a case, it has been shown that polydispersity could not be extracted from FCS data [12.39]. On the other hand, when mean diffusion times are very different from each other (Fig. 12.9), reasonably good results can be obtained.

It is interesting to compare the potential capabilities of FCS and PCS to work with polydisperse systems. From a mathematical point of view, the main difference is the polyexponential decay of the PCS correlation function [12.11], as opposed to a polyhyperbolic decay (12.4) for FCS. Since an exponential dependency changes faster than a hyperbolic one, PCS should be more sensitive to polydispersity than FCS. To test this point, FCS and PCS autocorrelation curves were generated for normal Gaussian distributions of variable polydispersities. The sum of the mean-squared differences between the autocorrelation function of a monodisperse or polydisperse system as a function of the polydispersity ratio is shown in Fig. 12.10. This sum is indeed larger for PCS than for FCS. On the other hand, the two curves are not very different and in both cases, the sum of the mean squared differences depend linearly on polydispersity, which indicates that there is no functional difference between the two methods in resolving polydispersity. The difference in the sum of squares between a noisy correlation function and the smooth "best" fit reflects the sensitivity limit of each method. Experimental measurement of this difference using 100 nm latex particles is shown by the solid (FCS) and dashed (PCS) horizontal lines in Fig. 12.10. Below these lines, the difference of the sum of squares due to polydispersity is less than the statistical errors. From this graph, it is apparent that it is impossible to determine size distributions with polydispersity ratios less than 0.03 using FCS, whereas for PCS this limit is lower (~ 0.01). On the other hand, because the

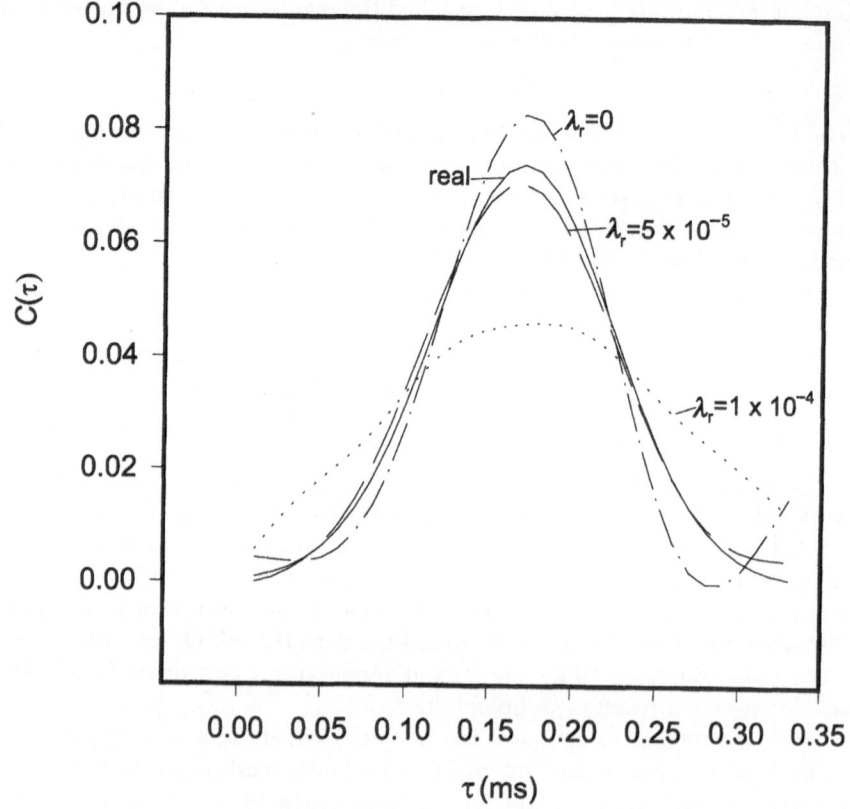

Fig. 12.8. Size distribution determined from simulated FCS autocorrelation curves fitted using different values of λ_τ [12.33]

signal-to-noise ratio in FCS can be enhanced by using longer accumulation times, both methods can be considered to have reasonably similar capabilities. In addition, it is likely that FCS could be more useful for measurements of small particles and molecules (size < 50 nm) which scatter light weakly, in which case the signal-to-noise ratio in FCS is much better than in PCS.

12.4 Example: Determination of the Diffusion Coefficients of Humic Substances as a Function of Solution Conditions

As mentioned above, the determination of diffusion coefficients for humic substances (including fulvic or humic acids and denoted below as "HS") is of great practical interest. Humic macromolecules are useful "model" environmental compounds in that they are a group of compounds which, like most

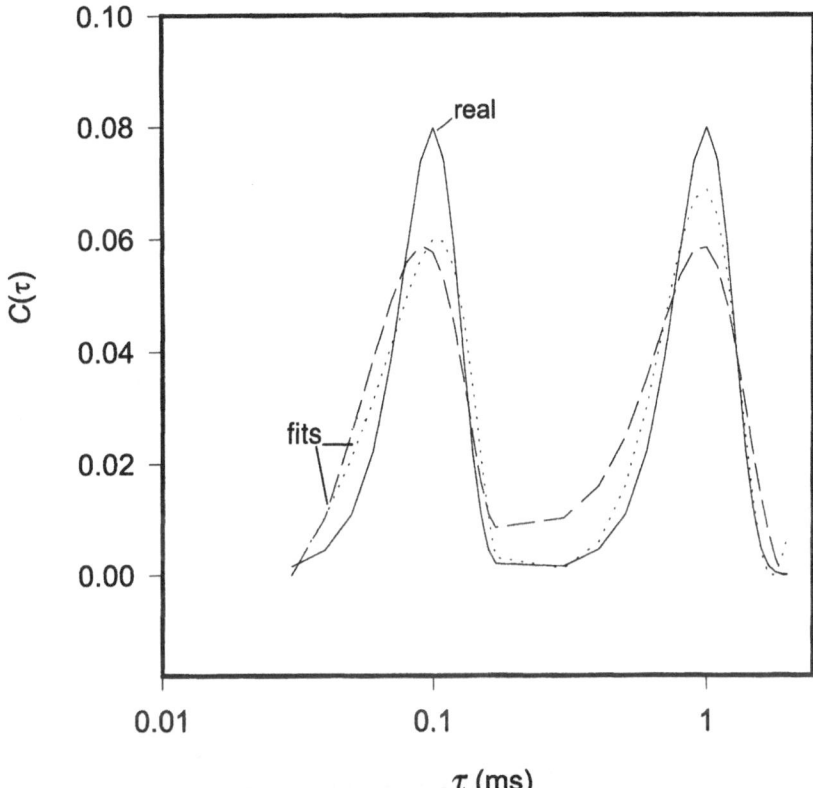

Fig. 12.9. Fitting of a bi-modal distribution with peak positions which differ by a factor of 10. The **dotted line** is obtained by using an initial approximation of a bi-modal distribution with sharp peaks in their correct positions. The **dashed line** is obtained with the initial approximation of a large peak in the middle of the scale [12.33]

environmental colloids, are chemically heterogeneous and physically polydisperse. Furthermore, diffusion coefficients of humic macromolecules have been measured by other independent techniques, so that the results obtained by FCS can be validated. In the following, we will briefly discuss three factors which must be taken into account when interpreting FCS results with humic substances, and then present some of the results which have been obtained for humics as a function of solution conditions.

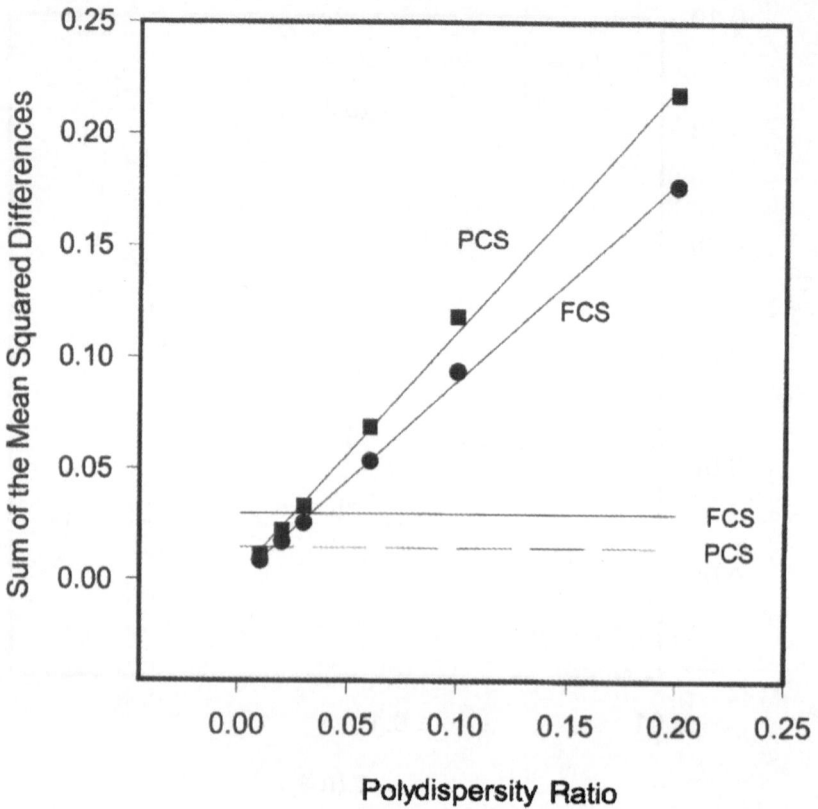

Fig. 12.10. Sum of the mean squared differences between the autocorrelation functions for monodisperse and polydisperse suspensions with different polydispersity ratios: PCS (**squares**) and FCS (**circles**). The noise limit for the PCS is given by a **horizontal dashed line** and for the FCS by a **horizontal solid line**

12.4.1 Factors Distinguishing Humic Substances From Model Compounds

At least three considerations should be taken into account when studying humic substances:

- Dependence of the fluorescent signal on the chemical heterogeneity of the humics,
- Polydispersity of the humic substances,
- Photodecomposition of the humic substances.

(i) *Heterogeneity of the fluorophore distribution.* Excitation and emission spectra of humic substances are typically quite broad compared to chemically homogeneous fluorophores such as the R6G used to calibrate the system. The broad peaks are due to the presence of chemically diverse fluorophores among and within the humic macromolecules, as well as diverse chemical environments around the fluorophores. The FCS signal will therefore depend upon the homogeneity of the fluorophore distribution among the various HS size fractions, at the measured laser excitation wavelength used in the experiment. In this respect, studies have been made on the role of laser excitation wavelengths on the fluorescence of NOM components isolated by capillary electrophoresis (separation based upon the ratio of charge / hydrodynamic size ratio). Under identical separation conditions, fluorescence electropherograms were obtained at 325, 454, 488 and 514 nm [12.40]. Spectra were identical qualitatively, indicating that over a wide range of fluorescence wavelengths, similar fluorophores were excited in the humic mixture, albeit to different extents. This was seen in the larger emission intensities at the lower wavelengths closer to the excitation maxima. The validity of these results was confirmed by measurements of diffusion coefficients of standard fulvic acids of the International Humic Substances Society (IHSS) across a wide range of pH values. At each pH value, the diffusion coefficient using 488 nm laser excitation was the same as that determined at 514 nm [12.41]. In spite of these promising results, further research is needed in this direction.

(ii) *Polydispersity of humic substances.* As discussed above, interpretation of the FCS autocorrelation function with respect to the polydispersity is possible for broad size distributions. Hydrodynamic diameters of IHSS standard fulvic acids isolated from the Suwannee river (SRFA) were determined using FCS and nuclear magnetic resonance spectrometry (NMR) (Fig. 12.11; [12.42]). Distributions obtained by FCS were similar for an identical sample of SRFA under similar conditions (low ionic strength and circumneutral pH). Furthermore, in another set of experiments, Balnois et al. [12.43] determined polydispersity indices (average weight size / average number size) on U.K. geological survey humic substances using atomic force microscopy (AFM). The absolute values of the indices were not identical due to important differences in the nature of the techniques. AFM measures the hights of the macromolecules absorbed on mica while FCS measures hydrodynamic diameters. Despite this, the observed trends in the AFM data across the pH range of 3.5 to 9.6 (ionic strength of 5 mM NaCl) were similar to the FCS results. The similarity of the FCS polydispersity measurements with those obtained by AFM, NMR and fFFF suggests that the FCS technique will be a useful means by which to determine polydispersity of environmental samples.

(iii) *Role of photodecomposition.* Humic substances are photosensitive, especially under relatively intense laser photoexcitation. Nonetheless, for the experimental conditions described here, photodecomposition is likely to be minimized by exciting the molecules at wavelengths far from their excitation

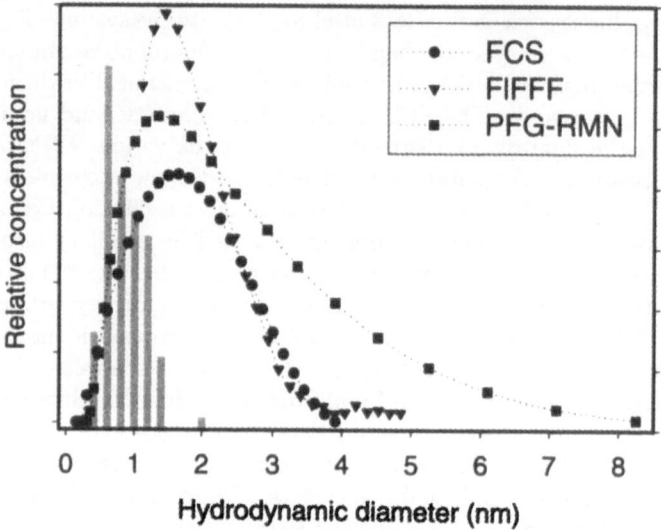

Fig. 12.11. Distribution of the hydrodynamic diameters of a standard fulvic acid (neutral pH, 5 mM ionic strength) isolated from the Suwannee River as determined by FCS, pulsed field gradient nuclear magnetic resonance spectrometry and flow field flow fractionation. The adsorbed heights of the fulvic acid as determined by atomic force microscopy are given in the bar graph [12.42]

maxima (≈ 350 nm). In a preliminary experiment designed to estimate the maximum possible effect of photodecomposition due to the passage of a humic macromolecule through the confocal volume, a 3 ml humic solution was irradiated with 5.2 mW for three hours at 488 nm ± 20 nm [12.41]. These conditions corresponded to 18.7 J/cm^2 or approximately a 0.075 ms permanence time in the confocal volume. A decrease of 2–46% of the fluorescence emission was observed depending upon the experimental conditions examined (including solution pH). These values are high with respect to the FCS measured triplet fraction of 29–35% observed for the IHSS fulvic acids. Furthermore, as demonstrated in [12.41], excellent agreement was found between the FCS results and those obtained by the non-fluorescence-based techniques, fFFF and NMR. In any case, the possible role of humic photodecomposition, and especially the ability of the data treatment to isolate the exponential decay of the triplet fraction from the normal hyperbolic signal which is obtained for Brownian motion, will need to be examined in greater detail in future experiments.

The previous three examples illustrate the continuing need for fundamental research to systematically examine the factors which influence the FCS signal under realistic environmental conditions. As illustrated below, the potential environmental applications of FCS are numerous, and the time and effort spent validating the technique is well-justified.

12.4.2   The Role of Solution Conditions (pH, Ionic Strength, Concentration) on the Diffusion Coefficients of Humic Substances

(i) *Effect of pH.* Humic macromolecules are polyelectrolytes, subject to conformational changes which are a function of the physico-chemical conditions of the medium. The binding of protons will reduce the charge of the humic macromolecule, especially at pH values below six (e.g. the Suwannee River humic acid has $6.1\,\mathrm{mmol\,g^{-1}}$ of carboxyl groups [12.44]). Proton binding may lead to two important effects which could affect the diffusion of the humic macromolecule: reduction of the intra-molecular repulsion of the anionic (primarily carboxylic) functional groups and increased aggregation due to a reduction of inter-molecular repulsion. While the first effect will lead to a compression of the humic molecule, leading to an increase in the diffusion coefficient, aggregation will have the opposite effect, i.e. a reduction of the average diffusion coefficient. Standard IHSS fulvic (SRFA) and humic (SRHA) macromolecules have been examined under a wide range of pH conditions. For example, the diffusion coefficient, D, of the SRFA was observed to decrease as the pH decreased from 7 to 4 (Fig. 12.12, [12.42]. The order of magnitude of the reduction in D represented an average aggregation of two or three humic macromolecules corresponding to an approximate increase in hydrodynamic diameter from 1.6 (SRFA) or 1.7 nm (SRHA) at high pH to approximately 2 nm at low pH. The absence of an important aggregation of these humic substances at low pH is in close agreement with AFM images and height measurements which were made on the same samples. The lack of aggregation might be related to the elaborate extraction procedure of the IHSS fulvics which may eliminate the most hydrophobic fractions [12.45].

(ii) *Effect of ionic strength.* As the ionic strength is increased, the diffuse layer around the humic substances is expected to be compressed, resulting in the simultaneous reduction of hydrodynamic diameter and repulsive inter-molecular forces due to a shielding of the charge of the humic macromolecule leading to partial aggregation. In this case the observed effects were less clear than in the case of a decrease in pH. Although a small but systematic reduction in the diffusion coefficient was observed across a wide range of pH values for 50 mM of added NaCl compared to 5 mM (Fig. 12.12, [12.42]), diffusion coefficients were not significantly different for ionic strength variations of 1 to 100 mM NaCl at pH 4.5 or 6.8 (Fig. 12.12, inset). The observed slight or negligible aggregation due to ionic strength is in accordance with the pH effect discussed above. Although experiments were also performed at higher ionic strengths (100 to 500 mM) where aggregation could be more important, the FCS structural parameter (ω_2/ω_1) was nonsensical, indicating destruction of the confocal volume due to the high refractive index of the medium.

272   K. Starchev, K.J. Wilkinson, and J. Buffle

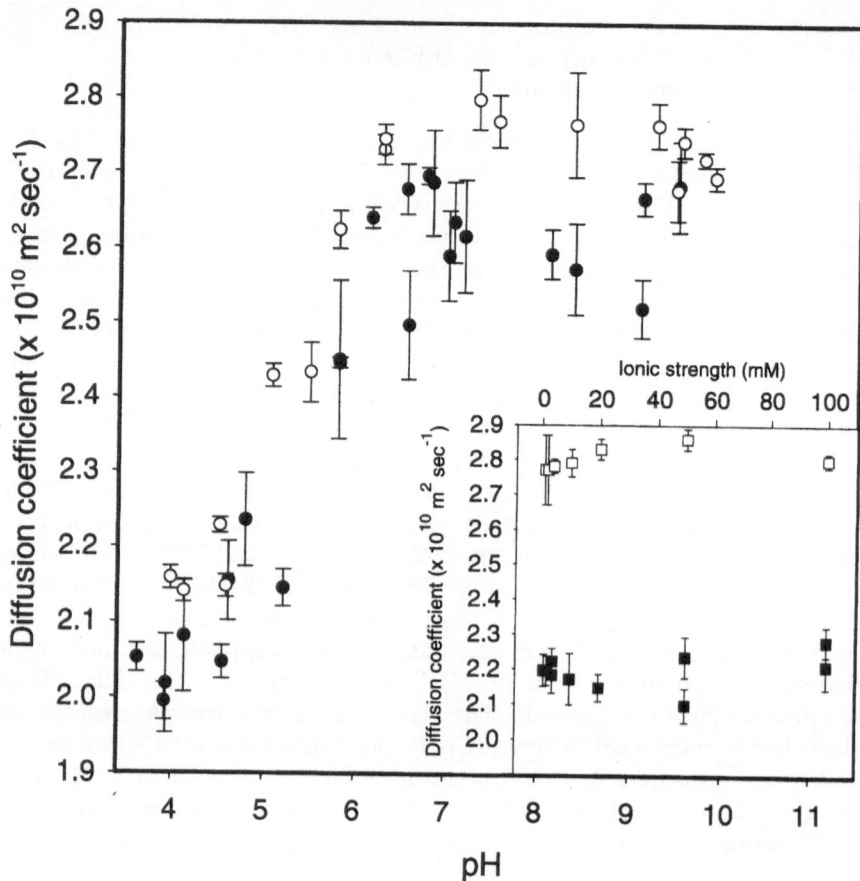

Fig. 12.12. Diffusion coefficients of the Suwannee River fulvic acids determined across a wide range of pH values in the presence of 5 mM (○) or 50 mM (•) of added NaCl. *Inset*: Diffusion coefficients of the Suwannee River fulvic acids as a function of the ionic strength at pH 4.5 (■) and 6.8 (□) [12.41]

(iii) *Concentration and nature of the humic macromolecules.* Aggregation is more likely to occur at high humic concentrations, especially in the presence of hydrophobic humic components. At the present time, the concentration range which can be studied is limited by the FCS technique: it is difficult to examine concentrations lower than $1\,\mathrm{mg\,l^{-1}}$ due to the weak fluorescence signal which is obtained, whereas, at concentrations exceeding $50\,\mathrm{mg\,l^{-1}}$ (approximately 5×10^{-5} M), the large quantity of fluorescing macromolecules may limit the signal quality.

For humic samples which are more hydrophobic than the standard aquatic fulvic or humic acids examined here, very large aggregates have been observed at low pH and high ionic strengths [12.43]. For example, for the U.K. geological survey humic substance at pH 4, interpretation of FCS data with a two-component model gave hydrodynamic diameters of 1 and 150 nm. This somewhat surprising bi-modal distribution was in good agreement with a size distribution observed by cascade tangential flow ultrafiltration [12.46] and by height observations obtained by AFM [12.43]. These results imply that the exact role of sample hydrophobicity will need to be examined in the future if we are to determine the precise environmental roles of these complex macromolecules.

12.5   Conclusions and Future Perspectives

The usefulness of the FCS technique for environmental applications is obvious. Nonetheless, the complexity of the samples and the necessity to label the majority of environmental components using non-perturbing labeling techniques will (temporarily) limit the development of the technique.

Therefore, in our opinion, future work in environmental systems should concentrate on:

- developing FCS applications for particles for which the number of adsorbed fluorophores varies with the particle size,
- improving the labeling techniques, especially for applications in the presence of fluorescence quenchers,
- determining the role of photobleaching of the humic substances.

Nonetheless, the potential applications of FCS are numerous. They include:

- the study of the conformation of fibrillar biopolymers,
- the study of aggregates and the formation of flocs,
- the determination of diffusion in complex systems, e.g., gels resembling the cell walls of micro-organisms.

To facilitate these kinds of studies in the future, it will be extremely useful to further develop:

- FCS using UV excitation (especially useful for the humic substances),
- cross-correlation FCS so as to facilitate studies examining the interaction of several components in aggregation or other environmentally relevant problems.

Acknowledgements

Funding for this work was provided by the Fonds National Suisse (Project number 2000-050629.97/1) and Zeiss/Evotec. We would like to acknowledge S. Assemi, E. Balnois, R. Beckett, S. Canonica, B. Cutak, C. Larive, J. Lead and E. Pérez for their contributions to this paper.

References

12.1 W. Stumm and J.J. Morgan: In: Aquatic Chemistry. Chemical Equilibria and Rates in Natural Waters (3rd ed.; Wiley: New York, 1996)

12.2 W. Stumm: In: Chemistry of the Solid-Water Interface (Wiley-Interscience: New-York, 1992)

12.3 L. Guo and P.H. Santschi: Reviews of Geophysics **35**, 17–40 1997

12.4 L.-M. Martin and M.-H. Dai: Limnol. Oceanogr. **40**, 119–131 1995

12.5 M.R. Jekel: Water Res. **20**, 1543–1554 1986

12.6 E. Tipping and D.C. Higgins: Colloid Surf. **5**, 85–92 1982

12.7 K.J. Wilkinson, J.C. Nègre, and J. Buffle: J. Contamin. Hydrol. **26**, 229–243 1997

12.8 K.J. Wilkinson, A. Joz-Roland, and J. Buffle: Limnol. Oceanog. **42**, 1714–1724 (1997)

12.9 J. Buffle and G.G. Leppard: Env. Sci. Techn. **29**, 2169–2175 (1995)

12.10 M. Filella, J. Zhang, M. Newman, and J. Buffle: Colloids and Surfaces **120**, 27–46 (1997)

12.11 J. Buffle and G.G. Leppard: Env. Sci. Techn. **29**, 2176–2184 (1995)

12.12 K.J. Wilkinson, E. Balnois, G.L. Leppard, and J. Buffle: Coll. Surf. A. **155**, 287–310 (1999)

12.13 J. Buffle, K.J. Wilkinson, S. Stoll, M. Filella, and J. Zhang: Environ. Sci. Technol. **32**, 2887–2899 (1998)

12.14 J. Buffle: In: Complexation Reactions in Aquatic Systems (An Analytical Approach, Ellis-Horwood: Chichester, 1988)

12.15 M. Filella and J. Buffle: Colloid Surf. **73**, 255–273 (1993)

12.16 A. Lerman: In: Geochemical Processes (Wiley-Interscience: New York, 1979)

12.17 E.R. Morris, D.A. Rees, and G. Robinson: J. Mol. Biol. **138**, 349 (1980)

12.18 P.H. Santschi, E. Balnois, K.J. Wilkinson, J. Zhang, J. Buffle, and L. Guo: Limnol. Oceanogr. **43**, 896 (1998)

12.19 J. Lyklema: In: Fundamentals of Interface and Colloid Science. (Vol. 1 and 2; Academic Press: London, 1991)

12.20 M. von Smoluchowski: Z. Physikal. Chem. **92**, 129 (1918)

12.21 C.R. O'Melia: In: Aquatic Chemical Kinetics, (W. Stumm (ed.) Wiley: New York. 1990), 447–474

12.22 D. Perret, M. Newman, J.-C. Negre, Y. Chen, and J. Buffle: Wat. Res. **28**, 91-106 (1994)

12.23 S. Stoll and J. Buffle: J. J. Coll. Int. Sci. **180**, 548–563 (1996)

12.24 H. Qian, E.L. Elson, and C. Frieden: Biophys. J. **63**, 1000–1010 (1992)

12.25 B.A. Scalettar, J.E. Hearst, and M.P. Klein: Macromolecules **22**, 4550–4559 (1989)

12.26 E.L. Elson, J. Schlessinger, D.E. Koppel, D. Axelrod, and W.W. Webb: Prog. Clin. Biol. Res. **9**, 137–147 (1976)

12.27 N. Senesi: Analytica Chimica Acta. **232**, 77–106 (1990)

12.28 K. Starchev, J. Zhang, and J.J. Buffle: Coll. Int. Sci. **203**, 189–196 (1998)

12.29 S.R. Aragon and R. Pecora: J. Chem. Phys. **64**, 1791 (1976)

12.30 R. Rigler, J. Widergreen, and U. Mets: In: Fluorescence Correlation Spectroscopy (O.S. Wolfbeis (ed.) Springer: Berlin, 1992) pp. 13–24

12.31 N.L. Thompson: In: Topics in Fluorescence Spectroscopy, volume 1: Techniques (Joseph R. Lakowicz (ed.) Plenum Press: New York, 1991)

12.32 D. Koppel: J. Chem. Phys. **57**, 4814 (1972)

12.33 K. Starchev, E. Pérez, and J. Buffle: J. Coll. Int. Sci. **213**, 479 (1999)

12.34 S. Provencher: Computer Physics Communications **27**, 213 (1982)

12.35 R.C. Smith and W.T. Grandy (eds.) In: Maximum.Entropy and Bayesian Methods in Inverse Problems (Reidel: Dortrecht, 1985)

12.36 M. Cernik, M. Borkovec, and J. Westall: Environ. Sci. Technol. **29**, 413 (1995)

12.37 E. Pérez, J. Lang: J. Phys. Chem. **103**, 2072 (1999)

12.38 U. Meseth, T. Wohland, R. Rigler, and H. Vogel: Biophys. J. **76**, 1619 (1999)

12.39 E. Gulari, E. Gulari, Y. Tsunashima,and B. Chu: J. Chem. Phys. **70**, 3965 (1979)

12.40 J.R. Lead, E. Balnois, M. Hosse, R. Menghetti, and K.J. Wilkinson: Environ. Int. **25**, 245–258 (1999)

12.41 J.R. Lead, K.J. Wilkinson, K. Starchev, S. Canonica, and J. Buffle: Environ. Sci. Technol. **34**, 1365 (2000)

12.42 J.R. Lead, K.J. Wilkinson, E. Balnois, B. Cutak, C. Larive, S. Assemi, and R. Beckett: Environ. Sci. Technol. **34**, 3508 (2000)

12.43 E. Balnois, K.J. Wilkinson, J.R. Lead, J. Buffle: Environ. Sci. Technol. **33**, 3911 (1999)

12.44 E.C. Bowles, R.C. Antweiler, and P. MacCarthy: In: Humic Substances in the Suwannee River, Georgia: Interactions, Properties, and Proposed Structures (R.C. Averett, J.A. Leenheer, D.M. McKnight, K.A. Thorn (eds.,) U.S. Geological Survey, Water Supply Paper 2373. Denver). (1995)

12.45 G.G. Leppard, J. Buffle, and R. Baudat: Water Res. **20**, 185–196 (1986)

12.46 K.J. Wilkinson, R. Menghetti, and E. Balnois: personal communication.

13 Photophysical Aspects of FCS Measurements

Jerker Widengren

13.1 Introduction

Photophysical and photochemical aspects of dyes are of great importance for applications of fluorescence spectroscopy where a high read out rate or a high sensitivity is required. The need to optimize the fluorescence, both with respect to the total number of photons that can be extracted as well as to the fluorescence rate itself, is especially pronounced in applications regarding single molecule detection or fluorescence correlation spectroscopy (FCS), where a low fluorescence signal can not be compensated by an increased fluorophore concentration. In the past, photophyscial and photochemical limitations have also, to a large extent, restricted the use of FCS. Although the concept and the first experimental results of FCS were presented more than 25 years ago [13.1–13.3], FCS has not, until the last five to ten years, started to find a more extensive use. In the early applications of the FCS technique, low detection efficiency combined with limited photostability of the fluorophores restricted the use of the method. High background intensities, dominated by Raman scattering of the water molecules, made it impossible to reduce the fluorophore concentrations without having a dominating background over the fluorescence signal. As a consequence, the concentration of the fluorescent molecules under investigation had to be high, leading to lower relative fluctuations in fluorescence. To be able to distinguish and analyze the low amplitudes of the relative fluctuations, long measurement times were often necessary. Therefore, the optical and electronical parts of the instrumentation had to fulfil rather strict requirements concerning stability and absence of systematic noise. The limited instrumental stability over long times in combination with long time range photochemical degradation restricted the possible applications of FCS (see [13.4,13.5] for a further discussion).

By the introduction of very small open detection volumes (fractions of a femtoliter), objectives with high numeric apertures, and highly selective band-pass filters, a very high spatial and spectral discrimination has been reached. This, and the use of silicon avalanche photodetectors with high detection quantum yields, has made it possible to increase signal-to-noise ratios considerably compared to early FCS experiments.

A small detection volume, at its extreme restricted radially by diffraction and by spherical aberration along the beam axis, minimizes background

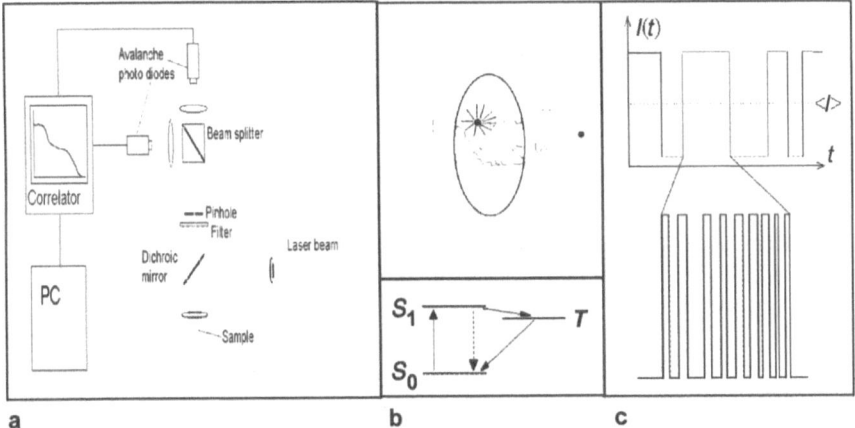

Fig. 13.1. a Experimental setup for FCS experiments, **b** schematic view of the detection volume (**black line**), given by the dimensions of the focus of the excitation laser beam and the collection efficiency function of the fluorescence microscope, **c** additional fluorescence fluctuations (*below*) generated from the presence of a transient non-fluorescent state, for instance a triplet state (**b**, *bottom*), which are superimposed on those due to translational diffusion (top) of the fluorophores into and out of the detection volume (**b**, *top*)

due to Raman scattering, since at constant excitation intensities, the detected Raman scattering will be proportional to the sample volume, i.e. the number of solvent molecules. Also, from a photophysical point of view, a small volume element facilitates a fast exchange of fluorophores in and out of the sample volume element. Due to the short passage times, higher excitation intensities can be applied without photobleaching the molecules, thus allowing higher fluorescence emission rates per molecule. Compared to previous setups, a high-numerical aperture objective in combination with single photon counting avalanche photodiodes (SPAD) with high detection yields has increased the fraction of emitted fluorescence photons that are detected. With the introduction of highly selective band-pass filters, the background counts remaining are mainly due to dark current from the detectors and can be reduced to about one thousand times less than what can be obtained from a single rhodamine molecule. With this modified experimental set-up (Fig. 13.1a), measurement times have been shortened drastically [13.6] compared to those needed in previous FCS experiments and it has become possible to perform measurements at the level of single molecules [13.7–13.9]. Many of the obstacles previously encountered in FCS measurements could thus be circumvented, or at least strongly alleviated, making FCS a useful tool in biological and medical research (see [13.10] for a review).

Nonetheless, photophysical properties of the fluorescently labeled molecules still set fundamental limits for the sensitivity and overall performance of the present day FCS measurements. Important figures of merit for the performance of FCS measurements are the fluorescence flux, and the total number of photons emitted per molecule. The flux of the emitted fluorescence photons will be limited by the finite deexcitation rate of the excited fluorophores, given by the fluorescence lifetime, τ_f, and by the extent to which the fluorophores are transformed into transient non-fluorescent states, such as the triplet state. The total number of photons emitted by a fluorophore molecule will ultimately be limited by the photochemical lifetime and is, in the absence of nonlinear effects, given by $n_f = \Phi_f / \Phi_D$, where Φ_f is the fluorescence quantum yield and Φ_D is the photodestruction quantum efficiency. Φ_f and Φ_D can be defined as the fractions of excitations to the excited singlet state, S_1, that lead to fluorescence emission and photodegradation, respectively.

Although of fundamental importance in many fields, many of these photophyscial processes are not sufficiently understood or characterized. Even for many of the most common fluorophores information is sparse. This reflects, to some extent, experimental difficulties in determining the relevant parameters in sufficient detail. Also, many of these parameters tend to be sensitive to the environment, which in some cases makes it difficult to compare different investigations and also makes it difficult to precisely predict the photophysical properties specific for a certain environment.

In the following it will be shown how FCS can be used to characterize photophysical processes relevant for ultrasensitive fluorescence spectroscopy, taking place over a time scale of nanoseconds to milliseconds. First, reversible processes are discussed, which take place on a time scale typically fast enough to let the molecules occupy the transient state several times when diffusing through the detection volume. In the second part, photodegradation will be treated, an irreversible process that can be treated as a quasi-stationary process in the FCS measurements. Finally, some approaches will be briefly discussed for how to improve the photophysical and photochemical conditions under which FCS measurements are performed.

13.2   Photophysics in the Fast Time Range

In the absence of any photophysical process affecting the dye molecules, the translational diffusion is the dominating process yielding fluorescence fluctuations in the microsecond to millisecond time scale as the fluorescent molecules diffuse into and out of the detection volume (Fig. 13.1B, top). The dye concentration is governed by the diffusion equation:

$$\frac{d\delta C(\bar{r}, t)}{dt} = D\nabla^2 \delta C(\bar{r}, t).$$
(13.1)

The time-dependent normalized intensity autocorrelation function can be written as:

$$G(\tau) = \lim_{T\to\infty} \frac{1}{T} \int\limits_0^T \frac{\delta I(t+\tau)\delta I(t)}{\langle I\rangle^2}\,\mathrm{d}t$$

$$= \frac{1}{N}\left(\frac{1}{1+4D\tau/\omega_1^2}\right)\left(\frac{1}{1+4D\tau/\omega_2^2}\right)^{1/2} + 1\,.$$

(13.2)

Here $\delta I(t)$ denotes the fluctuations of the detected fluorescence around its time-averaged mean $\langle I\rangle$, i.e. $\delta I(t) = I(t) - \langle I\rangle$, N is the mean number of molecules within the sample volume element, and D is the translational diffusion coefficient of the fluorescent molecules. Brackets denote time average. ω_1 and ω_2 denote the distances in the radial and axial dimensions, respectively, at which the detected fluorescence per unit volume has decreased by a factor of e^2. The total detected fluorescence is given from:

$$I(t) = \int \mathrm{CEF}(\bar{r})c(\bar{r},t)k_{21}qS_1(\bar{r},t)\mathrm{d}V\,.$$

(13.3)

Here, q accounts for the quantum efficiency of the detectors, fluorescence quantum yield of the fluorophore as well as attenuation of the fluorescence in the passage from the sample volume to the detector areas, $\mathrm{CEF}(\bar{r})$ is the collection efficiency function of the confocal microscope setup and $c(\bar{r},t)$ denotes the concentration of fluorophores. $S_1(\bar{r},t)$ denotes the fraction of fluorophores being in their excited singlet states. Typically, the generation rate of fluorescence is assumed to vary linearly with the excitation intensity, and the concentration of dye molecules in their ground and excited singlet states is assumed to be constant over time. However, at higher excitation intensities, where the excitation rate is of the same magnitude as that of the decay of the excited singlet states of the dyes, fluorescence saturation and different photo-induced transient non-fluorescent states can come into play. In the FCS measurements, transient non-fluorescent states present themselves as fluorescence fluctuations superimposed on those caused by concentration changes due to translational motion of the fluorophores into and out of the sample volume element (Fig. 13.1C). These superimposed fluorescence fluctuations originate from changes in the excited singlet state population. One may note that the typical time range for formation and decay of many photo-induced transient states is much faster, or at least can be made much faster, than the typical decay time of the fluorescence correlation function, as given by (13.2), and that the diffusion properties of the compounds are typically not changed upon generation of a photo-induced transient state.

Generally, under these conditions, i.e. if:

1. diffusion is much slower than the chemical relaxation time(s) ($\tau_\mathrm{D} \gg \tau_\mathrm{chem}$)

and/or

2. the diffusion coefficients of all fluorescent species are equal,
   the fluorescence correlation function can be separated into two factors, the

first, $G_D(\tau)$, which depends on transport properties (diffusion or flow) and the second, $X(\tau)$, which depends on the reaction rate constants [13.11]:

$$G(\tau) = G_D(\tau)X(\tau) + 1,\tag{13.4}$$

where

$$X(\tau) = \frac{\displaystyle\sum_{i,j=1}^{M} Q_i Q_j S_{ij}(\tau)}{\displaystyle\sum_{i=1}^{M} Q_i{}^2 \overline{C}_i}.\tag{13.5}$$

Here, Q_i denotes the fluorescence capacity of compound i and is given by the product of its fluorescence quantum yield and its excitation cross section, $Q_i = (\Phi_f \sigma)_i$. $S_{ij}(\tau)$ is the solution to the following set of differential equations and initial conditions:

$$dS_{ij}(\tau)/d\tau = \sum_{j=1}^{M} T_{ij} S_{jk}(\tau)$$
$$S_{ik}(0) = \overline{C}_i \delta_{ik}\tag{13.6}$$

where T_{ij} corresponds to the reaction rate matrix of a reaction involving M different species. From (13.5) and (13.6) one can see that $X(\tau)$ is a normalized sum of the terms $S_{ij}(\tau)$. Given an initial mean concentration, $\overline{C}_i$, of the compound i at $\tau = 0$ the terms $S_{ij}(\tau)$ state the average concentration of the state j at time τ.

When applying the above formalism to the case of photophysically generated transient non-fluorescent states, one can note from (13.3) that the detected fluorescence is proportional to the population of the excited singlet state, S_1, of the fluorophores. Fluctuations in fluorescence thus reflect fluctuations only in the population of S_1, and the factor $X(\tau)$ only contains one term, that of $S_1(\tau)/\overline{S}_1$, since the fluorescence capacity, Q_i, of the other states will be zero. Here, $\overline{S}_1$ is the mean fraction of fluorophores in the detection volume that are in their excited singlet states, and $S_1(\tau)$ represents that fraction at time τ.

For the full correlation function, using (13.4–13.6) above, one will arrive at:

$$G(\tau) = G_D(\tau) \times \frac{S_1(\tau)}{\overline{S}_1} + 1.\tag{13.7}$$

In the following subsections, examples will be given for how different transient non-fluorescent photo-induced states can be taken into account in FCS experiments, making use of the approach above, and how FCS can be used to characterize these states with respect to their kinetic properties.

13.2.1   Triplet State Formation

Formation of triplet states can, for many fluorescent dyes, be observed in FCS experiments already at relatively modest excitation intensities. As discussed above, triplet states are of considerable importance in fluorescence spectroscopy and a knowledge of their character is needed in order to find optimal excitation conditions. Triplet states are also believed to be involved in the process of photodegradation. Therefore, fluorophores having moderate triplet quantum yields are generally preferred. However, triplet properties of such dyes have, in the past, proven to be comparatively difficult to investigate. Several investigations of the triplet state properties of Rh6G and Fluorescein have been undertaken. Most investigations were performed by flash photolysis and transient absorption techniques, where after a short intense laser flash the absorption of triplet state fluorophores was measured. Under practically identical experimental conditions, reported triplet parameters differ by orders of magnitude between different investigators (see [13.12,13.13] and references therein). In the following, an outline will be given of a study of the triplet state properties of Rhodamine 6G (Rh6G) and Fluorescein Isothiocyanate (FITC) [13.13,13.14], which was undertaken to demonstrate the use of FCS for triplet state monitoring in solutions.

The relevant electronic states of Rh6G involved in the processes of fluorescence at 514.5 nm excitation wavelength can be modeled as shown in Fig. 13.2A. S_0 denotes the ground singlet state, S_1 is the excited singlet state and T is the lowest triplet state. k_{12}, k_{21}, k_{23} and k_{31} are the rate constants for excitation and deexcitation of the singlet state, intersystem crossing and deexcitation of the triplet state, respectively. The rate of ground singlet state excitation, k_{12}, may be written as $\sigma_{\mathrm{exc}}\Phi$, where σ_{exc} is the excitation cross section of the ground singlet state and Φ is the excitation intensity. Since vibrational relaxation takes place in the picosecond time scale [13.15], it can be neglected as it will occur much faster than the other transitions. Also, the excitation cross section from the first excited singlet state to higher singlet states, σ_{sn}, can be neglected. At the above excitation wavelength σ_{sn} will be as high as one fourth of that from the ground singlet state [13.16,13.17]. However, since the relaxation from the higher singlet states (denoted by k_{n1} in Fig. 13.2a) takes place in the sub-picosecond time scale, excitation to higher singlet states is not considered in our model. Furthermore, the effects of stimulated emission, σ_{21}, or triplet-triplet absorption, σ_{tn}, need not be taken into consideration. Stimulated emission and triplet-triplet absorption [13.18] will be negligibly small when exciting with a wavelength out of the emission band and when the concentration of the fluorophores is in the nanomolar range. The excitation and emission dipole moments of the fluorophore molecules can be treated as isotropic since the rotation of the molecules will take place in the sub-nanosecond time scale. Due to the fact that the triplet state lifetime $(1/k_{31})$ is quite long compared to that of the singlet state $(1/k_{21})$ and that

282   J. Widengren

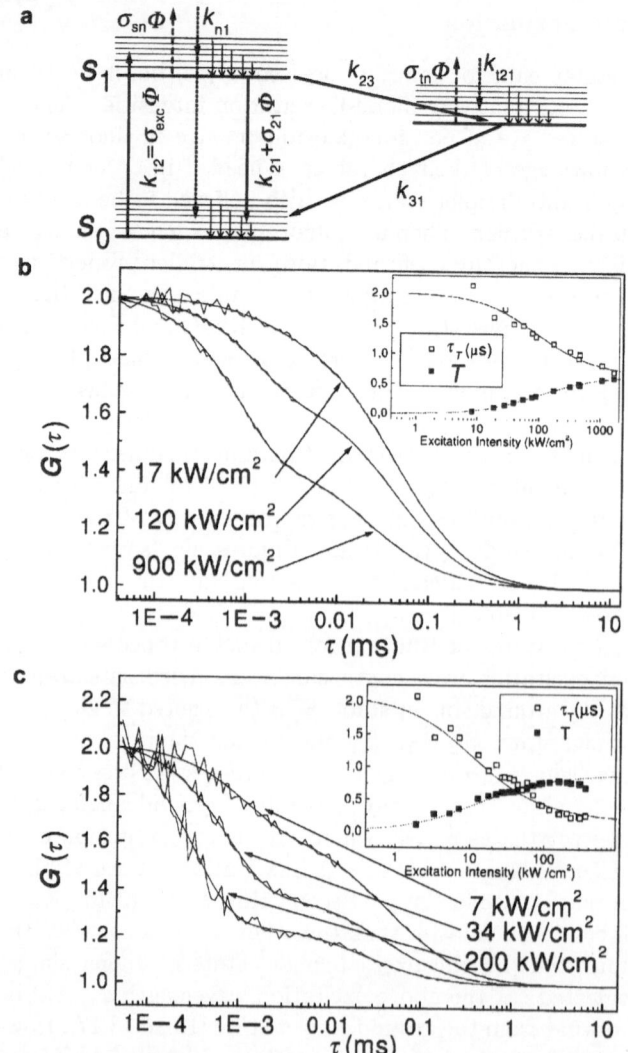

Fig. 13.2. a Electronic state model for Rh6G in water (see text for details), **b** and **c** FCS curves of Rh6G **b** and FITC **c** in aqueous solution measured at different excitation intensities. Fitting the curves to (13.14), and plotting the obtained triplet state population, T, and relaxation times, $\tau_T = 1/\lambda_3$, as functions of the excitation intensity (inserts of Figs. 13.2b and 13.2c) the rates of intersystem crossing, k_{23}, and triplet decay, k_{31}, could be determined to $1.1 \times 10^6\,\mathrm{s}^{-1}$ (k_{23} Rh6G) and $5.7 \times 10^6\,\mathrm{s}^{-1}$ (k_{23} FITC) and $0.5 \times 10^6\,\mathrm{s}^{-1}$ (k_{31} Rh6G and FITC)

the phosphorescence quantum yield from the triplet state is quite low, the triplet state can be considered as a non-luminescent state [13.19].

According to the electron state model, the fluorophore concentration can be written as:

$$c(t) = c(t) \left[S_1(t) + S_0(t) + T(t) \right] , \tag{13.8}$$

where S_0, S_1, and T are the fractions of the fluorophores being in their ground singlet, excited singlet or the lowest triplet states, respectively. The sum of the states are normalized to unity.

For a given fluorophore, subject to constant excitation intensity, the probability of occupying one of the three different states as a function of time can be expressed in terms of the rate constants by solving the following system of three first order differential equations:

$$\frac{d}{dt} \begin{pmatrix} S_0(t) \\ S_1(t) \\ T(t) \end{pmatrix} = \begin{bmatrix} -k_{12} & k_{21} & k_{31} \\ k_{12} & -(k_{23} + k_{21}) & 0 \\ 0 & k_{23} & -k_{31} \end{bmatrix} \begin{pmatrix} S_0(t) \\ S_1(t) \\ T(t) \end{pmatrix} , \tag{13.9}$$

where $k_{12} = \sigma_{exc} \Phi$. One can assume that the decay of the singlet state by fluorescence or internal conversion is much faster than either of the processes of intersystem crossing or triplet state decay, i.e.

$$k_{21} \gg k_{23}, k_{31} . \tag{13.10}$$

In the process of fluorescence, a photon is released and the fluorophore molecule enters its ground singlet state. Defining the time of fluorescence emission as $t = 0$ one can write:

$$\frac{d}{dt} \begin{pmatrix} S_0(t) \\ S_1(t) \\ T(t) \end{pmatrix} = \begin{pmatrix} 1 \\ 0 \\ 0 \end{pmatrix} . \tag{13.11}$$

Applying the assumption of (13.10) and the boundary condition of (13.11) to the above system of differential equations, one obtains the probabilities of occupying the different electronic states as a function of time, t, after photon release:

$$S_0(t) = \frac{k_{21}k_{31}}{k_{12}(k_{23} + k_{31}) + k_{21}k_{31}} e^{\lambda_1 t} +$$
$$+ \frac{k_{12}}{k_{12} + k_{21}} e^{\lambda_2 t} + \frac{k_{12}k_{21}k_{23}}{(k_{12} + k_{21})[k_{12}(k_{23} + k_{31}) + k_{21}k_{31}]} e^{\lambda_3 t} , \tag{13.12a}$$

$$S_1(t) = \frac{k_{12}k_{31}}{k_{12}(k_{23} + k_{31}) + k_{21}k_{31}} e^{\lambda_1 t} -$$
$$- \frac{k_{12}}{k_{12} + k_{21}} e^{\lambda_2 t} + \frac{k_{12}{}^2 k_{23}}{(k_{12} + k_{21})[k_{12}(k_{23} + k_{31}) + k_{21}k_{31}]} e^{\lambda_3 t} , \tag{13.12b}$$

$$T(\bar{r}, z, t) = \frac{k_{12}k_{23}}{k_{12}(k_{23} + k_{31}) + k_{21}k_{31}}e^{\lambda_1 t} -$$
$$- \frac{k_{12}k_{23}}{k_{12}(k_{23} + k_{31}) + k_{21} + k_{31}}e^{\lambda_3 t}, \qquad (13.12c)$$

where the eigenvalues are given by:

$$\lambda_1 = 0 \qquad (13.13a)$$

$$\lambda_2 = -(k_{21} + k_{12}) \qquad (13.13b)$$

$$\lambda_3 = -\left[k_{31} + \frac{k_{12}k_{23}}{k_{12} + k_{21}}\right]. \qquad (13.13c)$$

The first eigenvalue, λ_1, will be zero indicating that the populations in the three states will approach a steady state as $t \to \infty$. This follows from the fact that the triplet state model of Fig. 13.2 constitutes a closed system assuming a constant total population and no photobleaching. The first terms in (13.12a–13.12c) then expresses the steady-state concentrations of the S_0, S_1 and T states, respectively. The second eigenvalue, λ_2, will be of high magnitude and represents the so called "antibunching" term [13.20–13.22]. Antibunching refers to the phenomenon that a fluorescent dye can not, at the moment it has emitted a photon, directly emit a subsequent photon, since it first has to undergo another excitation-emission cycle. The relaxation rate of λ_2 is related to the exchange between S_1 and S_0, which takes place in the nanosecond time range for Rhodamine dyes. Antibunching is treated elsewhere in this book. The magnitude of the third eigenvalue, λ_3, is roughly related to the rate at which the build-up of the triplet state population takes place. Its inverse will be referred to as the bunching time, τ_T, which for many dyes in air-saturated aqueous solutions typically is in the microsecond time range. In this time range, the e^{λ_2} terms related to dye antibunching will vanish. Disregarding the dye antibunching, and using (13.4–13.7) and (13.12b), the fluorescence correlation function can be expressed as:

$$G(\tau) = G_D(\tau)\frac{S_1(\tau)}{\overline{S}_1} + 1 = G_D(\tau)\left[\frac{\overline{T}}{(1 - \overline{T})}e^{\lambda_3 \tau} + 1\right] + 1, \qquad (13.14)$$

where $\overline{T}$ is the mean fraction of fluorophores in the detection volume that are in their triplet states and is given by the first term of (13.12c):

$$\overline{T} = \frac{k_{12}k_{23}}{k_{12}(k_{23} + k_{31}) + k_{21}k_{31}}. \qquad (13.15)$$

For determining the rate constants k_{23} and k_{31}, as well as the excitation cross section σ_{exc}, fluorescence autocorrelation (AC) functions at different

excitation intensities were measured. These data were numerically fitted to
(13.14). In this way, one obtains two equations, (13.13c) and (13.15), for each
one of the AC functions relating the experimentally determined values of $\tau_T = 1/\lambda_3$ and $\overline{T}$ to the three unknown parameters of k_{23}, k_{31} and σ_{exc}. By plotting
these experimental values of $\overline{T}$ and τ_T as a function of excitation intensity
and by making a simultaneous nonlinear least squares fit to the expressions
of (13.13c) and (13.15), respectively, the three unknown parameters could be
determined (k_{21} was fixed to a known value). This corresponds to solving an
overdetermined equation system of independent expressions of k_{23}, k_{31} and
σ_{exc}. In this way, a more accurate determination of the parameters could
be obtained. In Figs. 13.2B and 13.2C correlation curves and corresponding
plots of $\overline{T}$ and τ_T as a function of excitation intensity are shown for Rh6G
and FITC.

Environmental Effects on the Triplet State Properties. Measurements performed on Rh6G in ethanol and ethylene glycol show that the triplet
state kinetics can vary substantially between different solvents (Fig. 13.3a).
The differences are likely to reflect the different polarities, viscosities and
oxygen solubilities of these solvents.

Triplet states of dyes are, largely due to the relatively slow rates of inter-
sytem crossing (k_{23}) and triplet state decay (k_{31}) influenced by the presence
of various compounds that would affect these rates. Iodide ions belong to
this category and are known to affect the singlet-triplet transitions by the
so-called heavy atom effect, where the rate of intersystem crossing and the
triplet decay rate are promoted by collisional interactions with the iodide ions
leading to a linear increase of these rates with the concentration of iodide.
AC functions were measured at different intensities and at different concentrations of potassium iodide, ranging from 0–10 mM. For each concentration,
the obtained values of τ_T and $\overline{T}$ were plotted as functions of excitation intensity and analyzed as described above. The effects of iodide could be seen on
the AC functions as a strong build-up of the triplet state population as well
as a reduced bunching time (Fig. 13.3B). The measured values of k_{31} and k_{23}
each display a linear dependence on the iodide concentration.

In Fig. 13.3c the effect of triplet state quenching by oxygen is shown. Oxygen is known to be a potent quencher of the triplet state of Rh6G. Although,
in principle, having the same impact on the rate parameters as iodide the
presence of oxygen has an opposite effect on the triplet population due to a
higher relative quenching effect on k_{31} than on k_{23}. Consequently, for Rh6G
equilibrated with a pure oxygen atmosphere, the triplet population decreases
considerably, while in an argon atmosphere, where the oxygen is purged, a
tremendous increase in triplet population can be seen. In the latter case, the
lifetime of the triplet state of Rh6G far exceeded the time of diffusion into
and out of the sample volume element. For this reason the sample volume

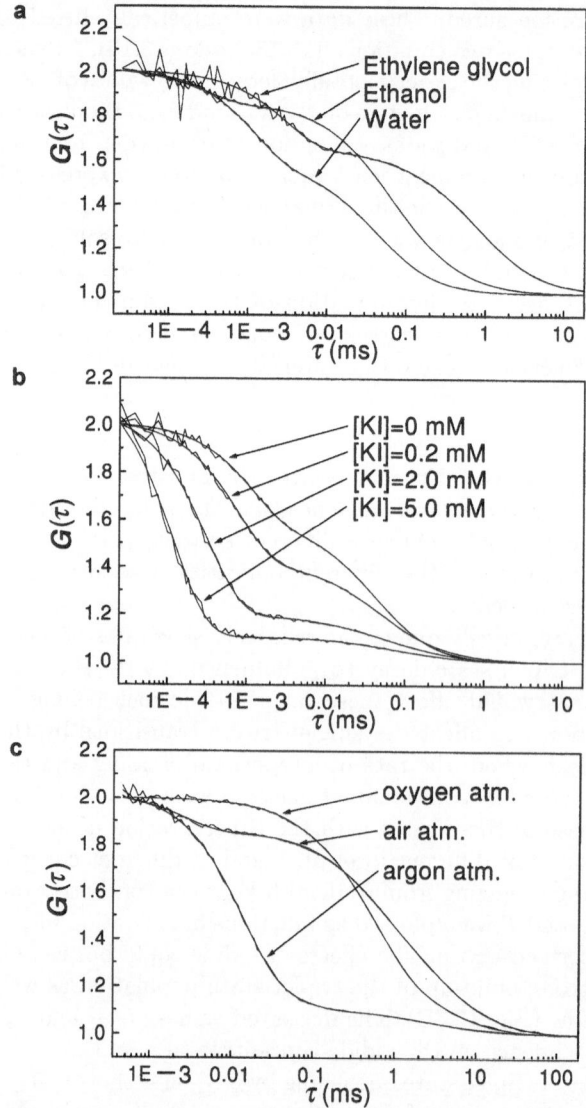

Fig. 13.3 a–c. FCS curves of Rh6G in different solvents **a**, with different concentrations of potassium iodide added **b**, and in argon, air and oxygen atmospheres **c** at equal excitation intensities applied (approx. $100\,kW/cm^2$). In Fig.13.3c, the detection volume was expanded by defocusing the laser beam (beam radius $2.5\,\mu m$ in the focal plane) and by introduction of a pinhole, matching the size of the focused laser beam. In this way the long triplet relaxation times found in argon atmosphere could be observed within the transit times of the Rh6G molecules through the detection volume

element was increased by reducing the focusing of the laser beam and by using a larger pinhole radius.

The presence of oxygen thus strongly reduces the population of the triplet state. On the other hand, in an oxygen atmosphere the FCS curve can be seen to decay faster than that measured in air (Fig. 13.4c). This is most likely caused by a higher photodecomposition yield. One can conclude that for the effects of oxygen on fluorescence, there is a trade-off between a reduced triplet population and a higher photodecomposition rate. For the experimental conditions presented here, the optimum oxygen concentration for Rh6G was found to be close to that in air (0.2 atm oxygen).

13.2.2   Charge Transfer Reactions

As seen above, interaction with the environment can result in changes of the triplet state properties. In this section we will show how FCS can be used to monitor the generation of additional photo-induced transient states of a dye following interaction with its neighboring molecules. In particular, the electron-transfer-induced quenching of Rh6G by the nucleotide dGTP will be discussed. Electron transfer can, in a simplified manner, be described in terms of electronic motion between a sensitizer and a quencher – an electron jumps from an occupied orbital of one reactant to an unoccupied orbital of the other (see [13.23] for an introductory review). The quenching of a number of different polynuclear aromatic dye fluorophores by different DNA bases is believed to involve electron transfer mechanisms if the redox potentials are of the appropriate relative magnitudes [13.24]. From investigations on dye-DNA base interactions in water in general [13.25–13.29], the electron transfer reactions can be assumed to take place upon hydrophobic complex formation and be fast enough for the hydrophobic interaction to be the rate-limiting process.

In [13.30] the feasibility of FCS in the investigation of processes involving electron-transfer was investigated by adding different nucleotides to an aqueous solution of Rh6G. Among the nucleotides added (dGTP, dATP, dCTP and dTTP), guanosine is known to be most readily oxidized and quenching by electron transfer should, for Rh6G, only be possible by this nucleotide (generating a Rh6G radical anion and a guanosine cation) [13.29]. For rhodamine dyes, their hydrophobic nature promotes aggregate formation and likewise aggregation of rhodamine molecules with other hydrophobic compounds, like nucleotides, can be observed. Upon aggregation with guanosine the quenching efficiency is a result of a redox reaction between the rhodamine dye and the guanosine, promoted by the proximity between the molecules and the matching redox potentials, whereby the guanosine acts as an electron donor and the rhodamine as an electron acceptor [13.31].

In agreement with the predictions, it was found that only the presence of dGTP caused a significant change in the fluorescence correlation functions (Fig. 13.4a). Four distinct dynamic processes could be identified. They were

288   J. Widengren

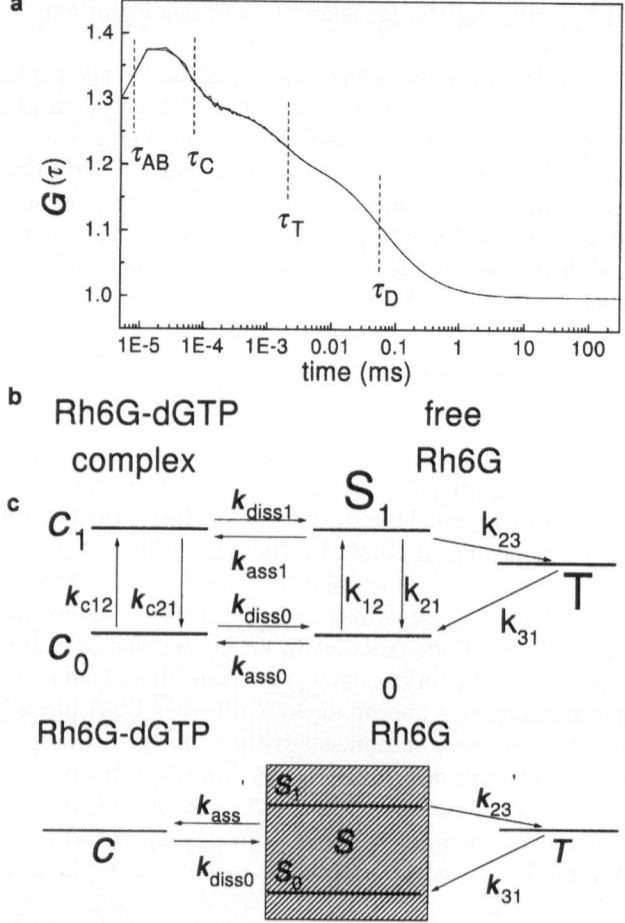

Fig. 13.4. a FCS curve of Rh6G in aqueous solution with $2\,\mathrm{mM}$ dGTP added. Excitation intensity $200\,\mathrm{kW/cm^2}$. Four distinct relaxation processes can be noticed: τ_{AB}, τ_C, τ_T, and τ_D denote the relaxation times due to antibunching, interactions with dGTP, singlet-triplet transitions, and translational diffusion, respectively. In this curve, one may note the broad time range over which different dynamic processes can be simultaneously monitored with FCS, **b** Kinetic scheme of the different states involved in the interaction of Rh6G and dGTP in water. An electron-transfer reaction is assumed to take place upon complex formation between Rh6G and dGTP leading to a complete quenching of the Rh6G fluorescence. $k_{12} = \sigma_{\mathrm{exc}} \Phi$ is the excitation rate from the ground singlet (S_0) to the first excited singlet (S_1) state, k_{21} is the decay rate of the excited singlet state, k_{23} is the intersystem crossing rate from S_1 to the lowest triplet state (T), and k_{31} is the triplet state decay rate. k_{ass1} and k_{ass0} are the rates of complex formation (leading to fluorescence quenching via electron transfer) from the excited and ground singlet state, respectively. k_{diss1} and k_{diss0} are the dissociation rates of the complex from the excited complex and ground state complex state of Rh6G, **c** Simplified scheme used in the treatment of the time dependence of the fluorescence intensity. Here, $k'_{\mathrm{ass}} = k_{\mathrm{ass0}}k_{21}/(k_{21} + k_{12}) + k_{\mathrm{ass1}}k_{12}/(k_{21} + k_{12})$ and $k'_{23} = k_{23}k_{12}/(k_{21} + k_{12})$

attributed to: 1. antibunching caused by transitions (excitation and deexcitation) between the ground and the excited singlet state of the Rh6G molecules (approximately 5 ns) [13.22,13.32]. 2. dGTP specific interactions (approximately 50 ns). 3. singlet-triplet interactions (μs) [13.13,13.14]. 4. Translational diffusion into and out of the sample volume element (approximately 50 μs). The typical time ranges for each of the processes are given within brackets. Disregarding the fastest process due to antibunching the experimental curve can be fitted to a two-exponential function superimposed onto the correlation function arising as a result of translational diffusion, $G_D(\tau)$:

$$G(\tau) = G_D(\tau)\frac{1}{1 - \overline{T} - \overline{C}}\left[1 - \overline{T} - \overline{C} + \overline{T}\,e^{-\tau/\tau_T} + \overline{C}\,e^{-\tau/\tau_C}\right] + 1, \quad (13.16)$$

where the first exponential, characterized by $\overline{T}$ and τ_T, is caused by singlet-triplet interactions in the Rh6G molecules, as outlined above, and the second exponential, described by $\overline{C}$ and τ_C, is due to an effect specific for dGTP.

Upon variation of the excitation intensity in the FCS measurements, the amplitude of $\overline{C}$, unlike the mean fraction of molecules being in their triplet states, given by $\overline{T}$, did not vanish at low excitation intensities. This indicates that a complex formation leading to quenching takes place from the ground singlet as well as from the excited singlet state. Additionally, from fluorescence decay measurements, the fluorescence decay of Rh6G in the presence of different concentrations of dGTP showed a biexponential decay, with a lifetime of the complex approximately 25 times shorter than that of the free Rh6G, which in turn was reduced with higher concentrations of dGTP. dGTP thus generated both static and dynamic quenching.

From these observations, a kinetic scheme could be set up including singlet-triplet interactions of the free dyes, and complex formation and the following electron-transfer-induced quenching originating from both the excited as well as the ground singlet state (Fig. 13.4b). Since the transitions between the singlet states take place on a time scale much faster than any of the other dynamic processes, the S_0 and S_1 states will, for the longer time ranges of these processes not be resolved and can therefore be regarded as one entity, that of S. Additionally, due to the very short deactivation time of the dye-dGTP complex, the fraction of molecules in the excited state of the complexed dyes could be neglected, leading to a simplified reaction scheme (Fig. 13.4c).

Based on this reaction scheme, the following system of coupled first order linear differential equations can be set up:

$$\frac{d}{dt}\begin{pmatrix} S(t) \\ T(t) \\ C(t) \end{pmatrix} = \begin{bmatrix} -(k'_{12} + k'_{ass}) & k_{31} & k_{diss0} \\ k_{23} & -k_{31} & 0 \\ k'_{ass} & 0 & -k_{diss0} \end{bmatrix} \begin{pmatrix} S(t) \\ T(t) \\ C(t) \end{pmatrix}. \quad (13.17)$$

where $k'_{23} = k_{23}k_{12}/(k_{12} + k_{21})$ and $k'_{ass} = k_{ass0}k_{21}/(k_{12} + k_{21}) + k_{ass1}k_{12}/(k_{12} + k_{21})$. The correlation function expresses the probability of a photon emission at a time τ given a photon emission at time 0. This probability is proportional to $S(\tau)$, which is obtained from the solution of (13.17) with the boundary condition:

$$\begin{pmatrix} S(0) \\ T(0) \\ C(0) \end{pmatrix} = \begin{pmatrix} 1 \\ 0 \\ 0 \end{pmatrix}. \tag{13.18}$$

In analogy to the analysis of the singlet-triplet interactions overviewed above, the correlation function is given by:

$$G(\tau) = G_D(\tau) \frac{S(\tau)}{\overline{S}} + 1, \tag{13.19}$$

where $S_1(\tau)$ in the triplet analysis above has been replaced by $S(\tau) = S_1(\tau) + S_0(\tau)$ and (13.19) takes the form given by (13.16).

From the solution of (13.17) and (13.18), the variables $\overline{T}$, $\overline{C}$, τ_T and τ_C can be expressed as functions of the rate parameters k_{ass0}, k_{ass1}, k_{diss0}, k_{23}, k_{31}, k_{21} and $k_{12} = \sigma_{exc} \Phi$ in the rate schedules of Figs. 13.4b and 13.4c. These rate parameters can be determined in a similar fashion as the triplet state parameters in the previous section: fluorescence correlation functions are measured at different excitation intensities, Φ. For each correlation curve, four experimental values $(\overline{T}, \overline{C}, \tau_T$ and $\tau_C)$ and hence four equations for the rate parameters are obtained. From the resulting overdetermined equation system the different rate constants can be extracted (Fig. 13.5a). The same evaluation procedure was repeated for several different concentrations of dGTP. In (Fig. 13.5b) the rate constants of k_{ass0}, k_{ass1} and k_{diss0} are plotted versus the concentration of dGTP. From these plots, the quenching efficiencies from the ground as well as from the excited singlet state can be extracted.

The above experimental example shows that FCS offers a complementary technique to study inter-molecular interactions between dyes and nucleotides. From the correlation data the dissociation rate from the complex ground state and the association rates from the ground singlet as well as from the excited singlet state can be extracted. The influence of the quencher on the triplet state can be obtained as well. In the above study, the decay rate of the triplet state was found to decrease with increasing dGTP concentrations, which could be indicative of shielding of oxygen upon complex formation.

In contrast to traditional techniques used to measure electron transfer reactions, such as transient absorption spectroscopy, electron-transfer quenching can, with FCS, be studied on a microscopic scale. The degree of electron-transfer-induced quenching can reflect proximities and accessibilities between possible reactants, giving information about inter- as well as intra-molecular dynamic processes.

13.2.3 Photo-Induced Isomerization

Apart from the use of cyanine dyes in fluorescence spectroscopy and microscopy, cyanine dyes are widely used in the photographic industry and as

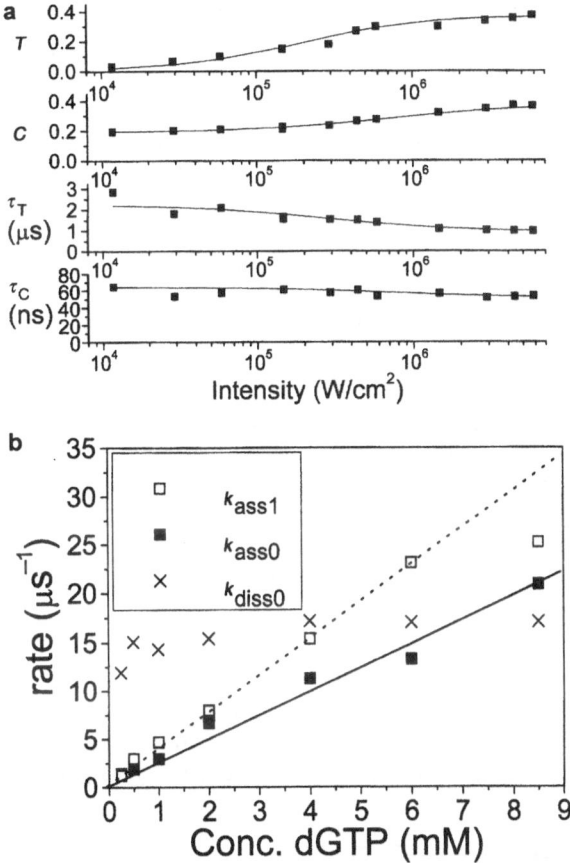

Fig. 13.5. a Outcome of a general analysis of Rh6G in a 2 mM aqueous solution of dGTP at different excitation intensities. The measured values of $\overline{T}, \overline{C}, \tau_T$ and τ_C are plotted as well as their values, fitted according to the solution of (13.17) and (13.18), where $\overline{T}, \overline{C}, \tau_T$ and τ_C are expressed as functions of the rate parameters contained in the rate schedule of Fig. 13.4b, **c** The calculated rate constants of association from the ground and excited singlet states, k_{ass0} and k_{ass1}, and that of dissociation, k_{diss}, plotted as a function of dGTP concentration. The quenching efficiencies from the ground as well as from the excited singlet state can be extracted and determined to be $2.4 \times 10^9 \, M^{-1} s^{-1}$ and $3.8 \times 10^9 \, M^{-1} s^{-1}$, respectively. According to Smoluchowski's relation this corresponds to approximate reaction radii of 0.6 and 0.9 nm with reactions taking place in approximately every third and second dye-dGTP collision, respectively. The dissociation rate remained more or less constant $(1.5 \times 10^7 \, s^{-1})$. The dissociation constants, defined as k_{diss0}/k_{ass1} and k_{diss0}/k_{ass0} for the ground and the excited singlet state are then estimated to $135 \, M^{-1}$ and $210 \, M^{-1}$, respectively

passive modelockers in lasers [13.33]. They may also find use as potential
sensitizers in tumors treatment by photodynamic therapy [13.34,13.35].

A characteristic of many compounds of this dye family is their tendency,
upon light exposure, to undergo trans-cis isomerization in their conjugated
hydrocarbon chains, which link the two heads of the cyanine dyes. The details
of this photoisomerization can largely determine their photophysical properties. The rate and character of the isomerization can vary greatly from one
dye to another. Among parameters that influence the isomerization, viscosity of the medium, temperature, solvent polarity and absence or presence of
steric hinderances can be mentioned [13.19,13.36,13.37].

To demonstrate the feasibility of the FCS technique to investigate trans-
cis isomerizations, the cyanine dye Cy-5 [13.38,13.39] was studied
[13.40,13.41]. In Fig. 13.6, correlation curves of Cy-5 in water excited by a He-
Ne laser emitting at 632 nm at different excitation intensities are shown. For
all curves, even at very low excitation intensities, an exponential process could
be observed, in addition to the correlation due to translational diffusion. The
relaxation time of this exponential decreased markedly with increasing excitation power. Plotting the inverse relaxation time versus excitation power
a linear dependence could be obtained. The amplitude of this exponential
remained almost constant (insert, Fig. 13.6). At higher laser intensities an
additional exponential gradually appeared, whose inversed relaxation times
showed a more sigmoidal-shaped intensity dependence and with an amplitude
dropping to zero with decreasing intensities. When adding glycerol to increase
the solvent viscosity, the first relaxation process was found to be slowed down
considerably. Similarly, slower relaxation rates were found for dyes having a

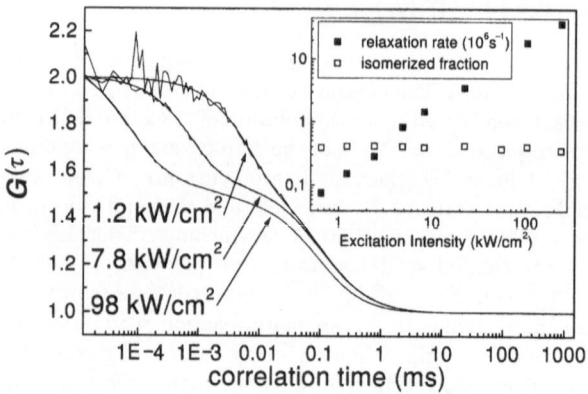

Fig. 13.6. Excitation intensity dependence for Cy-5 in aqueous solution. The main
effect seen is likely to be attributed to trans-cis isomerization. For the highest excitation intensity applied one can note some degree of photobleaching, as indicated by
the shortened decay in the 100 μs time scale. Insert: The relative amplitude and the
relaxation rate of the trans-cis isomerization relaxation process in the correlation
functions as a function of excitation intensity

nucleotide or a DNA strand linked to it. By addition of potassium iodide, the second exponential was found to be more pronounced, as for rhodamine dyes. Based on these observations, the first-mentioned exponential process can most likely be attributed to trans-cis isomerization and the latter ones to triplet formation. The fact that the amplitude in the correlation curves of the isomerization process remained constant, while the relaxation rate scaled with the excitation intensity, indicates that the isomerization as well as the back-isomerization process are photo-induced. This would be in agreement with a model that has been proposed for, DTCI, a cyanine dye similar to Cy-5 [13.42], where photo-induced back-isomerization is the main deactivation pathway of the cis state back to the trans state, and where back-isomerization over the ground states of the isomers is strongly disfavored due to a very high activation energy.

13.2.4   Effects of Non-Uniform Excitation

In the examples discussed above, no consideration has been taken of the effects of non-uniform excitation within the detection volume. With a non-uniform excitation profile over the sample volume element, the intensity-dependent exponentials of photophysical origin in the experimental autocorrelation (AC) function will appear as weighted integrals over the sample volume element. Therefore, instead of a single exponential decay, a continuous distribution of exponentials over an interval of relaxation rates can be expected. However, due to moderate S/N ratios in the microsecond time range, where many of the photophysical processes take place, and when working with a narrow $CEF(\bar{r}, z)$, the span of excitation intensities over the sample volume element will be restricted, and the distribution of exponentials is usually not resolved. However, a wider $CEF(\bar{r}, z)$, as obtained with a larger pinhole diameter in the image plane, may have a projection in the sample focal plane which clearly exceeds the dimensions of the excitation laser beam. Then effects of a non-uniform excitation intensity should be considered in order to more correctly estimate the photophysical parameters. The effects of non-uniform excitation intensities within the sample volume element have been treated in detail for the case of singlet-triplet interactions [13.13].

Another effect of non-uniform excitation intensities is that of saturation broadening of the sample volume element. With increasing excitation intensities, saturation of the fluorophores will first appear in the center of the beam in the focal plane, while in the periphery the local fluorescence generated will still show a linear dependence on the excitation intensity. This will lead to a broadening of the fluorescence intensity profile (proportional to $\bar{S}_1$, given by the first term in (13.12b)), thereby increasing the sample volume element and decreasing the amplitude of $G(\tau)$. However, with a pinhole size matching the dimensions of the excitation laser beam, the sample volume element will be determined mainly by the $CEF(\bar{r}, z)$ at high intensities.

13.3   Photophysics in the Slow Time Range-Photodegradation

In the previous section, the effects of photobleaching on the FCS measurements could be neglected for the case of a small molecule, quickly diffusing through a small detection volume, limited in size by diffraction and spherical aberration. The passage time through the detection volume would typically be much shorter than the photochemical lifetime of the molecule, even at excitation intensities close to saturation. However, if the time for translational diffusion through the detection volume is long, due to an increased size of the volume element or due to more slowly moving fluorescent molecules, the photochemical lifetime of the molecules may very well be shorter than the passage time. This can be seen in FCS as a faster overall decay of the experimental AC curves, especially when higher excitation intensities are applied. An influence from photobleaching can normally be avoided, also for longer passage times of the fluorophores, by reducing the excitation intensity, at the price of an increased measurement time. However, one may also deliberately increase the sample volume element in order to clearly monitor and characterize the photobleaching [13.43].

The possibility that a dye molecule is photobleached is usually assumed to be proportional to the number of excitation-emission cycles, or to the total time a molecule has spent in its excited singlet (or triplet) state. The photostability of a fluorescent compound is normally expressed as the quantum yield of photobleaching, Φ_D, given by:

$$\Phi_D = \frac{\text{number of photodestructed molecules}}{\text{total number of absorbed photons}} \tag{13.20}$$

and, with the assumption that photon absorption leads to excitation to the excited singlet state, the average number of photons emitted from a fluorophore before it bleaches is the ratio of emission quantum efficiency to bleaching quantum efficiency, or Φ_f / Φ_D. Then, the rate of photobleaching of the fluorescent molecules can be expressed as:

$$k_D = \Phi_D (k_{21} + k_{23}) \overline{S}_1 , \tag{13.21}$$

where $\overline{S}_1$ denotes the mean fraction of fluorophores in the detection volume populating the excited singlet state. For a three-level electronic model, given by the ground and the excited singlet state and the triplet state (Fig. 13.2a), $\overline{S}_1$ is given by the first term of (13.12b).

In the evaluation of FCS curves, photobleaching can be assumed to proceed as a first-order process at a constant rate during the whole passage through the detection volume. This applies to the case of a uniform excitation intensity within the detection volume, where the photobleaching is proportional to the average population of S_1 and T. Although photobleaching is fundamentally a non-equilibrium process the dye concentration within the detection volume can normally be regarded to be in a steady-state when exposed to CW excitation. The bulk sample volume in FCS experiments are often sufficiently large to assure that only a very small fraction of the fluorophores in the solution is within the laser beam at the same time. The rate of the overall depletion of the concentration due to bleaching is then orders of magnitude slower than the passage times of the fluorophores through the detection volume and does not influence the correlation function within the time range of the passage times.

By the assumption of an average effective photobleaching rate constant over the whole sample volume element, the actual space-dependent homogeneous distribution of bleaching reactions is changed into an overall bleaching reaction affecting only a fraction, A, of all excited molecules within the sample volume element [13.44]. In the presence of photobleaching, with the relaxation rate of singlet-triplet transitions, λ_3, being typically much faster than the average rate of photobleaching, k_D, the correlation function can be expressed as:

$$G(\tau) = G_D(\tau) \left[1 - A + A\,e^{-k_D\tau} \right] \left[1 + \frac{T}{1-T} e^{-\lambda_3\tau} \right] + 1\,, \qquad (13.22)$$

where $G_D(\tau)$ is the time-dependent correlation function for translational diffusion.

Figure 13.7 shows a set of correlation curves and the corresponding photobleaching rates and quantum yields, as given by (13.20) and (13.21), for Rh6G in aqueous solution. The photodestruction quantum yield of Rh6G was

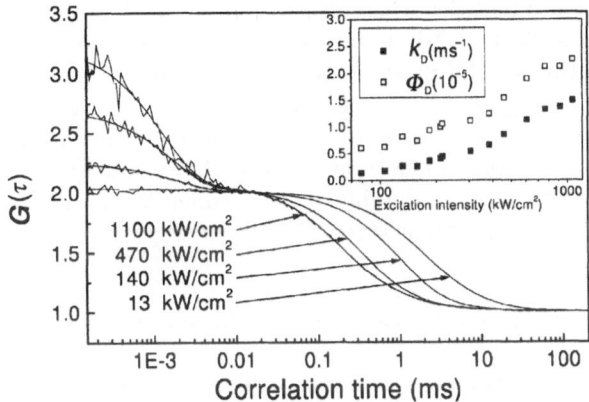

Fig. 13.7. FCS curves of Rh6G in water measured at different excitation intensities, Φ, with an extended detection volume ($\omega_1 = 1.7\mu m$). With increasing Φ the FCS curves decay with significantly shorter times and the degree of triplet state build-up increases strongly. Insert: Photobleaching rates, k_D, and photodestruction yields, Φ_D measured and calculated according to (13.21–13.22), and plotted versus excitation intensity. A prominent increase in Φ_D can be observed at excitation intensities at or close to saturation

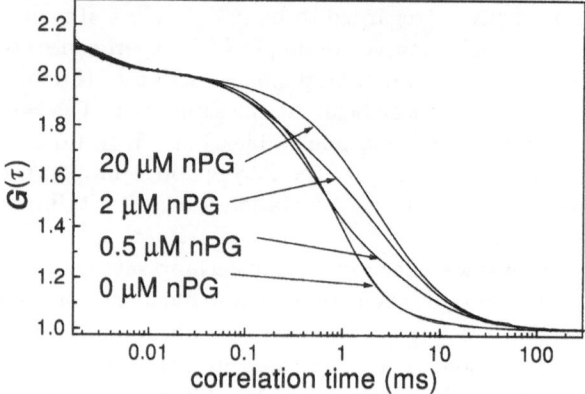

Fig. 13.8. FCS curves of Rh6G in water at $200\,\mathrm{kW/cm^2}$ excitation intensity. With increasing concentrations of n-propyl gallate (n-PG) the correlation curves decay more slowly, and a relaxation process can be identified consistent with a reduction of photo-induced dye radicals by n-PG, which thereby acts as an electron donor

found to be considerably lower than that for FITC. This is in line with the common view that the photobleaching is proportional to the triplet state population of the dyes, considering the much higher rate of intersystem crossing found for FITC.

The photodestruction quantum yield was found to be markedly increased at or close to excitation intensity levels, where the rate of excitation, $\sigma_{\mathrm{exc}}\,\Phi$, was comparable to or higher than the deexcitation rate of the excited singlet state, k_{21}, i.e. at excitation intensities where saturation of the fluorophores was reached. This is most likely attributable to multi-photon absorption effects, leading to excitation of the fluorophores to higher excited singlet and triplet states [13.45]. Indeed, the non-constant photobleaching quantum yields found in FCS measurements can be taken into consideration by the application of a five-level electronic model, including higher excited singlet and triplet states, which show a considerably higher tendency to undergo photobleaching than their lower neighbor states [13.44].

13.4   Strategies to Improve Photophysical Conditions

Different strategies have been proposed to improve fluorescence rates and diminish photobleaching, including deoxygenation and addition of anti-oxidants, oxygen scavengers, singlet oxygen quenchers or triplet state quenchers [13.45]. However, in many cases the reports are contradictory about the effects of the compounds in question [13.46]. Additionally, for compounds having a retarding effect on the bleaching rate itself a corresponding or even stronger fluorescence quenching often occurs [13.47,13.48]. This situation indicates that the underlying mechanisms of photobleaching are not fully understood and that many of the data on photophysical parameters of fluorophores, such as intersystem crossing rates, can be very inconsistent between different reports. Still, from the measurements reviewed above it is clear that the non-constant, increased yield of photodestruction at or close to saturating excitation intensities should be borne in mind. At very low background levels, excitation intensities low enough not to generate fluorescence saturation can be recommended to avoid multi-photon excitation to higher triplet and singlet states, which are generally highly photolabile. This effect probably underlies the higher photodestruction quantum yields found for rhodamine dyes [13.49] with two-photon excitation. This might indicate that two-photon excitation is a less favorable choice for many applications in ultrasensitive fluorescence spectroscopy, and should be regarded as an alternative only when the specific merits of the technique [13.50] lead to distinct advantages. From a photophysical point of view, the inherent confinement of the excitation volume in two-photon excitation microscopy is just such an advantage. Even with a higher photodestruction quantum yield, two-photon excitation can lead to a lower overall depletion of the fluorophores in constrained volumes, such as within cells.

Oxygen is known to be able to act as a promotor of photobleaching. Deoxygenation has therefore been proposed as a strategy to prevent photodegradation. However, oxygen is also a potent quencher of dye triplet states and has been long known to increase lasing intensities in dye lasers. As noted previously, there is a trade-off between a reduced triplet population due to the quenching ability of molecular oxygen on the triplet states, and a higher photodecomposition yield, due to oxygen molecules reacting with transient photo-induced states, thereby forming stable non-fluorescent products. For Rh6G measured by FCS in aqueous solution, purging of oxygen slightly diminished the photobleaching rates at excitation intensities below saturation. However, due to increased triplet populations as a consequence of the slowed deactivation of the triplet states, the excited singlet state population decreases, which may even indicate higher photobleaching quantum yields (see (13.20) and (13.21)). At excitation intensities at or close to saturation the bleaching rates were quite comparable to those in air atmosphere. Reduction of oxygen as a strategy to prevent bleaching should therefore be practiced with caution.

Anti-oxidants have been proposed to be of value to prevent photobleaching in confocal fluorescence microscopy. We have investigated the anti-oxidant n-propyl-gallate (n-PG) [13.51]. Retarding effects on photobleaching when adding this compound have been reported for fluorescein protein conjugates in glycerol [13.47,13.52] and for phycobiliproteins in phosphate buffer [13.53]. According to these reports, relatively high n-PG concentrations of around 10 mM should be necessary in order to obtain a retarding effect on the photobleaching. However, from our FCS measurements n-PG was found to strongly quench the Rh6G molecules at those concentrations. The fluorescence autocorrelation curves obtained strongly resembled those obtained from the electron-transfer-induced quenching of dGTP, as described above, showing an additional exponential process. By analyzing the two exponentials of the FCS curves in a similar way a close to identical behavior was found. Considering that n-PG is known as a reducing agent it is likely that n-PG acts as an electron donor, in a similar manner to dGTP, and that n-PG upon hydrophobic interaction with the Rh6G molecules induces charge-transfer. In other words, at millimolar concentrations n-PG did not show an overall beneficial effect on the fluorescence due to the strong static as well as dynamic fluorescence quenching. However, at concentration levels in the micromolar range, when most of the quenching effects were no longer traceable, the retarding effect of n-PG was still pronounced, in the sense that addition of increasing amounts of n-PG increased the correlation decay time. This effect could clearly be seen when expanding the sample volume element so that the passage times of the Rh6G molecules were of the order of a few milliseconds (Fig. 13.8). From the correlation curves for different excitation intensities and for different concentrations of n-PG in the micromolar range it was clear that the effect could be treated as a bimolecular kinetic reaction, where the on-rate depends on the excitation intensity and the back-rate can be enhanced by collisional queching with n-PG. This would indicate that the retarding effect is not related to a quenching of the triplet state but rather depends on a collisional deactivation of a reduced form of the Rh6G dye, which in turn is formed in proportion to the population of the triplet and the excited singlet state. By the addition of n-PG in micromolar concentrations, an increase of count rates per molecule of more than 50% was reached at excitation intensities at or close to the saturation level. This difference in optimal concentration of n-PG to those previously suggested illustrates the necessity to modify the retardant concentration to the specific dye used, as well as to the dye concentration and its specific environment.

In FCS measurements, CW excitation intensities close to the saturation limit are often applied. The consequence is that populations of photo-induced states with low quantum yields of formation but with long lifetimes can be quite significant. This will lower the fluorescence flow. Also, these photo-induced states may form precursor states for photobleaching. The triplet state is considered by many investigators to be the main precursor state of photobleaching. For this reason it is of interest to find triplet state quenchers, which then ideally would improve the fluorescence rate as well as retard photobleaching. We have investigated mercapto-ethylamine (MEA) [13.51], which has been reported to quench the triplet state of fluorescein without quenching the fluorescence [13.54]. In our experiments, addition of MEA to an air-saturated aqueous Rh6G solution led to a strong increase in fluorescence count-rates per dye molecule (approx. 100%). The effect of triplet state quenching of MEA is even more pronounced if the solution is deoxygenated, since the dyes would accumulate very strongly in their triplet states in the absence of oxygen. However, MEA also seems to have the effect of reducing the Rh6G molecules, just like n-PG, although not quite as strongly. This additional effect of MEA on the Rh6G molecules has the consequence that if too high a concentration of MEA is added, the fluorescence output will be reduced, just as observed for n-PG. This effect was most evident for de-oxygenated solutions since the reduced form was then not quenched by the oxygen molecules, but also for air-saturated solutions the fluorescence output, in terms of count-rates per molecule, did not increase at concentrations of MEA above a few millimolar.

13.5   Concluding Remarks

In this chapter, the fundamental photophysical parameters which determine the total number and rate of photons emitted per molecule have been discussed. Many photophysical processes have been found to be very sensitive to environmental parameters, which in part can explain the low common knowledge about their nature. Here, an outline has been given how to take photophysical processes into account in FCS measurements, and how FCS can be used to characterize these processes. In FCS measurements, with continuous excitation intensities close to the saturation level, attention should be paid to the build-up of transient states with a low quantum yield of formation and long lifetimes, as well as to an enhanced yield of photobleaching. Use of anti-fading compounds has proven to be useful also in FCS measurements. However, the optimal concentrations to be added seem to be considerably different from those found to be optimal for other fluorescence techniques. The FCS technique in general, offers a very convenient tool to monitor dynamic processes on a molecular level, which take place over time scales from nanoseconds up to seconds. In contrast to relaxation techniques, no perturbation of the system under study is needed. In the photophysics field, this gives FCS the ability to measure simultaneously both the extent of formation of non-fluorescent transients, such as triplet states, as well as quantitatively analyze the photodegradation process. With these properties in mind, it is likely that FCS can develop into a standard technique in the photophysics field, as an alternative or complementary method to flash photolysis and transient absorption techniques. Since FCS is a microscope-based technique, the environment sensitivity of many photophysical parameters can be analyzed on a microscopic scale, and be exploited as a means for microenvironmental probing.

Acknowledgements

The author would like to thank R. Rigler for cooperation and for providing the facilities where the work has been performed, and Ü. Mets, C. Seidel, C. Eggeling, J. Dapprich, and P. Schwille for cooperation on different parts of the work presented here. This study was financially supported by grants from the Swedish Foundation for International Cooperation in Research and Higher Education, the Swedish National Science Research Council, the Swedish Technical Research Council, Magnus Bergwall Foundation, and the Swedish Society of Medicine.

References

13.1 D. Magde, E.L. Elson, and W.W. Webb: Physical Review Letters **29** 705 (1972)

13.2 E.L. Elson and D. Magde: Biopolymers **13** 1 (1974)

13.3 D. Magde, E.L. Elson, and W.W. Webb: Biopolymers **13** 29 (1974)

13.4 E.L. Elson: Annu. Rev. Phys. Chem. **36** 379 (1985)

13.5 N.O. Petersen and E.L. Elson: Methods Enzymol **130** 454 (1986)

13.6 R. Rigler and J. Widengren: In: Bioscience (B. Klinge and C. Owman (eds.) Lund University Press, Lund. p. 180 1990)

13.7 R. Rigler, J. Widengren, and Ü. Mets: In: Fluorescence Spectroscopy (O.S. Wolfbeis (ed.) Springer Verlag, Berlin, p. 13 1992)

13.8 R. Rigler and Ü. Mets: SPIE **1921**, 239 (1992)

13.9 Ü. Mets and R. Rigler: J. Fluores. **4** 259 (1994)

13.10 M. Eigen and R. Rigler: Proc. Natl. Acad. Sci. USA **91**, 5740 (1994)

13.11 A.G. Palmer and N.L. Thompson: Biophys. J. **51**, 339 (1987)

13.12 E. Thiel and K.H. Drexhage: Chem. Phys. Lett. **199**, 329 (1992)

13.13 J. Widengren, Ü. Mets, and R. Rigler: J. Phys. Chem. **99**, 13368 (1995)

13.14 J. Widengren, R. Rigler, and Ü. Mets: Journal of Fluorescence **4**, 255 (1994)

13.15 A. Penzkofer, W. Falkenstein, and W. Kaiser: Chem. Phys. Lett. **44**, 82 (1976)

13.16 V.V. Rylkov and E.A. Cheshev: Opt Spectrosc (USSR) **63**, 462 (1987)

13.17 P.C. Beaumont, D.G. Johnson, and B.J. Parsons: J. Chem. Soc. Faraday Trans. **89**, 4185 (1993)

13.18 V.E. Korobov, V.V. Shubin. and A.K. Chibisov: Chem. Phys. Lett. **45**, 498 (1976)

13.19 V.E. Korobov and A.K. Chibisov: Russ. Chem. Rev. **52**, 27 (1983)

13.20 M. Ehrenberg and R. Rigler: Chem. Phys. **4**, 390 (1974)

13.21 P. Kask, P. Piksarv, and Ü. Mets: Eur. Biophys. J. **12**, 163 (1985)

13.22 Ü. Mets, J. Widengren, and R. Rigler: Chem. Phys. **218**, 191 (1997)

13.23 G.J. Kavarnos and N.J. Turro: Chem. Rev. **86**, 401 (1986)

13.24 G. Loeber and L. Kittler: Stud. Biophys. **73**, 25 (1978)

13.25 P. Lianos and S. Georghiou: Photochem. Photobiol. **29**, 13 (1979)

13.26 N.E. Geacintov, R. Zhao, V.A. Kuzmin, S.K. Kim, and L.J. Pecora: Photochem. Photobiol. **58,2**, 185 (1993)

13.27 S.J. Atherton and A. Harriman: J. Am. Chem. Soc. **115**, 1816 (1993)

13.28 J.M. Kelly, W.J.M. van der Putten, and D.J. McConnel: Photochem. Photobiol. **45**, 167 (1987)

13.29 C. Seidel: In: Dissertation (Heidelberg University 1992)

13.30 J. Widengren, J. Dapprich, and R. Rigler: Chem. Phys. **216**, 417 (1997)

13.31 M. Sauer, K.T. Han, R. Mueller, S. Nord, A. Schulz, S. Seeger, J. Wolfrum, J. Arden-Jacob, G. Deltau, N.J. Marx, C. Zander, and K.H. Drexhage: J. Fluoresc. **5** 247 (1995)

13.32 M. Ehrenberg and R. Rigler: Chem. Phys. Lett. **14**, 539 (1972)

13.33 C.V. Shank and E.P. Ippen: In: Dye Lasers (F.P. Schäfer (ed.) Springer Verlag, Berlin, **1**, p. 139, 1990)

13.34 A.C. Benniston and A. Harriman: J. Chem. Soc. Faraday Trans. **90**, 953 (1994)

13.35 M. Krieg and R.W. Redmond: Photochem. Photobiol. **57**, 472 (1993)

13.36 P.F. Aramendia, R.M. Negri, and E. San Roma'n: J. Phys. Chem. **98**, 3165 (1994)

13.37 D. Noukakis, M. Vanderauweraer, S. Toppet, and F.C. Deschryver: J. Phys. Chem. **99**, 11860 (1995)

13.38 L.A. Ernst, R.K. Gupta, R.B. Mujumdar, and A.S. Waggoner: Cytometry **10**, 3 (1989)

13.39 R.B. Mujumdar, L.A. Ernst, S.R. Mujumdar, and A.S. Waggoner: Cytometry **10**, 11 (1989)

13.40 J. Widengren: In: Fluorescence correlation spectroscopy, photophysical aspects and applications (Dissertation, Karolinska Institute, Stockholm 1996)

13.41 J. Widengren and P. Schwille: J. Phys. Chem. **104**(27), 6416 (2000)

13.42 R.E. Dipaolo, L.B. Scaffardi, R. Duchowicz, and G.M. Bilmes: J. Phys. Chem. **99**, 13796 (1995)

13.43 J. Widengren and R. Rigler: Bioimaging **4**, 149 (1996)

13.44 C. Eggeling, J. Widengren, R. Rigler, and C. Seidel: Anal. Chem. **70**, 2651 (1998)

13.45 R.Y. Tsien and A. Waggoner: In: Handbook of Biological Confocal Microscopy (J. Pawley (ed.) Plenum Press, New York, p. 169, 1989)

13.46 G. Boeck, M. Hilchenbach, K. Schauenstein, and G. Wick: J. Histochem. Cytochem. **33**, 699 (1985)

13.47 A. Longin, C. Souchier, M. French, and P.A. Bryon: J. Histochem. Cytochem. **41**, 1833 (1993)

13.48 T. Hirschfeld: J. Histochem. Cytochem. **27** 96 (1979)

13.49 E.J. Sanchez, L. Novotny, G.R. Holtom, and X.S. Xie: J. Phys. Chem. **101**, 7019 (1997)

13.50 W. Denk, J.H. Strickler, and W.W. Webb: Science **248**, 73 (1990)

13.51 J. Widengren, C. Eggeling, and S. Seidel: manuscript in preparation

13.52 H. Giloh and J.W. Sedat: Science **217**, 1252 (1982)

13.53 J.C. White and L. Stryer: Anal. Biochem. **161**, 442 (1987)

13.54 L.L. Song, C.A.G.O. Varma, J.W. Verhoeven. and H.J. Tanke: Biophys. J. **70**, 2959 (1996)

Part V

New Developments and Trends

14 Fluorescence Correlation Spectroscopy: Genesis, Evolution, Maturation and Prognosis

Watt W. Webb

> *The brightest flashes in the world of thought are incomplete until they have been proved to have their counterparts in the world of fact.*
> John Tyndall, Scientific Materialism

14.1 Introduction

Fluorescence correlation spectroscopy (FCS) [14.1] was conceived thirty one years ago with the objective of observing the dynamics of denaturation of DNA. Our strategy was to extend the elegant concepts of fluctuation correlation spectroscopy to the dynamics of chemical interactions, which had not been accessible to previous fluctuation indicators such as quasi-elastic light scattering. Its first realization in studying interaction of a fluorescent drug with DNA proved the concept and developed the first experimental method [14.1–14.3]. During the ensuing thirty years, fluorescence correlation spectroscopy has evolved into a user-friendly procedure for measurements of molecular diffusion in very dilute solutions and on membranes, for analyzing chemical kinetics and conformational dynamics, for sensitive analysis of sparse molecular and particulate species, and recently for observations of the dynamics of molecular processes in and on living cells and tissues.

Here I first report on the genesis of the concept and its realization in my laboratory at Cornell University. Then the evolution of FCS toward today's user-friendly status is discussed. Its maturation in some of our current research is summarized to illustrate our recent advances, and finally I venture a prognosis on the future of FCS. This report is not intended as a comprehensive history of FCS. It is expected that the other chapters of this book on FCS will recount the important contributions of other laboratories during these years and especially additional recent applications.

14.2 Genesis

Fluorescence correlation spectroscopy (FCS) was conceived [14.1] within a scientific milieu that was pervaded by research utilizing fluctuations to measure the fundamentals of the statistical thermodynamics and dynamics of continuous phase transitions, critical phenomena and quantum superfluids. Using optical techniques, but not fluorescence, we had measured for the first time the diffuse interfaces in critical fluid mixtures created by diverging critical concentration fluctuations [14.4] and the dynamics of the thermal excitations of these critical fluid interfaces [14.5]. The quantum statistical physics

of fluctuations in superfluidity and superconductivity were being studied in
our laboratory by fluctuation correlation spectroscopy using electrical and
magnetic signals, for example: the non-equilibrium annihilation instability in
hard superconductors was recognized by magnetic flux jumps [14.6]; intrinsic quantum fluctuations of one-dimensional superconductors were detected
by microKelvin transition temperature shifts [14.7,14.8]; non-equilibrium flux
creep in type II superconductors was detected by magnetic flux quanta displacement [14.9]; and quantum conductance fluctuations were measured in
Josephson oscillators [14.10]. Quasi-elastic light scattering by isotopic mixtures ^{3}He $+^4$ He near the tricritical point at 0.866 ± 0.001 K with ΔT stabilized to $\pm 1\,\mu$K was the most difficult of the optical experiments [14.11,14.12].
Our experimental research on fluctuations was nourished in those days by
a stimulating background of related theoretical research by our colleagues
Ben Widom, Michael Fisher, Viney Ambegaokor, Jim Langer, John Wilkens,
Mark Nelkin and Ken Wilson, and their students.

With this background, it seemed only natural to consider measuring the
statistical thermodynamics of a chemically reacting system when Elliot Elson presented his exciting new problem [14.1]. Use of fluorescent molecules
to study chemical reactions at room temperature promised, at the time, to
be easier than our measurements of fluctuations of light scattering by refractive index fluctuations of a few parts per million in the ^{3}He $+^4$ He isotopic
mixtures. Nevertheless, it is the theoretical basis for understanding concentration fluctuations in mixtures that provided the starting point for developing FCS. This theoretical background is nicely summarized by Berne and
Pecora [14.13]. The cooperative concentration fluctuations in critical fluids
produced spatial correlations on the scale of visible light wavelengths that
provided readily detectable light scattering.

All of these previous fluctuation measurements in critical solutions and
quantum cooperative systems had features that made them experimentally
accessible. Their statistical physics was so tightly coupled that the behavior of
individual atoms or molecules was not as important as the cooperative interactions of many atoms, molecules or electron pairs that created strong signals.
Quantum fluctuations in superconductors created fluctuations that we could
measure at a few femtovolts electrically, or magnetically at $< 10^{-15}$ Wb, or
by temperature shifts of a few microKelvin.

But why invoke fluorescence to study chemically reactive systems? Purely
diffusive or thermal fluctuations were being measured with great precision by
quasi-elastic light scattering, but extension to chemical reaction kinetics had
failed for the simple reason that the refractive index changes induced by
chemical reaction were too small to measure. How about using light absorption by molecular chromophores; it is often altered by chemical reactions?
The problem is that the absorption by a few molecules in the background
of many was not usefully measurable. However, molecular fluorescence can
be changed dramatically by chemical reaction, and an individual fluorophore

can, in principle, emit fluorescence at up to $\sim 10^9$ photons per second; detecting $\sim 0.1\%$ of the emitted photons yields $\sim 10^6$ photons per second per molecule, which suggests temporal fluctuation resolution to $\sim 10^6$ Hz. Often, 10^3 to 10^4 photons per molecule can be detected before irreversible photobleaching occurs. Thus, fluorescence is suitable to study ensembles containing only a few molecules. Because the relative mean square amplitude of fluctuations of independent random molecular processes in a system is equal to the reciprocal of the average number $\langle N \rangle$, the fluctuations of the number of fluorophores in a small system can yield a large change of the fluorescence relative to the time average fluorescence: $\langle \Delta F^2 \rangle / \langle F \rangle^2 = 1/\langle N \rangle$ if the fluorescence per molecule is a constant.

The problem motivating development of FCS was introduced by Elliot Elson, then the expert on DNA chemistry in Cornell's chemistry faculty. Elson wanted to understand how the double helix of DNA became denatured for transcription. At that time, the genome was visualized as a ball of twisted double-stranded string that had to be untangled for transcription. Elliot Elson had already developed a theory of denaturation of DNA based on twist diffusion while still a post-doc in association with Bruno Zimm. Elliot Elson noted that the binding to DNA of ethidium, a dye that inhibits nucleic acid synthesis *in vivo* and that becomes fluorescent only if intercalated in the double helix of DNA, appeared to offer an interesting probe based on kinetics measurements by Don Crothers at Yale. This proved to be an ideal archetype to develop FCS. The paucity of successful applications of FCS to chemical kinetics in the ensuing decade indicates the wisdom of his choice. Emphatically acknowledged here is the generation of the FCS concept in an enduring and delightful collaboration with Elliot Elson and its realization through the talented persistence of Douglas Magde participating in our research.

In FCS, one usually considers the dynamics of number fluctuations in an open sampling volume of a macroscopic system which fluctuates around average equilibrium concentrations determined by the surrounding medium and its thermodynamics. The fluctuations of concentration of each species can occur by *in situ* chemical reactions and by diffusion of each species into and out of the sampling volume. Thus, one can describe the local statistical thermodynamics within the sampling volume as an ensemble with average temperature and average concentrations fixed by the surrounding reservoir but with linear dynamics of small fluctuations around equilibrium determined by the kinetics of chemical reactions and diffusion. In general, this is the concept associated with using the Onsager reciprocal relations and reflects microscopic reversibility on the appropriate timescales and a phase space satisfying the quasi-ergodic hypothesis to accommodate the usual formulations of irreversible thermodynamics [14.14].

One can thus specify a set of coupled differential equations for every spatial and temporal point for this class of system in the form:

$$\frac{\partial}{\partial t}\delta C_j(\overline{r},t) = D_j\nabla^2\delta C_j(\overline{r},t) + \sum_k T_{jk}\delta C_k(\overline{r},t)\,, \tag{14.1}$$

where $\delta C_j(\overline{r},t) \equiv C_j(\overline{r},t) - \langle C_j(\overline{r},t)\rangle$ are the local concentration fluctuations of the reactants; D_j are the corresponding diffusion coefficients, and the T_{jk} are the elements of the matrix of linear chemical interaction coefficients representing the relevant chemical reactions. These matrix equations generally provide a sufficient description for most linearizable processes.

However, the existing fluorescence experiments cannot measure fluorescence fluctuations at a single point unless there is only a single point fluorophore present. Instead, the fluorescence emitted by a small sampling volume determined by the optical geometry of illumination and fluorescence detection (usually diffraction-limited) is measured as a function of time. The fluorescence fluctuations around the time average $\delta F(t) = F(t) - \langle F(t)\rangle$ are supposed to reflect the fluctuations of the number of molecules in the volume. (This condition holds in the absence of complications due to saturated excitation or photobleaching which depend on local excitation intensity.) Unfortunately, the focal volume is never illuminated or sampled uniformly. Thus, there is invariably a non-uniform spatial sampling profile composed of the product of the effective local illumination intensity which excites the fluorescence times a spatial profile for selective detection. For example, confocal optics, first used for FCS by Koppel et al. [14.15] as reported in 1976, are still being used today to select a small (femtoliter) focal volume from the double cone of focused laser illumination. For simplicity we can define the focal volume in terms of a spatial "detectivity" function $\phi(\overline{r})$ such that one can identify a local fluctuation in the concentration of a fluorescent species $\delta C(\overline{r},t)$ with its corresponding contribution to fluctuation of fluorescence signal; thus

$$\delta F(\overline{r},t) = \phi(\overline{r})\delta C(\overline{r},t)\,. \tag{14.2}$$

Integration of these point contributions over the sampling volume defines the fluorescence fluctuation signal; thus

$$\delta F(t) = \int \phi(\overline{r})\delta C(\overline{r},t)\mathrm{d}^3r\,, \tag{14.3}$$

and, of course, the fluorescence signal itself is

$$F(t) = \int \phi(\overline{r})C(r,t)\mathrm{d}^3r\,. \tag{14.4}$$

The focal volume V defined by the spatial distribution of the "detectivity" function for various cases can be expressed in terms of the moments of the detectivity function as

$$V = \left[\int \phi(\overline{r})\mathrm{d}^3r\right]^2 \bigg/ \int [\phi(\overline{r})]^2\,\mathrm{d}^3r\,. \tag{14.5}$$

We have developed illustrative forms for $\phi(\bar{r})$ for various optical geometries and calculated the corresponding volumes for one- and two-photon fluorescence excitation [14.16–14.19]. We are currently engaged in detailed analysis of the serious systematic errors in the results of FCS experiments that arise in sensitive experiments due to variations in $\phi(\bar{r})$. Note that measurements of fluorophore concentration depend on knowledge of V since:

$$\langle C \rangle = \langle N \rangle / V \,. \tag{14.6}$$

As is well known, the autocorrelation function of the relative fluorescence fluctuations

$$G(\tau) \equiv \langle \delta F(t) \cdot \delta F(t+\tau) \rangle / \langle F(t) \rangle^2 \tag{14.7}$$

provides the significant basis for FCS analysis. The zero time correlation $G(0)$ which is the normalized variance of $F(t)$ yields $\langle N \rangle$ and thus $\langle C \rangle$

$$G(0) \equiv \frac{\left\langle [\delta F(t)]^2 \right\rangle}{\langle F(t) \rangle^2} = \frac{1}{\langle N \rangle} = \frac{1}{\langle C \rangle V} \,. \tag{14.8}$$

The form of $G(\tau)$ reflects the dynamics of the fluctuating system, represented, for example, by (14.1), as modulated by the spatial detectivity function $\phi(\bar{r})$ in (14.3); thus for any fluorescent species the convolution of the spatial concentration fluctuations with the detectivity parameter yields the fluorescence correlation functions:

$$G_f(\tau) = \int \phi(\bar{r}) \delta C(\bar{r}, t) \phi(\bar{r}') \delta C(\bar{r}', t+\tau) \mathrm{d}^3 r \mathrm{d}^3 r' \,. \tag{14.9}$$

It has been customary to approximate $\phi(\bar{r})$ by a Gaussian cylinder in two-dimensions or as a Gaussian prolate ellipsoid in three-dimensions. This approximation simplifies the algebra and identifies simple diffusion factors in the autocorrelation functions of the form in two-dimensions here, $(1+\tau/\tau_D)^{-1}$ where $\tau_D \sim w^2/4D$, with w a radius of the sampling volume at $\phi(w)/\phi(0) \sim e^{-2}$ or e^{-4} depending on the form of $\phi(\bar{r})$. Note that $\phi(\bar{r})$ usually represents a product of two factors that appear in the optical point spread function (see, for example, the appendix of [14.18]). For two-photon excitation with no confocal aperture, $\phi(\bar{r})$ is simply the square of the illumination intensity profile.

The FCS analysis of the original DNA-ethidium reaction in the excess DNA limit and the principles of FCS have been discussed in detail in our early papers [14.1–14.3,14.20,14.21]. One cautionary note should be expressed about application of the original detailed discussion to other systems [14.2]: it is necessary to determine the form of the matrix equations to set up the eigenvalue problem and determine the eigenvectors and eigenvalues in the appropriate limits for each reactive system to be studied. The correlation functions differ strongly depending on whether, for example, there is a fluorescent product or a fluorescent reagent in the case of chemical reactions.

A variety of interesting phenomena in FCS has been studied in the 1980s by Elliot Elson and his colleagues, and during the 1990s by Rudolph Rigler and colleagues in Sweden and Germany, as is evident in other chapters of this book. One of the first FCS projects in Europe was the work of Ehrenberg and Rigler [14.22] on rotational diffusion. Some more complex cases have been studied by Icenogle and Elson [14.23,14.24].

To my mind, one of the most interesting discoveries in our DNA-ethidium study was the strong coupling between the fast diffusion of ethidium and the slow diffusion of the larger molecules of DNA. Because a local concentration fluctuation of free ethidium could relax either by diffusion or by binding to slow-moving DNA, the ethidium diffusion is slowed by its reactivity. One now recognizes a common biological appearance of this effect in research involving the effective diffusivity of Ca^{++} in the presence of buffers. If a high concentration of buffer is mobile, it can carry Ca^{++} along with itself, apparently speeding Ca^{++} diffusion, but if the buffer is immobile, it arrests the motion of Ca^{++} for part of the time, slowing diffusion. A somewhat different coupling phenomenon can occur in electrophoresis of high concentrations of membrane proteins on cell surfaces [14.25,14.26].

14.3   Evolution

Initially, a typical FCS experiment required about 30 min, and only a few per cent worked. Today, the measurement times for FCS may be as short as 10 s, and virtually every experiment works. In fact, FCS is now so efficient and so sensitive that it can generate misleading data (if one is not meticulously careful about experimental design) faster than any other spectroscopy I know.

Our first experimental realizations of FCS had demanded heroic measures from Douglas Magde and later Dennis Koppel to translate techniques learned from our experience in correlation spectroscopy based on quasi-elastic light scattering and to overcome the technical limitations of the day. Koppel's analysis of the statistical physics of the noise in FCS helped to squeeze useful information out of our marginal signals [14.27]. In early FCS, the limiting noise on the fluctuation correlations was primarily the photon shot noise due to signal plus background; this problem motivates choosing small values of average particle number $\langle N \rangle$ and suppression of extraneous background. The available lasers in the early 1970s were so noisy that compensation for their fluctuations was often necessary. Confocal optics have improved signal-to-noise ratios by defining a small focal volume, thus improving ratios of fluctuation signal to ambient background noise [14.15]. Koppel's early analysis of confocal optics for FCS was the first of a long series of studies [14.28] that continue today as we now aim to eliminate systematic errors in FCS experiments.

The technical developments that made these evolutionary advances possible are still in progress and continue to illustrate the synergistic interplay between scientific discovery and technological progress. We can list some of the recognizable advances that have made FCS generally feasible today. This list is surely incomplete and still expanding, but clearly a synergistic combination of developments rather than a single advance is responsible for the remarkable improvement.

(1) Confocal optics to restrict focal volume to less than 1 femtoliter.
(2) Correlator electronics of high efficiency and convenience. The introduction of the personal computer to accommodate dedicated PC correlator boards for FCS, notably available from ALV. Now, superb custom-designed correlators are available commercially as custom-programmed PC boards or external reconfigurable chips coupled through a fast port into a modern PC from Correlator.com.
(3) Avalanche photo diodes (APD) of small area, available from EG&G, have provided the high quantum efficiency needed for fast correlators. However, as this is written, new photomultipliers with comparable quantum efficiencies, larger sensitive areas, and selectable spectral sensitivities are becoming available from Hamamatsu. The larger areas are needed for ideal fluorescence collection optics in FCS with multi-photon excitation microscopy.
(4) Cross-correlation of two detectors in two-channel correlators enables rejection of random detector noise to access faster processes.
(5) Stable lasers in a variety of types and wavelengths. However, when stability fails, modern "noise eaters" can monitor laser output and feed back a modulating correction with wide bandwidth.
(6) Ultra purification of water by HPLC and use of clean room techniques to minimize (but not yet eliminate) impurity background fluorescence.
(7) Commercial availability of FCS apparatus from Evotec-Zeiss.
(8) Multi-photon excitation (MPE) of fluorescence for FCS to define a reproducible sampling volume, separate Raman scattering background of water, and minimize photobleaching problems. MPE is particularly preferable for FCS in biological preparations.

These technological advances have made possible substantial improvements of experimental design. Smaller focal volumes allow reduction of the average fluorophore number $\langle N \rangle$ while retaining chemically controllable nanomolar fluorophore concentrations. Understanding of the noise analysis has focused design on small $\langle N \rangle$ and low background fluorescence, and efficient correlators have allowed data acquisition in reasonable collection times. Better understanding of the photophysical limitations on photons detectable per molecule has focused attention on high quantum efficiency detectors to extend measurements to shorter times. FCS measurements of fast dynamics require fast photon emission rates per molecule since data contributions to correlation function at a correlation time τ requires detection of two photons per molecule during time τ. Thus, efficient measurement of a microsecond correlation requires detection of $\sim 10^6$ photons/s/molecule. This in turn requires emission of $\sim 10^9$ fluorescence photons/s since typically only $\sim 0.1\%$ are detected. Both intersystem crossing and photobleaching can defeat attempts to reach this objective. Pathological fluorophore photophysics seen at the single molecule level also remains a fundamental problem for FCS, as will be illustrated later in a discussion of green fluorescent proteins (GFP).

In the early 1990s, much of the progress that enabled this evolution of methods took place under the leadership of Rudolf Rigler in Stockholm and Manfred Eigen in Göttingen. Amongst the participating students, P. Kask, Ü. Mets and J. Widergren have come to my attention, but proper acknowledgements appear in chapters from representatives of that group. As the technical capability of FCS has improved, applications with sensitivities reaching the single molecule level have succeeded. Qian and Elson [14.29] extended the analysis of the statistics of FCS in confocal geometry to try to improve the precision and to analyze for higher moments; Thompson analyzed FCS in the optical geometry of localized wave excitation [14.30]; Nils Petersen, who had worked in FCS with Elson, developed a spatial correlation extension of the technique to observe molecular clustering on surfaces [14.31]; this technique may evolve too, as intensive computational technologies become more accessible. An excellent summary of the thinking about the capabilities of FCS was published by Eigen and Rigler in 1994 [14.32]. Developments in the 1990s brought FCS into a modern era in which commercial instruments can provide a user-friendly technology. Details of much of this research appear in other chapters of this book. We will concentrate here on some related research in our own laboratory.

While the aforementioned advances in FCS proceeded in Europe, our laboratory refocused temporarily on other targets of molecular fluorescence research that are now reflected in new concepts in FCS. Nanometer tracking of individual molecules, especially on the surfaces of living cells, was developed to understand the restraints on diffusion of cell membrane proteins [14.33–14.38]. This research eventually led to the discovery of ubiquitous anomalous subdiffusion of cell membrane macromolecules, a phenomenon based on intrinsic fluctuation processes due to molecular interactions, which we now find and study by FCS in cell cytoplasm as well as on living cell surfaces [14.38,14.39]. Technical lessons of this single molecule research guide our present research on FCS. Our second diversion from FCS research was the development of laser scanning multi-photon microscopy (MPM) and its adaptation for deep tissue and ultraviolet fluorescence imaging [14.40–14.42]. We have now found that MPM improves FCS, especially for effective applications in living cells and tissues and as a means to eliminate systematic errors due to detectivity function irregularity [14.18].

Meanwhile, our correlation spectroscopy research had again turned from fluorescence to the use of different indicators of fluctuations to study other fundamental problems. The mechanism of auditory transduction in vertebrate animals was studied by Differential Interference Contrast (DIC) optical measurements of the mechanical fluctuations of the sensory hair bundles which are cross-correlated with the cell membrane potential fluctuations that lead to propagation of the auditory nerve signals. Cross-correlation spectroscopy showed that mechanical Brownian noise limits the threshold of hearing [14.43]. Stochastic modulation of these same signals during evanescent binding and release of antibiotic molecules proved that the mechanism of auditory transduction was actually based on a mechano-sensory ion channel that served as the linear mechano-electrical transducer [14.44–14.46]. Note that this experiment measures chemical kinetics by a different indicator of the fluctuations.

Using ion current signal amplitude distribution analysis of the stochastic switching of conductance of individual ion channels reconstituted in lipid membranes, we first showed that the acetylcholine channel was a discrete molecular structure independent of cellular cofactors [14.47]. Our later measurements of conductance fluctuations of open channels showed that channel structures fluctuated rapidly [14.48,14.49]. Similarly, the statistical thermodynamics of mechano-sensitivity of an isolated trans-membrane ion channel [14.50] was established using signal amplitude probability distribution analysis methods. This statistical thermodynamics problem is quite analogous to FCS studies of reactive systems. It is interesting to note that these techniques are precursors of recently described fluorescence-intensity distribution analysis methods that can complement FCS as analyzed by Qian and Elson [14.51,14.52], see for example [14.53].

Yet, the applications of correlation spectroscopy are even broader: concurrent with the research noted above, we were studying resistive fluctuations in fabricated nanostructures, in which the wave mechanics of electron interference in metals and electron tunneling in composites causes conductance to fluctuate wildly with a broad spectrum with an amplitude proportional to the correlation time, usually over more than six decades of this "$1/f$ noise" [14.54–14.57]. The curvature energies of biological lipid membranes were determined by optically imaging and measuring shape fluctuations and analyzing the dynamics [14.58,14.59]. Fluctuations of lipid liquid crystal structures were studied by coherent scattering of synchrotron X-rays [14.60]. Turbulent boundary layers in fluids provided a challenging problem for correlation spectroscopy that we approached by developing a method for measuring vorticity fluctuations [14.61,14.62].

These developments and applications of correlation spectroscopies emphasize that fluorescence correlation spectroscopy is one member of a great family of experimental strategies from which one can seek the appropriate indicator of fluctuations of the physically relevant phenomenon and probe the fundamental questions with it. Fluorescence is a particularly useful fluctuation probe for biophysical processes where the relevant numbers of molecules may be very small.

14.4   Maturation of FCS at Cornell

Most of the accompanying chapters of this book report current achievements of FCS in their authors' laboratories. Thus, we discuss a few topics of interest in our own laboratories. Here, we employ FCS as a research tool which is closely coupled with many other research tools, and we find ourselves continuously trying to push the limits of capability of each of these tools. We are particularly motivated to develop the research tools to solve a long list of fundamental biophysical problems. Here are some tasks that can in principle be approached by development of variations of FCS concepts:

A. Measuring the dynamics of protein folding and structure fluctuations.
B. Characterizing the dynamics, thermodynamics and photophysics of mutants of green fluorescent proteins (GFPs) and other autofluorescent proteins of biological importance.
C. Analyzing and understanding the biophysical restraints responsible for the anomalous mobility of macromolecules on membranes and in living cells and tissues.
D. Detecting and measuring the restraints on the trajectories of individual macromolecules measured by nanometer tracking and recognition of mechanisms of molecular signaling interactions in and on living cells and tissues.
E. Sequencing of single template-DNA molecules by our new time series fluorescence correlation method.
F. Development of reliable methods that solve the sampling problem for systematic, sparse picomolar particle detection and characterization in biological fluids without using DNA amplification.
G. Characterizing the photophysical pathology of fluorophores at the single molecule level to facilitate molecular and cellular biophysics research on functions at the level of individual molecules.
H. Search for better fluorescent markers for biological research, and their evaluation. Understand the problems of nanometer semiconducting molecular markers and attempt to control them sufficiently well to replace organic dyes.
I. Develop the optics defining "detectivity" profiles that determine the effective shapes, profiles and volumes of the focal volumes of FCS in order to eliminate systematic errors in FCS and techniques to extend the scope of FCS experiments.
J. Understand and accommodate or circumvent the photophysical problems of fluorophores that limit the range of applicability of FCS measurements.

Our interest in measuring the dynamics of protein folding revived our use of fluorescence for correlation spectroscopy. The concept is to utilize fluorescent labels which are environmentally or FRET sensitive at appropriate sites on interesting proteins to follow the dynamics of protein folding and denaturation. Since the folding process is supposed to follow diverse pathways, often including an intermediate molten globule state, it is quite likely that conformational fluctuations occur over a wide range. Our problems in extracting data that we consider sufficiently reliable have focused our attention on some of the remaining limitations and artifacts in FCS; but we can report some good news that shows the power of FCS.

This section summarizes some discoveries made using FCS in our laboratory. Most of the results have recently been published, are in press, or are in preparation, so this presentation relies on that literature for details. These topics are discussed:

A. Statistical thermodynamics and photophysics of mutants and analogs of green fluorescent protein.
B. Molecular diffusion and phase characterization in lipid bilayers of giant unilamellar vesicles (GUVs).
C. Two-photon molecular excitation of fluorescence for FCS.
D. FCS in cells and tissues observing anomalous subdiffusion in membranes and cytoplasm and measuring cytoplasmic pH.

14.4.1   Green Fluorescent Proteins in FCS

Green fluorescent proteins (GFP) may provide the most useful advance in fluorescence probes in biophysics, biochemistry, and molecular and cell biology (!) research of the 1990s. The gene for this fluorescent protein, native to a Pacific jellyfish, can be expressed in the genome of a cell or an animal so that it labels a selected protein as an intrinsic fluorescent marker that is expressed with the selected protein.

Many mutants of GFP have been developed to enhance the properties of the fluorophore. The chromophore is formed autocatalytically from three amino acids, in the wild type Ser-65/Tyr-66/Gly-67, which are embedded in a large hydrophobic beta-barrel structure, the interior of which includes an intricate network of hydrogen bonds that interact with the chromophore. Mutations can favor either a protonated neutral state or a deprotonated anionic state of the tyrosyl hydroxyl group and can enhance desired fluorescence properties.

The mutant called EGFP $\equiv$ S65T/F64L can exist in either state over a pH range from about pH 4 to pH 11, with $pK_a \sim 5.8 \pm 0.1$, and is chemically stable over this range. The dominant deprotonated anionic state absorbs strongly with a peak at $\sim 490\,\mathrm{nm}$ and fluoresces with high quantum efficiency, but the protonated state or states absorb and emit weakly with this excitation (see Fig. 14.1) [14.63]. Therefore, the fluorescence intensity provides a useful and sensitive pH indicator for circumstances where it can be calibrated. As we shall see, FCS can enhance the pH measurement capability.

We have studied the dynamics of fluorescence fluctuations in EGFP by FCS, aiming in this research to measure the dynamics and statistical thermodynamics of protonation and deprotonation [14.63]. Correlation functions of

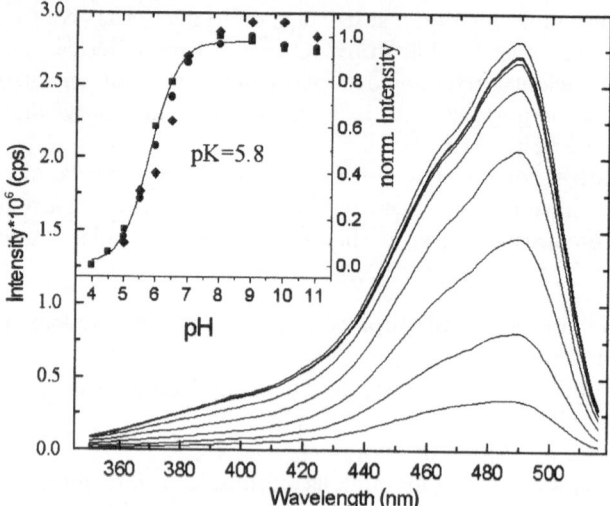

Fig. 14.1. Fluorescence excitation spectra of EGFP in 100 mM potassium phosphate, 10 mM citrate buffer at various pH values (*top-down*): 9.0; 8.0; 11.0; 7.0; 6.5; 6.0; 5.5; 5.0. *inset*: fluorescence emission intensity at 510 nm (**squares**), fluorescence excitation intensity at 490 nm (**circles**) and absorption at 488 nm (**diamonds**) versus pH. *Solid line*: fit to the emission intensity from pH 4.0 to 9.0 yielding a pK_a of 5.8 ± 0.1

EGFP as a function of pH are shown in Fig. 14.2. They clearly show the kinetics of the proton-binding reaction at low pH superimposed on the diffusion factor of the correlation function. The rate constant is $k_c = \tau_c^{-1} = k_p[H^+] + k_d$, where k_p = protonation rate constant and k_d = deprotonation rate constant. Since the GFP concentration is much smaller than the proton concentration $[H^+]$, it drops out. Our best fit values are $k_p = 3.45 \times 10^9 \, M^{-1} s^{-1}$ and $k_d = 9.05 \times 10^3 \, s^{-1}$ yielding $pK_a = \log(k_p/k_d) = 5.5 \pm 0.3$. To prove that we have correctly identified the origin of the exponential kinetics, we also measured the pH dependence of FCS for another GFP mutant, Y66W, which has no proton-binding site, as seen in Fig. 14.2. As expected, FCS data on Y66W are independent of pH.

The FCS fluorescence flicker data yield the time average fraction F_p of the protonated state as the weighting factor of the exponential correlation due to proton binding; this reflects the equilibrium statistical thermodynamics. Measurement of its temperature dependence yields $\Delta H° = -11.5 \pm 0.7 \, kJ/mol$ and $\Delta S° = -150 \pm 7 \, J/mol \, K$. Thus, FCS spectra can provide an internally calibrated ratiometric measure of pH, which can be implemented as a pH indicator at EGFP concentrations $\sim 10 \, nM$ without needing independent measurements of EGFP concentration.

However, our FCS studies of EGFP revealed an additional small but disturbing feature in its photophysics. Superimposed on the external protonation

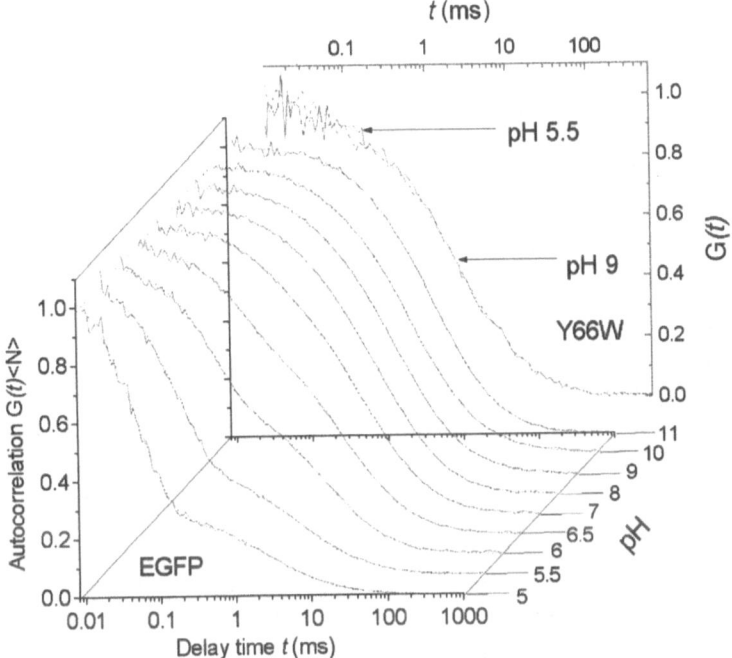

Fig. 14.2. Autocorrelation $G(t)$ curves (normalized to 1 at 10 ms) for EGFP at different pH values. back to front: pH 11, 10, 9.0, 8.0, 7.0, 6.5, 6.0, 5.5 and 5.0. Appropriate theoretical fits are shown as **dotted lines**. The decay of $G(t)$ at high pH is dominated by diffusional relaxation, while upon decreasing pH chemical relaxation at shorter time is growing in amplitude and speed. Autocorrelations of GFP-Y66W at pH 9.0 (**solid line**) and 5.5 (**broken line**) show absence of pH-dependent fast fluorescence flicker in this mutant, which lacks a protonatable hydroxyl on the chromophore

dynamics was an additional pH-independent fluorescence flicker with a dark fraction $\sim 13\%$ and a time constant $\sim 340 \pm 50\,\mu s$. This additional flicker process was attributed to an internal protonation rearrangement via transfer of a proton from an internal proton-binding site predicted by structural studies to the hydroxyl group of the chromophore; since this transfer quenches the fluorescence, it is detected by an additional flicker in the FCS. The disturbing feature of this process is a small but noticeable shift of its kinetics and partition with excitation intensity, suggesting that excitation provides an additional pathway for transition from fluorescent to "dark" states of the molecule. These experiments, however, have not yet established the identity of the additional dark state. The practical consequence is that use of EGFP as a pH indicator should take into account this small excitation intensity dependent perturbation. This complex photophysical question has led us to study excitation-driven processes in other mutants in which slow transitions had been reported [14.64,14.65] and to continuing studies of EGFP by Ahmed Heikal.

In 1997, Moerner and his associates reported that observation of immobilized single molecules of two yellow-shifted GFP mutants S65G/S72A/T203Y and S65G/S72A/T203F, called here YFPs, could slowly "blink" between brightly fluorescent and "dark" states by weak illumination, with the transitions occurring on a timescale of $\sim 1\,$s [14.64]. Motivated by weak effects of excitation of EGFP, we initiated collaboration to apply FCS to these mutants [14.65]. The FCS experiments showed fast fluorescence flicker at 10^{-6} to $10^{-3}\,$s overlying the usual diffusion dynamics. This effect was clearly distinguishable from intersystem crossing flicker. Protein binding from the external medium was recognized by a pH-sensitive flicker quite like that of EGFP but with $pK_a \sim 5.3 \pm 0.2$ for T203Y and 5.2 ± 0.2 for T203F.

The most interesting feature of these YFPs, however, is an additional flicker term in the fluorescence correlation functions that has a strong dependence on the excitation intensity. This is not due to singlet-triplet intersystem crossing, which can also be detected at much higher excitation intensities. As seen in Fig. 14.3, analysis of this phenomenon showed that this flicker rate increases linearly with excitation intensity and extrapolates toward zero $\pm 100\,$Hz at zero intensity. It is not entirely clear whether this excitation-driven fluorescence flicker process observed by FCS is identical with the blinking previously observed for immobilized YFP. The flicker fraction F_2 for the driven process remains constant, independent of excitation

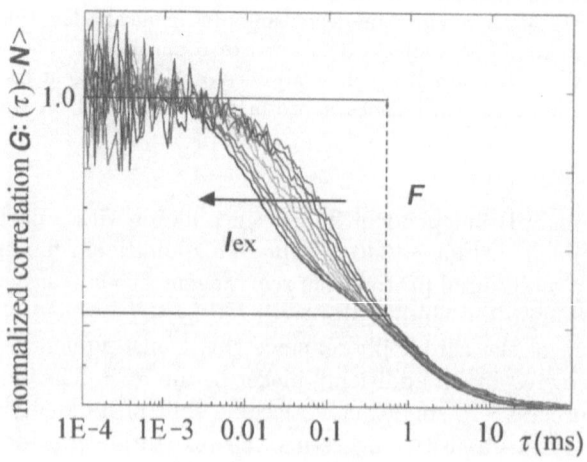

Fig. 14.3. Excitation intensity I_{ex} dependence of normalized fluorescence correlation curves for T203Y at pH 8 with 514 nm excitation. *Right to left:* I_{ex} increases from 5 to $50\,$kW/cm^2 by 0.1 OD steps. The average dark fraction $F(\sim 60\%)$, determined by the fitted amplitude of the fast flicker process, remains constant although the time constants have decreased by an order of magnitude as the intensity decreases

intensity. This indicates that transitions between bright and "dark" systems are equally facilitated by excitation in both directions. The dark state is thought to be associated with a theoretically anticipated isomerization transition of the chromophore.

This research has led us to consider the photophysics of a wider selection of GFP mutants and further experiments that combine FCS measurements at 200 ns temporal resolution with separate fluorescence lifetime measurements at 20 ps resolution. This research, plus single molecule photophysics observations, leads us to think that photophysical pathology is a ubiquitous characteristic of GFP mutants and probably of most fluorophores. For example, time-resolved fluorescence lifetime experiments by Ahmed Heikal on EGFP excited at shorter wavelengths to access the protonated "dark" states indicate a plurality of dark states. Studies in progress by Sam Hess and Ahmed Heikal on a recently developed "ecliptic" mutant [14.66] have revealed excitation-driven transitions between a bright state and multiple dark states that do shift the bright-dark equilibria in patterns yet to be understood. FCS has turned out to be a surprisingly useful tool for understanding the photophysical pathology of fluorophores.

It would be useful to develop efficient methods to explore the experimental time gap between photophysical data on immobilized, individual molecule data which appear to be limited by photobleaching and the FCS timescale in solution of $< 10^{-3}$ s, the usual diffusion time. Actually, both regimes are limited by photobleaching to $\sim 10^5$ protons detected per molecule. FCS in its usual format suffers from another limitation of the fluorophore, which reduces its effective time resolution with continuous illumination; intersystem crossing contributes to depleting the ground state so that the photon emission rate is reduced far below the singlet system saturation levels $\sim 10^9\,s^{-1}$ which would, with 0.1% detection efficiency, allow detection of $\sim 10^6\,s^{-1}$ photons per second per molecule. It is interesting to note that this phenomenon is not a problem in laser scanning microscopy because the dwell time of the scanning beam is so short.

14.4.2   Molecular Diffusion in Lipid Membranes of Giant Unilamellar Vesicles

We first reported measurements by FCS of molecular diffusion in planar black lipid membranes (BLM) in 1976 [14.67], and our research soon showed that diffusion in BLM was perturbed by retained solvent. We have recently reported interesting measurements of diffusion in pure lipid mixture giant unilamellar vesicle (GUV) membranes by FCS [14.68,14.69]. These membranes are formed as GUVs that form excellent stable preparations for measurements of two-dimensional phase equilibria and two-dimensional diffusion by FCS. Employing dilute fluorescent markers that preferentially partition into the various coexisting lipid phases, the phase segregation patterns and equi-

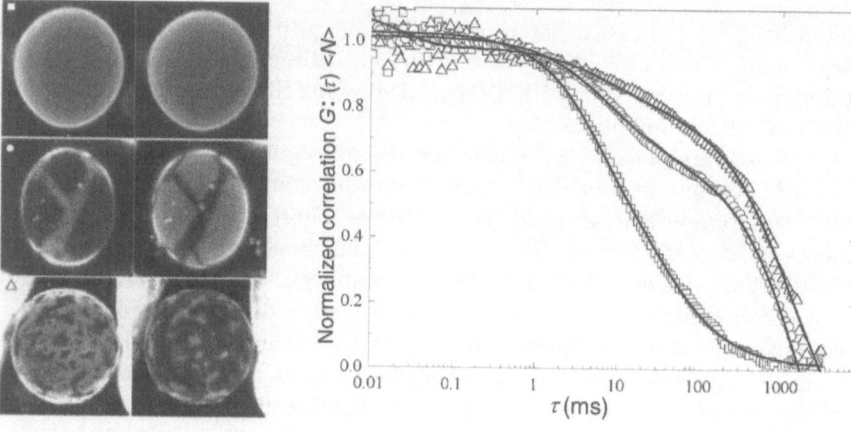

Fig. 14.4. Confocal fluorescence imaging and FCS applied to characterize lipid bilayer models for biomembranes. Giant unilamellar vesicles (GUVs) composed of dilauroylphosphatidyl choline (DLPC) and dipalmitoyl phosphatidylcholine (DPPC) were imaged by confocal fluorescence microscopy, using two fluorescent probes, DiI-C20 (*left images*) and Bodipy-PC (*right images*). Shown are three-dimensional image reconstructions of confocal z-scans through GUVs, revealing the morphology of coexisting phase domains on the surface of these vesicles. Phase separation is visualized by differential probe partitioning into coexisting phases, with diI-C20 partitioning into the spatially ordered phase (slow diffusion), and Bodipy-PC into fluid phases (fast diffusion). FCS autocorrelation curves were obtained for samples with a DLPC/DPPC composition of 1/0 (**squares,** *top image pair*), 0.60/0.40 (**circles,** *middle*), and 0.20/0.80 (**triangles,** *bottom*). At DLPC/DPPC 1/0, a single fluid phase is present, whereas ordered-fluid two-phase coexistence is present in the two other cases. This is reflected by the shape of the corresponding correlation curves, which have sampled both phases. FCS could thus accurately determine the phase-state of the lipid mixture, as well as distinguish two different diffusion coefficients in a region of two-phase coexistence on a single vesicle. *Solid curves* are data fitting curves for diffusion theory from which diffusion coefficient values were determined. Correlation amplitudes are normalized to 1 by multiplying $G(t)$ by the average number of fluorescent molecules in the focal area $\langle N \rangle$ in order to compare the shapes of the curves for different compositions. Fluorescent probes were added to the lipid mixture at a concentration of 10^{-1} mol% for confocal imaging and 10^{-4} mol% for FCS

librium coexistence diagrams can be observed by laser scanning microscopy, and viscosities can be estimated by molecular diffusivities obtained by FCS.

Figure 14.4 illustrates some of our results (see [14.69] for color figures, noting captions are corrected on line) which have several interesting features. Note that the FCS correlations in membranes have been extended significantly to > 10 s in cell membranes and are typically > 1 s for diffusion coefficients $D \sim 10^{-8}$ to 10^{-10} cm²/s without difficulty in GUVs. Since phase boundaries are mobile, FCS can sometimes measure the diffusivity in two phases, so that a stationary focal area can actually sample both phases as the boundaries move through. Our fluorescence-intensity measurements have shown that our ordered phases incorporate both leaflets of the lipid bilayer and not a single leaflet. Other FCS data reveal the high viscosity in phospholipid+cholesterol single-phase mixtures where extended short-range order due to solution cooperativity leads to a slower diffusing entity. We have found some tentative evidence for diffusion anomalies near a line of critical points in some mixtures where continuous phase transitions can occur. The "critical slowing down" theoretically predicted and observed near three-dimensional critical points has not been studied in the two-dimensional lipid bilayers, but could be with sufficient effort and care.

Although these membrane FCS experiments were carried out using confocal laser scanning microscopy for the imaging and confocal optics in a separate instrument for the FCS, we have found that multi-photon molecular excitation offers important advantages for FCS on membranes of living cells and within living cells and would presumably be favored for these experiments.

14.4.3   Two-Photon Molecular Excitation (2PE) for FCS

Two-photon molecular excitation can occur by simultaneous absorption of two low energy photons of the excitation radiation by a molecule which is then excited to a state of energy equal to the sum of the energies of the two absorbed photons. In this way, the usual excited state bands for fluorescence emission are reached and the usual fluorescence is emitted [14.70]. Absorption is simultaneous within $\sim 10^{-15}$ s, the appropriate uncertainty principle timescale, and the photons must impact a single molecular cross-section area $\sigma_1 \sim 10^{-16}$ cm^2.

For simultaneous absorption of two photons, the rate is proportional to the product of the absorption probabilities for each photon multiplied by the time scale defining "simultaneous" so the two-photon cross-sections have units cm^4 s with appropriate limiting magnitude $10^{-16} \times 10^{-16} \times 10^{-15} \sim 10^{-47}$ cm^4 s. Usual units are 10^{-50} cm^4 s = 1 GM named after Maria Goeppert-Mayer, who developed the theory of MPE in 1931. To conveniently obtain for FCS rapid and recurrent multi-photon excitation in small volumes, the infrared pulses of ~ 100 fs pulsewidth from a mode-locked laser with a repetition rate of ~ 80 MHz are focused through the usual microscope optics to produce photon flux densities in the focal volume at 10^{27} to 10^{30} photon/cm^2 s. Since the two-photon excitation rate is proportional to the square of the illumination intensity, which drops as $(\Delta z)^{-4}$ with the axial distance Δx above and below the plane of focus, this effect provides an optical resolution essentially equivalent to confocal optics. Since resolution scales with wavelength, one might expect lower 2PE resolution than with confocal optics, but it turns out that the necessary opening of the confocal aperture for optimal resolution reduces the confocal resolution to essentially the same resolution as the corresponding two-photon excitation [14.71].

Two-photon, or multi-photon, laser scanning fluorescence microscopy (MPM) does not require a confocal aperture, since the three-dimensional resolution provided by 2PE or MPE with the focused laser beam is inherent. Note that absorption of the infrared radiation or excitation is negligible outside the focal volume, so that out-of-focus background fluorescence, photodamage and photobleaching are eliminated. Imaging deep into strongly scattering tissue is facilitated since all of the emitted fluorescence originates in the focal volume, and it can still be collected even if scattered since it does not have to pass a confocal aperture. Scattering of the illumination attenuates the illumination in the focal volume, but this can be compensated by increasing the laser power because the out-of-focus light does not excite fluorescence or damage, unlike the conventional one-photon case. Caution and knowledge are needed, however, since the powerful lasers that are available can be set to inappropriate overpowering levels within the focal volume. It is unfortunate that most of the literature on multi-photon photodamage has ignored or misunderstood this fact. It is important to keep in mind that the rate of fluorophore excitation in the focal volume must be the same for one- and two-photon excitation for the same fluorescence emission.

FCS in a complex structure, such as biological cells or tissues, at a location to be selected by microscopy, is possible using a stable laser scanning microscope adapted for MPM. It is convenient to form a fluorescence image, select a desired position for an FCS measurement, and park the laser beam for either 2PE or confocal FCS. This method requires that the beam scan control be sufficiently stable to hold the parked beam stationary. Fortunately, this is the case with the Bio-Rad laser scanning microscopes but is not necessarily so with others. Alternatively, a conventional full-field image can be observed and the specimen moved into position of beam focus.

The optics of 2PE provide a significant advantage for reproducible precision FCS that is seen by considering the detectivity function discussed with (14.2–14.9) [14.28]. For 2PE, the detectivity function which corresponds to the point spread function of the optics becomes the square of the illumination profile with no further adjustment in the absence of an (unnecessary) confocal aperture. Standard optical algorithms can compute $\phi(\bar{r})$ yielding the volume V and thus $\langle N \rangle / \langle C \rangle$ as defined in (14.5). We have discussed this concept recently [14.16,14.19]. In principle, analogous but more complex calculations can be carried out for confocal optics. The appropriate expressions for the detectivity functions with 2PE and 1PE and their specialization to the Gaussian approximation are discussed in the appendix of [14.18]. Unfortunately, careful analysis of correlation functions recorded with confocal optics and analyzed assuming Gaussian profiles can show serious errors that yield fictitious extra terms in the correlation functions that tend to resemble exponential flicker kinetics. Current studies in our laboratory are addressing this issue.

14.4.4   FCS in Cells and Tissues

We have recently carried out a detailed comparison of 2PE and conventional 1PE confocal optics for applying FCS in living cells and tissues [14.18]. Although I would never have expected 30 years ago for FCS to work satisfactorily in such a hostile environment as cell cytoplasm, we have found it quite satisfactory, especially with 2PE, in several cultured cell lines. However, in thick multi-cellular tissues [14.72] it has worked only with 2PE. One cannot, however, be certain of universal success, because some tissues have such strong autofluorescence from mobile fluorophores that it may be difficult to avoid background interference with the desired fluorophores. But how do we escape the deleterious effects of even mild autofluorescence? It turns out that most autofluorescence in many cell types is primarily due to immobile or localized fluorophores. Thus, the first few seconds of exposure to the parked laser beam produces photobleaching which avoids much of the background. The presence of any residual background fluorescence does produce an error in the normalization denominator $\langle F(t) \rangle^2$, which will contain an extra contribution F_B, so that the actual factor is $\langle F(t) + F_B(t) \rangle^2$ and the value of $\langle N \rangle$ is overestimated by a factor $\langle [F(t) + F_B(t)] / F(t) \rangle^2$ unless a control experiment is included to determine background.

Optical correspondence between 2PE and confocal 1PE optics is crudely adjusted by selection of an appropriate pinhole size for confocal optics. The value $25\,\mu\mathrm{m}$ noted for the optics of [14.18] includes a hidden effect of the fiber optic coupling. The axial resolution obtained with 2PE is significantly superior, but even with relatively low axial resolution due to a large confocal aperture in the 1PE case, the axial resolution is quite useful, as illustrated by Fig. 14.5.

In the cell cytoplasm, there is a major advantage in using 2PE to minimize photobleaching. Recall that photobleaching is recognized in FCS as a decrease in the apparent diffusion time, because the dwell time of molecules in the photovolume is curtailed by photobleaching. Since the 2PE photobleaching is more sharply focused as the square of illumination while 1PE photobleaching spreads, there is a different direct effect of photobleaching with 2PE. This advantage is evident in the control limits found for illumination power in cells [14.18]. Because 2PE does not photobleach along the laser beam outside of the focal volume, the available fluorophore in restricted volumes such as cell cytoplasm is not seriously depleted. Photobleaching within the focal volumes appears to be about the same for equivalent signals excited by one or two-photon absorption.

Fluorescence photobleaching recovery methods for measurements of diffusion in solution, that is with three-dimensional diffusion using 2PE, have recently been developed in our laboratory and compared with FCS measurements in solutions and in cells, and it was satisfying to find that the results of the two methods agreed very well [14.39].

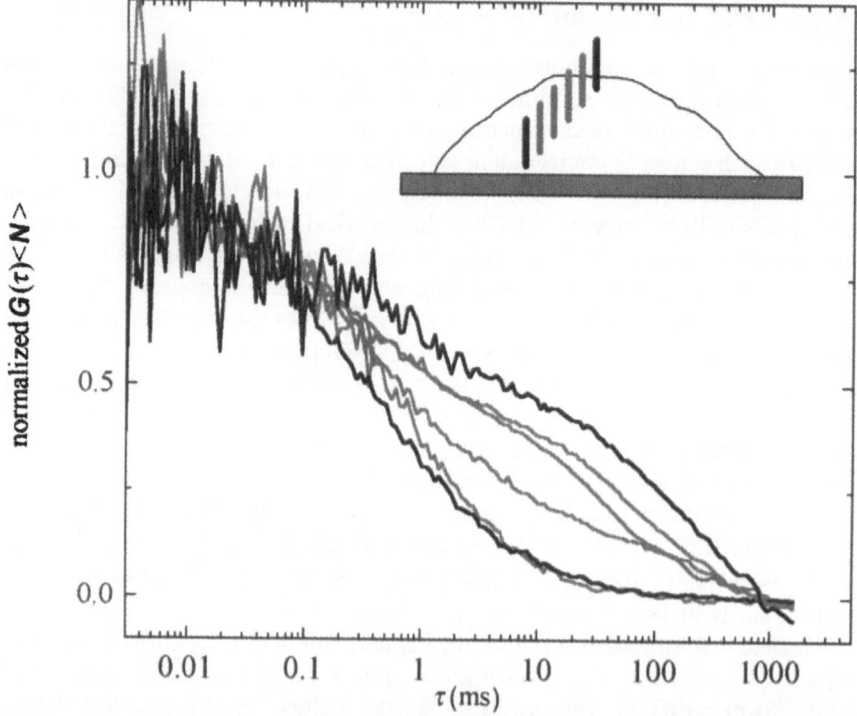

Fig. 14.5. Coexisting cytoplasmic and membrane diffusion spatially resolved by FCS with quasi-confocal optics: z-stepping with $1\,\mu$m steps ($100\,\mu$m pinhole, low resolution) through a RBL cell expressing GFP-labeled palmitoylated Lyn analogs distributed in cytoplasm and on membranes. Close to the cell surface, both intracellular (fast) and membrane (slow) mobilities are represented; in deeper z positions, the membrane contribution can be avoided. The fast intracellular part is always present

An interesting, and to us surprising, result of our measurements of diffusion in both cellular cytoplasm and on cell membranes was the observation of anomalous subdiffusion instead of conventional Brownian motion diffusion [14.38]. Ordinarily in diffusion fluctuations, the mean square molecular displacements grow linearly in time $\langle(\Delta x)^2\rangle \propto Dt$. Anomalous subdiffusion is characterized by power law growth of mean square displacements with a time exponent $\alpha < 1$; thus, $\langle(\Delta x)^2\rangle \propto \Gamma t^\alpha$. The only applicable physical model that yields this phenomenon heuristically assumes that diffusion does not occur on a "level playing field"; instead, the potential energy fluctuates in both time and space with bandwidths greater than the experimental bandwidth. Under these conditions, the non-linear power law t^α is expected. One hypothetical origin of the effect can be a weak evanescent attractive interaction amongst mobile molecules. We have first observed this

EGFP flicker (pK=5.8) measures pH in cells

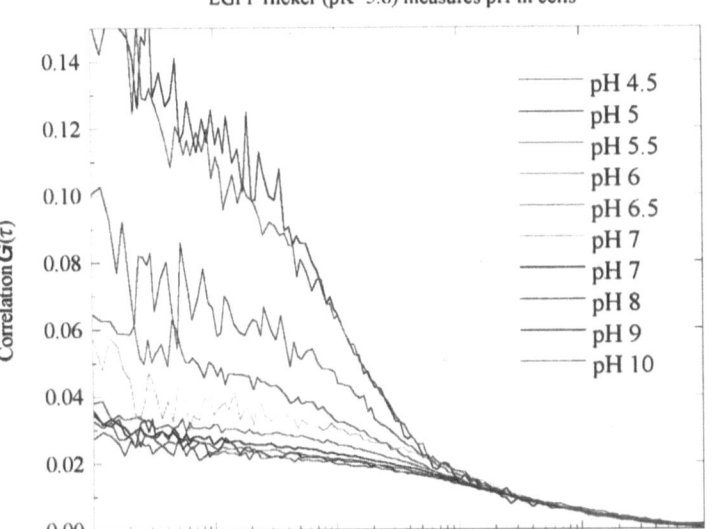

Fig. 14.6. Measurements of pH in cellular cytoplasm using FCS correlations of proton binding dynamics by EGFP to determine the "dark fraction" corresponding to the mean fraction of EGFP anionic molecules neutralized by binding external protons. External calibrations of pH are quite uncertain in these illustrative experiments, so the results are not precise

phenomenon in nanometer tracking of individual macromolecules on cell surfaces over 3 orders of magnitude in time and have extensively analyzed the phenomenon [14.37,14.38]. It surprised us to find anomalous subdiffusion of lipids in cell membranes, and even more surprising to us, anomalous subdiffusion of lipoproteins in cell cytoplasm by FCS [14.18]. The results were confirmed by 2PE-fluorescence photobleaching recovery [14.39]. The FCS experiments were controlled to test for perturbations due to unusual intensity profiles by measurements in pure solutions, which invariably showed conventional normal diffusion with both methods.

Some unpublished preliminary research carried out by Paul Pyenta and Petra Schwille combined our techniques for FCS in cytoplasm with the pH measurement capability of EGFP using FCS to test the feasibility of measuring pH in the cytoplasm. Figure 14.6 shows a set of illustrative but rather hasty pH measurements in the cytoplasm of mucosal mast cells with the cellular pH externally adjusted (but not necessarily carefully calibrated). This indicates feasibility for pH values within about 2 pH units of pK_a at modest EGFP concentrations here with $\langle N \rangle \sim 50$, [EGFP]$\sim 50$ nM.

14.5   Prognosis for FCS

I believe that the current concept of fluorescence correlation spectroscopy will generate progeny in the form of related fluorescence correlation methods that will proliferate and expand the capabilities of FCS just as have the earlier correlation spectroscopies that preceded FCS. Fluorescence as the indicator of fluctuation dynamics is ideally suited for the demands of biological research and measurement where small numbers of molecules commonly amplify their signals by enormous factors in the responses. For example, we can aim to detect and measure the dynamics of signals generated by a single polyvalent antigen molecule binding to a cell surface receptor that may stimulate a global response of the cell, yielding the release of 10^{10} times as many molecules. That represents enormous gain in the signaling process! But what new measurement capabilities do we need?

Can we measure dynamics by FCS at high concentrations by using smaller volumes? The problem is simply to define smaller open volumes. Can one use near field microscopy or nanostructured systems to define attoliter or even milli-attoliter volumes instead of the typical femtoliter focal volumes? We have preliminary success in this endeavor.

How do we use FCS to detect picomolar or sub-picomolar concentrations of fluorescent molecules or particles such as the concentrations of β amyloid clusters sometimes associated with degenerative neural diseases? The problem here is one of sampling, which has a canonical solution of scanning the detection volume and perhaps utilizing cross-correlation between separate positions. This approach has repeatedly been demonstrated but needs some technological solutions. I believe this kind of measurement to be entirely feasible, but we have not had the occasion to address the problem.

Fluorescence pattern correlation techniques, initiated by Nils Petersen, and even used with pattern photobleaching in our laboratory [14.73], have the potential to become a powerful tool of cell biology but would benefit from much greater spatial resolution, multichannel detection for speed and instant cross-correlation, as well as orders of magnitude extension of computational analysis to address the challenging problems. Could a focused effort solve the problems?

The photophysical limitations of fluorophores is the most serious limitation of FCS. Organic fluorophores improve continuously but incrementally. However, for 2PE, orders of magnitude increases in cross-section have been achieved [14.74]. Perhaps inorganic nanoparticles can help also, but the literature is worrisome, as it reports much of the same photophysical pathology we know for organic fluorophores. Specifically, can we eliminate photobleaching of fluorophores in useful environments? Can we reduce ground state depletion of fluorophores on strong excitation by reducing excited state lifetimes? It is an interesting physical trick that intersystem crossing is a serious problem in reducing fluorescence emission in FCS but not in laser scanning microscopy, at least not in MPM laser scanning microscopy. This occurs because the dwell time of a molecule in the illumination volume is short, $\sim 1\,\mu s$, and the repeat rate is slow enough that triplet states cannot be heavily populated. Can we incorporate this trick into FCS?

The recent reconsiderations of intensity distribution analysis of fluorescence appears promising for distinguishing dilute concentrations of sufficiently disparate particles, but the very long measurement times seem discouraging [14.53]. Accuracy and sensitivity might be significantly enhanced by improving the illumination optics of the inhomogeneous focal volume. As in previous very successful manifestations of "intensity" distribution analysis with other fluctuation indicators, technological improvements seem to be indicated. Two-photon excitation and some systematic calibration might help.

No doubt fluorescence correlation spectroscopy will evolve many progeny that are not anticipated here as measurement problems present new and more demanding challenges. It would be of benefit to measure conveniently the dynamics of molecular processes over the continuous time range from 20 ps to 20 s in the same apparatus with the same specimen. Would that be too much to hope for or to work for? Can FCS be conveniently combined with other indicators of fluctuations such as quasi-elastic light scattering or impedance fluctuations for cross-correlation to detect the coupling of basic processes? Can FCS be successfully utilized to analyze non-linear processes or dynamics far from equilibrium? Can one use atomic fluorescence to study atomic correlations; how about diagnosis of fluctuations in Bose–Einstein condensates? Will FCS become a useful technique in the chip technologies utilizing human genome data; will it help in aiming to characterize the proteome, and will it become routine in medical diagnostics? One can at least predict useful applications of FCS to additional interesting problems in the cellular biophysics of signaling interactions [14.75].

Acknowledgements

This manuscript was prepared in the Developmental Resource for Biophysical Imaging Opto-electronics with funding provided by the National Science Foundation (DIR 88002787), and the National Institutes of Health (P41-RR04224), and a grant from the Department of Energy (066898-0003891). Recent research mentioned here also benefited from a laser loaned by Spectra Physics, a Confocor FCS machine loaned by Zeiss, and support of Bio-Rad Laboratories. The individuals contributing to the reported research are generally identified by the references to our publications, but recent unpublished contributions by Samuel Hess, Michael Levene, and Warren Zipfel have also been mentioned.

References

14.1 D. Magde, E. Elson, and W.W. Webb: Phys. Rev. Lett. **29**, 705 (1972)

14.2 E.L. Elson and D. Magde: Biopolymers **13**, 1 (1974)

14.3 D. Magde, E. L. Elson, and W.W. Webb: Biopolymers **13**, 29 (1974)

14.4 G.H. Gilmer, W.C. Gilmore, J.S. Huang, and W.W. Webb: Phys. Rev. Lett. **14**, 491 (1965);
J.S. Huang and W.W. Webb: J. Chem. Phys. **50**, 3677 (1969)

14.5 J.S. Huang and W.W. Webb: Phys. Rev. Lett. **23**, 160 (1969)

14.6 M.R. Beasley, W.A. Fietz, R.W. Rollins, J. Silcox, and W.W. Webb: Phys. Rev. A **137**, 1205 (1965)

14.7 W.W. Webb and R.J. Warburton: Phys. Rev. Lett. **30**, 461 (1968)

14.8 J.E. Lukens, R.J. Warburton, and W.W. Webb: Phys. Rev. Lett. **25**, 1180 (1970)

14.9 M.R. Beasley, R. Labusch, and W.W. Webb: Phys. Rev. **181**, 682 (1969)

14.10 W.H. Henkels and W.W. Webb: Phys. Rev. Lett. **26**, 1164 (1971)

14.11 P. Leiderer, D.R. Watts, and W.W. Webb: Phys. Rev. Lett. **33**(8), 483 (1974)

14.12 P. Leiderer, D.R. Nelson, D.R. Watts, and W.W. Webb: Phys. Rev. Lett. **34**, 1080 (1975)

14.13 B.J. Berne and R. Pecora: Dynamic Light Scattering (John Wiley & Sons, Inc., NY, 1976)

14.14 S.R. De Groot: Thermodynamics of Irreversible Processes (North-Holland Publishing Company, Amsterdam, Interscience Publishers, Inc., New York 1952)

14.15 D.E. Koppel, D. Axelrod, J. Schlessinger, E.L. Elson, and W.W. Webb: Biophys. J. **16**, 1315 (1976)

14.16 J. Mertz, C. Xu, and W.W. Webb: Optics Lett. **20**, 2532–2534 (1995)

14.17 S. Maiti, U. Haupts, and W.W. Webb: PNAS **94**, 11753–11757 (1997)

14.18 P. Schwille, U. Haupts, S. Maiti, and W.W. Webb: Biophys. J. **77**, 2251–2265 (1999)

14.19 C. Xu and W.W. Webb: In: Topics in Fluorescence Spectroscopy: V5., Nonlinear and Two-Photon-Induced Fluorescence (J. Lakowicz (ed.) Plenum Press, NY, 1997)

14.20 E.L. Elson and W.W. Webb: Ann. Rev. Biophys. Bioeng. **4**, 311 (1975)

14.21 W.W. Webb: Quart. Rev. Biophys. **9**, 49 (1976)

14.22 M. Ehrenberg and R. Rigler: Chem. Phys. **4**, 390–401 (1974)

14.23 R.D. Icenogle and E.L. Elson: Biopolymers **22** (8), 1919–1948 (1983)

14.24 R.D. Icenogle and E.L. Elson: Biopolymers **22** (8), 1949–1966 (1983)

14.25 T.A. Ryan, J. Myers, D. Holowka, B. Baird, and W.W. Webb: Science **239**, 61–64 (1988)

14.26 T. Ryan, J. Myers, and W.W. Webb: Biol. Bull. **176** (S), 164–169 (1989)

14.27 D.E. Koppel: Phys. Rev. A **10**, 1938 (1974)

14.28 D.R. Sandison, D.W. Piston, R.M. Williams, and W.W. Webb: Appl. Opt. **34**, 3576–3588 (1995)

14.29 H. Qian: Biophys. Chem. **38**, 49 (1990): H. Qian, E.L. Elson: Appl. Opt. **30** (10), 1185 (1991)

14.30 N.L. Thompson: In: Topics in Fluorescence Spectroscopy V1. (J.R. Lakowicz (ed.), Plenum Press, NY. 1991)

14.31 N.O. Petersen, D.C. Johnson, and M.J. Schlesinger: Biophys. J. **49**, 817 (1986)

14.32 M. Eigen and R. Rigler: PNAS **91**, 5740 (1994)

14.33 L.S. Barak and W.W. Webb: J. Cell Biol. **95**, 846 (1982)

14.34 D.W. Tank, W. J. Fredericks, L.S. Barak, and W.W. Webb: J. Cell Biol. **101**, 148 (1985)

14.35 D. Gross and W.W. Webb: In: Spectroscopic Membrane Probes, Vol II. (L.M. Loew (ed.), CRC Press, Inc., Boca Raton, FL, 1988)

14.36 D. Gross and W.W. Webb: Biophys. J. **49**, 901–911 (1986)

14.37 R.N. Ghosh and W.W. Webb: Biophys. J. **66**, 1301–1318 (1994)

14.38 T.J. Feder, I. Brust-Mascher, J.P. Slattery, B. Baird, and W.W. Webb: Biophys. J. **70**, 2767–2773 (1996)

14.39 E. Brown, E. Wu, and W.W. Webb: Biophys. J. **77** (5), 2837–2849 (1999)

14.40 W. Denk, J.H. Strickler, and W.W. Webb: Science **248**, 73–76 (1990)

14.41 S. Maiti, J.B. Shear, R.M. Williams, W.R. Zipfel, and W.W. Webb: Science **275**, 530–532 (1997)

14.42 P. Kloppenburg, W.R. Zipfel, W.W. Webb, and R.M. Harris-Warrick: In press: Neuroscience (1999)

14.43 W. Denk and W.W. Webb: Phys. Rev. Lett. **63**, 207–210 (1989)

14.44 W. Denk, W.W. Webb, and A.J. Hudspeth: PNAS **86**, 5371–5375 (1989)

14.45 W. Denk and W. Webb: Appl. Opt. **29**, 2382–91 (1990)

14.46 W. Denk, R.M. Keolian, W.W. Webb: J. Neurophysiology **68**, 927–932 (1992)

14.47 D.W. Tank, R.L. Huganir, P. Greengard, and W.W. Webb: PNAS **80**, 5129 (1983)

14.48 D-O.D. Mak and W.W. Webb: Biophys. J. **72**, 1153–1164 (1997)

14.49 D-O.D. Mak and W.W. Webb: Biophys. J. **69**, 2337–2349 (1995)

14.50 L.R. Opsahl and W.W. Webb: Biophys. J. **66** 71–74, (1994)

14.51 H. Quan and E.L. Elson: PNAS **87** (14), 5479–5483 (1990)

14.52 H. Quan and E.L. Elson: Biophys. J. **57** (2), 375–380 (1990)

14.53 P. Kask, K. Palo, D. Ullmann, and K. Gall: PNAS **96**, 13756–13761 (1999)

14.54 J.H. Scofield and W.W. Webb: Phys. Rev. Lett. **54**, 353 (1985)

14.55 J.V. Mantese and W.W. Webb: Phys. Rev. Lett. **55**, 2212–2215 (1985)

14.56 J.V. Mantese, W.A. Curtin and W.W. Webb: Phys. Rev. B **33**, 7897–7901 (1986)

14.57 N.M. Zimmerman and W.W. Webb: Phys. Rev. Lett. **65**, 1040–43 (1990)

14.58 M.B. Schneider, J.T. Jenkins, and W.W. Webb: Biophys. J. **45**, 891 (1984)

14.59 M.B. Schneider, J.T. Jenkins, and W.W. Webb: J. Phys. **45**, 1457 (1984)

14.60 D.C. Wack and W.W. Webb: Phys. Rev. A **40**, 1627–1636 (1989)

14.61 M.B. Frish and W.W. Webb: J. Fluid Mech. **107**, 172 (1981)

14.62 R.D. Ferguson and W.W. Webb: In: Eighth Biennial Symposium on Turbulence, University of Missouri-Rolla (1983)

14.63 U. Haupts, S. Maiti, P. Schwille. and W.W. Webb: PNAS **95**, 13573–13578 (1998)

14.64 R.M. Dickson, A.B. Cubitt, R.Y. Tsien, and W.E. Moerner: Nature **388**, 355–358 (1997)

14.65 P. Schwille, S. Kummer, A. Heikal, W.E. Moerner, and W.W. Webb: PNAS **97**, 151–156 (2000)

14.66 D.A. DeAngelis, G. Miesenböck, B.V. Zemelman, and J.E. Rothman: PNAS **95**, 12312–12316 (1998)

330   W.W. Webb

14.67 P.F. Fahey, D.E. Koppel, L.S. Barak, D.E. Wolf, E.L. Elson, and W.W. Webb: Science **195**, 305 (1977)
14.68 P. Schwille, J. Korlach, and W.W. Webb: Cytometry **36**, 176–182 (1999)
14.69 J. Korlach, P. Schwille, W.W. Webb, and G.W. Feigenson: PNAS **96**, 15 8461–8466 (1999)
14.70 C. Xu and W.W. Webb: J. Opt. Soc. Am. B **13** (3), 481–491 (1996)
14.71 R.M. Williams, D.W. Piston, and W.W. Webb: FASEB Journal **8**, 804–813 (1994)
14.72 R. Köhler, P. Schwille, W.W. Webb, and M. Hanson: Submitted (1999)
14.73 J. Thomas and W.W. Webb: In: Non-Invasive Techniques in Cell Biology. (S. Grinstein and K. Foskett (eds.) Wiley-Liss, Inc., 1990)
14.74 M. Albota, D. Beljonne, J. Brédas, J.E. Ehrlich, J-Y. Fu, A.A. Heikal, S. Hess, T. Kogej, M.D. Levin, S. Marder, D. McCord-Maughon, J.W. Perry, H. Röckel, M. Rumi, G. Subramaniam, W.W. Webb, X-L. Wu, and C. Xu: Science **281**, 1653–1656 (1998)
14.75 J.L. Thomas, D. Holowka, B. Baird, and W.W. Webb: J. Cell Bio. **125**, 795–802 (1994)

15 ConfoCor 2 – The Second Generation of Fluorescence Correlation Microscopes

Tilo Jankowski and Reinhard Janka

15.1 Introduction

The initial idea to develop a commercial instrument for fluorescence correlation spectroscopy (FCS) was given to Carl Zeiss by Manfred Eigen and Rudolf Rigler in 1993. In cooperation with EVOTEC BioSystems Hamburg, Germany), Zeiss designed a system under the trademark ConfoCor to measure autocorrelation functions to study molecular interactions in solutions. The first FCS workshop by Zeiss with the new system was made in cooperation with Rudolf Rigler on October 24th, 1995, at the Karolinska Institute. Two years later a first prototype of a ConfoCor based on a two channel detection system for the measurements of FCS cross-correlation was developed at Zeiss. Two systems were given to Manfred Eigen and Rudolf Rigler's laboratories for evaluation. The two channel FCS system, ConfoCor 2, will be discussed in this Chapter. It is based on their experiences but is designed using a large part of the technological solutions developed originally for the Zeiss laser scanning microscope LSM 510 The result is a more flexible handling of multiwavelength excitation and detection for cross-correlation studies, as well as a principal experimental setup for combining both of the technologies in the near future.

15.2 Instrumental Setup

A simplified view of the optical setup of the ConfoCor 2 fluorescence correlation microscope (FCM) is shown in Fig. 15.1. The instrument consists of the following different modules: the FCS detection unit which is coupled directly onto the Axiovert 100 M microscope, the laser unit, avalanche photodiode (APD) detector box, the electronic controls unit and a PC for data analysis. This modular design has several advantages: it provides specific interfaces such as fiber coupling devices for the laser illumination and detection and a software interface for the storage of the original data. The optical path of the FCM is folded to achieve a very compact and therefore stable device. The combination of the FCS detection together with laser scanning technology is highly advantageous in the studies of molecular interactions in cells.

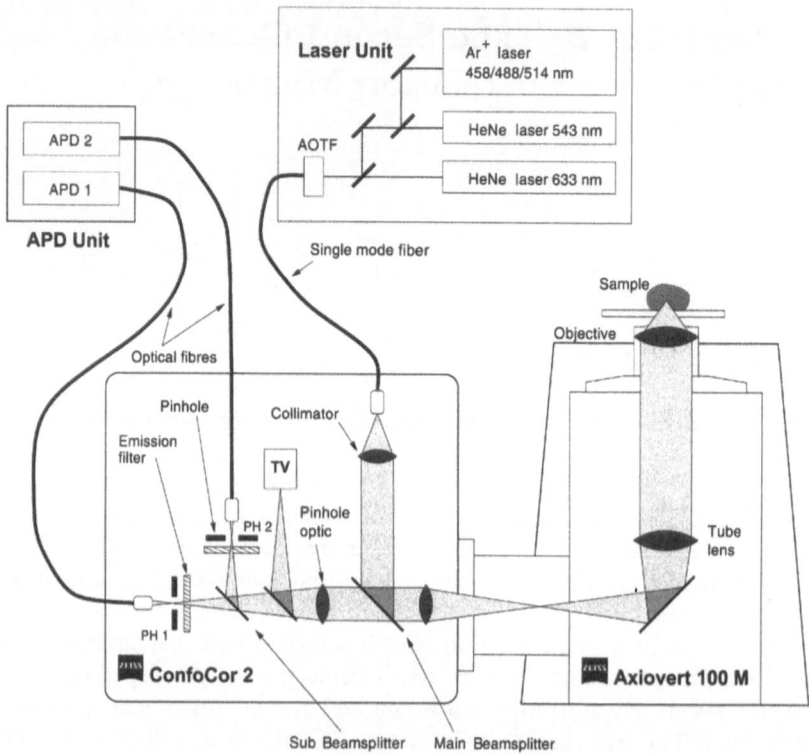

Fig. 15.1. The ConfoCor 2 beampath

The Axiovert 100 M has two identical optical ports: a sideport and a base-port, which can be used to mount simultaneously an FCM unit and an LSM 510 laser scanning unit on the same microscope stand. The control of both units can be done with the same integrated LSM and FCS software package using the same laser unit and an integrated electronic control.

15.2.1 Laser Module

For flexible research or routine experiments using a large range of fluorescent dyes it is convenient to have access to a variety of laser excitation lines. This applies as well to FCS and LSM work. The well-established laser unit of the LSM 510 suits both techniques, combining up to three different laser wave-lengths via dichroic mirror sets into one light fiber. Usually an argon-ion laser with emission lines at 458, 488 and 514 nm can be combined with two HeNe lasers, adding the excitation lines of 543 nm and 633 nm. An acousto-optic tunable filter (AOTF) mounted on the front side of the fiber is used to se-lect the laser lines and simultaneously control the transmitted laser intensity of each line individually. The operation of the AOTF is based on the inter-action of light with a travelling acoustic wave within an anisotropic crystal medium [15.3]. The incident light passes a narrow band-pass depending on the intensity of the acoustic wave, as controlled by software. In the case of FCS experiments, one can select one or two wavelengths to illuminate the sample simultaneously and adjust their intensities *completely independent* of each other by an attenuation factor of up to 1000. This can be achieved in LSM experiments in the same way with four laser lines.

15.2.2 Detection Unit

The FCS detection unit includes the main parts for laser light illumination and for the detection of fluorescence light. The unit is shown in Fig. 15.1. A collimator generates a parallel beam which is reflected by the main beam splitter towards the microscope. The 40-fold water-corrected C-Apochromat $40 \times /1.2$ Korr lens with high aperture gives a good working distance of about $0.2\,\mu m$ into the sample and has a high efficiency. The focal spot size is in the femtoliter volume range. Fluorescent light emitted by molecules from within this focus is collected by the same high aperture lens. Long wavelength fluorescence light is transmitted by the main beam splitter, whereas all the reflected light from the laser excitation is blocked by the dichroic mirror. The pinhole optics consist of another highly corrected lens, and it projects the fluorescence light from the focused image of the sample into the pinhole which acts as the field diaphragm. This type of confocal setup allows effective suppression of the background signal coming from non-confocal sample regions: only light which comes from within the confocal focus section of the sample can pass the pinhole aperture and enter the fiber light guide to the APD detector. A sub-beamsplitter behind the pinhole lens divides the fluorescent light beam into two detection channels with a pinhole for each channel. The beamsplitter can be used for further spectral division in cross-correlation applications. A mirror can be switched into the beam for autocorrelation measurements in the second detection channel. One of the channels is optimized for the long wavelength range, greater than 550 nm.

The pinholes are software controlled in xyz axes for optimal count-rate adjustment and their diameter can be adjusted between $10\,\mu m$ and 1 mm. Using two pinholes in the detection section is one of the main differences between ConfoCor 2 and the prototypes for cross-correlation based on ConfoCor [15.9]. This simplifies the placement of the pinholes for each color exactly in its optimal position. This is very helpful in eliminating residual nonideal aspects of the optical components. The adjustment of the motorized pinholes can be done individually for each channel but is fully automated by software with a precision of about $2\,\mu m$. Another major difference is that the pinhole is not placed in the image plane of the tube lens. Additional optical components form a smaller cross section of the beam in the detection unit. Here, the instrument is more compact with the advantage being that it can be made even less sensitive to environmental influences (vibrations or temperature drifts). Smaller filters are more optically flat, resulting in a higher quality optical wavefront. The total magnification is given by:

$$M_{\text{tot}} = \frac{10}{3} M_{\text{ob}}, \tag{15.1}$$

where M_{ob} is the magnification of the objective.

In addition to the beamsplitters, emission filters are placed in front of each pinhole to reduce residual scattered laser light and to select the spectral detection range. Both dichroic mirrors and the emission filters have to be optimized for the laser excitation wavelength and for the emission spectrum of the dyes used. Thus, a variety of different filter sets is needed to cover the full range of FCS applications. All filters inside the detection unit are placed on motorized filter wheels containing up to eight filter positions. The pinhole coordinates are saved for different filter combinations, therefore the pinhole adjustment must not be repeated when another combination of dichroic mirrors and emission filters is selected.

For dual color cross-correlation experiments a special characteristic of the main beamsplitter is needed: it must reflect two laser wavelengths for excitation and transmit two separate spectral bands of fluorescence light. The high quality of this optical component is very important for the accurate detection of cross correlation signals. A special filter for cross correlation measurements using the wavelengths 488 and 633 nm is included in the basic filter set. As an example, the spectral detection efficiency of both detection channels is shown in Fig. 15.2. This filter set is optimal for the use of Rhodamine Green and Cyanin 5 (Cy5) dyes. The cross talk intensity of Rhodamine Green fluorescence into channel 1 (red) is only a few per cent of the intensity detected in channel 2 (blue).

Table 15.1. Basic filter set for the two channel version of the ConfoCor 2 (LP: longpass filter, BP: bandpass filter, U: user position)

Main beamsplitter	Sub-beamsplitter	Emission filter 1	Emission filter 2
458 nm	Mirror	BP 560–615 nm	BP 475–525 nm
488 nm	635 nm	BP 585–615 nm	BP LP 475 nm
514 nm		LP 560 nm	BP 505–530 nm
543 nm		LP 585 nm	BP 505–550 nm
633 nm		LP 650 nm	LP 505 nm
488/633 nm			BP 530–600 nm
		(U)	LP 530 nm

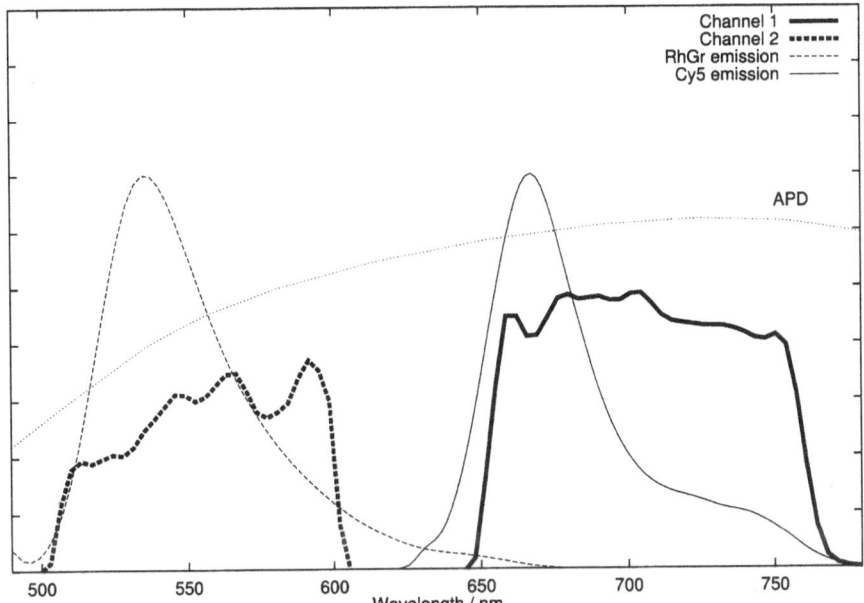

Fig. 15.2. Spectral sensitivities of the two detection channels. Dichroic mirror 488/633 nm, dichroic mirror 635 nm, LP 650 nm for channel 1 and BP 505–600 nm for channel 2. The detection efficiency of the APDs (*thin dotted line*) is included in the calculation of the resulting channel sensitivity. For comparison, the emission spectra of Rhodamine Green and Cyanin 5 dyes are shown. There is a clear overlap between the Rhodamine Green emission and the channel 2 spectral detection efficiency

15.2.3 Detection Efficiency Profile

The correlation function of photon counts measured in FCS experiments depends on the spatial characteristics of the detection probability in the sample volume. The diffusion of fluorescent particles is not traced with constant molecule detection efficiency (*MDE*). The real *MDE* depends on both the excitation intensity profile $I(r, z)$ and the photon collection efficiency function $CEF(r, z)$:

$$MDE(r, z) = \varepsilon I(r, z) CEF(r, z) , \qquad (15.2)$$

where the constant ε is determined by the quantum yield of the fluorophore, the detector's efficiency and the transmittance of the filter set used. The vector r denotes the position of a sample point in the plane perpendicular to the optical axis and z is the distance between this plane and the focal plane. For a three-dimensional Gaussian profile of the molecule detection efficiency,

$$MDE(r, z) = \varepsilon \exp\left(\frac{-2r^2}{r_0^2}\right) \exp\left(\frac{-2z^2}{z_0^2}\right) , \qquad (15.3)$$

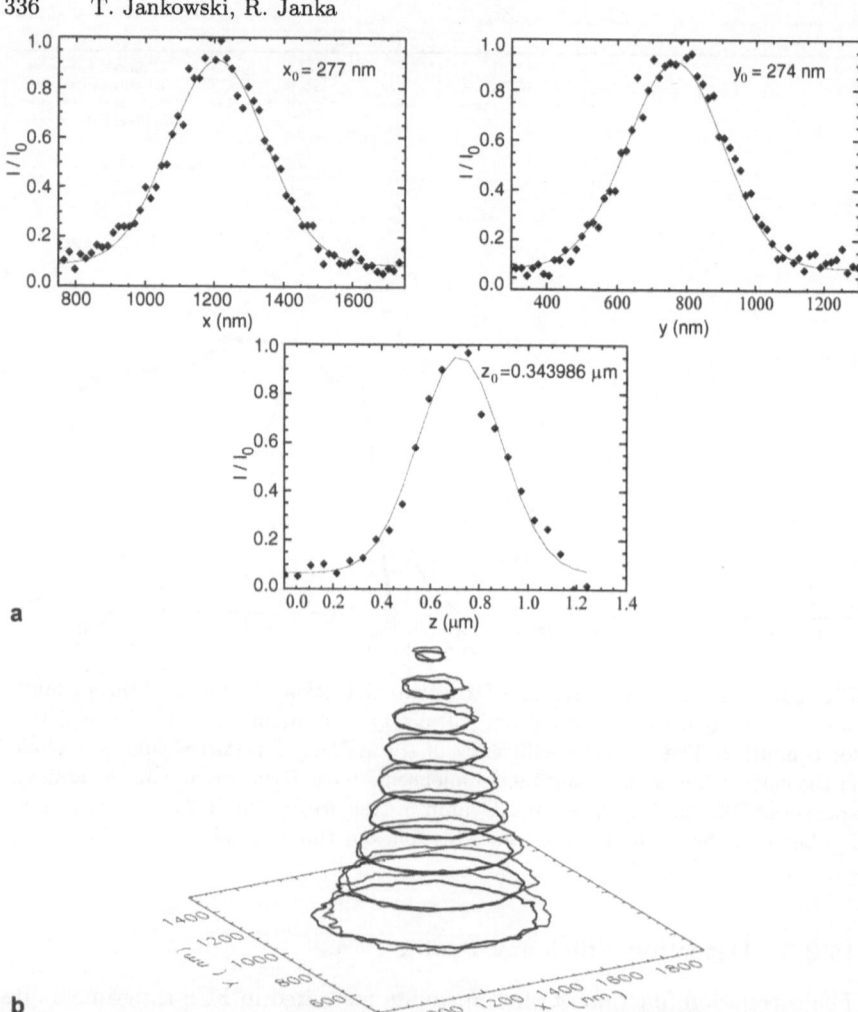

Fig. 15.3 a,b. Focus scan results. Measured intensities are normalized to their maximal value. The excitation profile is well described by a three-dimensional Gaussian **a**. The overlap of excitation spots is essential to detect cross correlation for particles being excited by two laser wavelengths. For the combination of 633 nm and 488 nm laser lines, the distance between the two foci is less than the resolution of the measurement **b**. For visualization, data smoothing was applied in **b** at a scale of 60 nm. The lowest contour level plotted corresponds to the $1/e^2$ level, the difference between two levels is 0.1

the calculation of the expected correlation function of fluorescence photon counts of translational diffusing particles can be carried out easily (15.5). Computations become rather complicated for other *MDE* profiles. Despite this, (15.3) is a good approximation for a Gaussian laser beam in a confocal detection setup [15.10,15.8].

In contrast to the arguments given by Qian and Elson [15.8] the ConfoCor 2 produces a *diffraction limited* focal spot in the sample. The back aperture of the objective lens is filled by the $1/e^2$ radius of the collimated laser beam. The larger effective focusing angle of the laser produces a beam waist which is small compared to the ConfoCor setup. The external Airy rings produced by diffraction of light at the circular aperture of the objective (see [15.2], Chap. 8) are strongly reduced due to the inhomogeneous illumination of this aperture. Another effect of the truncated Gaussian intensity profile is a slight broadening of the central spot. Furthermore, the $1/e^2$ radius r_0 of the focus hardly varies with z. The major effect of the field diaphragm is reduction of the background, and its size is usually set approximately as an Airy unit:

$$d_{\mathrm{PH}} = M_{\mathrm{tot}} \frac{1.22\,\lambda}{\mathrm{N.A.}}\,. \tag{15.4}$$

Smaller pinhole diameters reduce the thickness of the observed optical slice, giving smaller values for the ratio of axes $S = z_0/r_0$. Although the theory of Gaussian beams does not apply for the ConfoCor 2 beampath, (15.3) remains valid. The loss of excitation intensity can be neglected but the smaller beam cross section allows the use of smaller dichroic mirrors which are available with higher optical quality.

To prove that the *MDE* can be approximated using (15.3), a scanning near field optical microscope (SNOM 210, Zeiss & Digital Instruments) was used to measure the excitation intensity profile directly with a spatial resolution of ≈ 80 nm. The results are shown in Fig. 15.3. The spatial superposition of the different laser foci is very important for cross-correlation measurements and can be checked exactly using this near field technique.

15.2.4 Detection of Fuorescence Correlation Signals

Fluorescence light passing through the emission filters and the pinholes is coupled into multimode fibers. This allows the detectors to be placed outside of the detection unit to avoid heating of the optical components. The connection to the detectors is established by standard FC fiber connectors. Thus, the fluorescence light collected by the ConfoCor 2 detection channels can be guided very easily to different instruments, like spectrometers. For each detection channel an avalanche photodiode SPCM-AQR-13-FC (EG&G, Canada) with a low background signal (<250 Hz) is used. The pulse length of this type of detector is ≈ 25 ns, and the dead time is 50 ns. The detection efficiency of the APDs varies with wavelength. The maximum in efficiency of 65% lies at

700 nm, while at 500 nm only 45% of incoming photons are detected. This strongly affects the spectral sensitivity of the detection channels (Fig. 15.2) and the cross talk of the fluorescence light from the blue laser excitation to the red detection channel.

The APD signals are forwarded to the electronic control unit where the first step of the analysis is performed. Time intervals between the pulses are measured in discrete units and transmitted to the computer via a fast SCSI connection. The normalized correlation function is calculated by a specially developed software correlator. Up to three correlation curves are calculated simultaneously: the two autocorrelation functions and the correlation between the two detection channels. In addition, it is possible to save the raw data (the detection time for each photon) for further analysis. Although the software correlator can process a few million photon detections per second in each channel, during measurements the mean count-rate should not go beyond 1.5 MHz. At this intensity level the data acquisition is automatically stopped to avoid damaging the detectors. In addition, there is a hardware temperature safety circuit for the APDs included. If raw data storage is activated, the mean count-rate in both channels should not go beyond ≈500 kHz (depending on the disk speed).

15.2.5   FCS Data Analysis

If the molecule detection efficiency function can be described by a three-dimensional Gaussian profile (15.3), and translational diffusing particles are observed, the correlation function of fluorescence photons is given by [15.1]:

$$G(t) = \frac{1}{N}\left(1 + \frac{T}{1-T}e^{-\tau/\tau_{\mathrm{T}}}\right)\sum_{i=1}^{M}\frac{f_i}{1+t/\tau_{\mathrm{Di}}}\frac{1}{\sqrt{1+t/(S^2\tau_{\mathrm{Di}})}}\,. \qquad (15.5)$$

The total number of particles within the effective detection volume $V_{\mathrm{e}} = \pi\sqrt{\pi}r_0^2 z_0$ is denoted by N. The M fractions of the individual components with the diffusion times τ_{Di} have to fulfill the additional constraint $\sum_i f_i = 1$. The parameter S is the ratio of axes of the effective volume $S = z_0/r_0$. A fraction T of the N particles is in the triplet state, which has a typical lifetime τ_{T}.

The parameters of this model are derived by a Levenberg Marquard Fit [15.7] after automatic determination of the starting values. Only the number of present components has to be selected by the user. If there is only a poor correlation signal or different components have similar diffusion times or a particular component fraction is very low, the algorithm fails to give reliable results. By setting some of parameters to values which were pre-determined by calibration measurements, often one is able to avoid this problem.

The fit algorithm was tested using simulated measurements with known model parameters. If the signal-to-noise ratio SNR of the correlation function for 1 μs correlation time is 20, the scattering of the fit results is 1% for N and

4% for the diffusion time τ_D, but 20% for the ratio of axes and even more for the triplet parameter. For a two-component model with a $SNR \approx 250$ and diffusion times differing by a factor of two, the fraction of each component has to be above 20% to be detected. When the ratio of axes S is fixed to the true value during the fit routine, fractions of 10% can be measured. For larger differences in the diffusion times this problem becomes more controllable.

15.3   Autocorrelation Measurements

The quality of ConfoCor 2 autocorrelation results was tested by measuring the translational diffusion of rhodamine 6G (Rh6G) at room temperature. The 488 nm line of the argon$^+$ laser was the excitation wavelength, and the emission filter BP505-550 nm was used for detection.

A typical correlation function of 2 nM Rh6G is shown in Fig. 15.4a. The measured correlation function is well fitted by the one component model (15.5). The number of fluorescent particles within the confocal volume is $N = 0.24$. This is consistent with the measured size of the detection volume. To illustrate the effect of a very small confocal volume, a typical result of a FCS measurement using the ConfoCor is included in the graph. The correlation signal increases when the effective detection volume is reduced. Therefore the ConfoCor 2 has a higher signal-to-noise ratio of the correlation function. Differences in the size of the focal spot also cause differences in the diffusion time: for Rh6G a diffusion time of only 22 µs is measured, while 60 µs is a typical result for FCS measurements using the ConfoCor (Fig. 15.4b). In the range of short correlation times ($t < 3\,\mu s$) the normalized correlation curve does not depend on the size of the detection volume because the fraction of particles in the triplet state and the triplet state lifetime are

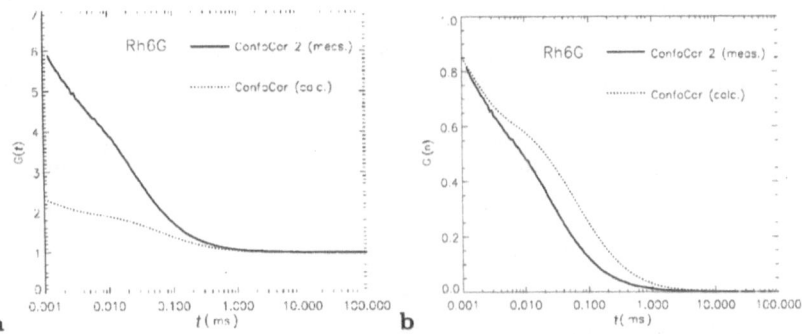

Fig. 15.4 a,b. Typical autocorrelation curves for rhodamine 6G ($\approx 1\,nM$). a: Comparison with a typical ConfoCor result. Due to the smaller detection volume, the correlation amplitude is much higher. b: The same curves, but normalized to $G(0)$. The diffusion time is smaller for the ConfoCor 2 result. The triplet state lifetime is the same, therefore both curves are identical in the region of short correlation times

independent of the size of the focal spot. To check the stability a number of measurements were repeated under the same conditions, and the results are very reproducible. Even the ratio of axes S, which is difficult to derive from the correlation curve by fitting a model, can be determined with considerable accuracy (typical: $S = 5.5 \pm 1$). In Table 15.2, the results of fitting a series of 100 subsequent measurements are listed for different filter settings.

The signal-to-noise ratio SNR for FCS measurements is proportional to the detection efficiency and the square root of the total measurement time [15.6]:

$$SNR \propto Q\sqrt{T}. \tag{15.6}$$

Here, Q denotes the mean count-rate per detected molecule. If this count-rate is increased by a factor of two, only one quarter of the measurement time is needed to achieve the same quality correlation signal. The count-rate per molecule is influenced by the properties of the dye used and the total sensitivity of the FCS instrument. Selection of the optimal filter set is essential to archive a higher value for Q. To avoid the use of a fit routine for the calculation of the current value of Q, the following approximation is used by the ConfoCor 2 software: the correlation function in the range from $2\,\mu s$ to $10\,\mu s$ is averaged. Then this value is considered to be approximately $1/N$ and multiplied by the mean count-rate. Errors due to triplet state excitation and very short diffusion times are of the order of a few per cent and can be neglected. To investigate the correlation signal quality, Q was measured in this manner for a wide range of excitation laser intensities and dye concentrations (Fig. 15.5).

The maximum value of Q is above $100\,\mathrm{kHz/molecule}$, which is slightly higher than the ConfoCor value. This value can be further improved by the use

Table 15.2. Examples for FCS measurements of Rh6G (≈ 1 nM) using different laser excitation lines and emission filters. The time for a single measurement was only 10 s. Different samples were used, therefore the particle number is not the same in all experiments

Excitation:	488 nm	488 nm	488 nm	514 nm
Emission Filter:	505–550 nm	505–600 nm	530–600 nm	530–600 nm
N	0.220 ± 0.006	0.314 ± 0.008	0.318 ± 0.007	0.110 ± 0.003
τ_D (μs)	23.4 ± 1.1	23.6 ± 1.1	22.7 ± 1.3	30.0 ± 2.1
S	5.6 ± 1.2	5.4 ± 1.2	5.5 ± 1.0	5.3 ± 1.4
T (%)	39 ± 7	36 ± 4	37 ± 7	27 ± 7
t_0 (μs)	1.2 ± 0.3	1.2 ± 0.3	1.2 ± 0.2	1.7 ± 0.7
count rate (kHz)	16.4 ± 0.3	31.7 ± 0.5	24.9 ± 0.4	12.8 ± 0.3
Q (kHz(Molec.)$^{-1}$)	77 ± 2	103 ± 2	80 ± 2	121 ± 3

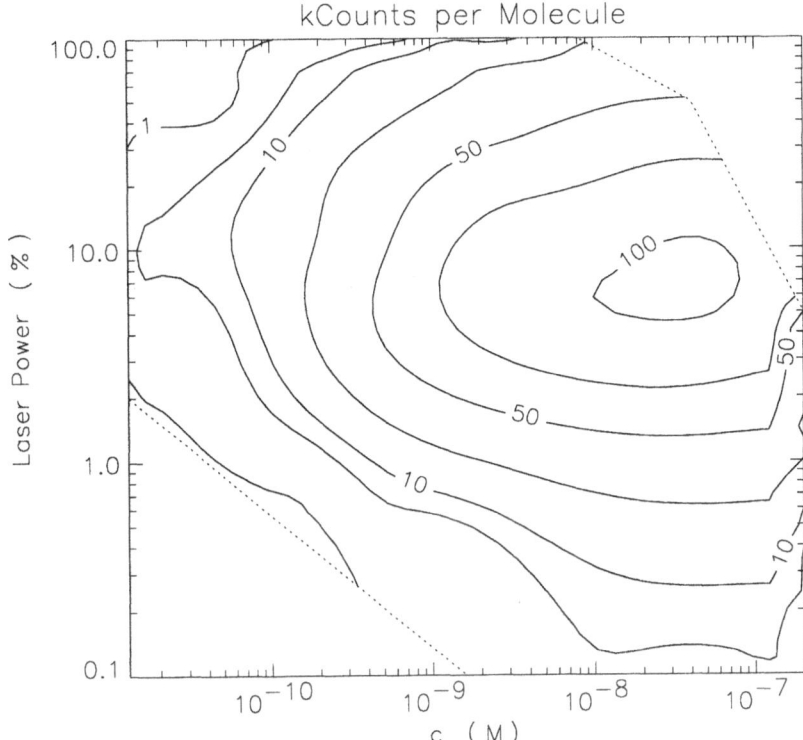

Fig. 15.5. The count-rate per molecule depends on the excitation laser power and the concentration of fluorescent particles. For very low and very high excitation intensity, the background signal dominates the detected signal, i.e. the count-rate per molecule becomes smaller

of a wider band-pass emission filter. Althought Q clearly becomes smaller for lower concentrations, there is a wide range of concentrations with high values. For concentrations above 1 nM one can achieve $Q > 50$ kHz Molecule. But the count-rate per molecule depends strongly the laser excitation intensity. Starting at very low laser power, the value of Q increases with intensity. For laser power settings above 10% (100% corresponds to ≈ 1.5 mW at the back aperture of the objective) Q decreases again, because the background signal is amplified. For low concentrations and low intensities the increasing noise is due to the dark count-rate of the detector. Compared to the signal quality of ConfoCor measurements, the maximum value of Q is shifted towards higher dye concentrations because the number of detected particles is determined by the size of the focal spot and the concentration of the fluorescent particles. A large particle number corresponds to a small correlation signal.

These tests showed that the ConfoCor 2 allows very precise, reproducible FCS measurements with tiny detection volumes and improved signal-to-noise

342   T. Jankowski, R. Janka

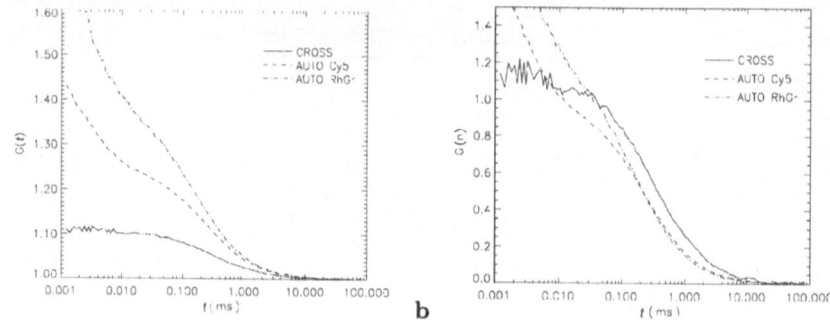

a b

Fig. 15.6 a,b. Cross-correlation measurements. A 10 nM solution of twice-labeled
DNA shows a cross-correlation signal with a low noise level. The total time of measurement was 100 s. For illustration the same data are shown in **b** with normalized
amplitudes.

ratios. Choosing an appropriate laser power is important in achieving maximal correlation signal quality.

15.4 Cross Correlation Measurements

Very frequently, FCS is used to investigate samples containing more than one
fluorescent species. The data analysis by modeling the measured correlation
function by (15.5) requires a large difference in the diffusion times of the
different species (a factor of ≈ 2 for the diffusion times leading to a factor
of at least five for the weights of the molecules). To avoid the degeneracy
of model parameters, either the diffusion time or the fraction of particles
for one component has to be set to a fixed value as determined by other
measurements. Another approach is to separate the different components
by their spectral properties. This leads to the concept of dual-color cross-
correlation FCS [15.12,15.9,15.5]. Two dyes with different emission profiles
are used to label different molecule species. FCS measurements for both dyes
are performed simultaneously using two detection channels. After the binding
a third species, double-labeled with both dyes, can be detected by cross-
correlation analysis of the photon counts in the two detection channels.

When there is only one species of products, (15.5) can be used to model
the cross-correlation FCS result. Triplet excitation can be neglected because
it is rare to find fluorescent particles with both dyes being in the triplet state
simultaneously. The meaning of cross-correlation amplitude differs from the
autocorrelation analysis:

$$G(0) = \frac{1}{N} = \frac{N_{br}}{(N_r + N_{br})(N_b + N_{br})} . \tag{15.7}$$

N_r, N_b and N_{br} denote the number of red, blue and double-labeled particles,
respectively.

As an example, for cross-correlation experiments using the ConfoCor 2, two complementary single stranded DNA (NAPS, Göttingen, Germany), one labeled with Rhodamine Green (RhGr) and the other with Cyanin 5 (Cy5) were renaturated at $1\,\mu$M concentration in the presence of $100\,$mM NaCl, $10\,$mM TRIS (pH8). The product is a double-labeled dsDNA with approximately twice the mass of each of the ssDNA. The FCS measurement was done using a $10\,$nM solution.

The results of cross-correlation FCS experiments are shown in Fig. 15.4. The diffusion time of the product is a factor of two larger than that of the unbound particles [15.12]. In the autocorrelation the double-labeled particles are not visible because their fraction of the total number of particles is too small, i.e.

$$N_{br} \ll N_{b}, N_{r} \, . \tag{15.8}$$

Although the cross-correlation amplitude is much smaller than the autocorrelation amplitudes due to (15.7), the reaction product is detected with a good signal-to-noise ratio.

If the sample is diluted, all particle numbers in (15.7) are reduced by the same amount, therefore the cross-correlation amplitude increases. During the renaturation experiment, N_b and N_r do not change but N_{br} increases. The approximation (15.8) leads to

$$G(0) \propto N_{br} \, . \tag{15.9}$$

This behavior was found in another experiment, in which FCS measurements were performed at different reaction times. As shown in Fig. 15.4 b, the reaction can be followed directly by the dual-color FCS setup.

Fluorescent light detected in the wrong channel produces an uncorrelated background and therefore reduces the cross correlation signal quality. The emission characteristics of both dyes and the spectral sensitivity of the detection channels for the filter settings used in this experiment are shown in Fig. 15.2. The cross talk from the RhGr emission into the red detection channel can not be neglected. For the single stranded DNA labeled with RhGr the cross talk is $\approx 10\%$, twice as high as that for pure Rhodamine Green. After the renaturation experiment, the fluorescense intensity caused by $488\,$nm excitation and detected in the red channel is $\approx 15\%$ of the value in the blue detection channel. The presence of cross talk complicates the analysis [15.9], and the simple interpretation (15.7) is only valid when the amount of cross talk is insignificant.

Cross-correlation experiments can be done using the ConfoCor 2 without any additional experimental effort. The two channel setup allows several new applications such as the detection of small particle fractions, and discrimination between components with similar diffusion times, including reaction kinetic studies.

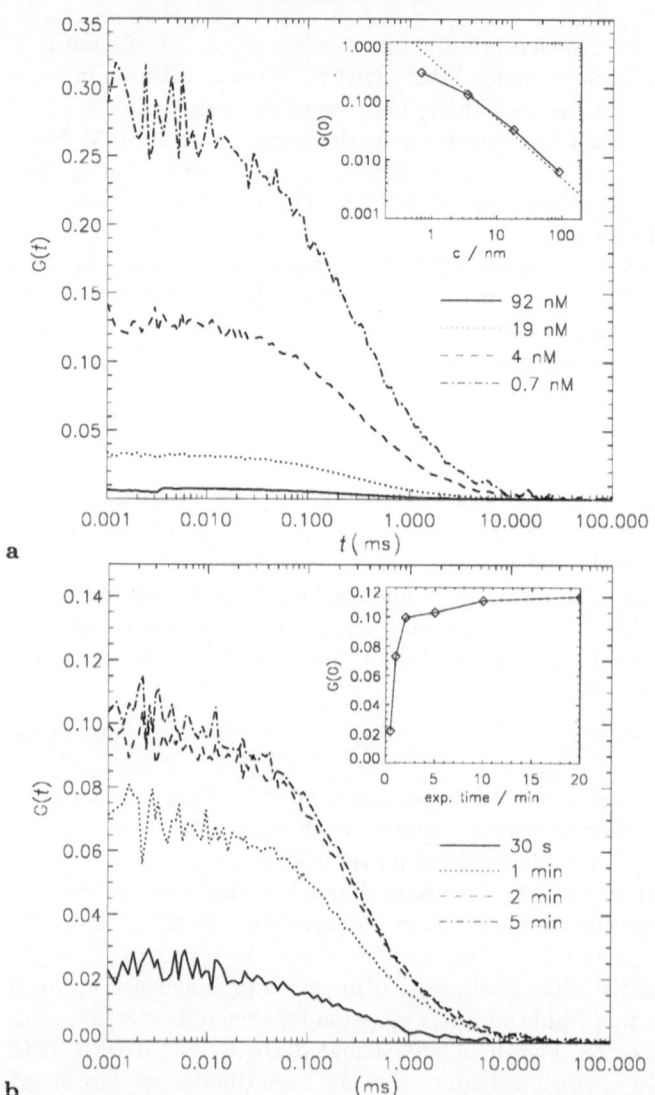

Fig. 15.7 a,b. Experiments using dual-color FCS. If the sample is diluted, all concentrations change by the same amount, therefore the cross-correlation amplitude grows with decreasing concentration **a**. Due to the renaturation process, the number of twice-labeled particles increases with time but the concentration of single labeled particles stays almost the same, so the cross-correlation amplitude increases when the reaction is followed **b**

15.5   Summary

The setup of the ConfoCor 2 was introduced in this Chapter. Due to its modular and flexible design, this new instrument is convenient for diverse applications. High quality autocorrelation measurements can be done using different dyes and laser excitation lines. A reduction of the detection volume leads to higher correlation signals and shorter diffusion times. In addition, the two channels of the ConfoCor 2 can easily be used to perform cross-correlation measurements. This is useful in situations where the diffusion times differ by only a small amount, or when small fractions are to be detected.

References

15.1 S.R. Aragon and R. Pecora: J. Chem. Phys. **64**, 1791 (1976)

15.2 M. Born and E. Wolf: In: Principles of Optics (6th ed. Cambridge University Press, Cambridge 1997)

15.3 I.C. Chang: Optical Engineering **20**, 824–929 (1981)

15.4 E.L. Elson and D. Magde: Biopolymers **13**, 1–27 (1974)

15.5 U. Kettling, A. Koltermann, P. Schwille, and M. Eigen: Proc. Natl. Acad. Sci. USA **95**, 1416–11420 (1998)

15.6 D.E. Koppel: Phys. Rev. A **10**, 1938 (1991)

15.7 W.H. Press, S.A. Teukolsky, W.T. Vetterling, and B.P. Flannery: In: Numerical Recipes in C (2nd ed. Cambridge University Press, Cambridge 1995)

15.8 H. Qian and E.L. Elson: Appl. Optics **30**, 1185–1195 (1991)

15.9 R. Rigler, Z. Földes-Papp, F.J. Meyer-Almes, C. Sammet, M. Völcker, and A. Schnetz: J. Biotechnol. **63**, 97–109 (1998)

15.10 R. Rigler, Ü. Mets, J. Widengren, and P. Kask: Eur. Biophys. J. **22**, 169–175 (1993)

15.11 P. Schwille: In: Fluoreszenz-Korrelations-Spektroskopie (Dissertation TU Braunschweig 1996 ISBN 3-930333-14-7)

15.12 P. Schwille, F.J. Meyer-Almes, and R. Rigler: Biophys. J. **72**, 1878–1886 (1997)

15.13 M. Völcker, A. Schnetz, and R. Grub: Technisches Messen **63**, 128–135 (1996)

16 Antibunching and Rotational Diffusion in FCS

Ülo Mets

16.1 Introduction

The topics of this chapter, photon antibunching and rotational diffusion, have found far fewer practical applications than translational diffusion and chemical or photophysical effects in FCS. In spite of the interesting features and unique information that these phenomena provide, they have attracted merely academic interest. One reason for this could be the added complexity of the experimental technique. Also, the physical interpretation of the results of real experiments is more involved than the basic theory suggests.

The physical basis of both phenomena is different. Rotational diffusion is well known as the reorientational component of the Brownian motion. Antibunching in fluorescence can be viewed as a property of individual fluorescent molecules to emit photons which are separated in time by a characteristic minimum interval. The reason for treating these phenomena together is that they both appear usually in the nanosecond time range of the autocorrelation function. Therefore, when attempting to measure rotational diffusion, antibunching must also be taken into account.

Although photon antibunching can be observed in the emission of various sources (e.g., specific laser configurations, nonlinearly downconverted light), this chapter will be restricted to the fluorescence of organic dye molecules. We shall describe the basic theory, illustrated with some experimental examples and potential applications of FCS concerning antibunching and rotational diffusion.

16.2 Antibunching

To our knowledge, the first prediction of the photon antibunching in fluorescence was made by Ehrenberg and Rigler [16.1] in their treatment of rotational diffusion in FCS. Independently, antibunching in resonance fluorescence was theoretically predicted by Carmichael and Walls [16.2]. The first experimental observation of photon antibunching was performed by Kimble et al. [16.3] in the fluorescence of a low number of sodium atoms. In the more familiar medium for FCS, i.e. water-based solution, antibunching in the fluorescence of dye molecules was first measured by Kask et al. [16.4].

The phenomenon has also been observed in the fluorescence of dye molecules trapped in solid matrices at liquid helium temperatures [16.5], and on the dye molecules adsorbed on the air-glass interface [16.6]. An overview of photon antibunching and sub-Poisson photon statistics can be found in [16.7].

When detecting individual photons in a light beam, one can see that photon arrival times appear to be random. This is due to the stochastic nature of photon emission from the source: the exact moment of photon emission cannot be predicted.

Some important statistical properties of the photon flux are described by the second order autocorrelation function of intensity

$$G(\tau) = \frac{\langle I(t)I(t+\tau)\rangle}{\langle I(t)\rangle^2} = 1 + \frac{\langle \delta I(t)\delta I(t+\tau)\rangle}{\langle I(t)\rangle^2} , \tag{16.1}$$

where $I(t)$ is the light intensity at time t, $\langle\,\rangle$ denotes time or ensemble average, and $\delta I(t) = I(t) - \langle I(t)\rangle$.

This correlation function can be viewed as describing the evolution of the conditional probability of detecting a photon at time τ, provided the previous photon was detected at $\tau = 0$. The behavior of this conditional probability is the criterion for defining antibunching: if $G(\tau) > G(0)$, then the probability of detecting a pair of photons separated by an interval τ decreases as τ approaches zero, and antibunching is observed. On the other hand, if $G(\tau) < G(0)$, then short intervals between photons are favored, i.e. photons appear in "bunches," hence the phenomenon is called photon bunching. Photon bunching is characteristic of thermal light, where the classical intensity fluctuates in time. If the classical intensity does not fluctuate, e.g., in coherent laser light, then the photon detection times obey Poisson statistics and $G(\tau) = G(0) = 1$. In the case of antibunching, photons have a tendency to stay apart in time, and such a field cannot be described by classical electrodynamics.

In normal experiments on organic dye fluorescence, a large number of molecules is simultaneously under observation, and the relative fluctuations of the fluorescence intensity are negligibly small. This is not the case when a single fluorescent molecule is under observation. When the molecule as a quantum electronic system emits a photon, it undergoes a transition from a higher energy level (excited state) S_1 to the ground state S_0 at a lower energy. Consequently, detection of a photon at time $t = 0$ reflects the fact that the molecule is in the ground state. Before the molecule can emit the following photon, it must first be excited. Then it stays in the excited state for a random time, before it again emits a photon and returns to the ground energy level. The times that the molecule spends in the excited state are distributed exponentially, and their average is the excited state lifetime τ_d.

Here we shall treat mathematically only the simplest possible model of a fluorescent molecule, consisting of two energy levels. We neglect the vibrational relaxation since it is much faster than the time resolution of our experimental apparatus. We can also neglect the stimulated emission, since the Stokes' shift at room temperature is so large that we can excite outside of the emission band, i.e. we are effectively very far from resonance. The build-up of the triplet population, on the other hand, is much slower than the delay time range in antibunching experiments, therefore it only modifies the apparent number of molecules.

According to the remaining two level model, we obtain the following differential equation:

$$\frac{dS_1}{dt} = -k_d S_1(t) + k_e[1 - S_1(t)], \tag{16.2}$$

where S_1 is the population probability of the singlet excited state, $k_d = 1/\tau_d$ is the decay rate of the excited state, $k_e = I_e \times \sigma_{abs}$ is the rate of excitation of the ground state of the molecule, I_e is the intensity of the exciting beam in photons per second per unit area, and σ_{abs} is the absorption cross section of the molecule in the ground state. We have also assumed that the sum of populations of both states $S_0 + S_1 = 1$.

In FCS experiments in the nanosecond time range, the frequency of the detection of photon pairs with different delays between the photons is measured. The fact of the detection of the first photon in a pair, which defines the zero point on our time axis, means that the molecule responsible for that photon has just entered the ground state. With this boundary condition

$$S_1(0) = 0, \tag{16.3}$$

the special solution of (16.2) is:

$$S_1(t) = \frac{k_e}{k_e + k_d}\{1 - \exp[-(k_e + k_d)t]\}. \tag{16.4}$$

The fact that the rate at which the population of the excited state approaches its equilibrium depends on the excitation rate k_e, can, in principle, be utilized for estimating the absorption cross section of fluorescent molecules [16.8].

In the case of observing a single fixed molecule, the normalized autocorrelation function of the fluorescence intensity defined in (16.1) becomes

$$G(\tau) = 1 - \exp[-(k_e + k_d)\tau]. \tag{16.5}$$

$G(0)$ is exactly zero, since a fluorophore cannot emit two photons at the same time, and $G(\infty) = 1$ due to normalization.

If the number of molecules in the observation volume is larger than one, then $G(0)$ is not zero. The reason is that the electronic states of different molecules are independent and absorb and emit energy in an uncorrelated manner. This makes it possible to detect two photons even with zero time delay between them. The corresponding correlation function is [16.5]

$$G(\tau) = 1 - \frac{1}{N}\exp[-(k_e + k_d)\tau], \tag{16.6}$$

where N is the number of molecules that are simultaneously observed. In these two cases, $G(\tau) < 1$ over its time dependent part (theoretically the whole function). This condition defines the sub-Poisson statistics, i.e. the variance of photon numbers $var(n)$ is lower than n, the value predicted by Poisson statistics.

When the molecules are in continuous random motion, as is the case with the Brownian motion in water solutions, then additional fluctuations appear in the fluorescence intensity due to molecules entering and leaving the observation volume. These additional fluctuations are of a classical nature and contribute to a component with a positive amplitude in the correlation function. This translational diffusion component is the subject of the majority of FCS applications. In the correlation function the translational diffusion contribution is superimposed on antibunching [16.1]

$$G(\tau) = 1 + \frac{1}{N(1 + [\tau/\tau_D])} - \frac{1}{N} \exp[-(k_e + k_d)\tau], \tag{16.7}$$

where τ_D is the translational diffusion time of the molecules across the Gaussian laser beam profile. N is now the average number of molecules. The functional form of the translational component in (16.7) corresponds to the case of diffusion between two walls perpendicular to the excitation beam [16.9]. It can be seen from (16.7) that in this case $G(\tau) \geq 1$ for any τ and an interesting situation will be observed where antibunching is clearly present, but the statistics are not sub-Poisson. Another interesting property of this situation is that $G(0) = G(\infty) = 1$. This can be understood when observing that the "delayed" photon contributing to the correlation function both at $\tau \to 0$ and at $\tau \to \infty$ originates from a different molecule than the photon which defined $\tau = 0$. At $\tau \to 0$ the same molecule cannot emit since it is in the ground state, and at $\tau \to \infty$ it cannot emit either since it has left the observation volume. Therefore the detected photons in pairs both with $\tau \to 0$ and with $\tau \to \infty$ are completely uncorrelated and contribute only to the delay independent background of the correlation function.

If the build-up of the triplet population occurs, then the correlation function contains an additional term

$$G(\tau) = 1 + \frac{1}{N} \left[\frac{1}{1 + (\tau/\tau_D)} + \frac{T}{1 - T} \exp\left(-\frac{\tau}{\tau_T}\right) \right]$$

$$- \frac{1}{N(1 - T)} \exp[-(k_e + k_d)\tau], \tag{16.8}$$

where T is the stationary population of the triplet state, and τ_T is the associated time constant which is close to the lifetime of the triplet state [16.10]. In this expression, the apparent number of molecules that contribute to antibunching is $N(1 - T)$. This is the average number of molecules in the singlet state at any moment of time and therefore the number of molecules capable of emitting simultaneously. Here again,

$$G(0) = G(\infty) = 1.$$

Now we note that the number of independently emitting fluorophore molecules is not necessarily equal to the number of independently diffusing fluorescent particles. An example could be multiply labeled biomolecules, where each particle (e.g., a protein molecule) carries several fluorescent labels. Let us denote the number of fluorescent labels on each diffusing particle as m. Now (16.8) obtains a modified form

$$G(\tau) = 1 + \frac{1}{N(\tau/\tau_D)} + \frac{T}{Nm(1-T)} \exp\left(-\frac{\tau}{\tau_T}\right)$$

$$- \frac{1}{Nm(1-T)} \exp[-(k_e + k_d)\tau]. \qquad (16.9)$$

The terms for the triplet relaxation and for antibunching contain the total number of fluorophores in the observation volume Nm, since both are electronic processes. It is the characteristic feature of the correlation function of multiply labeled particles that while at long delays still $G(\infty) = 1$, at zero delay now $G(0) = 1 + [(m-1)/Nm]$, which is larger than 1. The component describing the contribution of the triplet population, although modifying the amplitude of the antibunching signal, does not affect the values of the amplitude of the translational component and that of $G(0)$. Therefore, we omit the triplet term in the following expressions.

This property of the antibunching signal looks very attractive for determining the multiplicity of labeling. However, it is very rare that the number of labels per particle is fixed. Usually this number has some statistical distribution $p(m)$, with $\sum_{m=1}^{\infty} p(m) = 1$. Due to the fact that FCS relies on the second order intensity correlation function (16.1), the molecular contributions are weighted as $p(m)m^2$ in the amplitude of the translational component of the autocorrelation function, and without a priori knowing the form of $p(m)$, even the average number of diffusing particles N cannot be determined in this case

$$G(\tau) = 1 + \frac{\sum_{m=1}^{\infty} p(m)m^2}{N\left(\sum_{m=1}^{\infty} p(m)m\right)^2} \frac{1}{1 + (\tau/\tau_D)}$$

$$- \frac{1}{N \sum_{m=1}^{\infty} p(m)m} \exp[-(k_e + k_d)\tau]. \qquad (16.10)$$

The triplet term has been left out of this expression for simplicity.

We have used the following considerations regarding the amplitudes of different terms in (16.10). The denominator in the normalized correlation function (16.1) is the square of the average detected intensity

$$\langle I(t) \rangle^2 = \left(N \sum_{m=1}^{\infty} p(m)m \right)^2 .$$

The variances of classical intensities for different species are proportional to the average number of molecules of the given species $Np(m)$, as well as to the square of the molecular brightness m. The intensity variances of different species sum as

$$\langle \delta I(t) \delta I(t+\tau) \rangle = N \sum_{m=1}^{\infty} p(m)m^2 , \quad \tau_D \gg \tau \gg (k_e + k_d)^{-1} .$$

This expression gives us the amplitude of the translational diffusion term. For understanding the effect of antibunching, the fluorescence intensities with a delay of $\tau \ll (k_e + k_d)^{-1}$ should be considered. The two intensities in the numerator of (16.1) are physically measured with two different detectors in order to overcome the effect of the detector dead time. Moreover, this scheme relieves us from the shot noise contribution in the correlation function. The fact of the detection of a photon at $t = 0$ means that one of the m labels on the particle is in the ground state and is therefore unable to emit immediately. Consequently, for the second detector at $\tau \to 0$ only $m - 1$ labels are visible on the particle which carries the first emitter. Therefore at very short delays the product of correlated intensities is $m(m-1)$ instead of m^2

$$\langle \delta I(t) \delta I(t+\tau) \rangle = N \sum_{m=1}^{\infty} p(m)m(m-1) , \quad \tau \to 0 .$$

The amplitude of the antibunching term in (16.10) is the difference of these two amplitudes and turns out to be the inverse of the average number of fluorophores.

In cases when the time constants in the correlation function depend on the excitation intensity $I_e(x,y)$, (e.g., photon antibunching at high intensities, photoisomerization) the averaging over the Gaussian excitation intensity profile

$$I_e(x,y) = \frac{2P}{\pi \omega_0^2} \exp\left(-2\frac{x^2 + y^2}{\omega_0^2} \right) \tag{16.11}$$

has to be taken into account. P is the power of the laser beam, and ω_0 is its radius at $1/e^2$ intensity. The intensity that is effective in this averaging is not the excitation intensity I_e, but the detected intensity, which is (when neglecting the triplet state) proportional to $\langle S_1(x,y) \rangle = [k_e(x,y)/k_e(x,y) + k_d]$, where $k_e(x,y) = \sigma_{abs} I_e(x,y)$. Taking into account that the contribution to the correlation amplitude is proportional to the square of the emission intensity, and designating the time dependent correlation function by $F\{\tau[I_e(x,y)]\}$

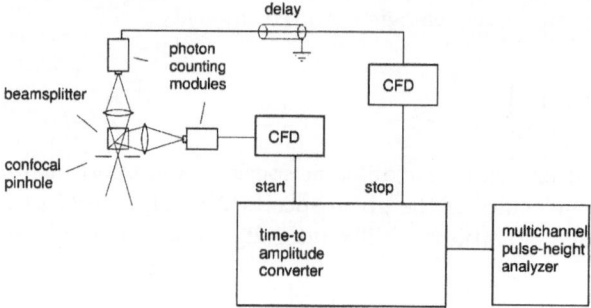

Fig. 16.1. Schematic diagram of the detection and electronics section of the nanosecond FCS spectrometer. CFD: constant fraction discriminator

and the local concentration by $c(x, y)$, the general form of the averaged correlation function is

$$G(\tau) = 1 + \frac{\iint \langle S_1(x,y) \rangle^2 c(x,y) F\{\tau[I_e(x,y)]\} \mathrm{d}x \mathrm{d}y}{(\iint \langle S_1(x,y) \rangle c(x,y) \mathrm{d}x \mathrm{d}y)^2} . \qquad (16.12)$$

The local concentration can vary due to photobleaching.

The example experiment on antibunching was performed to investigate the feasibility of measuring the excitation rate of dye molecules via the intensity dependence of the antibunching relaxation rate $k_e + k_d$. The optical setup for antibunching experiments is identical to that for standard diffu-

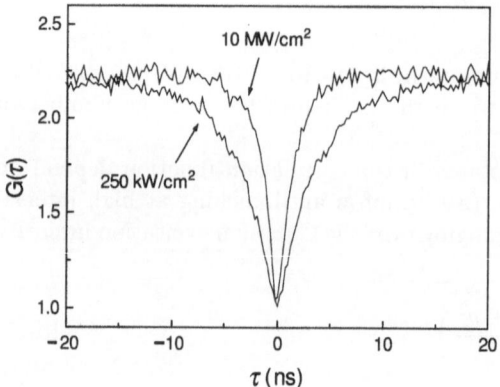

Fig. 16.2. Example of measured correlation functions of rhodamine 6G at different laser excitation intensities. The decrease of the relaxation time at higher excitation intensity is clearly seen. The negative delay times correspond to the reversed time order of detection by the two detectors. The photons are divided between the detectors randomly by the beamsplitter. If both photons of a pair fall onto the same detector, then that pair will be lost due to the detector dead time. The data collection time was 30 min at $10\,\mathrm{MW/cm^2}$ and 70 min at $250\,\mathrm{kW/cm^2}$

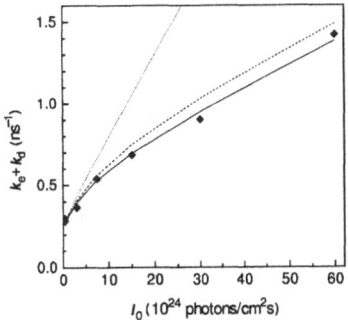

Fig. 16.3. Decay rate of the antibunching component in the correlation function $k_e + k_d$ for Rh6G, as a function of the peak laser excitation intensity at 514.5 nm. The *squares* represent the experimental values, the lines show the calculated dependencies. **Solid line**: considering both singlet and triplet saturation; **dashed line**: considering singlet saturation only; **dotted line**: neglecting any saturation

sion measurements in a bulk solution. The specificity of the equipment is related to very short delay times (on the order of nanoseconds) and resides behind the confocal pinhole (Fig. 16.1). In order to avoid the detector dead time effect, the fluorescent light was split between two detectors, and the cross-correlation of the two photon count signals was calculated, similar to Hanbury Brown and Twiss [16.11]. To achieve a subnanosecond resolution, a time to amplitude converter (TAC) was used, which performed the function of a one-bit clipped correlator: the output pulse is generated only if a pulse arrives (i.e. the count number is ≥ 1) at both inputs, start and stop. The amplitude of the output pulse of the TAC, which is proportional to the time difference between the start and stop input pulses, determines the memory channel, the content of which is incremented by 1. The resulting histogram presents the unnormalized correlation function $\langle I(t)I(t+\tau)\rangle$. The TAC can process only one photon pulse on each input during the active time range (200 ns in our experiment). For this reason, in order to avoid the distortions of the measured correlation function, the probability of detecting two or more photons by one detector per time range (200 ns) should be kept negligible. This is not difficult, since the average count rates are in the order of $10^5\,\mathrm{s}^{-1}$.

An example of the measured correlation functions at two different laser excitation intensities is presented in Fig. 16.2. The fitted time constants of the antibunching signal are plotted against the peak laser excitation intensity in Fig. 16.3. The solid line and the dotted line demonstrate the strong effect of singlet saturation on the average over the Gaussian excitation profile. At higher excitation intensities, the singlet population $S_1(x,y)$ saturates, and the relative contributions to the average from the peripheral regions of the sample volume with lower excitation intensities become significantly enhanced.

16.3   Rotational Diffusion

As compared to the translational diffusion coefficient, the rotational diffusion coefficient is much more sensitive to the changes in the size and shape of molecules. A stronger dependence on the geometry of molecules can be crucial for studying conformational changes or interactions between the molecules of similar size. Additionally, rotational diffusion can offer information about the segmental flexibility of sample molecules. In spite of these principal values, the FCS experiments on rotational diffusion are extremely rare. The main reasons restricting the use of FCS in rotation measurements are the complexity of interpretation of the data from real experiments as well as the unfavorable signal-to-noise ratio due to necessarily short dwell times.

The first theoretical treatment of rotational diffusion in FCS [16.1] was published soon after the introduction of the method itself [16.12]. The theory was developed further by Aragón and Pecora [16.13,16.14], who analyzed the influence of the actual geometrical configuration of the optical setup on the expected correlation function. The case of driven rotational motion in FCS has also been analyzed theoretically [16.15,16.16]. Experimentally, the rotational motion of labeled myosin in muscle fibers has been studied, using the fluctuations of the fluorescence polarization ratio [16.17]. The rotational diffusion of fluorescently labeled protein molecules (Bovine carbonic anhydrase) in water-based solution has been measured by Kask et al. [16.18]. A modification of the theory for the separation of the isotropic factor and the rotational factor in the correlation function, together with an experimental realization, was presented by Kask et al. [16.19]. Recently, the rotational diffusion of green fluorescent protein (GFP) was observed in FCS studies of the photodynamic processes in the molecule [16.20].

In contrast to the more well-known method for rotational diffusion measurements, time-resolved fluorescence depolarization (TRFD), where pulsed excitation is used, the illumination in FCS is continuous.

Let us consider an experiment where all the photons emitted by the sample are detected with equal efficiency, irrespective of their propagation direction or polarization. Sample molecules are excited by a parallel laser beam propagating along the laboratory y axis, and the absorption and emission properties of the sample molecule are described by the respective transition dipole moments μ_a and μ_e. In such an experimental configuration where no detection polarizer is used, orientation dependent changes in the fluorescence intensity still arise. The only reason for these changes will be the reorientation of the absorption dipole moment of the molecule. The changes of fluorescence intensity occur even if the excitation laser beam is unpolarized: with the absorption dipole moment parallel to the propagation of the laser beam, there will be no absorption. This example of an unpolarized experiment illustrates the difficulty of eliminating the rotational contribution in FCS experiments. In order to achieve a completely isotropic experiment, the laser excitation beams should propagate in at least two orthogonal directions.

The correlation function $G(\tau)$ with rotational effects included is rather complex in the general case [16.14,16.19]. The expression can be greatly simplified if the rotational time constants and the time constants of orientation independent processes are well separated in time. In particular, the fluorescence lifetime $\tau_d \to 0$, and all other processes (diffusion, singlet-triplet relaxation) are much slower than $1/6D_R$, where D_R is the rotational diffusion coefficient. Under these assumptions the correlation function factorizes [16.19]

$$G(\tau) \cong S(\tau)[1 - \exp(-\tau/\tau_d)]C(\tau). \qquad (16.13)$$

Here, $S(\tau)$ represents the contribution from the fluctuations of the classical intensity in a fully unpolarized experiment, the second factor describes the photon antibunching, and $C(\tau)$ describes the rotational correlations.

The rotational factor

$$C(\tau) = \langle A^2(0)E_1^2(0)A^2(\tau)E_2^2(\tau)\rangle \qquad (16.14)$$

depends on the angular factors of absorption

$$A^2\{\boldsymbol{\mu}_a[\boldsymbol{\Omega}(t)]\} = 3(\boldsymbol{\mu}_a \cdot \boldsymbol{a})^2 \qquad (16.15)$$

and emission detection

$$E^2\{\boldsymbol{\mu}_e[\boldsymbol{\Omega}(t)]\} = 3(\boldsymbol{\mu}_e \cdot \boldsymbol{e})^2 \qquad (16.16)$$

$\boldsymbol{\mu}_a(\boldsymbol{\Omega})$ and $\boldsymbol{\mu}_e(\boldsymbol{\Omega})$ are the directions of absorption and emission dipole moments of the chromophore, with the molecule having orientation $\boldsymbol{\Omega}$, $\boldsymbol{a}$ and $\boldsymbol{e}$ are the directions of excitation and emission detection polarizations respectively (Fig. 16.4). The detection polarization of the initial photon $\boldsymbol{e}_1$ and that of the delayed photon $\boldsymbol{e}_2$ can be different, which provides extra flexibility in choosing the experimental geometry. The corresponding cross-correlation functions provide additional information about the rotational dynamics of the molecule.

Let us assume now, for simplicity, that the diffusing molecule is a spherical rotor, and that the transition moments of absorption $\boldsymbol{\mu}_a$ and emission $\boldsymbol{\mu}_e$ are parallel to each other

$$\boldsymbol{\mu}_a = \boldsymbol{\mu}_e = \boldsymbol{\mu}. \qquad (16.17)$$

The behavior of the orientation of the molecule in solution is described by the equation of rotational diffusion

$$\frac{dP(\boldsymbol{\mu},t)}{dt} = D_R L^2 P(\boldsymbol{\mu},t), \qquad (16.18)$$

where L is the angular momentum operator, D_R is the rotational diffusion coefficient, and $P(\boldsymbol{\mu},t)$ is the probability density for the transition moment to

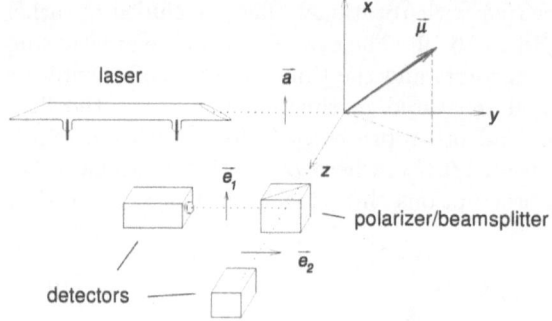

Fig. 16.4. Schematic diagram explaining the geometry of the rotational diffusion experiment. Fluorescence is collected along the z axis. In experiments with (x, x, x) and (x, y, y) geometries the beamsplitter is non-polarizing, and a separate polarizer is added. The electronic part is identical to that for the antibunching experiment

have orientation $\boldsymbol{\mu}$ at time t. The special solution of this differential equation corresponding to the initial orientation $\boldsymbol{\mu}_0$ is

$$P(\boldsymbol{\mu}, t | \boldsymbol{\mu}_0) = \sum_{l=0}^{\infty} \sum_{m=-1}^{l} Y_{lm}(\boldsymbol{\mu}_0) Y_{lm}^*(\boldsymbol{\mu}) \exp[-l(l+1)D_{\mathrm{R}}t] \,, \tag{16.19}$$

where Y_{lm} are the spherical harmonics.

According to the definition of the rotational factor (16.14)

$$C(\tau) = \\ \iint A^2(\boldsymbol{\mu}_0) E_1^2(\boldsymbol{\mu}_0) P(\boldsymbol{\mu}_0) P(\boldsymbol{\mu}, \tau | \boldsymbol{\mu}_0) A^2(\boldsymbol{\mu}) E_2^2(\boldsymbol{\mu}) \mathrm{d}^2 \boldsymbol{\mu}_0 \mathrm{d}^2 \boldsymbol{\mu} \,. \tag{16.20}$$

Substituting (16.19) into (16.20) and using $P(\boldsymbol{\mu}_0) = (4\pi)^{-1}$, the rotational factor of the correlation function for the spherical molecule becomes

$$C(\tau) = \sum_l B_l(\boldsymbol{a}, \boldsymbol{e}_1, \boldsymbol{e}_2) \exp[-l(l+1)D_{\mathrm{R}}\tau] \tag{16.21}$$

with $l = 0, 2, 4$,

$$B_l(\boldsymbol{a}, \boldsymbol{e}_1, \boldsymbol{e}_2) = \sum_{m=-l}^{l} b_{lm} b_{lm}^* \,, \quad \text{and} \tag{16.22}$$

$$b_{lm}(\boldsymbol{a}, \boldsymbol{e}) = \frac{1}{2\sqrt{\pi}} \int A^2(\boldsymbol{\mu}) E^2(\boldsymbol{\mu}) Y_{lm}(\boldsymbol{\mu}) \mathrm{d}^2 \boldsymbol{\mu} \,. \tag{16.23}$$

The coefficients B_l depend on several geometrical parameters of the experiment: the direction of the propagation of the excitation beam and that of the detected light, the directions of polarization of the excitation beam $\boldsymbol{a}$ and

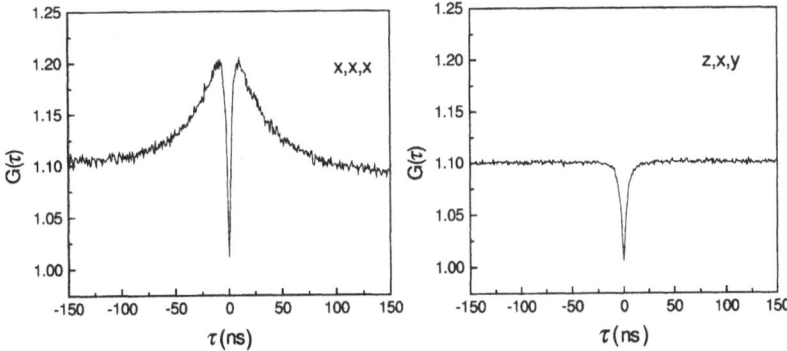

Fig. 16.5. Examples of the rotational diffusion measurement of Texas Red labeled porcine pancreatic lipase with different combinations of excitation and detection polarizations. In the (z, x, y) configuration the rotational components are still present and have small negative amplitudes. The data collection time with photomultiplier tube detectors was 10 h. With silicon detectors the data of the same quality might be collected in about 20 min

those of the emission analyzers e_1 and e_2, and also on the size of the solid angle over which the emission is collected. It was shown by Ehrenberg and Rigler [16.1] that in the case of collecting all emitted photons with all polarizations (i.e. the angular factor of emission $E^2 \equiv 1$), the term with $l = 4$ vanishes in (16.21), leaving only the constant and the $\exp(-6D_R\tau)$ terms. However, it is technically extremely difficult to perform such an experiment. Aragón and Pecora [16.14] have calculated the values of B_l for a number of experimental geometries and different solid angles of emission collection, offering a convenient guidance for designing FCS experiments on rotational diffusion. These results, along with two new configurations with $e_1 \neq e_2$, were used by Kask et al. [16.19] to obtain a combination of correlation functions where the rotational contribution canceled out. This isotropic correlation function was then used for dividing the result of a polarized experiment, yielding a purely rotational correlation function $C(\tau)$. As examples, the experimental correlation functions, obtained with two different geometries, are presented in Fig. 16.5. The geometry with all polarizations parallel $(a, e_1, e_2) = (x, x, x)$ yields the strongest rotational contribution with $B_0 = 3.24$; $B_2 = 5.29$; $B_4 = 0.47$, while the correlation function with the geometry with all polarizations orthogonal to each other $(a, e_1, e_2) = (z, x, y)$ contains only weak rotational contributions with negative amplitudes: $B_0 = 0.36$; $B_2 = -0.0735$; $B_4 = -0.0294$. Experimentally, the latter geometry can be realized only in the setup with orthogonal excitation and detection directions, which is not possible in modern confocal FCS devices. However, the three different geometries realizable with confocal optics, i.e. (x, x, x), (x, x, y), and (x, y, y), also allow simultaneous canceling of rotational contributions in a linear combination of the corresponding correlation functions.

16.4   Discussion

Photon antibunching and rotational diffusion in FCS are examples of the possibility of measuring the dynamics of processes in the nanosecond range with continuous wave excitation. A brief discussion of some ideas relevant to these applications is as follows.

In particular, antibunching can yield information about the excitation rates of fluorophores under strong laser excitation. It thereby provides an independent way of estimating the absorption cross section. However, we must note that for practical applications, the fluorescence efficiency of molecules should be very high. A good example is rhodamine 6G, and other laser dyes are also suitable. With molecules of lower efficiency the duration of the experiment could become impractically long.

The amplitude of the antibunching term compared to that of translational motion allows detection of the presence of particles carrying more than one fluorescent label. The quantitative measurement of the labeling ratio is complicated by the distribution of the number of labels per diffusing particle.

FCS has clear advantages over the pulsed anisotropy method when the rotation under study is slow. FCS works best if the rotational diffusion time is long compared to the fluorescence lifetime, while the pulsed method can observe rotation only during the excited state of fluorescence. A drawback of FCS is the relative complexity of identifying and separating the rotational contribution from the isotropic one in the correlation function.

At the expense of the increased complexity of the equipment (each detector replaced by a beamsplitter and two detectors, and using three TACs) the correlation functions of three configurations (x, x, x), (x, x, y), and (x, y, y) could be measured simultaneously. This would both reduce the duration of the experiment and avoid some artifacts caused by drifts in concentrations and optical alignment.

For the validity of the theory of rotational diffusion in FCS, the excitation intensity must be low enough to avoid the saturation of the excited state. As was pointed out in [16.1], the presence of saturation would flatten the angular dependency of excitation A^2 (16.15), resulting in serious errors.

The sensitivity to rotation in FCS is the result of the inherent anisotropy of the absorption of molecules. A factor which can significantly reduce this anisotropy is multiple labeling of sample molecules. If non-specific labeling is used, then the respective orientations of different labels attached to a particle are random. Therefore, when the absorption moment of one label turns perpendicular to the excitation polarization, another label on the same particle could take over, and the fluorescence fluctuations vanish.

The segmental flexibility in very large molecules can be one reason for the difficulty in observing slow rotation with FCS. If the small fluorescent label or its binding site have sufficient orientational freedom with respect to the whole molecule, then only the faster local reorientation will be reflected in the correlation function. In this context the green fluorescent protein could be a promising label due to its larger size and rigid structure.

FCS in the nanosecond time range is capable of extracting unique information from the sample concerning slow rotational diffusion, actual excitation rates of fluorophores, and the multiplicity of labeling of biomolecules.

References

16.1 M. Ehrenberg and R. Rigler: Chem. Phys. **4**, 390–401 (1974)

16.2 H.J. Carmichael and D.F. Walls: J. Phys. B **9**, 1199–1219 (1976)

16.3 H.J. Kimble, M. Dagenais, and L. Mandel: Phys. Rev. Lett. **39**, 691–695 (1977)

16.4 P. Kask, P. Piksarv, and Ü. Mets: Eur. Biophys. J. **12**, 163–166 (1985)

16.5 T. Basché, W.E. Moerner, M. Orrit, and H. Talon: Phys. Rev. Lett. **69**, 1516–1519 (1992)

16.6 W.P. Ambrose, P.M. Goodwin, J. Enderlein, D.J. Semin, J.C. Martin, and R.A. Keller: Chem. Phys. Lett. **269**, 365–370 (1997)

16.7 L. Davidovich: Rev. Mod. Phys. **68**, 127–173 (1996)

16.8 Ü. Mets, J. Widengren, and R. Rigler: Chem. Phys. **218**, 191–198 (1997)

16.9 E.L. Elson and D. Magde: Biopolymers **13**, 1–27 (1974)

16.10 J. Widengren, Ü. Mets, and R. Rigler: J. Phys. Chem. **99**, 13368–13379 (1995)

16.11 R. Hanbury Brown and R.Q. Twiss: Nature **177**, 27–29 (1956)

16.12 D. Magde, E. Elson, and W.W. Webb: Phys. Rev. Lett. **29**, 705–708 (1972)

16.13 S.R. Aragón and R. Pecora: Biopolymers **14**, 119–138, (1975)

16.14 S.R. Aragón and R. Pecora: J. Chem. Phys. **64**, 1791–1803, (1976)

16.15 H. Hoshikawa and H. Asai: Biophys. Chem. **22**, 167–172 (1985)

16.16 B.A. Scalettar, M.P. Klein, and J.E. Hearst: Biopolymers. **26**, 1287–1299 (1987)

16.17 J. Borejdo, S. Putnam, and M.F. Morales: Proc. Natl. Acad. Sci. USA **76**, 6346–6350 (1979)

16.18 P. Kask, P. Piksarv, Ü. Mets, M. Pooga, and E. Lippmaa: Eur. Biophys. J. **14**, 257–261 (1987)

16.19 P. Kask, P. Piksarv, M. Pooga, Ü. Mets, and E. Lippmaa: Biophys. J. **55**, 213–220 (1989)

16.20 J. Widengren, Ü. Mets, and R. Rigler: Chem. Phys., **250**, 171–186 (1999)

17 Cross-correlation analysis in FCS

Petra Schwille

17.1 Introduction

The concept of cross-correlation spectroscopy actually follows as a logical consequence if one wants to extend the common established fluorescence autocorrelation analysis to achieve a larger detection specificity. Instead of mathematically comparing, i.e. convoluting, a single fluorescence time signal with itself at different times, two different signals of whatever origin are now compared with each other, again as a function of time. Consequently, while autocorrelation analysis of a recorded signal provides us with information about its self-similarity in time, thus in particular about characteristic repetitive processes of significant duration, cross-correlation of two different signals reveals underlying mechanisms that relate the measured quantities with each other. These two signals do not necessarily both have to be fluorescence traces; it is possible and potentially attractive to compare fluctuations in the emission behavior of fluorescent probes with any other parameter that can be measured sensitively enough to reveal fluctuations on a single molecule or single particle scale. In this respect, cross-correlation is a valuable concept if one wants to follow different parameters at once but also if long-distance or long-timescale effects have to be investigated. Although there exist manifold ways to apply the cross-correlation concept, in particular two quite simple modes have now been experimentally established: two-volume fluorescence cross-correlation, which is discussed in detail elsewhere in this book (Chap. 18), and dual-color fluorescence cross-correlation which we want to highlight here in particular.

Before cross-correlation was first employed for fluorescence correlation spectroscopy by Ricka and Binkert in 1988 [17.1] to measure pairwise Coulomb interactions between deionized diffusing latex beads with different emission characteristics, it had had a rather long tradition in the methodological predecessor technique to FCS, dynamic light scattering (Berne and Pecora 1976 [17.2]). Griffin and Pusey [17.3] introduced a so-called "anticorrelations" concept which enabled them to determine rotational dynamics of nonspherical particles in a setup with two detectors recording the scattered signal under different angles. The same principle had further been used to measure internal dynamics such as conformational relaxation of random coils and association-dissociation dynamics [17.4]. By locating the two detectors at different regions in the scattered field with different wave vectors, the first approach towards a two-detection-volume variant was made, promising exciting applications, such as to probe the local structure of disordered systems [17.5,17.6]. Later, Tong et al. [17.7] established incoherent cross-correlation concepts to separate internal particle motion and translational and rotational dynamics from relative particle motions that induce rapid phase fluctuations by detecting the scattered light at two separate wavelengths. This enabled them to measure magnitudes of local velocities and vorticity in laminar flow. This technique was extended by using two spatially separate laser beams of different wavelengths to improve the discrimination of distance-insensitive effects such as particle rotation, and to become sensitive to the flow direction [17.8]. The same principle, but with two excitation beams of the same wavelength was little later established for fluorescence correlation spectroscopy on the single molecule level. The theory was extended to separate diffusive contributions from active transport in flow systems on microstructures [17.9,17.10].

The use of dual-color fluorescence cross-correlation schemes for ultrasensitive diagnostic applications in the biosciences was first proposed by Eigen and Rigler [17.11]. By cross-correlation, they suggested to improve the signal-to-background ratio in reaction schemes with fluorescently labeled probe substances, where in favor of extremely low concentrations of target molecules a large excess of unbound probe ought to be suppressed. One prominent example is the specific binding of DNA probes or antibodies to infectious agents such as bacterial or viral DNA or core proteins in clinical tests. The proposed scheme (Fig. 17.1) consists of two different probes labeled by spectrally well separable dyes (e.g., red and green), both directed to the same target agent. By splitting the emission light into two separate channels and cross-correlating the two different detection signals from the same measurement volume, contributions only arise if particles of different color interact with or,

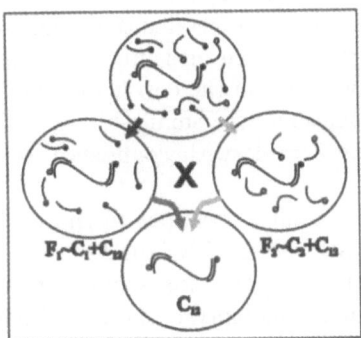

Fig. 17.1. Principle of background rejection in diagnostic applications of dual-color cross-correlation spectroscopy as proposed in [17.11]: each detector observes a certain color, only the doubly labeled species contributes to both detectors and can be cross-correlated. F_i: Fluorescence in channel i, C_j: concentration of species j

in particular, bind to each other. In a search for freely diffusing particles of low concentrations, this can be achieved by designing probes of different color specifically directed to different binding sites of a common target molecule. A doubly labeled unit, requirement for non-vanishing cross-correlation, could only be formed if target molecules with the two specific binding sites are provided. In clinical tests for certain agents, a positive cross-correlation signal would then be equivalent with a positive test result. Based on the simplicity of the proposed scheme, an attractive applicability also in biotechnological drug screening procedures was foreseen.

The first experimental realization of dual-color fluorescence cross-correlation with spectrally separable dyes on a single molecule level [17.12,17.13] was demonstrated with a system of differently labeled complementary DNA oligonucleotides that irreversibly hybridize to each other. Only the double-stranded hybrid enables positive cross-correlation while the autocorrelation signals of either color contains single strands as well as the hybrid products. In order to excite both dyes (Cy-5 and Rhodamine green) at their respective absorption maximum, two laser beams had to be employed which both illuminate the same focal volume element. Although, due to this requirement and to the remaining spectral overlap of the absorption and emission characteristics, several experimental hurdles had to be overcome, dual-color fluorescence cross-correlation proved to be extremely powerful in discriminating the product species from free reaction partners. It could be shown that in proper setups, the cross-correlation amplitude is directly proportional to the concentration of reaction products and thus enables a simple evaluation of reaction kinetics. This is in contrast to association or dissociation processes observed by single-color autocorrelation analysis on the basis of changes in diffusion times. On the other hand, the initial high expectations regarding a higher sensitivity in systems with minute amounts of doubly labeled molecules in large excesses of singly labeled ones could not be confirmed because of the reduced signal-to-noise ratios of cross-correlation functions due to the intrinsic normalization process. Because all the fluorescence measured by either one of the detectors contributes to the denominator of the correlation curve which reduces the strength of relative fluctuations, the smallest ratios between product and educt concentrations measurable with confidence were estimated to be around 1%.

As a conclusion, not only the desired capability of simple yes/no decisions in test schemes but also the specific analysis of doubly labeled product was demonstrated, opening up interesting new applications such as intermolecular reactions without large changes in the mobility properties of one partner, as required for conventional autocorrelation analysis [17.13]. In the following, a comprehensive outline for dual-color applications of fluorescence cross-correlation spectroscopy is given highlighting the proper experimental conditions and pointing out some of the potential artifacts in nonideal setups.

17.2   Theory

17.2.1   Fluctuation Correlations

Although correlation spectroscopy of time series signals can, in principle, reveal any repetitive process with characteristic duration that modulates the fluorescence emission signal – a feature that makes it susceptible to several kinds of systematic noise using nonideal equipment – its predominant mode of application is the observation of minute spontaneous fluctuations. This so-called fluctuation correlation analysis is concerned with non-coordinated stochastic deviations from an equilibrated system that can only be resolved in highly restricted ensembles down to the single molecule level. The relaxation processes following such a spontaneous molecular fluctuation carry the same relevant thermodynamic and kinetic information as conventionally obtained by relaxation techniques in bulk samples, but without the large disturbances of the equilibrium.

In presently established FCS setups regardless of whether autocorrelation or cross-correlation is performed, small volume elements are realized by illuminating a microscope objective with parallel laser light and imaging the so-defined focal spot onto surfaces of ultrasensitive detectors. Depth discrimination in sample space is obtained by inserting field diaphragms (pinholes) in the image plane or by employing multi-photon excitation [17.14]. Typical sizes of effective volume elements are then of the order of $10^{-16} - 10^{-15}$ l. Sufficiently small ensembles are usually guaranteed if the concentration is kept well below micromolar, most applications work best in the nanomolar range where only single molecules reside in the volume element at any time.

The general form for the normalized fluctuation correlation function reads:

$$G_{ij}(\tau) = \frac{\langle \delta F_i(t) \delta F_j(t+\tau) \rangle}{\langle F_i(t) \rangle \langle F_j(t) \rangle} , \tag{17.1}$$

with $i = j$ for autocorrelation, $i \neq j$ for cross-correlation. Fluctuations in the fluorescence signal $F_D(t)$ recorded by a certain detector D, to be processed by correlation analysis have the following form:

$$\delta F_D(t) = \sum_i \kappa_i \int_V I_{ex}(\underline{r}) CEF_i(\underline{r}) \delta \left(\sigma_i q_i C_i(r, t) \right) dV . \tag{17.2}$$

The index i here determines the observed molecular species contributing to $F_D(t)$ with specific excitation and detection features such as overall fluorescence detection efficiency κ_i, the spatial distribution of the excitation intensity $I_{ex}(\underline{r})$ with amplitude I_0, and the dimensionless optical transfer function of the objective-pinhole combination $CEF(\underline{r})$. The fluctuating term that describes the dynamics on single particle scale, $\delta \left(\sigma_i q_i C_i(\underline{r}, t) \right)$, can include fluctuations of the molecular absorption cross-section $\delta \sigma_i$, the fluorescence quantum yield δq_i, or in the simplest case just local fluctuations of particle concentration due to Brownian motion δC_i. Equation (17.1) can be simplified by combining the convolution factor of the two dimensionless optical transfer functions $I_{ex}(r)/I_0 * CEF(r)$ into a single distribution function $W(r)$ and by combining the excitation intensity amplitude I_0 with the factors κ, σ and q into a parameter that determines the count-rate per detected molecule per second: $\eta_i = \kappa_i, \sigma_i, q_i$. Equation (17.1) can now be rewritten as

$$\delta F_D(t) = \sum_i \int_V W(r)\delta\left(\eta_i C_i(\underline{r}, t)\right) dV, \tag{17.3}$$

with $\delta(\eta_i C_i) = C_i \delta\eta_i + \eta_i \delta C$. In this expression, intra-molecular dynamics that change the fluorescence characteristics of single molecules are contributing to $\delta\eta_i$, whereas particle motion determines the number fluctuation term δC_i. If intra-molecular dynamics occur on much faster time scales than the translational motions (in most cases, simply diffusion), and if no attractive or repulsive forces between molecules exist, we can simplify $\langle \delta C_i(\underline{r}, 0)\delta C_j(\underline{r}', t)\rangle = 0$ for $i \neq j$ and separate the fluctuation terms in autocorrelation functions:

$$\langle \delta F_D(t)\delta F_D(t+\tau)\rangle =$$

$$\iint W(r)W(r') \sum_i \left(\begin{array}{c} C_i^2\langle\delta\eta_i(t)\delta\eta_i(t+\tau)\rangle \\ +\eta_i^2\langle\delta C_i(\underline{r}, t)\delta C_i\left(\underline{r}', (t+\tau)\right)\rangle \end{array} \right) dVdV'. \tag{17.4}$$

For translational particle diffusion of a single fluorescent species i and without internal dynamics, the sum term is replaced by the so-called concentration correlation factor $\phi(\underline{r}, \underline{r}', \tau)$ which reads

$$\phi(r, r', \tau) = \langle \delta C_i(\underline{r}, 0)\delta C_i(\underline{r}', \tau)\rangle$$

$$= \langle C_i\rangle (4\pi D_i\tau)^{-3/2} \exp\left(-\frac{(\underline{r}-\underline{r}')^2}{4D_i\tau}\right). \tag{17.5}$$

This expression is equivalent to the classical solution of the diffusion equation and determines the joint probability of finding a molecule at a position $\underline{r}'$ at time τ that had been at position $\underline{r}$ at time zero, multiplied by the average concentration. D_i is the diffusion coefficient of species i. By introducing (17.4) in (17.3) and taking the spatial integrals, the physical meaning of the concentration correlation function becomes evident: it is the probability that a molecule that has been inside the detected volume at time zero is still inside at time τ, weighted with the spatial detection efficiency, thereby describing a decay function with τ its characteristic time constant resembles the average molecular residence time within the observed spatial regime.

17.2.2   The Effective Measurement Volume in FCS

An important issue for all FCS applications, for cross-correlation applications in particular, is the sensible definition of the actually observed, effective measurement volume. It is well-established that the usually applied confocal optics with diffraction-limited illumination and detection through field apertures or so-called pinholes in the image plane, strictly has to be described by Fourier optics [17.15,17.16], where the objective point spread function is convoluted with the geometrical (generally circular) pinhole function. In the diffraction-limited case, the illumination point spread function is described by Bessel functions while it can, for an underfilled back aperture, be approximated as Gaussian in radial and Lorentzian in axial directions. However, the convolution routine in image space that is inferred by the pinhole function can in no way be solved analytically, and for a better handling of the optical system, a three-dimensional Gaussian distribution for detection optics is assumed [17.17,17.18]. In the theoretical form of the detected fluorescence signal, (17.1), this means in principle that the normalized $W(\underline{r})$ as well as the full *spatial product* $E_i(\underline{r}) = Q_i \kappa_i I_{\mathrm{ex},i}(\underline{r}) \mathrm{CEF}_i(\underline{r}) = \eta_i W(\underline{r})$ with amplitude η_i have 3D Gaussian shape:

$$E_i(\underline{r}) = \eta_i W(\underline{r}) = \eta_i \exp\left(-\frac{2(x^2 + y^2)}{r_0^2}\right) \exp\left(-\frac{2z^2}{z_0^2}\right), \tag{17.6}$$

without making decisive assumptions about the single contributions that result from either excitation or detection. To distinguish this product from the pure illumination profile of the exciting laser beam, $E(\underline{r})$ is termed "emission intensity distribution"; it is proportional to the probablity to *detect* a fluorescent molecule of species i dependent on its location in the sample space. This probability is naturally nonzero throughout the whole space, consequently the concept of a strictly defined "hard sphere" volume element with fixed borders is not valid for our measurement geometry. Unless there are no mechanical constraints, the observed volume in confocal setups is principally of infinitesimal size. However, the convergence of the Gaussian functions' integrals allows us to assign a finite, effective volume V_{eff} that at a given average concentration contains exactly the number of molecules resembled by the inverse correlation amplitude $N = G(0)^{-1} = V_{\mathrm{eff}}^{-1} \langle C \rangle^{-1}$. This relationship between N and the normalized $G(0)$ follows from Poissonian statistics where the standard deviation equals the mean value. For autocorrelation with only one illuminating beam (for simplicity written for only one fluorescent species), the normalized $G(\tau)$ is given by

$$G(\tau) = \frac{\iint W(\underline{r})W(\underline{r}')\eta^2 \phi(\underline{r},\underline{r}',\tau)\mathrm{d}V\mathrm{d}V'}{(\langle C \rangle \eta \int W(\underline{r})\mathrm{d}V)^2}. \tag{17.7}$$

With $\phi(\underline{r},\underline{r}',0) = \langle C \rangle \delta(\underline{r} - \underline{r}')$ [17.19], the effective volume element V_{eff} is fully defined as

$$V_{\mathrm{eff}} = \frac{[\int W(\underline{r})\mathrm{d}V]^2}{\int W^2(r)\mathrm{d}V}, \tag{17.8}$$

and yields for integration over the whole space: $V_{\mathrm{eff}} = \pi^{3/2} r_0^2 z_0$, if r_0 and z_0 are the $1/e^2$ half-axes of the $E(\underline{r})$ Gaussian distribution. It is important to note that this volume is not identical with the 3D integral of $E(\underline{r})$ itself, which would be a factor of $\sqrt{8}$ smaller.

In dual-color cross-correlation, excitation profiles created by the two independent laser beams may differ in size. In any case, the different detection wavelengths modify the detection point spread functions $CEF(\underline{r})$ in different ways and inevitably lead to different effective detection volume distribution functions $W_1(\underline{r}) \neq W_2(\underline{r})$. The definition for an effective volume element reads in this case:

$$V_{\text{eff}} = \frac{[\int W_1(\underline{r})dV][\int W_2(\underline{r})dV]}{\int W_1(\underline{r})W_2(\underline{r})dV}, \tag{17.9}$$

resulting in

$$V_{\text{eff},x} = \pi^{3/2} \frac{(r_1^2 + r_2^2)}{2} \sqrt{\frac{(z_1^2 + z_2^2)}{2}}, \tag{17.10}$$

r_1 and r_2, respectively z_1 and z_2 are the $1/e^2$ half axes for the two different Gaussian distributions $W_1(\underline{r})$ and $W_2(\underline{r})$.

17.2.3 Autocorrelation and Cross-correlation Functions for Pure Diffusion

With a Gaussian distribution for $W(\underline{r})$, the effective volume elements as defined and if no internal dynamics takes place: $\langle \delta\eta_i(t)\delta\eta_i(t+\tau)\rangle \approx 0$, the expression for the normalized autocorrelation function of a detector signal F_D following (17.1) and (17.4) reads, for multiple species i (for simplicity, we assume equal η_i):

$$G_D(\tau) = \frac{\sum_i \langle C_i\rangle \eta_i^2 \text{Diff}_i}{V_{\text{eff}}(\sum_i \eta_i\langle C_i\rangle)^2} = N_{\text{tot}}^{-1} \sum_i Y_i \text{Diff}_i, \tag{17.11}$$

with

$$\text{Diff}_i \equiv \left(1 + \frac{\tau}{\tau_{d,i}}\right)^{-1} \left(1 + \frac{r_0^2\tau}{z_0^2\tau_{d,i}}\right)^{-1/2}, \tag{17.12}$$

the temporal fluctuation decay function in 3D for each diffusing species i. The relative fractions of each species i to the total number of molecules $N_{\text{tot}} = V_{\text{eff}} \sum_i \langle C_i\rangle$ are given as $Y_i = \langle C_i\rangle / \sum_i \langle C_i\rangle$, the parameters $\tau_{d,i} = r_0^2/4D_i$ are the characteristic residence or diffusion times for molecules of species i with diffusion coefficients D_i in/through the effective volume element V_{eff}.

For the simplest form of cross-correlation as schematized in Fig. 17.1, we assume that again no attractive or repulsive interactions between particles and no internal dynamics take place so that the only non-vanishing fluctuation correlation terms are $\langle \delta C_i(\underline{r}, 0)\delta C_i(\underline{r}', t)\rangle$. The system under observation consists of molecules of species 1 and 2, as well as some that bear both label types which we connote with species 12, and that have to be specifically selected by the cross-correlation procedure. The two detector signals are now given as

$$\delta F_{D1}(t) = \int W_1(\underline{r})\eta_1[C_1(r,t) + C_{12}(\underline{r},t)]dV,$$

$$\delta F_{D2}(t) = \int W_2(\underline{r})\eta_2[C_2(r,t) + C_{12}(\underline{r},t)]dV. \tag{17.13}$$

In the ideal case where the two detection channels can perfectly separate fluorescent species 1 and 2, the autocorrelation and cross-correlation functions read ($i = 1, 2$):

$$G_{Di}(\tau) = \frac{(\langle C_i \rangle \mathrm{Diff}_i + \langle C_{12} \rangle \mathrm{Diff}_{12})}{V_{\mathrm{eff},i}(\langle C_i \rangle + \langle C_{12} \rangle)^2}, \tag{17.14}$$

for autocorrelation and

$$G_{D1 \times D2}(\tau) = \frac{\langle C_{12} \rangle \mathrm{Diff}_{12}}{V_{\mathrm{eff},x}(\langle C_1 \rangle + \langle C_{12} \rangle)(\langle C_2 \rangle + \langle C_{12} \rangle)}, \tag{17.15}$$

for cross-correlation. Note that for the separation of doubly labeled species to be effective, it is not necessary that the measurement volumes be of the same size. In the case of different volume geometries, the average residence time of species 12 in $V_{\mathrm{eff},x}$, reads $\tau_{d,12} = (r_1^2 + r_2^2)/(8D_{12})$. D_{12} is the diffusion coefficient of species 12. A correct evaluation of parameters measured by cross-correlation requires good prior knowledge of the system or a careful calibration procedure of the two basic parameters resulting from FCS analysis, the effective volume element and the lateral characteristic residence time.

From (17.14) and (17.15) it can be seen that by cross-correlation, it is indeed possible to separate a species bearing two different fluorescent labels and specifically determine its dynamic features in a system composed of freely diffusing molecules. Moreover, since the denominator in (17.15) is constant if there is no loss of fluorescent material, the amplitude of the cross-correlation function is directly proportional to the concentration of doubly labeled species which gives a real advantage over the autocorrelation alternative in studies of slow reaction kinetics: While by autocorrelation, evaluation of product concentration is only possible if there are large changes in the diffusion behavior upon reaction, and has to be achieved by fitting the correlation curve, by cross-correlation, no changes of the diffusive behavior are required, and the relative amount of product at any time can be directly read from the $G(0)$ value of the curve. The denominator in (17.15) can be expressed in terms of amplitudes $G(0)$ with $\mathrm{Diff}_i(0) = 1$ of the two autocorrelation functions (17.14) so that the absolute amount of product can be easily derived as

$$\langle C_{12} \rangle = \frac{G_x(0)V_{\mathrm{eff},x}}{G_{D1}(0)V_{\mathrm{eff},1}G_{D2}(0)V_{\mathrm{eff},2}}. \tag{17.16}$$

17.2.4   Detector Cross-Talk

Unfortunately, the ideal case of total separability of the two distinct fluorophore species 1 and 2 is, in reality, hard to achieve. The absorption cross-section for the relatively red-shifted dye is usually nonzero for the short excitation wavelength, on the other hand there is always a small amount of leakage of the green-shifted dye's emission through the long wavelength detection window. In contrast, blue-shifted emission of the red dye and absorption of the red laser light by the blue dye are not likely and can be neglected in most cases. Consequently, it is in particular the red-shifted detector signal that suffers from optical "contamination". Due to the two possibilities, to excite and to detect either of the dyes with both lasers and by both detectors, eight different emission amplitude parameters η have to be defined, in principle. We abbreviate the absorption cross-section of dye i at the wavelength of laser j with excitation intensity amplitude $I_{0,j}$ as σ_{ij}, and the detection efficiency of dye i by detector k with κ_{ik} which results in a detected counts per molecule signal η_{ijk} at detector k: $\eta_{ijk} = \kappa_{ik}\sigma_{ij}I_{0,j}$. The two fluctuating emission signals in (17.13) have to be rewritten as

$$\delta F_{D1}(t) = \int \eta_{111}W_1(\underline{r})(\delta C_1 + \delta C_{12})dV$$

$$\delta F_{D2}(t) = \int [\eta_{212}W_1(\underline{r}) + \eta_{222}W_2(\underline{r})]\delta C_2(\underline{r},t)dV$$

$$+ \int [\eta_{112}W_1(\underline{r}) + \eta_{212}W_1(\underline{r}) + \eta_{222}W_2(\underline{r})]\delta C_{12}(\underline{r},t)dV$$

$$+ \int \eta_{112}W_1(\underline{r})\delta C_1(\underline{r},t)dV .$$

It can easily be verified that with this modified expression of δF_{D2}, the exclusive representation of species C_{12} in the numerator of (17.15) is no longer possible due to non-vanishing contributions of $\langle \delta C_1(\underline{r},0)\delta C_1(\underline{r}',\tau)\rangle$. The amplitude of the cross-correlation function has to be modified as follows:

$$G_x(0) = \frac{(r_1\langle C_1\rangle + r_2\langle C_2\rangle + r_{12}\langle C_{12}\rangle)\int W^2(\underline{r})dV}{\langle F_1\rangle\langle F_2\rangle} , \qquad (17.17)$$

if equal intensity distributions $W_1(\underline{r})$ and $W_2(\underline{r})$ are assumed for simplicity. The representation factors r_i of species i in the numerator of the cross-correlation function are given by: $r_1 = \eta_{111}\eta_{112}$, $r_2 = 0$ and $r_{12} = \eta_{111}(\eta_{112} + \eta_{212} + \eta_{222})$. Thus, the selective representation of doubly over singly labeled species or the so-called quality factor for cross-correlation Q_x is given by the relationship

$$Q_x = \frac{r_{12}}{r_1} = \frac{\eta_{112} + \eta_{212} + \eta_{222}}{\eta_{112}} . \qquad (17.18)$$

To yield an improvement over single-color autocorrelation in reaction schemes where both species are labeled with the same dye (and the product is just two-fold brighter), this factor has to be greater than 4 due to the square dependence on the η values in the numerator of the correlation curve. In setups with proper filters and dyes with the smallest possible spectral overlap integrals, values of $Q_x \geq 20$ can easily be obtained [17.13].

17.2.5   Not Completely Overlapping Detection Volumes

As mentioned above, the typically assumed ideal case of exactly matching detection volume elements $W_1 = W_2$, is experimentally difficult to achieve due to the dispersion of optical transfer functions. In reality, neither perfect equality of size nor the full spatial overlap can be guaranteed (Fig. 17.2). Small mismatches in laser alignment on the excitation and chromatic aberration effects of the microscope objective on the detection side cannot be ruled out. Insufficient overlap is very easy to test, as described below, and has a different effect on the measured curves than the cross-talk due to not ideally separable dyes. Strictly, the formalism of displaced detection volumes as used for the derivation of spatial cross-correlation functions [17.10] has to be applied to correct for these effects in the measured correlation curves. If the displacement is small compared to the size of the volume elements however, a simplified approach can be used as follows.

The fluorescence signals can be split into a part that contributes to both channels and a part that contributes to only one channel. Consider two effective detection volumes observed by either detector, $V_{\text{eff},1}$ and $V_{\text{eff},2}$, and an overlapping region that is contained by both, $V_{\text{eff},o}$. The relative size of $V_{\text{eff},o}$ compared to $V_{\text{eff},1}$ will be $v = V_{\text{eff},o}/V_{\text{eff},1}$, the relative size compared to $V_{\text{eff},2}$ will be $w = V_{\text{eff},o}/V_{\text{eff},2}$. Speaking in terms of concentrations, the molecular observed species $\langle C_i \rangle$ now have to be divided into an "inside-part" $\langle C_{i,o} \rangle = v\langle C_i \rangle$ for detector 1 and $\langle C_{i,o} \rangle = w\langle C_i \rangle$ for detector 2 inside and an "outside-part" $(\langle C_i \rangle - \langle C_{i,o} \rangle)$ with respect to the overlapping regime. Only the inside-parts of either species (ideally, only $\langle \delta C_{12,o} \delta C_{12,o} \rangle$) give nonzero contributions to the cross-correlation numerators, while all species contribute to the denominators. As a consequence, the measured cross-correlation amplitude is wrongly decreased and the concentration determination of doubly labeled species according (17.15) and (17.16) is no longer valid. Equation (17.15) at time zero has to be modified as follows:

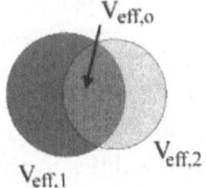

Fig. 17.2. Non-overlapping volume elements

$$G_{D1 \times D2}(0) = \frac{v \cdot w \cdot \langle C_{12} \rangle}{V_{\text{eff},x}(\langle C_1 \rangle + \langle C_{12} \rangle)(\langle C_2 \rangle + \langle C_{12} \rangle)}, \tag{17.19}$$

i.e. the ratios of volume overlap directly enter into the amplitude expression and can be determined experimentally. The autocorrelation functions' amplitudes naturally do not suffer from artifacts because the relative position of the volume elements do not play any role here.

17.2.6   Cross-correlation of Internal Fluctuations

All cases treated so far concern the situation that no fast molecular dynamics reversibly modulating the emission within the diffusion time window take place, i.e. that the only contributions to the fluctuation correlation term are number fluctuations on the time scale of particle diffusion. Consequently, cross-correlation is only possible if molecules are stationarily labeled with both dyes during the full particle observation time. If non-coordinated internal fluctuations in either of the observed species occur and $\delta\eta_i(t)\delta\eta_i(t+\tau) \neq 0$, as is the case for singlet-triplet transitions known for a large variety of dyes (see Chap. 13), the autocorrelation functions of either detector show additional contributions, typically represented as additive or multiplicative exponential decay factors $G_{\text{trip}}(\tau) = [1 - T + T\exp(-\tau/\tau_t)]$ with amplitudes T proportional to the fraction of time spent in the dark triplet state and decay constants τ_t denoting the relaxation times back to the bright state.

The probability to detect singlet-triplet transitions in the cross-correlation function is proportional to the joint probability for them to occur in both of the two observed dyes at the same instant. Following (17.4), the idealized expression for setups without cross-talk (everything vanishing but C_{12} contributions) is given by

$$\langle \delta F_{D1}(t)\delta F_{D2}(t+\tau) \rangle$$

$$= \iint W_1(r)W_2(r') \begin{pmatrix} C_{12}^2 \langle \delta\eta_1(t)\delta\eta_2(t+\tau) \rangle \\ +\eta_1\eta_2 \langle \delta C_{12}(t)\delta C_{12}(t+\tau) \rangle \end{pmatrix} dVdV' \tag{17.20}$$

and does not contain *self* terms $\langle \delta\eta_i\delta\eta_i \rangle$ so that the only fluctuation terms representing fast internal dynamics are the distinct expressions $\langle \delta\eta_1(t)\delta\eta_2(t+\tau) \rangle$. If the photophysical properties of the dyes are truly non-coordinated, this term representing the probability for both dyes to be in the dark triplet state within any time interval τ approaches zero at moderate intensities. Consequently, the cross-correlation functions in close-to-ideal setups lack triplet contributions in the fast time range. On the other hand, observation of non-vanishing triplet shoulders in the measured cross-correlation functions can be a valuable hint for large detector cross-talk or other nonidealities in the setup.

Evidently, a large variety of interesting cross-correlation schemes is opened up if the assumption $\langle \delta\eta_i(t)\delta\eta_j(t+\tau) \rangle \approx 0$ is no longer fulfilled. This is the

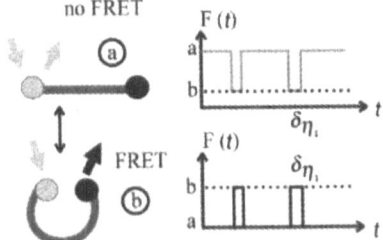

Fig. 17.3. Example for dual-color cross-correlation of internal e.g. conformational fluctuations with a FRET pair of dyes. The same scheme is possible for intermolecular dynamics between differently labeled molecules

case for, for example, inter- or intra-molecular dynamics that induce fast fluctuations in the emission behavior of both dyes at once with positive or negative cross-correlation. To date, no such cross-correlation application in conventional FCS has been reported although the measurement principle is rather simple. An example would be the combination of dyes with the non-vanishing possibility for Förster resonance energy transfer (FRET) within the same molecule, e.g. due to conformational dynamics, or bound to two different molecules that come close to each other for a short time to induce coordinated fluctuations in the emission signals of donor and acceptor, as schematized in Fig. 17.3. Due to the dependency of energy transfer efficiency on the distance between the donor and acceptor dye molecule, intervals of relative dye proximity induce fluctuations in both emission channels if only the do dye is excited.

This principle has recently become very popular to follow localization and dynamics of fixed single molecules [17.20], comparable experiments in bulk with a dye molecule and a quencher instead of a FRET pair have been carried out by single-color FCS [17.21] to study the intra-molecular dynamics of single stranded DNA molecules ("molecular beacons") with internal loops. However, in single-color applications, conformational and other internal fast dynamics may occur on time scales close to triplet dynamics mentioned above, which makes them difficult to study independently. Cross-correlation allows for a determination of FRET free from photophysical dynamics contributions. In such applications, one would expect a negative contribution from internal dynamics inducing a drop of the correlation amplitude at small τ.

17.3   Experimental Realization

Figure 17.4 shows an experimental setup of the dual-color cross-correlation setup realized first in 1996 [17.13]. The 488 nm line of an argon ion laser and the 647 nm line of a krypton-argon ion laser were combined on the optical table by a dichroic mirror. In order to realize similar focal spot sizes in the

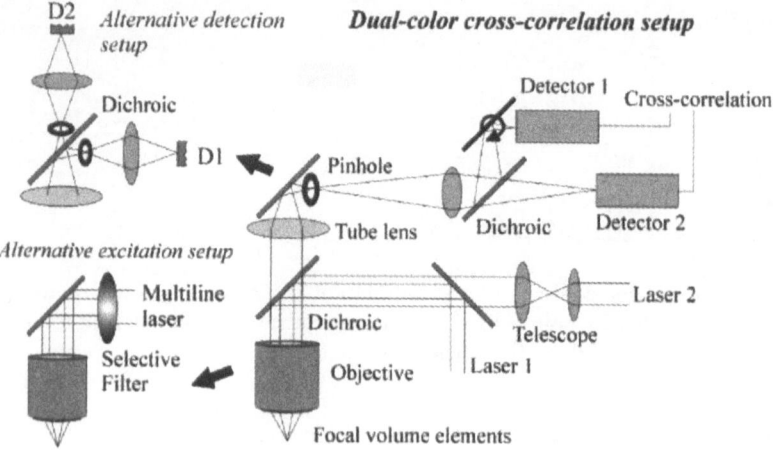

Fig. 17.4. Possibilities for an experimental realization of dual-color cross-correlation

objective, the 647 nm beam size had to be increased in diameter by a factor of 1.3 with respect to the 488 nm beam. This was accomplished by a parallel optics telescope system. After being reflected by a selective dichroic mirror, both beams epi-illuminate a water immersion 40 × 0.9 microscope objective. Beam diameter $1/e^2$ values at the objective back aperture were 3.5 mm (488) and 5 mm (647) resulting in a Gaussian focal spot diameter r_0 of 0.7 mm for both wavelengths. As usual in FCS setups, the axial dimension z_0 was determined in calibration measurements with pure dye of known diffusion properties and high photostability (e.g. Rhodamine derivatives) in order to avoid artifacts. The fraction z_0/r_0 for the given and related setups is typically between 3.5 and 5, yielding effective volume elements V_{eff} of approximately 1.3×10^{-15} l. Thus, at nanomolar concentrations the average number of molecules in the focal spot is of the order of 1. Fluorescence light is focused by a tube lens with focal length 164.5 mm onto a pinhole of 50 µm diameter in the image plane. Another 60 mm lens images the pinhole 1:1 onto the two avalanche photodiodes. Behind this lens, the fluorescence light is split by a dichroic mirror. Pinhole, lens and mirror are separately adjustable. Final detection specificity is accomplished by selective band-pass filters in front of each diode. Cross-correlation is performed on-line by the same hardware correlator used for conventional FCS.

In setups where two laser beams have to be combined into the same focal spot of the objective, the alignment is highly critical and has to be controlled indirectly by FCS calibration measurements [17.13] or directly by measuring the illumination profiles with a specifically designed focus scanner [17.22] offered by the designers of a commercial cross-correlation prototype. Although the determination of the full illumination point spread functions at different colors is very attractive from an optical point of view, the calibration measurements that can be performed in every FCS setup are adequate for practical purposes to guarantee a sufficient overlap of the measurement volumes. The idea is to illuminate a dye solution with comparable excitation cross-sections at both wavelengths successively with both lasers without changing the detection pathway. If the amplitudes and decay times of the measured correlation functions are equal or at least similar with deviations less than 10%, the alignment is close to perfect and sensible cross-correlation measurements are possible. De novo alignment of the two foci works in an equivalent way: first the detection path is adjusted with one illuminating beam, then this first beam is blocked and the second beam is aligned to yield the same correlation curve without changing the detection optics. The procedure is harder to perform, but also works if the excitation cross-sections differ considerably.

An interesting alternative to avoid alignment efforts for the illuminating laser beams is to use a laser in multiline mode. This has been successfully applied by Winkler et al. [17.23]. However, to achieve focal spots of the same size, selective filters that reduce the diameter of the green beam have to be employed. Additionally, it has to be guaranteed that the objective is perfectly corrected against chromatic aberration.

The alignment of the detection pathway is even more critical. The overlap of the sample volumes observed by either of the detectors has to be maximized in order to effectively suppress fluorescence contributing to only one of the signals. One critical feature for dual-color cross-correlation with largely different emission spectra is again the chromatic aberration of the microscope lens. If only one pinhole is used and the wavelength selection performed behind it, there is no possibility to correct against chromatic displacements of the focal spots in the z-direction at different wavelengths by adjustment of the detection optics. The objective's correction properties become very critical in this case [17.13]. An alternative setup with two pinholes, where the wavelength selection is already performed in the regime of parallel optics may be more flexible. Additionally, pinholes of different sizes are in principle required for the ideal case of equal effective detection volumes to cancel out the differences induced by the wavelength-dependent detection point spread functions.

17.4   Applications

17.4.1   Slow Association Reactions: Comparison Between Autocorrelation and Cross-correlation

The large majority of FCS applications to date have been concerned with analysis of diffusion properties of macromolecules in solution. In particular, changes in molecular diffusion constants upon association of small, labeled probes to larger reaction partners have been exploited. Because of the low concentrations necessary for FCS, irreversible association dynamics [17.12,17.13,17.24] could be conveniently followed within minutes to hours. The increasing amount of slowly diffusing reaction product induces a shift of the average diffusion time towards larger decay times; the product concentration has to be derived by careful fitting analysis of the overall correlation curve as described elsewhere in this book. Although the concept of kinetic studies by observation of diffusion times only is simple, versatile and even permits ultrasensitive diagnostic applications [17.25,17.26], it is especially the time-consuming fitting procedure as well the requirement of marked changes in molecular weight to indicate the reaction, that motivates the introduction of dual-color cross-correlation schemes.

Consequently, the first application was demonstrated for an association reaction between two differently labeled complementary nucleic acid strands of each 20 bases length [17.13]. The change in diffusion properties following the renaturation has, in previous studies, been determined to be considerably less than a factor of two in diffusion time, which does not allow reliable separation between free and bound molecules [17.12].

Figure 17.5 exemplifies the conceptual differences between the two methods by comparing results from two different renaturation experiments, the association of a 15 bases long labeled DNA-oligonucleotide to a 100 bases

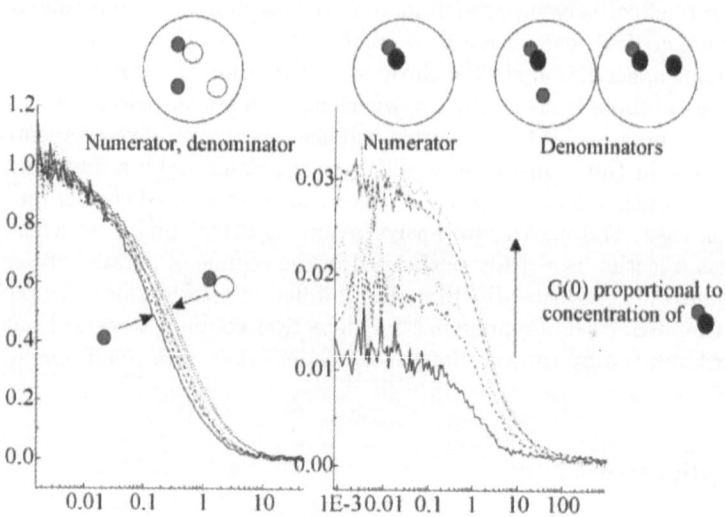

Fig. 17.5. Difference in measurement principle for association reactions: *left*, hybridization of a nucleic acid oligomer of 15 bases length to a target of five-fold size; the product ratio has to be determined by fitting the correlation curve with two different species (free and bound) and evaluating the partition of bound species. *Right*: Hybridization of two oligomers of the same size; evaluation of product can, in well calibrated setups, be performed simply by recording the amplitude values

RNA target, measured by conventional autocorrelation analysis (left), and the association of two 20 bases long oligonucleotides, each labeled with a distinctive dye, by dual-color cross-correlation (right). Although the change in molecular weight after binding to the RNA target in the autocorrelation case is a factor of 5–6, it can be easily seen that the dynamic frame for data analysis is rather narrow and that careful calibration processes are necessary in order to deconvolute the measured curve in contributions from free and bound probes.

On the other hand, the cross-correlation function as explained above, is predominantly sensitive to the relative amount of doubly labeled species regardless of how significant the changes in the diffusion characteristics. In proper setups, the amplitudes of the cross-correlation functions are directly proportional to the product concentration, the temporal decay represents the diffusion characteristics of product only, which enables easy data analysis.

17.4.2   Cross-correlation Applications in Various Biochemical Systems

Since the first demonstration of dual-color cross-correlation, not only a commercial prototype for dual-color FCS has been developed (see Chap. 15) but also a variety of interesting applications have been reported. In 1996, Eigen proposed substantial advantages for the employment of cross-correlation in discovering prospective early aggregation stages of pathogenic agents in neurodegenerative diseases such as BSE or Alzheimer's disease [17.27]. The first promising experimental results regarding the self-aggregation behavior of a homologue to the protease resistant, potentially infectious form of the prion protein PrP(27–30) were reported one year later [17.28]. Recently, dual-color detection of prion protein aggregates has shown to even fulfill the requirements for a rapid and specific test for CJD and related diseases [17.29].

In continuation of FCS investigations on nucleic acids, Rigler et al. [17.22] employed the cross-correlation concept to detect specifically amplified target DNA in multiple polymerase chain reactions (PCRs). In these experiments, the amplification primers were labeled with two different dyes which enabled a selective investigation of double-stranded PCR products and yielded information about their temporal accumulation. With only 10–25 initial copies of the template molecule, the reaction could be carried out in single compartments containing all reaction ingredients without any need for post-PCR purification and thus promises a simplification in identifying PCR targets in medical diagnostic applications. Moreover, the necessity of two primers to simultaneously bind the amplification product reduces the risk of detecting unspecific products.

The fact that cross-correlation analysis is not only sensitive to reactions or processes where doubly labeled species are formed out of singly labeled, but also to the opposite process, has been exploited for several studies on enzyme kinetics. Kettling et al. [17.30] followed the relative number of doubly labeled, 66-bp double-stranded DNA molecules and their singly labeled fragments upon restriction by EcoRI endonuclease, thereby yielded easy access to the kinetic quantitation of enzymatic activity by simply monitoring the cross-correlation amplitude with time. The enzyme activity could be quantified down to the low picomolar range, where a linear dependence of the rate constants on enzyme concentration was observed over two orders of magnitude. Based on these pioneering studies, good chances for high-throughput screening of enzyme assays with dual-color cross-correlation were proposed.

In a subsequent study, the measurement time for a yes/no decision about enzyme activity was further suppressed. Reduced signal-to-noise values of the recorded curves could be accounted for by focusing on the cross-correlation amplitudes $G(0)$ rather than the full functions in the evaluation [17.31]. Due to the slightly modified evaluation procedure, based on statistical analysis of $G(0)$ distributions, this application was termed RAPID FCS (rapid assay processing by investigation of dual-color fluorescence cross-correlation spectroscopy). Based on the scheme developed by Kettling et al. [17.30], a simulated high-throughput screening was carried out with homogeneous assays for different restriction endonucleases. The data recording time necessary to distinguish with less than 5% error between positive (specific) and negative (non-specific) enzyme samples was less than 800 ms.

As a consequence from the above findings that measurement times can be dramatically reduced by reduction of the readout parameters to $G(0)$ values, the dual-color detection scheme was modified to simply record correlation factors at time zero, i.e. to a mere dual-color coincidence analysis (CFCA, [17.23]). The instrumentation for such applications does no longer require hardware correlator boards for *on-line* analysis, but can be carried out by simply comparing count-rate traces recorded with two multichannel scalers. The comparison is done by multiplying independently recorded time bins of both detection channels at any instant where the preferable bin width is a function of photophysical and mobility parameters of the investigated molecules. To enhance the signal-to-noise ratios for the measurement of correlation factors at short measurement times and for small bin widths, the sample was additionally subject to rapid oscillations of the sample carrier to increase the number of events of doubly labeled particles entering the volume element. This feature allows us to improve the measurement statistics independently of the diffusion characteristics of the observed molecules. In endonucleolytic yes/no assays with EcoRI analogous to the ones tested by Koltermann et al. [17.31] the measurement time to distinguish positive from negative results (with an error $< 8\%$) was reduced to 50 ms. At this level, screening applications with throughput rates of 106 samples per day are opened up, with minute amounts of sample required in confocal detection setups. As a consequence, fluorescence cross-correlation and related schemes presently experience a growing popularity in new strategies for drug discovery and evolutionary biotechnology (see Chap. 9).

17.4.3 Outlook

Dual-color cross-correlation analysis yields significant advantages over auto-correlation analysis in two respects: first, it separates molecular characteristics such as diffusion and internal dynamics of singly labeled from doubly labeled molecules, and second, it simplifies the determination of absolute dual-color complex concentrations regardless of whether mobility changes are associated with complex formation or cleavage. The qualitative improvement over autocorrelation by distinguishing two different measurement parameters systematically enhances detection specificity, which is of particular importance in systems where two or more different molecular species interact with each other. The next step in cross-correlation application, besides further biotechnological developments, is certainly the investigation of molecular association or dissociation processes in living cells or on membranes. A large number of relevant cellular processes involving complexation of more than two molecules, i.e. within signaling cascades certainly motivate the employment of even more spectrally separable probes and multi-color detection. It has been shown [17.14] that two- or multi-photon excitation offers a number of important advantages for FCS on the cellular level such as lower background and reduced photobleaching of the dyes. With respect to multiple color systems, one additional advantage results from the broad and largely overlapping two-photon excitation spectra of many relevant fluorescent probes. In a recent publication [17.32], this attractive concept of dual-color, two-photon fluorescence cross-correlation (TPCC) analysis was experimentally demonstrated with an established assay probing the selective cleavage of dual labeled DNA substrates by restriction endonuclease *Eco*RI.

As a perspective, we conclude that the possibility to follow different reaction partners and their products at the same time, the compatibility with fast screening assays in biotechnological applications, and last but not least the availability of commercial excitation and detection modules promise a bright future for the dual-color cross-correlation concept.

References

17.1 J. Ricka and T. Binkert: Phys. Rev. A **39**, 2646–2652 (1989)

17.2 Berne and Pecora: J. Wiley & Sons, New York (1976)

17.3 W.G. Griffin, and P.N. Pusey: Phys. Rev. Lett. **43**, 1100–1104 (1979)

17.4 Z. Kam and R. Rigler: Biophys. J. **39**, 7–13 (1982)

17.5 N.A. Clark, B.J. Ackerson, and A.J. Hurd: Phys. Rev. Lett. **50**, 1459–1462 (1983)

17.6 B.J. Ackerson, T.W. Taylor, and N.A. Clark: Phys. Rev. A **31**, 3183–3193 (1985)

17.7 P. Tong, K.-Q. Xia, and B.J. Ackerson: J. Chem. Phys. **89**, 9256–9265 (1993)

17.8 K.-Q. Xia, Y.-B. Xin, and P. Tong: J. Opt. Soc. Am. A **12**, 1571–1778 (1995)

17.9 M. Brinkmeier and R. Rigler: Exp. Techn. Phys. **41**, 205–210 (1996)

17.10 M. Brinkmeier, K. Dörre, J. Stephan, and M. Eigen: Anal. Chem. **71**, 609–616 (1999)

17.11 M. Eigen and R. Rigler: Sorting single molecules: Proc. Natl. Acad. Sci. USA. **91**, 5740–5747 (1994)

17.12 P. Schwille, F. Oehlenschläger, and N. Walter: Biochemistry **35**, 10182–10193 (1996)

17.13 P. Schwille, J. Bieschke, and F. Oehlenschläger: Biophys. Chem. **66**, 211–228 (1997)

17.14 P. Schwille, U. Haupts, S. Maiti, and W.W. Webb: Biophys. J. **77**, 2251–2265 (1999)

17.15 H. Qian and E.L. Elson: Appl. Opt. **30**, 1185–1195 (1991)

17.16 D.R. Sandison, and W.W. Webb: Appl. Optics **33**, 603–615 (1994)

17.17 S.R. Aragon and R. Pecora: J. Chem. Phys. **64**, 1791–1803 (1976)

17.18 R. Rigler, Ü. Mets, J. Widengren, and P. Kask: Eur. Biophys. J. **22**, 169–175 (1993)

17.19 N.L. Thompson: Fluorescence correlation spectroscopy. In: Topics in Fluorescence Spectroscopy. Vol. 1. J.R. Lakowicz, Ed. (Plenum Press, New York 1991) pp. 337–378

17.20 S. Weiss: Science **283**, 1676–1683 (1999)

17.21 G. Bonnet, O. Krichevsky, and A. Libchaber: Proc. Natl. Acad. Sci. USA **95**, 8602–8606 (1998)

17.22 R. Rigler, Z. Földes-Papp, F.-J. Meyer-Almes, C. Sammet, M. Völcker, and A. Schnetz: J. Biotchnol. **63**, 97–109 (1998)

17.23 T. Winkler, U. Kettling, A. Koltermann, and M. Eigen: Proc. Natl. Acad. Sci. USA **96**, 1375–1378 (1999)

17.24 M. Kinjo and R. Rigler: Nucleic Acids Res. **23**, 1795–1799 (1995)

17.25 N. Walter, P. Schwille, and M. Eigen: Proc. Natl. Acad. Sci. USA **93**, 12805–12810 (1996)

17.26 F. Oehlenschläger, P. Schwille, and M. Eigen: Proc. Natl. Acad. Sci. USA. **93**, 12811–12816 (1996)

17.27 M. Eigen: Biophys. Chem. **63**, A1–A18 (1996)

17.28 J. Bieschke and P. Schwille: Aggregation of prion protein investigated by dual-color fluorescence cross-correlation spectroscopy. In: Fluorescent Microscopy and Fluorescent Probes. Vol. 2. J. Slavik, Ed. (Plenum Press, New York 1997) pp. 81–86

17.29 J. Bieschke, A. Giese, W. Schulz-Schaeffer, I. Zerr, S. Poser, M. Eigen, and H. Kretschmar: Proc. Natl. Acad. Sci. USA **97**, 5468–5473 (2000)

17.30 U. Kettling, A. Koltermann, P. Schwille, and M. Eigen: Proc. Natl. Acad. Sci. USA. **95**, 14116–1420 (1998)

17.31 A. Koltermann, U. Kettling, J. Bieschke, T. Winkler, and M. Eigen: Proc. Natl. Acad. Sci. USA **95**, 1421–1426 (1998)

17.32 K.G. Heinze, A. Koltermann, and P. Schwille: Proc. Natl. Acad. Sci. USA **97**, 10377–10382 (2000)

18 Cross-correlated Flow Analysis in Microstructures

Michael Brinkmeier

18.1 Introduction

In most of the FCS applications, diffusion is the physical parameter used in gaining translational information about the molecules observed. The underlying reason for this is that a change of the diffusion constant reveals a change of the molecule size, which can be coupled to chemical reactions: therefore opening the important field of chemical kinetics measurements. In all these applications, the measurements are performed in a time-spatially isotropic environment: No information about the structure of the environment and non-diffusional transport processes is needed or being used. But especially for biotechnological applications where complete reaction cascades take place on a small chip streaked with channels and chambers, these reactions and the active transport of the reactants have to be monitored. For this, flow parameters like velocity and its direction are determined via FCS on a micrometer scale. At the same time, other parameters, like diffusion or fluorescent intensity can be measured. The ability to measure flow is greatly enhanced by the so-called "two-beam FCS" where the fluorescent signals generated by two separated foci are being cross-correlated. In the following, the theory of two-beam FCS is outlined, and a typical setup and microchannel measurements are described. Finally, possible applications of FCS and single molecule detection in microstructures will be given.

In early studies of FCS, a theoretical analysis and corresponding FCS measurements of flowing particles was carried out [18.1,18.2]. Since then, the field of FCS has evolved tremendously through the use of a confocal epi-illuminated fluorescence microscope. Due to high excitation irradiances and sensitive detectors, it is now possible to detect single molecules in aqueous solutions at room temperature, i.e. under biological conditions. Nevertheless, flow measurements with a standard FCS setup (the standard setup is described in other chapters in this book) have two major disadvantages: First, one cannot measure slow flow velocities, that is, when the flow time through the volume element is larger than the characteristic diffusion time:

$$\frac{2w_0}{V} < \frac{w_0^2}{4D}, \tag{18.1}$$

with V being the velocity, D the diffusion constant of the particles and $2w_0$ describing the diameter of the volume element. Taking typical values in FCS experiments, $D = 3 \times 10^{-6}\,\mathrm{cm^2/s}$ and $2w_0 = 1\,\mu\mathrm{m}$, the lower limit of a measurable velocity lies at about a quarter of a centimeter per second. Lower values no longer change the autocorrelation curve. Second, it is not possible to determine the direction of the flow (with the theoretical exception of the z-axis which is irrelevant for most applications). Asai [18.3] proposed the use of an elliptically shaped laser beam in order to measure the flow direction, but still diffusion and flow are not effectively separated by this method.

However, the use of two volume elements in laser velocimetry to generate start and stop signals triggered by particles flowing or flying by already possesses the basic idea to expand the standard FCS setup towards a two-fold volume element detection apparatus. In addition, the powerful mathematical tool of cross-correlation analysis then delivers thorough transport information on a micrometer and microsecond scale.

18.2   The Experimental Setup

The experimental setup (Fig. 18.1) is an easy-to-build expansion of a standard FCS setup. Basically, the laser beam has to be doubled by a polarized beamsplitter in order to create two practically parallel laser beams. The $\lambda/2$ plate helps to adjust the relative intensities of the two beams. The distance between the laser beams before reaching the first focusing lens is usually

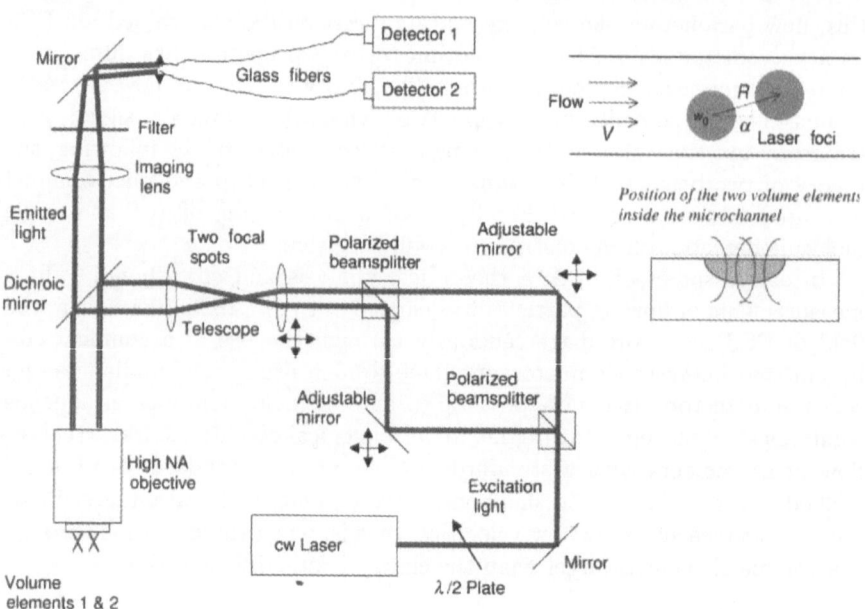

Fig. 18.1. Two-beam FCS setup as used in the experiments. The distance of the two volume elements is typically between 1 and 10 μm

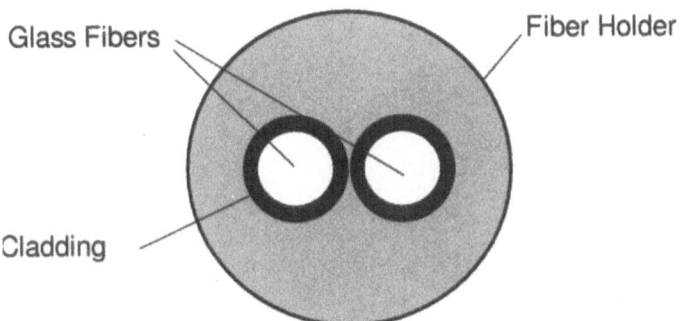

Glass Fibers

Fiber Holder

Cladding

Fig. 18.2. Glass fibers aperture for two-beam FCS

smaller than their diameter so that they appear as a single beam. Through minute alterations of the right angle deflection at the mirrors two (through the oculars) clearly visible separated focal spots below the objective can be created.

From these spots, the emerging fluorescent light is then collected by the objective, passes the dichroic mirror, a band-pass filter and the imaging lens. The pinhole which defines the exact dimensions of the volume element is replaced by the ends of two separate multimode fibers (Fig. 18.2). In the experiments shown here, the openings of the two fibers are separated by a fixed cladding (center-to-center distances are typically between 60–140 μm). This aperture is mounted in an x-y-z-α-controllable way on the optical rail. The other ends of the two fibers are directly coupled to two avalanche photon detectors (EG&G, version SPCM–AQ 131–FS). Their signals are directed to the PC-adapted correlator card (ALV–5000) which calculates the cross-correlation function (volume element #1 against volume element #2 and vice versa simultaneously).

In the experiments described here, the two volume elements are placed inside a transparent microchannel, either being a simple uncoated fused-silica capillary (inner diameter 2–25 μm, outer diameter 150 μm; the outer poly-acrylamide coating is removed by a flame) or custom-made microstructures made out of PMMA [18.4] or glass [18.5]. The channels are filled by capillary forces or slight pressure with an aqueous tetramethylrhodamine (TMR) solution (≈ 1 nM concentration), and the flow is induced either by electroosmotic force or hydrostatic pressure. Figure 18.3 shows a glass chip mounted with a plexiglass holder onto a disc. The disc itself is mounted onto a x-y-z table. A piezo element with sub-100 nm resolution (Physik Instrumente, P 721–10, controller 862 LVPZ) is added between the objective and the setup mount in order to adjust the height of the focus precisely.

When electroosmotic flow is used, one needs DC voltages up to 5 kV, depending on the length and shape of the channel. For security purposes, the

Fig. 18.3. Microchannel holder. The microchannel is inside a quartz strip which is fitted into the black holder. The mount on top also supplies the electric voltage via the thin silver wires. The holder and the disc have a large central hole in order to give the objective access to the channel and to let the laser light pass

objective should always be grounded and the maximum current should be limited to a value below 1 mA.

In free-hanging droplets under normal working conditions, signal-to-noise ratios of the order of 100:1 are normally achieved. Due to the microstructures, this ratio can decrease because of two reasons: intrinsic fluorescence of the bulk material and scattering effects on the surfaces. We found that SiO_2 and PMMA are suited equally well with respect to background minimization. Figure 18.4 shows the intensity signal as a function of the relative position of the laser focus in the channel.

A number of wall effects, especially adhesion, are also present in microstructures. A distinctive indication of such a phenomenon is the much longer diffusion time for molecules near the channel walls as determined by FCS (Fig. 18.5). Diffraction spots of single molecules can be seen by eye to remain for fractions of seconds in the laser focus indicating that the molecules diffuse slower close to the wall than in free solution.

18.3 Theory

The definition of the cross-correlation function is

$$G(\tau) = \frac{\iint W_1(\boldsymbol{r})W_2(\boldsymbol{r}')\phi(\boldsymbol{r},\boldsymbol{r}',\tau)\mathrm{d}^3r\,\mathrm{d}^3r'}{\bar{C}^2\int W_1(\boldsymbol{r})\mathrm{d}^3r\int W_2(\boldsymbol{r}')\mathrm{d}^3r'} \tag{18.2}$$

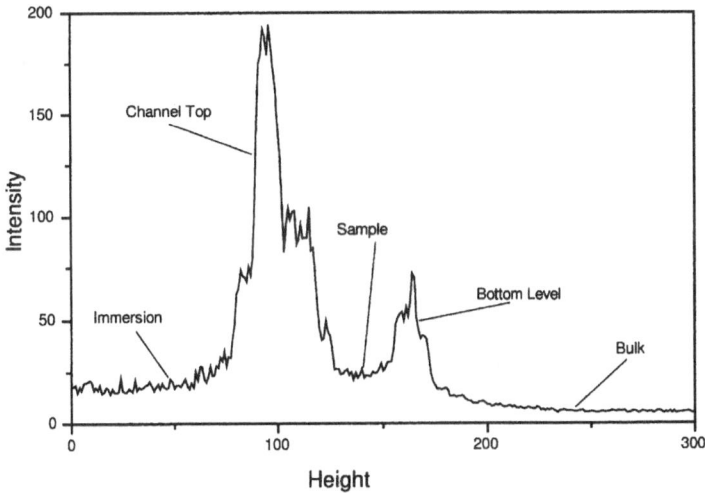

Fig. 18.4. Dependence of the measured intensity on the position of the laser focus in the microstructure. Height and intensity are given in arbitrary units. In this experiment the channel had a depth of $15\,\mu m$

where $\bar{C}$ is the average concentration of the fluorophores, ϕ the fluctuation function, and W_1 and W_2 are the description of the two separated volume elements, known as molecule detection efficiency function (MDE):

$$W_1(\boldsymbol{r}) = \kappa I_{01} \exp\left(-2\frac{\rho^2}{w_0^2}\right) \exp\left(-2\frac{z^2}{z_0^2}\right), \tag{18.3a}$$

$$W_1(\boldsymbol{r}') = \kappa I_{02} \exp\left(-2\frac{(\rho' - \boldsymbol{R})^2}{w_0^2}\right) \exp\left(-2\frac{z'^2}{z_0^2}\right), \tag{18.3b}$$

with $\rho = (x, y)$. As usual, the MDEs are approximated by a three-dimensional Gaussian distribution [18.6]. I_{01} and I_{02} are the excitation intensities of the beams, and κ is the system specific detection parameter. The two volume elements are separated by the vector $\boldsymbol{R} = (R_x, R_y)$; there is no shift along the optial axis.

The experiments are normally performed under conditions where the flow is constant during the measuring time and uniform in the region of the two volume elements. With this, the standard diffusion equation changes to

$$\partial_t \delta C(\boldsymbol{r}, t) = D\Delta\delta C(\boldsymbol{r}, t) - \boldsymbol{V}\nabla\delta C(\boldsymbol{r}, t), \tag{18.4}$$

with $\boldsymbol{V}$ being a uniform constant flow. In this case, there exists an exact solution of the fluctuation function, either generated by some tedious calculation or by a simple Galilei transformation $\boldsymbol{r}' \to \boldsymbol{r}' - \boldsymbol{V}t$ of the fluctuation function of the pure diffusion case:

$$\phi(\boldsymbol{r}, \boldsymbol{r}', \tau) = \bar{C}(4\pi D\tau)^{-3/2} \exp\left[-\frac{(\boldsymbol{r} - \boldsymbol{r}' + \boldsymbol{V}\tau)^2}{4D\tau}\right]. \tag{18.5}$$

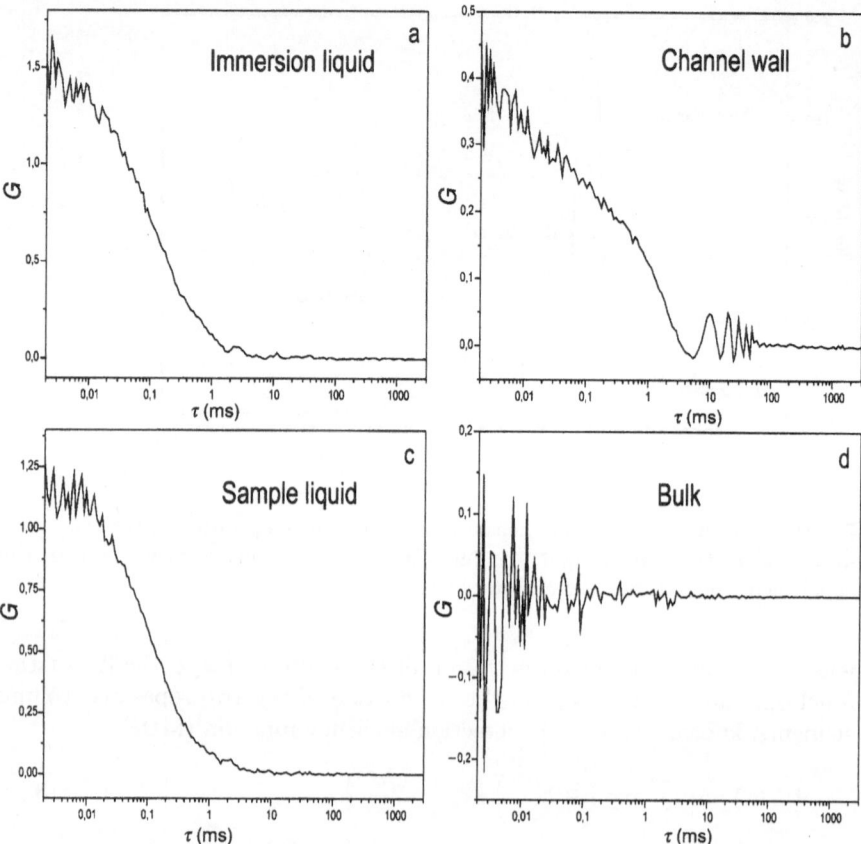

Fig. 18.5 a–c. Autocorrelation functions from different positions of the volume element with respect to the microstructure. The adhesion effect is clearly visible **b**. Due to the high reflectance of the channel wall (Fig. 18.4), the 50 Hz power supply effect on the laser beam is also visible. Note that the immersion liquid **a** shows a positive autocorrelation, too, which is due to the ever-present residues of dyes in the lab from former experiments sticking at the objective. This is not a problem as long as the focus is positioned inside the sample liquid

If (18.3) and (18.5) are used for the normalized correlation function (18.2), one obtains [18.7]

$$G(\tau) = \left[N \left(1 + \frac{\tau}{\tau_{\mathrm{D}}} \right) \sqrt{1 + \frac{w_0^2}{z_0^2} \frac{\tau}{\tau_{\mathrm{D}}}} \right]^{-1}$$

$$\exp \left[-\frac{1}{w_0^2} \frac{(\boldsymbol{V}_\rho - \boldsymbol{R})^2}{1 + (\tau/\tau_{\mathrm{D}})} - \frac{1}{z_0^2} \frac{V_z^2 \tau^2}{1 + (w_0^2 \tau/z_0^2 \tau_{\mathrm{D}})} \right] . \tag{18.6}$$

Here, $N = \bar{C} w_0^2 z_0 \pi^{1,5}$ is the weighted average number of fluorophores in one volume element, $\tau_{\mathrm{D}} = w_0^2/4D$ is the characteristic diffusion time and $\boldsymbol{V}_\rho$ the

x- and y-dependent term of the flow velocity, which is separated from $\mathbf{V}_z$, the z-dependent term. Since the flow in the z-direction was negligible in our experiments, the second term in the exponential becomes zero. We define the flow time $\tau_F = R/V_\rho$ as the ratio of the distance of the two foci and the two-dimensional flow velocity; α is the angle between the direction of the flow and the connecting vector of the two volume elements (Fig. 18.1). With these simplifications, (18.6) yields

$$G(\tau) = \left[N \left(1 + \frac{\tau}{\tau_D} \right) \sqrt{1 + \frac{w_0^2}{z_0^2} \frac{\tau}{\tau_D}} \right]^{-1}$$

$$\exp\left[-\frac{V^2}{w_0^2} \frac{\tau^2 + \tau_F^2 - 2\tau\tau_F \cos\alpha}{1 + (\tau/\tau_D)} \right]. \tag{18.7}$$

For the limiting case of zero diffusion, the cross-correlation function simplifies to

$$G(\tau) = \frac{1}{N} \exp\left[-\frac{(\mathbf{V}\tau - \mathbf{R})^2}{w_0^2} \right]$$

$$= 1 + \frac{1}{N} \exp\left[-\frac{R^2}{w_0^2} \left(\frac{\tau^2}{\tau_F^2} + 1 - 2\frac{\tau}{\tau_F} \cos\alpha \right) \right]. \tag{18.8}$$

For experiments with extremly slow diffusing particles, this equation is sufficient for the evaluation of flow parameters from the correlation function. In any case, the cross-correlation function increases from

$$G(\tau = 0) = \frac{1}{N} \exp\left(-\frac{R^2}{w_0^2} \right) \tag{18.9a}$$

to a maximum value of $G_c(t_{max})$ and decreases to zero for very high values of τ. For the case of zero diffusion, the relative maximum of the cross-correlation function is found at $\tau_{max} = \tau_F \cos\alpha$ with

$$G(\tau_{max}) = \frac{1}{N} \exp\left(-\frac{R^2}{w_0^2} \sin^2\alpha \right). \tag{18.9b}$$

If diffusion cannot be neglected, the maximum is found at

$$\tau_{max} = -\left[\tau_D + \frac{\tau_F^2}{2\tau_D} \left(\frac{R}{W_0} \right)^{-2} \right] \tag{18.10}$$

$$+ \sqrt{\left[\tau_D + \frac{\tau_F^2}{2\tau_D} \left(\frac{R}{W_0} \right)^{-2} \right]^2 + \tau_F^2 \left[1 + \left(\frac{R}{w_0} \right)^{-2} \right] + 2\tau_D\tau_F \cos\alpha}.$$

As long as the two volume elements are sufficiently separated, i.e. $R > w_0$, $G(\tau)$ increases from $G(0) = 1/N \exp(-R^2/w_0^2)$ to a maximum value and decreases to zero for $\tau \gg \tau_{max}$.

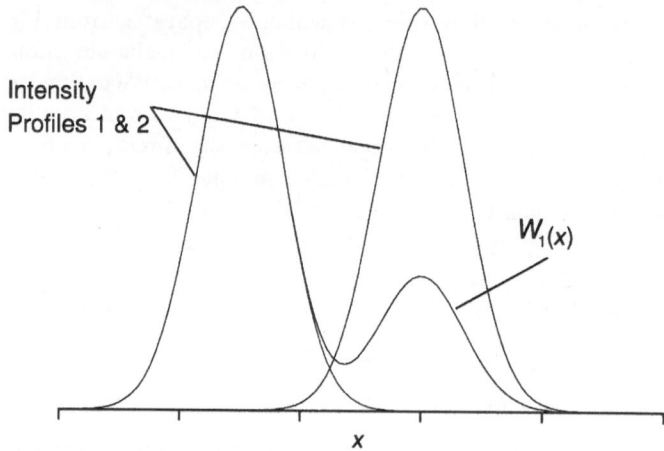

Fig. 18.6. Functional dependence of the molecule detection efficiency (MDE) perpendicular to the optical axis

18.3.1   Pseudo-Autocorrelation

For many experiments, (18.4) is a sufficient approximation to produce correct answers regarding the transport properties of the particles observed. These two equations state that each detector "sees" only those photons which have emerged from its assigned volume element. In reality, photons from the other focal spot also reach the detector; the closer the spots, the more "cross-talk" exists. When such a photon arrives at the "wrong" detector and is cross-correlated with a signal triggered by a photon from the same volume element (but which arrived at its assigned detector), one effectively gets an autocorrelation. Since, from the data processing point of view a cross-correlation has been performed, we term this effect "pseudo-autocorrelation".

Figure 18.6 scetches the real functional behavior of the MDE function. For the derivation of a quantitative description it is sufficient to perform a first-order approximation for the MDE function by introducing an overlap factor k:

$$W_1(\boldsymbol{r}) = I_1(\boldsymbol{r}) + kI_2(\boldsymbol{r}) \,, \tag{18.11a}$$

$$W_2(\boldsymbol{r}') = kI_1(\boldsymbol{r}') + I_2(\boldsymbol{r}') \,, \tag{18.11b}$$

with $k < 1$. Calculating $G(\tau)$ with this, one obtains [18.7]

$$G(\tau) = (1+k)^{-2}[G_{\mathrm{CC}}(\tau) + k^2 G_{\mathrm{CC}}(-\tau) + 2kG_{\mathrm{AC}}(\tau)] \,, \tag{18.12}$$

with G_{CC} as the original cross-correlation function, G_{AC} as the corresponding autocorrelation function

$$G_{\mathrm{AC}}(\tau) = G_{\mathrm{CC}}(\tau, R \equiv 0) \,, \tag{18.13}$$

and $G_{CC}(-\tau)$ is simply the cross-correlation for negative times, resulting from mathematically exchanging I_1 and I_2 in the k^2 term. Two things render this pseudo-correlation function useful for data evaluation. First, for the correlation signals generated by diffusion events, it does not matter whether signal #1 is correlated versus signal #2 or vice versa since diffusion is an isotropic transport process, resulting in $G(\tau) = G(-\tau)$. Second,

$$G_{CC}(\tau \le 0) < G_{AC}(\tau \le 0),\tag{18.14}$$

and one can neglect the k^2 term when estimating k. For $\tau = 0$ we obtain

$$\frac{G_{0AC}}{G_0} = \frac{(1+k)^2}{2k} > 1.\tag{18.15}$$

18.4   Experimental Procedures

18.4.1   Optimizing the Setup

Before carrying out flow experiments in microstructures, it is useful to optimize the setup via a free-hanging droplet measurement (i.e. pure diffusion). At the beginning, both beams should be adjusted singly to their respective detector so that both generated autocorrelation curves show the same shape, at the same time having an optimal number of detected photons per molecule $I_{1,2}/N$. In the next step, one of the beams is blocked and a one-minute cross-correlation measurement is performed. By this, one obtains the pure pseudo-autocorrelation function. After this follows the cross-correlation measurement with both beams unblocked (2–3 min measurement time). The overlap factor k is the ratio of the intensities measured by the two detectors during a pseudo-autocorrelation measurement; it should be equal, when using (18.15) and the amplitudes of the curves obtained by the experiments.

Since, in a pure diffusion experiment the cross-correlation direction does not matter, one should record both curves simultaneously (which is possible with the ALV correlator card) and average then the two curves. This effectively doubles the measurement time, thereby smoothing the function.

For data evaluation, the amplitude of the pseudo-autocorrelation curve has to be normalized to the G_0 value to the cross-correlation curve and subsequently subtracted. The resulting difference curve represents the pure cross-correlation function, which can be fitted to (18.7). Usually τ_D has been determined already during the adjusting experiments, so it can be held fixed for the fitting process. The main result of the fit is the separation of the two volume elements, R/w_0. For standard experiments, separations between 2.5 and 5.0 should be used.

18.4.2   Flow Measurements

This section shows typical experiments with TMR in microchannel structures. The channel used was 30 μm wide, about 10 μm deep and 30 mm long.

Voltages up to 4000 V were applied. Figure 18.7a shows a plot of a 4 kV measurement with $\alpha = 0$. The glass fibers used had a diameter of 100 μm and a center-to-center distance of 140 μm (or a 100/140 fiber, for short); the magnification of the objective was $M = 40$. As explained above, the back correlation curve is almost equal to the pseudo-autocorrelation, so it can be subtracted from the forward correlation curve in order to obtain the pure cross-correlation. The peak of the curve $G(\tau_{max})$ gives a good estimate of $\tau_F (\approx \tau_{max})$, but because of the strong influence of diffusion it is usually better to apply (18.7) (or the two-dimensional form) via a fit algorithm.

Depending on the optics, sometimes a distinct maximum cannot be seen at all before subtracting the pseudo-autocorrelation, as shown in Fig. 18.7b. Everything is the same as in the former experiment, except that a 50/60 fiber was used. So before doing the experiments, one should estimate the expected flow and diffusion times, choose, for a given objective, the appropiate glass fibers and then select the pre-focusing optics for optimal illumination of the two volume elements.

Diffusion becomes more important with decreasing flow velocities. The peak values $G(\tau_{max})$ decrease according to the flow (Fig. 18.8), due to a decreased probability that a molecule leaves the first volume element and subsequently enters the second one. Furthermore, diffusion also affects the shape of the curve: because of an increased random movement, the cross-correlation peak broadens. Connecting the peaks of all cross-correlation curves leads to the course of the autocorrelation function of diffusion within a volume element.

The velocity V is obtained via $V = \tau_F/R$. The distance R between the two volume elements can be calculated either via the center-to-center distance of the two apertures divided by the magnification M or via the focal spot diameter [18.7] multiplied by the separation value R/w_0, as obtained by the fitting procedure. For a well-aligned setup, both values should match.

The data of the determined flow velocity (Fig. 18.9) yield a linear dependence on the electric field, due to electroosmotic flow. The results are independent of the fibers and focusing lenses used. The slope of the resulting line is equivalent to the mobility of the molecules inside the channel, in this case $0.83 \, \text{cm}^2/(\text{kV s})$.

In addition to diffusion and flow, the angle α between the flow direction and the volume elements can be determined by two-beam FCS. Figure 18.10 shows the dependence of the (pure) cross-correlation curves on α. Obviously, an increasing angle results in a decrease of both the peak amplitude and its position, as predicted by the theory. τ_{max} is directly proportional to the component of the volume element connecting vector in the flow direction, whose value decreases with increasing angle α. On the other hand, the probability that a molecule leaves volume element #1 and subsequently is excited in volume element #2 decreases with increasing α. Taking, for example, the change

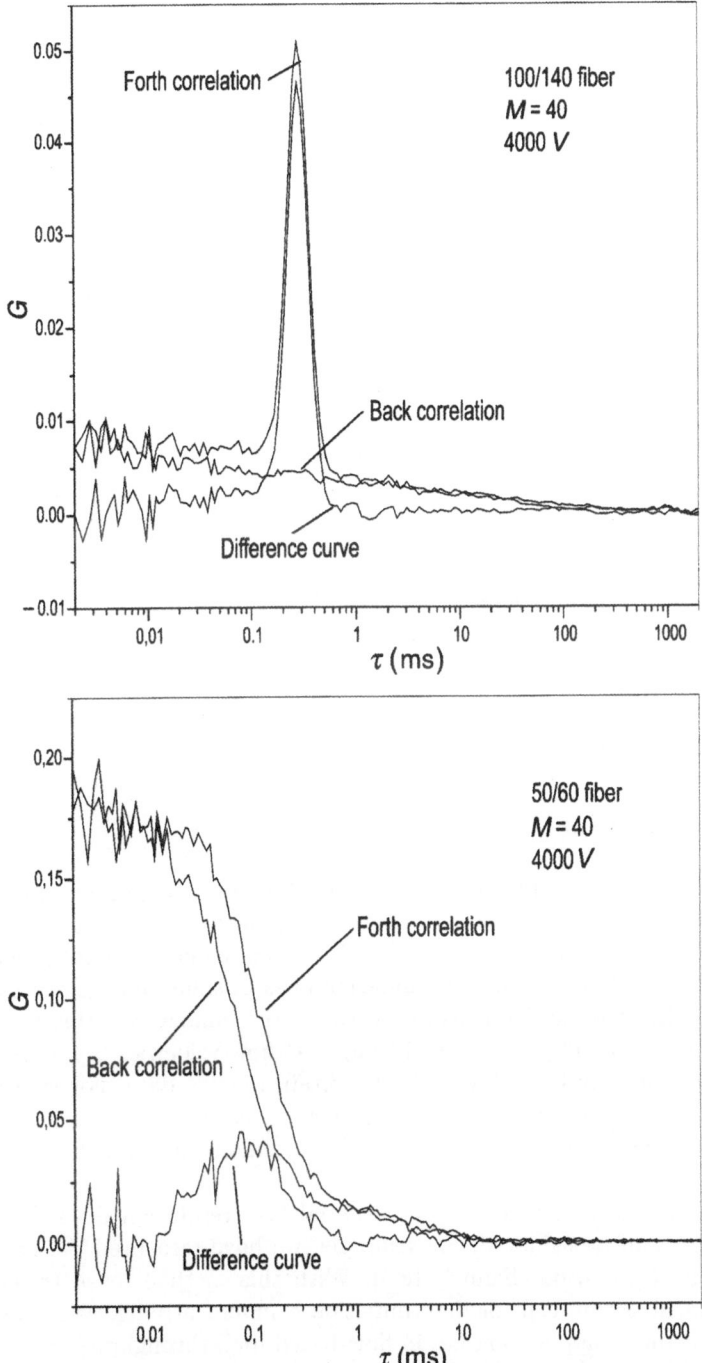

Fig. 18.7. Forth and back correlation as calculated by the ALV card. The difference curve is the actual cross-correlation curve to be used for the fitting process

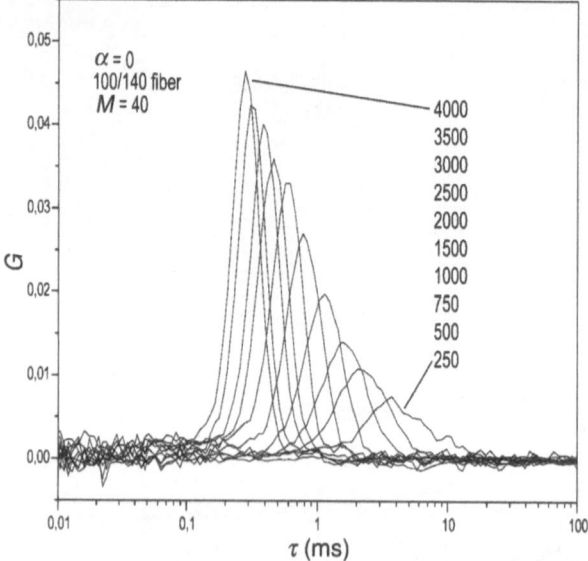

Fig. 18.8. Dependence of the cross-correlation on the applied voltage. The voltages in the legend are ordered according to the peaks

of the amplitude, theoretical calculations show a good agreement with the experimental data (Fig. 18.11).

18.5   Applications

The technique of two-beam FCS is a precise and simple method to determine different transport properties in microstructured channels or narrow capillaries on the single molecule level in solution with small amounts of sample. One major point is the fast and easy determination even of small flow velocities. The flow velocities that can be measured within a reasonable detection time of 10–100 s range from $100 \, \mu m/s$ up to $10 \, cm/s$. At flow velocities $> 1 \, mm/s$, the measuring time may be as low as 5 s to obtain reliable data. For values of $R/w_0 > 3$ at high velocities, the curves show a distinct maximum, and the influence of the third dimension as well as random diffusion becomes negligible.

Furthermore, due to the small foci, a spatial flow resolution down to a few micrometers can be achieved. For example, we have measured the flow profile inside a $15 \, \mu m$ wide channel [18.8]. With this method, even transport measurement within cells or determination of microturbulences should be possible. The main application lies in flow-based high-throughput screening techniques like single-molecule selection in a Y-shaped channel structure [18.7]. In the following, two possible applications, where transport measure-

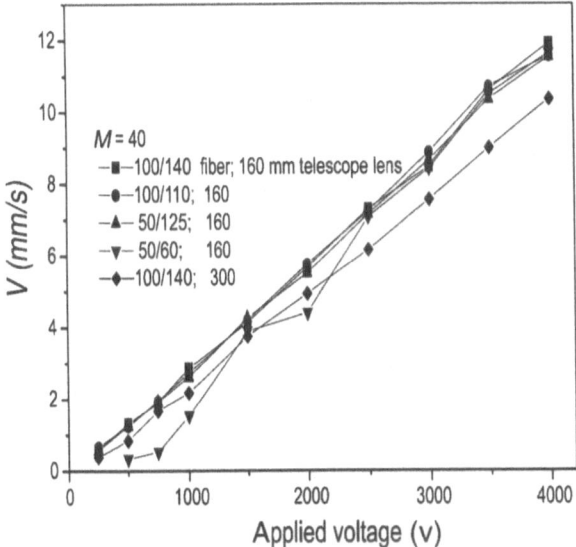

Fig. 18.9. Dependence of the measured flow velocity on the applied field. Five experiments with different fiber setups are shown. Ideally, the results should be independent of the fibers used. Here one can see that it is important to align the pre-focusing lens with the fiber aperture: in the 50/60;160 and the 100/140;300 experiments the focal spots generated by the pre-focusing lenses were actually too large so that there were no sharply defined and well-separated volume elements

ments are necessary for setting up the system or for the actual experiments, are presented.

18.5.1   Continuous Flow Kinetics

Imagine a Y- or T-shaped channel structure where two chemical components are brought in separately via two branches and are mixed and analyzed in the third branch. This resembles the first simple kinetic experiments: two components react in one of the branches with a given flow velocity V. The reaction begins at the branch point and proceeds with time as the mixed components flow along the branch. Assuming that the fluorescence properties of the reaction product differ from that of the educts (by intensity, color, triplet state lifetime, etc.), one can monitor the fluorescence at a certain distance d from the intersection. The fluorescence changes with time along the direction of flow as a function of the distance according $t = d/V$, enabling one to determine the kinetics of the reaction and eventually to gain information about the reaction mechanism. The drawback of this method has been the poor lateral resolution ($> 5\,\mathrm{mm}$), which limited the time resolution to $50\,\mathrm{ms}$ even at a flow velocity of $10\,\mathrm{cm/s}$. Even worse, the faster the flow, the more the sample is wasted during the experiment. To overcome this,

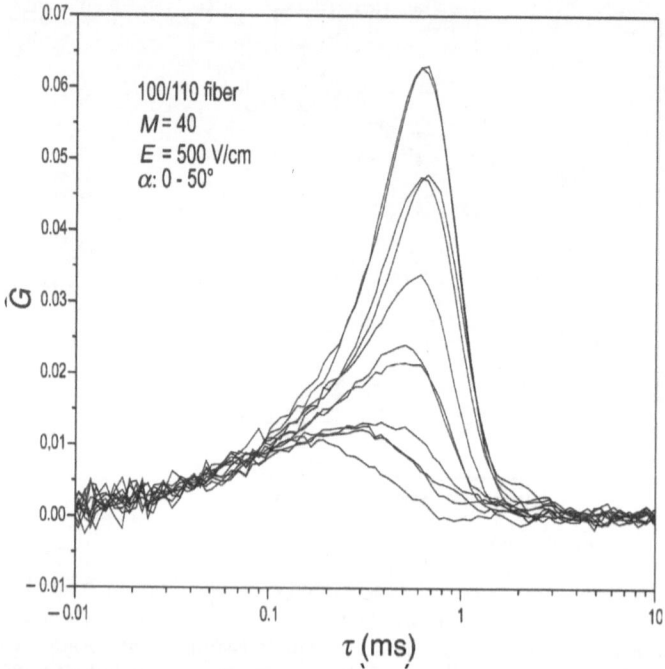

Fig. 18.10. Dependence of the cross-correlation function on the angle α. The highest peak corresponds to $\alpha = 0$, the lowest to $\alpha = 50°$

the method of "stopped flow" mixing experiments was developed [18.9]. It is still an important tool for measuring the kinetics of interactions between, for example, nucleic acids and proteins. However, with continuous flow mixing experiments within microstructures it becomes immediately obvious that one is dealing with a completely different order of magnitude. The size of the volume element is about $1\,\mu\text{m}$. Since it is no problem to achieve flow velocities on the order of $1\,\text{cm/s}$ in the microchannels, one can immediately obtain a time resolution of $100\,\mu\text{s}$ for non-equilibrium reactions, i.e. several orders of magnitude better than before. The amount of sample needed is a few μl of a nanomolar solution. Just by this simple estimation it becomes clear that the macroscopic setups for measuring kinetics can be scaled down to microstructure sizes.

One aim, for example, is to measure the exonucleic digestion of a DNA strand. For this, an intercalating dye is added to the DNA solution. The dye is non-fluorescent when it is not intercalated. Therefore the intensity of the fluorescent signal with respect to the distance d yields the length of the uncleaved DNA strand and thereby the cleavage rate. One major advantage over the macroscopic setups is that besides the average intensity additionally the fluctuations (and therefore changes in the diffusion constant or mobility) can be measured by FCS. With two-beam FCS, one can even measure the

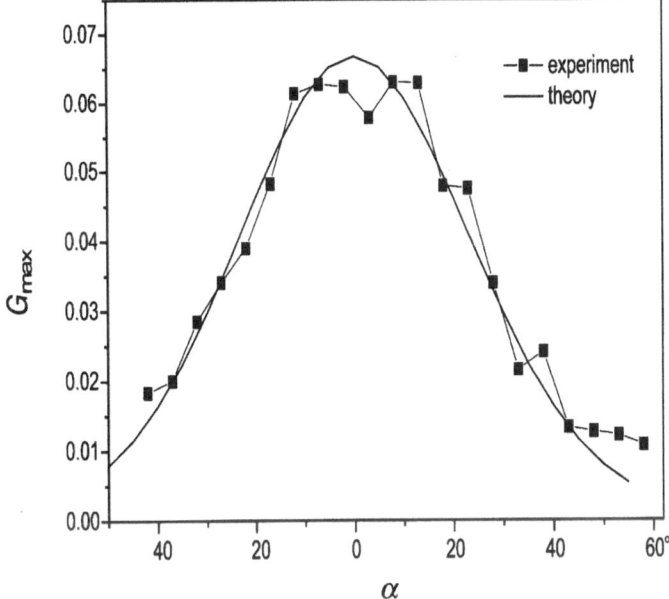

Fig. 18.11. Functional behavior of the cross-correlation amplitude with respect to the angle α; comparison between theoretical and experimental data

causal dependence between two reaction steps at times t_1 and t_2 by cross-correlating the fluctuations of the volume elements positioned at d_1 and d_2.

18.5.2 Rapid DNA Sequencing

Regarding commercial aspects, rapid DNA sequencing is by far the most interesting application of confocal spectroscopy in microchannel structures. It is a real single-molecule technique rather than a correlation, i.e. statistical, method: It uses a single DNA strand. The principle is as follows (Fig. 18.12; for a more detailed description, see [18.10]):

The first step is to label the four different bases via a PCR-like technique with four different dyes, say red, blue, yellow and green. One end of the DNA strand is attached to a bead of a few micrometer size. This bead is placed inside a microchannel, being held by laser trapping in the middle of the channel. Then an exonuclease enzyme is added to the solution. Once an enzyme attaches itself to the free end of the DNA, it starts to cleave the bases one-by-one from the strand. The free mononucleotides then flow to an elliptically shaped focal spot which covers the whole cross-section of the channel. In order to keep the background low, we use several glass fibers aligned in a row so that every detector sees only a part of the elliptical spot. With the help of dichroic mirrors, photons of different wavelengths are guided

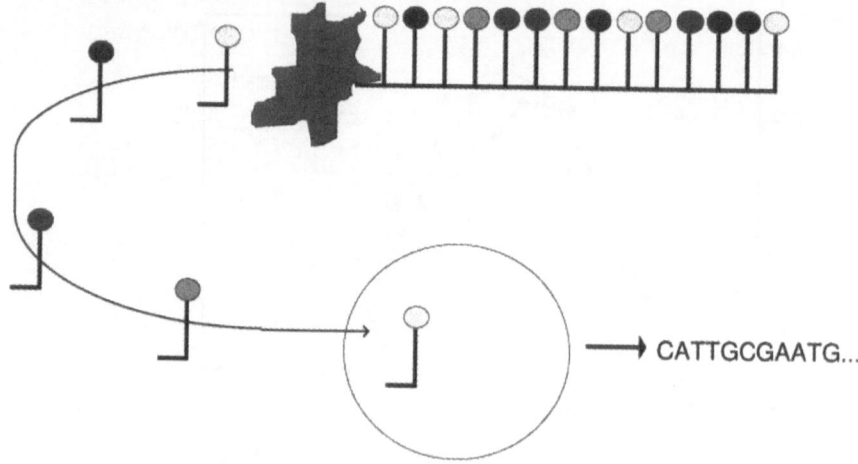

Fig. 18.12. Principle of single molecule DNA sequencing. The dye-tagged DNA strand is cleaved one-by-one by an exonuclease. After cleavage, the monomers are transported to the volume element (indicated by a circle) and detected. The whole process is performed inside a microchannel system

to different detectors leading to a colorwise distinction. A computer assigns the photon peaks to the respective bases, yielding the sequence of the DNA.

The speed of this sequencing method is basically determined by the cleavage rate of the exonuclease, about 100 bases per second. This means that a sequencing equipment of this type would be about thousand times faster than the best parallelized conventional sequencers today. It will still take much effort to achieve this formidable goal, but the results so far are very promising.

18.6 Conclusion

The application of FCS or, more general, of confocal spectroscopy in microstructures is not confined to measurements in microwells. The use of transparent microchannels will have a tremenduous affect on various analytical techniques like chemical kinetics and rapid DNA sequencing. Also, transportation measurements within living cells will become more and more important since one now is able to trace single molecules. In addition, the effect of combining other FCS techniques presented in this book with measurements in microstructures opens up a huge field of scientific research one could not have imagined just ten years ago and which still has to be explored.

Acknowledgement

Two-beam FCS and confocal spectroscopy in microstructures have been developed in a collaboration of the Max-Planck-Institute for Biophysical Chemistry, Göttingen, and the Karolinska Institute, Stockholm. Many thanks go to the heads of the two groups involved, Manfred Eigen and Rudolf Rigler, and to the numerous collaborators who supported this effort.

References

18.1 E.L. Elson and D. Madge: Biopol. **13**, 1–27 (1974)

18.2 E.L. Elson, D. Madge, and W.W. Webb: Biopol. **13**, 29–61 (1974)

18.3 H. Asai: Jpn. J. Appl. Phys. **19**(11), 2279–2282 (1980)

18.4 M. Lapczyna: Thesis, Vakuum-Ultraviolett-(VUV)-Laser-induzierte Mikrostrukturierung von Polymersubstraten für laserspektroskopische Anwendungen in der Bioanalytik, Univ. Kassel, Germany (1998)

18.5 J. Holm, H. Elderstig, O. Kristensen, R. Rigler: Nucleosides Nucleotides **16**, 557–562 (1997)

18.6 R. Rigler et al.: Eur. Biophys. J. **22**, 169–175 (1993)

18.7 M. Brinkmeier: Thesis, Fluoreszenz-Korrelations-Spektroskopie in Mikrostrukturen, Univ. Braunschweig, Germany (1996)

18.8 M. Brinkmeier, K. Dörre, K. Riebeseel, R. Rigler: Biophys. Chem. **66**, 229–239 (1997)

18.9 F. Bernasconi: Investigation of Rates and Mechanism of Reactions Part II, Tech. of Chem. Vol VI, 4th edn. (Wiley, New York)

18.10 K. Dörre, S. Brakmann, M. Brinkmeier, et al.: Bioimaging **5**, 139–152 (1997)

19 Introduction to the Theory
of Fluorescence Intensity Distribution Analysis

Peet Kask and Kaupo Palo

19.1 Introduction

The primary data of a fluorescence correlation experiment is a sequence of photon counts detected from a microscopic sample volume. An essential component of the fluorescence correlation analysis is the calculation of the second order autocorrelation function of detected photons. In this way a stochastic function of photon counts is transformed into a statistical function having an expected shape from which properties of the sample can be estimated. The calculation of the autocorrelation function, however, is not the only way to extract information about the sample from the sequence of photon counts. Another approach, based on collecting the distribution of the number of photon counts for a given time interval, was introduced into fluorescence fluctuation spectroscopy by Qian and Elson in 1990 [19.1].

The idea behind the fluorescence intensity distribution analysis (FIDA) can be well understood by imagining an ideal case in which a sample volume is uniformly illuminated and when there is almost never more than a single particle illuminated at a time, similar to the ideal situation in cell sorters. Under these circumstances, each time a particle enters the sample volume, the fluorescence intensity jumps to a value corresponding to the brightness of the particle. Naturally, the probability that this intensity occurs at an arbitrary time moment equals the product of the concentration of a given species and the size of the sample volume. Another fluorescent species which may be present in the sample solution produces intensity jumps to another value characteristic of this other species. In summary, the distribution of light intensity is, in a straightforward way, determined by the values of concentration and specific brightness[1] of each fluorescent species in the sample solution.

In reality, the intensity of fluorescence detected from a particle within a sample volume is not uniform but depends on the coordinates of the particle with respect to the focus of the optical system. Even though the calculation of a theoretical distribution of the number of photon counts is more complex for a bell-shaped profile than for a rectangular one, the distribution of the

---

[1] Under "specific brightness" we mean count rate of the detector from light emitted by a particle of given species situated in a certain point in the sample, conventionally in the point where the value of the spatial brightness profile function is unity.

number of photon counts sensitively depends on values of the concentration
and the specific brightness of fluorescent species. The measured distributions
of the number of photon counts can therefore be used for sample analysis.

The first successful realization of this kind of analysis was demonstrated
on the basis of moments of the photon count number distribution [19.2].
The k-th factorial moment of the photon count number distribution $P(n)$ is
defined as

$$F_k = \sum_n \frac{n!}{(n-k)!} P(n) \,. \tag{19.1}$$

In turn, factorial moments are closely related to factorial cumulants,

$$F_k = \sum_{l=0}^{k-1} C_l^{k-1} K_{k-l} F_l \,, \tag{19.2}$$

or

$$K_k = F_k - \sum_{l=1}^{k-1} C_l^{k-1} K_{k-l} F_l \,. \tag{19.3}$$

(C_l^ks are binomial coefficients, and K_ks are cumulants.) The basic expression used in moment analysis, derived for ideal solutions, relate k-th order
cumulant to concentrations (c_i) and specific brightness values (q_i)

$$K_k = \chi_k \sum_i c_i (q_i T)^k \,. \tag{19.4}$$

Here, χ_k is the k-th moment of the relative spatial brightness profile $B(r)^2$:

$$\chi_k = \int_{(V)} B^k(r) dV \,. \tag{19.5}$$

Qian and Elson used experimental values of the first three cumulants
to determine unknown parameters of the sample. The number of cumulants
which can be reliably determined from experiments is usually three to four.
This sets a limit to the applicability of the moment analysis.

In the pioneering publications on moment analysis, the idea that the count
number distribution could be directly fitted was also discussed. The present
chapter aims to present an adequate theory which has found a number of applications of the development in biochemical assays and drug screening [19.3].

---

[2] Usually in FCS, the unit of volume and the unit of B are selected which yield
$\chi_1 = \chi_2 = 1$. After selecting this convention, concentrations in our equations are
dimensionless, expressing the mean number of particles per "sample volume",
and the specific brightness of any species equals the mean count rate from a
particle if situated in the focus divided by the numeric value of $B(0)$. The value
of this constant is a characteristic of optical equipment. It can be calculated
from estimated parameters of the spatial brightness profile (see Sect. 19.2.4); it
is about 3.8 for our equipment.

19.2   Photon Count Number Distribution Corresponding to a Rectangular Sample Profile

A key to a successful realization of the photon count number distribution analysis is an adequate calculation of the expected count number distribution function. Let us first consider a simple theoretical case in which the light intensity reaching the detector from a particle as a function of coordinates of the particle is constant over the whole active volume of the sample, and zero outside it. Also, we assume that the diffusion of a fluorescent particle is negligible during the counting interval T. In this case, the distribution of the number of photon counts emitted by a single fluorescent species can be analytically expressed as a double Poisson: the distribution of the number of particles of a given species within this volume is Poisson, and the conditional probability of the number of detected photons corresponding to a given number of particles is also Poisson. The double Poisson distribution has two parameters: the mean number of particles in the active sample volume, c and the mean number of photons emitted by a single particle per dwell time, qT. The distribution of the number of photon counts n corresponding to a single species is expressed as

$$P(n; c, q) = \sum_{m=0}^{\infty} \frac{c^m}{m!} e^{-c} \frac{(mqT)^n}{n!} e^{-mqT} , \qquad (19.6)$$

where m runs over the number of molecules in the active volume. If $P_i(n)$ denotes the distribution of the number of photon counts from species i, then the resultant distribution $P(n)$ is expressed as

$$P(n) = \sum_{\{n_i\}} \prod_i P_i(n_i) \delta(n, \sum n_i) . \qquad (19.7)$$

This means that $P(n)$ can be calculated as a convolution of the series of distributions $P_i(n)$.

19.3   Photon Count Number Distribution Corresponding to an Arbitrary Sample Profile: The Convolution Technique

As in FCS, the rectangular sample profile is a theoretical model which can hardly be applied in experiments. The issue of which particular shape of the sample profile $B(r)$ should be selected as a model will be studied later in this chapter. Here, we will only study how the photon count number distribution of a single species can be calculated for a given sample profile. We may divide the sample into a large number of volume elements and assume that within each of them, the intensity of a molecule is constant. The contribution to the photon count number distribution from a volume element is therefore a double Poisson with parameters cdV and $qTB(\boldsymbol{r})$. (Here q denotes count-rate from a molecule in a selected standard position where $B = 1$, and $B(\boldsymbol{r})$ is the profile function of coordinates.) The overall distribution of the number of photon counts can be expressed as a convolution integral over double Poisson distributions. Note that integration is a one-dimensional rather than a three-dimensional problem here, because the result of integration does not depend on actual positions of volume elements in respect to each other. Figuratively, we may rearrange the three-dimensional array of volume elements into a one-dimensional array, for example in decreasing order of the value of B.

In a number of preliminary experiments in our laboratory, the photon count number distribution was indeed fitted, using the convolution technique. Our sample model consisted of twenty spatial sections, each characterized by its volume V_j and brightness B_j. However, a convenient and much faster computation technique exists, as described in the next subsection.

19.4   Photon Count Number Distribution Corresponding to an Arbitrary Sample Profile: The Technique of the Generating Function

The formal definition of the generating function of a distribution $P(n)$ is as follows:

$$P(\xi) = \sum_{n=0}^{\infty} \xi^n P(n) \,. \tag{19.8}$$

In particular, if one selects $\xi = e^{i\varphi}$, then the distribution $P(n)$ and its generating function $G(\varphi)$ are inter-related by the Fourier transform. What makes the generating function attractive in count number distribution analysis is the additivity of its logarithm: logarithms of generating functions of photon count number distributions of independent sources, like different volume elements as well as different species, are simply added for the calculation of the generating function of the combined distribution. (This is so because the transformation (19.8) maps distribution convolutions into the products of the corresponding generating functions.) Applying the definition (19.8) to formula (19.6) with $c \rightarrow cdV$ and $q \rightarrow qB(\boldsymbol{r})$, the contribution from a particular species and a selected volume element dV can be written as

$$G_i(\xi; dV) = \exp\left[c_i dV \left(e^{(\xi-1)q_i TB(\boldsymbol{r})} - 1\right)\right] \,. \tag{19.9}$$

Therefore, the generating function of the total photon count number distribution can be expressed in a closed form

$$G(\xi) = \exp\left[\sum_i c_i \int \left(e^{(\xi-1)q_i TB(\boldsymbol{r})} - 1\right) dV\right] \,. \tag{19.10}$$

Numeric integration according to (19.9) followed by a fast Fourier transform is, to our knowledge, the most effective means of calculating the theoretical distribution $P(n)$ corresponding to a given sample (i.e. given concentrations and specific brightness values of fluorescent species).

19.5   Sample Profile Models

The most widely used spatial profile model in FCS is the three-dimensional Gaussian profile with a single parameter of shape, the axial dimension ratio in longitudinal and radial directions. Logically, the first step in photon count number distribution analysis would be fitting a count number distribution obtained for single species like tetramethylrhodamine, see Fig. 19.1. The residuals curve with open circles illustrates the fit quality obtainable with the Gaussian profile. There are large systematic deviations in residuals. What is a sufficiently flexible model for fitting FCS data has turned out to be an absolutely inflexible and inadequate model for FIDA[3].

A model of the sample profile which has yielded a better fit of the measured distribution $P(n)$ is Gaussian–Lorentzian (see Fig. 19.1, open squares), but this model still lacks flexibility. Note that according to (19.8), a certain function of the spatial brightness B is integrated over the volume. In other words, it is a relationship between B and V, characterizing a given spatial brightness profile in FIDA. For example, the Gaussian profile yields the relationship

$$\frac{dV}{dx} \propto \sqrt{x}, \tag{19.11}$$

where we have denoted $x = -\ln B$. The Gaussian–Lorentzian profile yields the relationship

$$\frac{dV}{dx} \propto e^{x/4}\sqrt{\sinh \frac{x}{2}}. \tag{19.12}$$

Both of the relationships are inflexible, i.e. they do not provide any spatial shape parameters to adjust the theoretically calculated distribution to fit the measured data.

When looking for sufficiently flexible models to fit experimental data, we arrived at the following relationship:

$$\frac{dV}{dx} \propto (x + a_1 x^2 + a_2 x^3). \tag{19.13}$$

There is a formal rather than a physical model behind (19.13). The fit quality obtainable with (19.13) is illustrated by the filled squares curve of Fig. 19.1.

---

[3] Indeed, moments of the three-dimensional Gaussian profile do not depend of axial dimension ratios but only on their product which is an absolute measure of the sample volume.

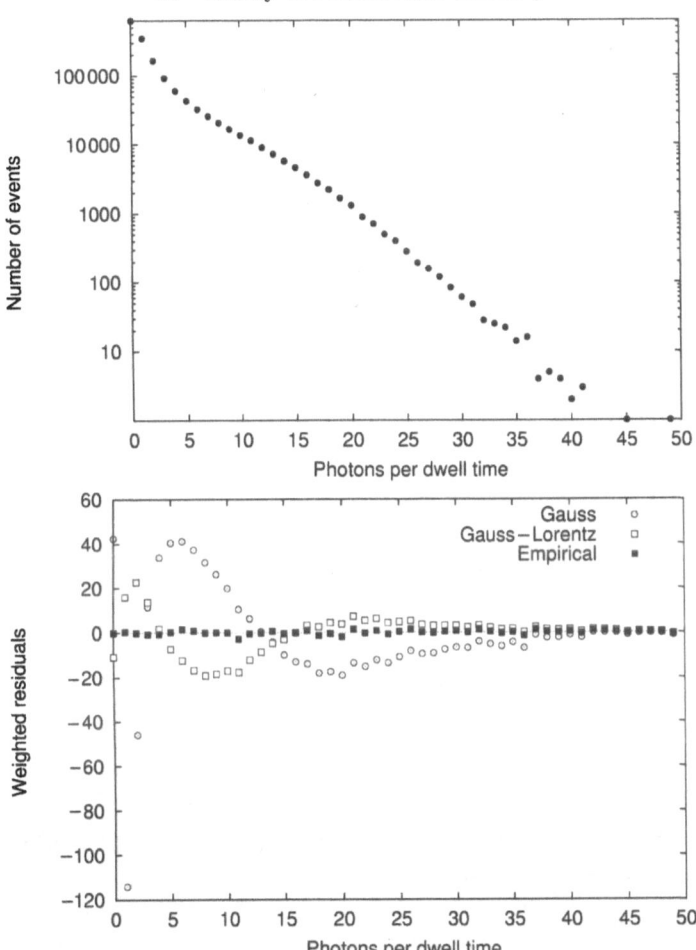

Fig. 19.1. A distribution of the number of photon counts measured at $T = 40\,\mu s$ for about 10^{-9} M solution of tetramethylrhodamine in water, data collection time 60 s. Three residuals curves are presented, corresponding to the best fit obtained with (19.11), (19.12), and (19.13)

19.6 Distribution of the Specific Brightness Within a Species

Some fluorescent species may have significantly wide distribution of specific brightness. For example, vesicles, which are likely to have a significantly broad size distribution and a random number of receptors, may have trapped a random number of labeled ligand molecules. In order to fit count number distributions for samples containing such species, we have modified (19.10) in the following manner. We assume that the distribution of brightness of particles q within a species is mathematically expressed as follows:

$$\rho(q) \propto q^{a-1}e^{-bq} .$$ (19.14)

This expression has been selected for the sake of convenience: all moments of this distribution can be analytically calculated, using the following formula:

$$\int_0^\infty x^a e^{-bx}\,\mathrm{d}x = \frac{\Gamma(a+1)}{b^{a+1}} .$$ (19.15)

It is straightforward to derive the modified generating function of a photon count number distribution. Following the arguments of Sect. 19.2.3 one can rewrite (19.9) as follows:

$$G(\xi) = \exp\left[\sum_i c_i \int \mathrm{d}V \int_0^\infty \mathrm{d}q \rho(q; a_i, b_i)\left(e^{(\xi-1)qTB(\mathbf{r})} - 1\right)\right],$$ (19.16)

where

$$\rho(q; a, b) = \frac{b^a}{\Gamma(a)}q^{a-1}e^{-bq} .$$ (19.17)

The integral over q can be performed analytically:

$$G(\xi) = \exp\left\{\sum_i c_i \int \mathrm{d}V\left[\left(\frac{b_i}{b_i - (\xi-1)TB(x)}\right)^{a_i} - 1\right]\right\},$$ (19.18)

The parameters a_i and b_i are related to the mean brightness $\bar{q}_i$ and the width of the brightness distribution σ_i^2 by

$$a_i = \frac{\bar{q}_i^2}{\sigma_i^2}, \quad b_i = \frac{\bar{q}_i}{\sigma_i^2} .$$ (19.19)

19.7 Weighting in FIDA

In the range of obtained count numbers, the probability to obtain a particular count number usually varies by many orders of magnitude, see for example the distribution of Fig. 19.1. Consequently, the variance of the number of events with a given count number has a strong dependence on the count number. To determine weights for least squares fitting, let us assume for simplification that light intensities in all counting intervals are independent. Under this assumption, we have a problem of distributing M events over choices of different count numbers n, each particular outcome having a given probability of realization, $P(n)$. Covariance matrix elements of the distribution can be expressed as follows:

$$\langle \Delta P(n)\Delta P(m)\rangle = \frac{P(n)\delta(n,m) - P(n)P(m)}{M},$$ (19.20)

where M is the number of counting intervals per experiment.

For a further simplification, one may ignore the second term on the right side of (19.20), which can be interpreted as a consequence of normalization (see Appendix). In this case, the weights simply equal the inverse values of the diagonal covariance matrix elements

$$W_n = \frac{M}{P(n)}.$$ (19.21)

19.8   Data Simulation Algorithms

Data simulation is a convenient means of testing FIDA algorithms and debugging the corresponding computer programs. As an illustration, in Fig. 19.2, a count number distribution is simulated for a case which models a binding reaction of a labeled ligand to vesicles. The distribution with open circles corresponds to the free ligand alone; the distribution with open squares corresponds to labeled vesicles; and the distribution with filled squares corresponds to their mixture. Concentrations and specific brightness values have been selected to model a realistic situation in drug screening.

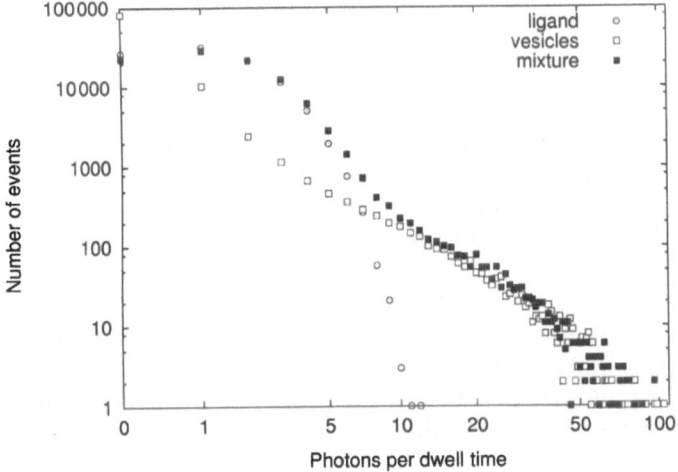

Fig. 19.2. Simulated distributions of the number of counts corresponding to $T = 40\,\mu s$, values of the spatial parameters of (19.13) $a_1 = -0.4$; $a_2 = 0.08$, background count rate $b = 1.0\,kHz$, data collection time 4 s. Curve *"ligand"* corresponds to a species of $c = 6.0$; $q = 6.0\,kHz/particle$; $\sigma_q = 0$. Curve *"vesicles"* corresponds to a species of $c = 0.05$; $\bar{q} = 300.0\,kHz/particle$; $\sigma_q = 150.0$. Curve *"mixture"* corresponds to their mixture. Fitting of curve (c) mixture returns the values of the five parameters characterizing the given "sample" with statistical errors which are mostly between 3.5 and 6%, except the error of σ_q of vesicles which is 13%. If, however, σ_q of vesicles is fixed in fitting, all the statistical errors are below 4%

For the fastest data simulation algorithm, we calculate the expected distribution and generate a random Poisson number of events for each value of n independently[4]. As a cosmetic error, the total number of events $\sum_n S(n)$ may slightly deviate from the pre-given number M.

A slower but a straightforward data simulation algorithm is the generation of a random configuration of particles in volume elements contributing to fluorescence, the calculation of the classical light intensity corresponding to the given configuration of particles, and the generation of a random Poisson number corresponding to this intensity, as a simulated number of photon counts. This procedure is repeated M times to obtain a simulated count number distribution.

19.9 Statistical Errors of Estimated Parameters

In general, a linear or linearized least squares fitting returns not only the values of the estimated parameters, but also their covariance matrix, provided the weights have been meaningfully set. It may turn out to be possible to express the statistical errors of the estimated parameters analytically in some simple cases (e.g., for the rectangular sample profile and single species) but in applications at least two-component analysis is usually of interest. Therefore, we have been satisfied with the numerical calculations of statistical errors. In addition to the "theoretical" errors with the assumption of uncorrelated measurements (19.20), in some cases we have determined statistical errors empirically, making a series of about a hundred FIDA experiments on identical conditions. As a rule, empirical errors are higher than theoretical ones by a factor of three to four. Empirical errors appear to be closer to the theoretical ones in scanning experiments. Therefore, we are convinced that the main reason for the underestimation of theoretical errors is the assumption of uncorrelated measurements.

Figures 19.3–19.8 illustrate how theoretical errors depend on the counting interval T, concentrations and brightness values. Table 19.1 compares statistical errors of parameters estimated by fitting a photon count number distribution (FIDA) and by the moment analysis. FIDA is overwhelmingly better than the moment analysis if the number if estimated parameters is higher than three.

---

[4] Generation of a random Poisson number is the following: For a given expected value of events E, a simulated number of events S is determined from a routinely generated random number R between 0.0 and 1.0 through inequations
$$\sum_{i=0}^{S-1} \text{Poisson}(i; E) < R \leq \sum_{i=0}^{S} \text{Poisson}(i; E).$$

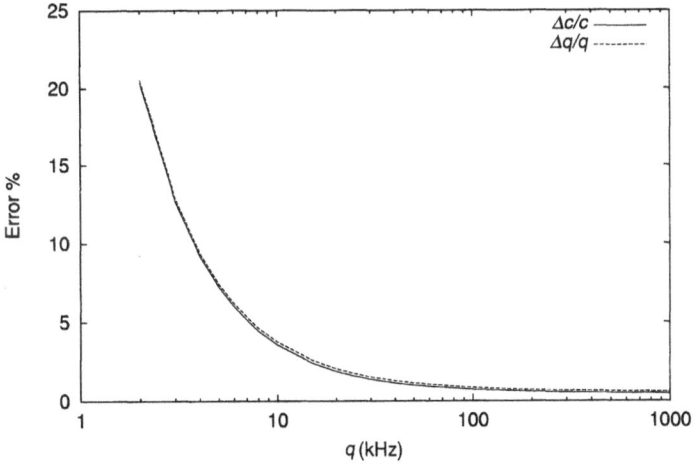

Fig. 19.3. Theoretical errors of the estimated parameters c and q of a solution of single species, depending on the value of q. The following values of experimental parameters were selected: $c = 1.0$; $T = 20\,\mu s$; $a_1 = -0.4$; $a_2 = 0.08$; $b = 1.0\,kHz$; data collection time 2 s

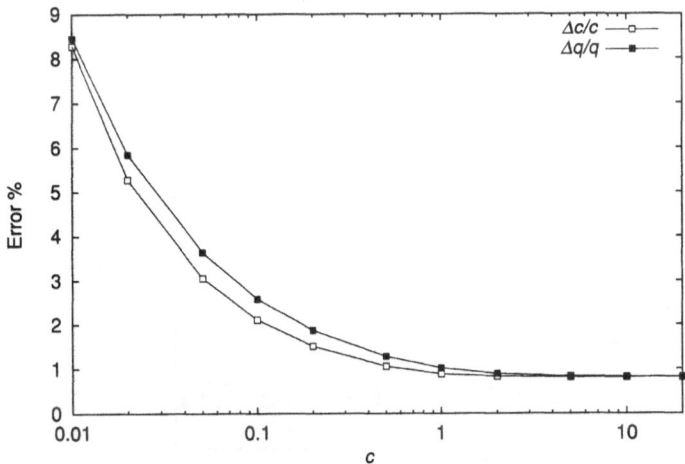

Fig. 19.4. Theoretical errors of the estimated parameters c and q of a solution of single species, depending on the value of c. The following values of experimental parameters were selected: $q = 60\,kHz/particle$; $T = 20\,\mu s$; $a_1 = -0.4$; $a_2 = 0.08$; $b = 1.0\,kHz$; data collection time 2 s

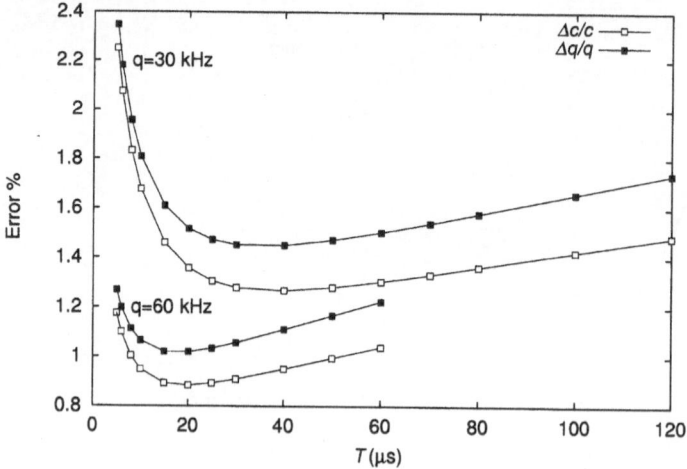

Fig. 19.5. Theoretical errors of the estimated parameters c and q of a solution of single species, depending on the value of T. The following values of experimental parameters were selected: $c = 1.0$; $q = 30.0\,\text{kHz/particle}$ (*upper graphs*); $q = 60.0\,\text{kHz/particle}$ (*lower graphs*); $a_1 = -0.4$; $a_2 = 0.08$; $b = 1.0\,\text{kHz}$; data collection time 2 s

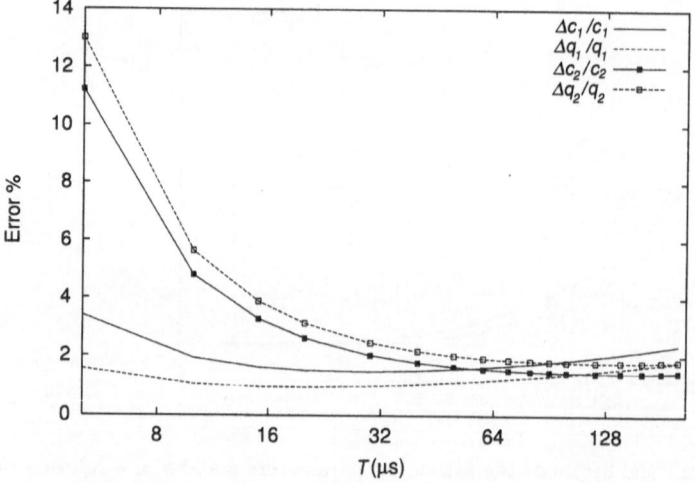

Fig. 19.6. Theoretical errors of the estimated parameters c and q of a mixture of two species, depending on the value of T. The following values of experimental parameters were selected: $c_1 = 0.1$; $c_2 = 2.0$; $q_1 = 200.0\,\text{kHz/particle}$; $q_2 = 10.0\,\text{kHz/particle}$; $a_1 = -0.4$; $a_2 = 0.08$; $b = 1.0\,\text{kHz}$; data collection time 10 s. Note that the optimal value of T for the determination of the parameters of the brighter species is lower than that of the darker species

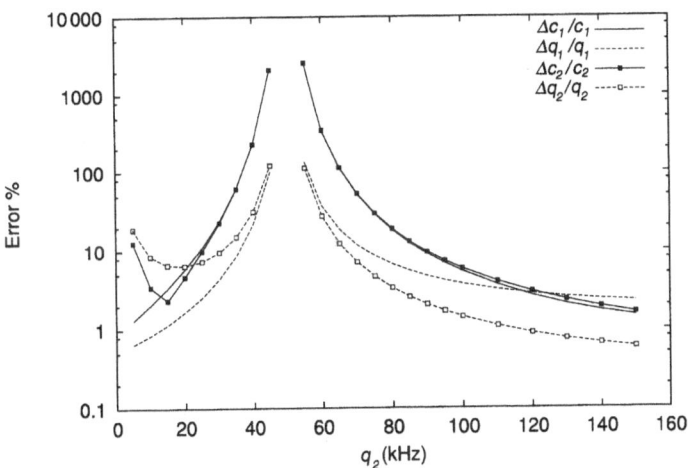

Fig. 19.7. Theoretical errors of the estimated parameters c and q of a mixture of two species, depending on the ratio of q_2 to q_1. The following values of experimental parameters were selected: $q_1 = 50.0\,\text{kHz/particle}$; $c_1 = c_2 = 0.5$; $a_1 = -0.4$; $a_2 = 0.08$; $b = 1.0\,\text{kHz}$; data collection time $40\,\text{s}$

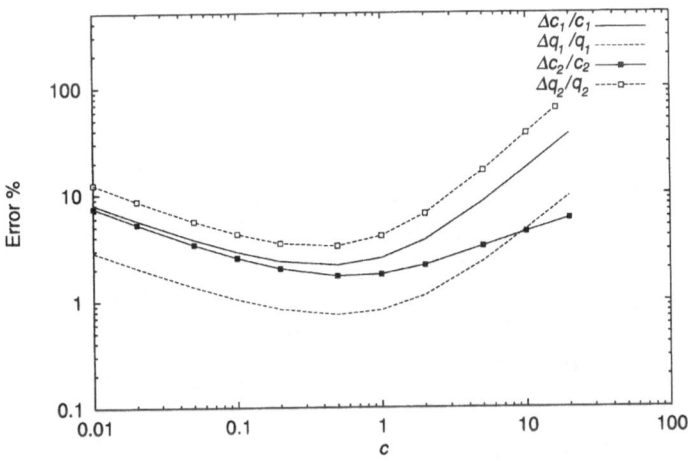

Fig. 19.8. Theoretical errors of the estimated parameters c and q of a mixture of two species, depending on concentrations. The concentrations were changed synchronously, $c_1 = c_2$. The following values of experimental parameters were selected: $q_1 = 75.0\,\text{kHz/particle}$; $q_2 = 25.0\,\text{kHz/particle}$; $a_1 = -0.4$; $a_2 = 0.08$; $b = 1.0\,\text{kHz}$; data collection time $60\,\text{s}$. Note that an optimal concentration exists at about one particle per sample volume. This is generally true, except if less than three parameters are to be determined

Table 19.1. Statistical errors of estimated parameters from least squares fitting of photon count number distributions (FIDA) and from moment analysis. Error values are determined through processing a series of simulated distributions

Data collection time (s)	Time window (μs)	Number of species	Number of estimated parameters	Parameter specification	Value (qs in kHz)	Error of FIDA (%)	Error of moment analysis (%)
10.0	40.0	1	2	c	0.5	0.59	0.54
				q	60.0	0.56	0.51
10.0	40.0	2	3	c_1	0.05	2.00	2.71
				q_1	150.0	1.54	1.89
				c_2	3.0	0.53	0.62
				q_2	5.0 (fixed)		
10.0	40.0	2	4	c_1	0.05	2.26	4.99
				q_1	150.0	1.63	2.53
				c_2	3.0	3.18	17.8
				q_2	5.0	3.35	14.9

Appendix

The dispersion matrix (19.20) corresponds to the multinomial distribution of statistical realizations of histograms. We will show here that the Poisson distribution, with the constraint that the total number of counting intervals M is fixed, will lead to the multinomial distribution. This is the rationale behind using Poisson weights as given in (19.21).

Let n_k be the expectation value of the number of events of counting k photons and let $N = \sum_k n_k$ be their sum. Let m_k be a statistical realization with $N = \sum_k m_k$. Assume that realizations $m_0, m_1, \ldots$ obey Poisson statistics

$$P(m_0, m_1, \ldots) = \frac{[n_0, n_1, \ldots]^{[m_0, m_1, \ldots]}}{[m_0, m_1, \ldots]!} e^{-N},\qquad(19.22)$$

where we have introduced the notation $n_0^{m_0} n_1^{m_1} \ldots \equiv [n_0, n_1, \ldots]^{[m_0, m_1, \ldots]}$. The probability of having the total of M events is

$$p(M) = \frac{N^M}{N!} e^{-N}.\qquad(19.23)$$

The conditional probability of having $m_0, m_1, \ldots$ events if there is a total of M events is

$$P(m_0, m_1, \ldots | M) \equiv \frac{P(m_0, m_1, \ldots)}{P(M)} = \frac{M! [n_0, n_1, \ldots]^{[m_0, m_1, \ldots]}}{N^M [m_0, m_1, \ldots]!},$$

or

$$P(m_0, m_1, \ldots | M) \equiv C^M_{m_0, m_1, \ldots} [P_0, P_1, \ldots]^{[m_0, m_1, \ldots]} .$$ (19.24)

This is the multinomial distribution where we have introduced:

$$P_k \equiv \frac{n_k}{M}$$

and $C^M_{m_0, m_1, \ldots}$ are multinomial coefficients.

References

19.1 H. Qian and E.L. Elson: Biophys. J. **57**, 375–380 (1990)
19.2 H. Qian and E.L. Elson: Proc. Natl. Acad. Sci. USA **87**, 5479–5483 (1990)
19.3 P. Kask, K. Palo, D. Ullmann, and K. Gall: Proc. Natl. Acad. Sci. USA **96**, 13759–13761 (1999)

20 Photon Counting Histogram Statistics

Joachim D. Müller, Yan Chen and Enrico Gratton

20.1 Introduction

In the macroscopic world fluctuations are typically exceedingly small and beyond the resolution power of most experiments. Only in special circumstances, such as near the critical point of a liquid, do fluctuations become visible to the naked eye. However, the importance of fluctuations increases once the macroscopic world is left behind and one starts to consider mesoscopic or microscopic systems, where fluctuation phenomena are readily observed. Fluctuation spectroscopy exploits this source of information [20.1] and embraces a diverse field of applications [20.2–20.4].

Fluctuations of fluorescent light emerging from a small region of the sample were first considered more than two decades ago [20.5,20.6]. The technique is known as fluorescence correlation spectroscopy (FCS) and was originally used to measure translational diffusion and chemical reactions [20.5–20.8]. FCS has been extended over the years to study a variety of processes, such as triplet state kinetics [20.9], surface processes [20.10–20.12], rotational correlations [20.8,20.13,20.14], and others [20.15–20.20]. The implementation of confocal [20.21–20.23] and two-photon microscopy [20.24] with their extremely small observation volumes greatly increased the sensitivity of FCS and helped to push the detection limit to the single molecule level [20.25,20.26].

Fluctuation experiments observe a physical process that is stochastic in nature. The time series of such a random signal can be rather complex. Consequently, an analysis based on statistical methods is required to examine stochastic processes [20.27]. Traditionally, FCS experiments determine the autocorrelation function $g(\tau)$ from the sequence of photon counts in order to examine the statistics of the time-dependent decay of the fluorescence intensity fluctuations to their equilibrium value. The determination of the autocorrelation function from the raw data is a data reduction technique. Hence, some information encoded in the time series of the photon counts is lost. While the autocorrelation approach is the method of choice for characterizing kinetic processes embedded in the stochastic signal, it lacks information regarding the amplitude distribution of the intensities [20.28]. Here we consider a different data analysis approach based on the amplitude distribution of the fluorescent intensities. Experimentally, photon counts rather than intensities are measured and the statistics of the photon count amplitudes must be considered. We are specifically interested in the probability to observe k photon counts per sampling time. This probability distribution of photon counts is experimentally determined by the photon counting histogram (PCH).

The analysis of fluorescence fluctuation experiments by PCH had been recently introduced by our group [20.29]. In this chapter, we focus on the theoretical foundation of the photon counting histogram and its experimental realization. We develop the theory for a single fluorescent species and confirm it experimentally. In addition, we apply the PCH method to separate heterogeneous sample compositions. Therefore, the theory is generalized to include multiple species. The autocorrelation function $g(\tau)$ describes the fluctuations in the time domain, but with the exception of $g(0)$ lacks amplitude information. The photon counting histogram, on the other hand, characterizes the amplitude distribution of the fluctuations, but lacks kinetic information. Thus, autocorrelation and PCH analysis are complementary techniques and PCH should be able to separate a mixture of species based on a brightness difference between the components regardless of their diffusion coefficients. We discuss the use of PCH to separate a mixture of components and consider sample conditions, such as the molecular brightness and the particle concentration. We demonstrate the technique by resolving a binary dye mixture from the photon count distribution. Furthermore, we apply PCH analysis to a mixture of biomolecules with either one or two fluorescent labels attached and resolve them experimentally.

20.2   Theory

To describe the PCH of a freely diffusing species we first consider a single, diffusing particle enclosed in a small box of volume V_0. The PCH of a single particle depends explicitly on the beam profile of the excitation light. In the next step, several particles are added to the box and the corresponding PCH function determined. To describe an open system with Poisson number fluctuations, the boundary condition of the volume is removed and particles are allowed to enter and leave the box. We expanded the model to cover multiple species and calculate the PCH for a number of beam profiles. However, first we consider the statistics of the photon detection process.

20.3   PCH and the Theory of Photon Detection

Fluorescence fluctuation experiments are typically performed using single photon counting techniques. Each detected photon gives rise to an electric pulse, which is sent to the data acquisition system. Thus, the fluorescence intensity reaching the photodetector is converted into photon counts. The photon detection process changes the statistics of the measured intensities by adding shot noise to the photon counts [20.30]. For example, a constant light intensity at the detector I_D gives rise to a Poisson distribution of photon counts k,

$$\text{Poi}(k, \langle k \rangle) = \frac{(\eta_I I_D)^k e^{-\eta_I I_D}}{k!} \,. \tag{20.1}$$

Shot noise is a random Poisson point process, and reflects the discreteness and statistical independence of the photoelectric detection process [20.31]. The factor η_I is the proportionality factor between the average photon counts $\langle k \rangle$ and the constant intensity I_D at the detector, $\langle k \rangle = \eta_I I_D$. It incorporates the detection efficiency and the sampling time interval. The variance $\langle \Delta k^2 \rangle$ of a distribution serves as an indicator of its width. For the Poisson distribution, the mean and the variance are equal, $\langle \Delta k^2 \rangle = \langle k \rangle$.

A Poisson distribution describes the photon count statistics for light of constant intensity. Fluctuations in the light intensity change the photon counting statistics and the corresponding distribution of photon counts was first described by Mandel [20.32],

$$p(k) = \int\limits_0^\infty \text{Poi}(k, \eta_I I_D) p(I_D) dI_D \,. \tag{20.2}$$

The probability of observing k photoelectron events is given by the Poisson transformation of the intensity probability distribution $rmp(I_D)$. However, to be more precise, photon detection involves a short, but finite sampling time Δt_s, which along with the detector area A needs to be integrated over to yield the energy W,

$$W(t) = \int\limits_t^{t+\Delta t_s} \int\limits_A I(r, t) dA dt \,. \tag{20.3}$$

Thus, technically we observe energy fluctuations of the collected light, rather than intensity fluctuations. However, if the time scale of the intensity fluctuations is longer than the sampling time Δt_s, then the energy fluctuations track the intensity variations. We will assume for the rest of the chapter that the sampling time Δt_s is chosen to be fast enough, so that the energy fluctuations track the intensity fluctuations of interest. We also assume a stationary process, so that there is no explicit time dependence to the statistical properties of the photon counts.

Mandel's formula (20.2) essentially describes a superposition of Poisson distributions scaled by the probability of the intensity distribution function $p(I_D)$. The photon count distribution $p(k)$ is now characterized by a variance $\langle \Delta k^2 \rangle$ greater than its mean value, $\langle \Delta k^2 \rangle > \langle k \rangle$, which classifies super-Poisson statistics [20.33]. Thus, intensity fluctuations lead to a broadening of the photon count distribution with respect to a pure Poisson distribution (Fig. 20.1). The variance $\langle \Delta I_D^2 \rangle$ of the intensity distribution $p(I_D)$ determines the variance of the photon counts $\langle \Delta k^2 \rangle = \langle k \rangle + \langle \Delta I_D^2 \rangle$ [20.34]. Thus, as the strength of the intensity fluctuations increases, so does the broadening of the photon counting histogram. The changes in the shape of the histogram from a Poisson distribution are characteristic for the intensity distribution $p(I_D)$ and it is possible to infer some characteristic properties of the light source from the photon counting histogram. This approach has been used in the past to investigate the scattering of light [20.35], the twinkling of stars [20.36] and the fluctuations of laser light [20.37]. Here, we follow the same approach and directly model the photon counting histogram to describe fluorescence fluctuation experiments of freely diffusing particles.

The experimental setup typically involves a microscope and an objective that is used to focus the excitation light in order to achieve a small spatial volume. A detector collects the fluorescent light emerging from the excitation volume. The effect of the microscope optics and the detector on the shape of the observation volume is characterized by the point-spread function (PSF) of the instrument. The shape of the PSF influences the photon

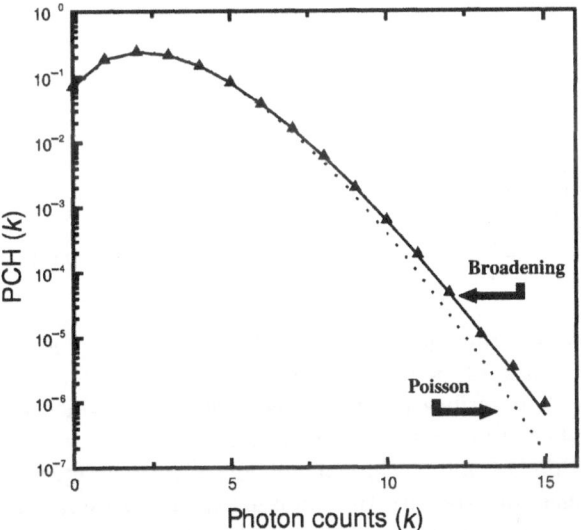

Fig. 20.1. The photon counting histogram. The normalized PCH (▲) of a solution of fluorophores is shown. The dye molecules are diffusing through the observation volume of the microscope and give rise to intensity fluctuations. A Poisson distribution (**dotted line**) describes the corresponding PCH in the absence of intensity fluctuations. Intensity fluctuations lead according to Mandel's formula to a broadening of the PCH compared to a Poisson distribution. The PCH in the presence of intensity fluctuations is characterized by super-Poisson statistics and described by the theory presented in this paper (**solid line**). The super-Poisson character of the PCH contains the information of the amplitude distribution of the intensity fluctuations

count distribution and will be considered for a few PSF conventionally used in confocal and two-photon detection [20.21,20.24]. In our context, it is more convenient to define a scaled PSF, $\overline{\text{PSF}}$, such that the volume of the scaled PSF, $V_{\text{PSF}} = \int \overline{\text{PSF}}(r)dr$, is equal to the volume defined for FCS experiments [20.38].

20.3.1   PCH of a Single Particle

We consider a single particle diffusing in an enclosed box with a volume V_0 large enough so that it essentially contains the PSF (Fig. 20.2a). Based on Mandel's formula we derived an equation which expresses the photon count distribution as a function of the PSF [20.39],

$$
\begin{aligned}
p^{(1)}(k; V_0, \varepsilon) &= \int \text{Poi}[k, \varepsilon\overline{\text{PSF}}(r)]p(r)dr \\
&= \frac{1}{V_0} \int_{V_0} \text{Poi}[k, \varepsilon\overline{\text{PSF}}(r)]dr ,
\end{aligned}
\tag{20.4}
$$

where $p(r)$ describes the probability of finding the particle at position r. For a freely diffusing particle, the probability equals $1/V_0$ inside the box and is zero outside of the box. The photon counting probability $p^{(1)}(k; V_0, \varepsilon)$ for a single particle enclosed in a volume V_0 depends on the shape of the PSF and the parameter ε. The meaning of this parameter is best illustrated by considering the average photon counts $\langle k \rangle$ of the PCH $p^{(1)}(k; V_0, \varepsilon)$,

$$
\langle k \rangle = \frac{\varepsilon}{V_0} \int_{V_0} \overline{\text{PSF}}(r)dr = \varepsilon\frac{V_{\text{PSF}}}{V_0} .
\tag{20.5}
$$

The average photon counts are thus determined by the product of ε and the probability of finding the molecule within the volume of the point spread function V_{PSF}. Therefore, ε describes the molecular brightness, which determines the average number of photon counts received during the sampling time Δt_s for a particle within the observation volume V_{PSF}. The average photon counts received $\langle k \rangle$ scale linearly with the sampling time. Therefore, the ratio $\varepsilon_{\text{sec}} = \varepsilon/\Delta t_s$ is independent of the somewhat arbitrary sampling time Δt_s. The parameter ε_{sec} expresses the molecular brightness in photon counts per second per molecule (cpsm) and allows a more convenient comparison between different experiments.

20.3.2   PCH of Multiple Particles

Now let us consider N independent and identical particles diffusing inside a box of volume V_0 (Fig. 20.2b). If one could follow a particular particle individually, the PCH of this particle would be given by $p^{(1)}(k; V_0, \varepsilon)$ according to (20.5). For N independent particles the corresponding PCH, $p^{(N)}(k; V_0, \varepsilon)$,

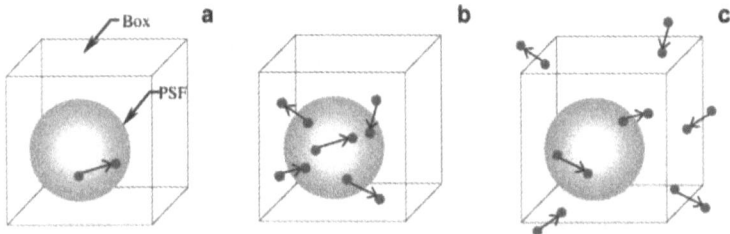

Fig. 20.2 a–c. Conceptual basis of PCH theory. **a** We consider a single particle confined to a box of volume V_0. The particle diffuses freely within the closed system of the box. We chose the box so that the point-spread function is contained within the volume of the box. The point-spread function maps the fluorescence intensity received at the detector to the spatial position of the particle. **b** In the next step, we consider several particles, which are independent and identical. The particles are allowed to diffuse within the volume of the reference box. **c** In the last step, we consider an open system. Particles are now allowed to enter and leave the box, which leads to fluctuations in the number of particles. For a small sub-volume in contact with a large reservoir, the particle number fluctuations are governed by Poisson statistics

is given by consecutive convolutions of the single particle PCH functions $p^{(1)}(k; V_0, \varepsilon)$ [20.40],

$$p^{(N)}(k; V_0, \varepsilon) = \underbrace{(p^{(1)} \otimes \cdots \otimes p^{(1)})}_{N-\text{times}}(k; V_0, \varepsilon). \tag{20.6}$$

20.3.3   PCH of Particles with Number Fluctuations

The assumption of a closed system, in which particles diffuse inside a box, does not describe the experimental situation, unless the reference volume includes the whole sample. But a macroscopic reference volume would require the evaluation of an astronomical number of convolutions according to (20.6). Instead, we choose to consider an open system in which particles are allowed to enter and leave a small sub-volume (Fig. 20.2c). The sub-volume is in contact with a much larger reservoir volume and the distribution of the number of particles inside the sub-volume is given by a Poisson distribution [20.41],

$$p_\#(N) = \text{Poi}(N, \overline{N}), \tag{20.7}$$

where $\overline{N}$ describes the average number of molecules within the reference volume V_0. Of course, if there are no particles in the reference volume, no photon counts are generated and we define the corresponding PCH as,

$$p^{(0)}(k; V_0, \varepsilon) = \delta(k), \quad \text{with} \quad \delta(k) = \begin{cases} 1, k = 0 \\ 0, k > 0 \end{cases}. \tag{20.8}$$

Now we can express the PCH for an open system $\hat{p}(k; V_0, \overline{N}, \varepsilon)$ as the expectation value of the N-particle PCH $p^{(N)}(k; V_0, \varepsilon)$ considering Poisson number statistics,

$$\Pi(k; \overline{N}_{\mathrm{PSF}}, \varepsilon) \equiv \hat{p}(k; V_0, \overline{N}, \varepsilon) = \langle p^{(N)}(k; V_0, \varepsilon) \rangle_N. \tag{20.9}$$

The PCH function $\hat{p}(k; V_0, \overline{N}, \varepsilon)$ describes the probability of observing k photon counts per sampling time for an open system with an average of $\overline{N}$ particles inside the reference volume V_0. The particular choice of the reference volume for an open system is irrelevant. It is intuitively clear that the properties of an open system have to be independent of the arbitrary reference volume V_0 [20.39]. Thus, the photon count distribution should be either referenced to an intensive quantity, like the particle concentration, or to some standard volume. We choose the convention used in FCS, where the volume of the PSF, V_{PSF}, connects the $g(0)$ value of the autocorrelation function to the average number of molecules $\overline{N}_{\mathrm{PSF}}$ [20.38]. Consequently, we drop the V_0 parameter dependence for the PCH of an open system and declare a new function $\Pi(k; \overline{N}_{\mathrm{PSF}}, \varepsilon)$, which characterizes the PCH of an open system referenced to the volume of the PSF. The average number of photon counts $\langle k \rangle$ can be calculated from (20.9) and is given by the product of the molecular brightness ε and the average number of particles $\overline{N}_{\mathrm{PSF}}$ inside the observation volume,

$$\langle k \rangle = \varepsilon \overline{N}_{\mathrm{PSF}}. \tag{20.10}$$

20.3.4   PCH of Multiple Species

So far, only identical particles have been treated. Often, more than one type of particle is present in the sample. It is straightforward to expand the theory under the assumption that the particles are non-interacting. Let us consider the case of two different species for simplicity. If we could distinguish the photon counts emerging from each species, we could directly determine the PCH of each species, $\Pi(k; \overline{N}_1, \varepsilon_1)$ and $\Pi(k; \overline{N}_2, \varepsilon_2)$. However, we cannot distinguish the origin of the photon counts. But as long as the photon emission of both species is statistically independent, the PCH of the mixture is given by the convolution of the photon count distributions of species 1 with the one of species 2,

$$\Pi(k; \overline{N}_1, \overline{N}_2, \varepsilon_1, \varepsilon_2) = \Pi(k; \overline{N}_1, \varepsilon_1) \otimes \Pi(k; \overline{N}_2, \varepsilon_2). \tag{20.11}$$

For more than two species all single species photon counting distributions are convoluted successively to yield the photon count distribution of the mixture.

20.3.5   PCH for Different PSFs

The photon count distribution depends on the PSF. Here we report the PCH of a single particle $p^{(1)}(k; V_0, \varepsilon)$ for a few PSFs of interest. The Gaussian–

Lorentzian squared PSF has been used to describe the two-photon excitation beam profile for our experimental conditions [20.24],

$$\overline{\mathrm{PSF}}_{2GL}(\rho, z) = \frac{4\omega_0^4}{\pi^2 \omega^4(z)} \exp\left[-\frac{4\rho^2}{\omega^2(z)}\right].$$ (20.12)

The PSF is expressed in cylindrical coordinates and the excitation profile has a beam waist ω_0. The inverse of the Lorentzian along the optical axis for an excitation wavelength of λ is given by

$$\omega^2(z) = \omega_0^2\left[1 + \left(\frac{z}{z_R}\right)^2\right], \quad \text{with} \quad z_R = \frac{\pi\omega_0^2}{\lambda}.$$ (20.13)

To calculate the PCH of a single particle for a reference volume V_0, (20.12) is inserted into (20.4), but integrated over all space. Integrating over all space is mathematically convenient and ensures the correct PCH for the open volume case, since from a mathematical point of view the PSF extends to infinity. However, the PCH of a closed volume is only approximately determined. The quality of the approximation depends on the size of the reference volume V_0. If the volume is chosen so that the contribution of the PSF to the photon counts is negligible outside of the reference volume, then the deviation between both functions is small. However, the PCH of a closed volume is from a practical point of view only of minor interest, because the PCH of an open volume describes the experimental situation. We refer the interested reader to a more detailed discussion of this point by Chen et al. [20.39]. The PCH of a single particle is then determined for $k > 0$ by a one-dimensional integral

$$p_{2GL}^{(1)}(k; V_0, \varepsilon) = \frac{1}{V_0} \frac{\pi^2 \omega_0^4}{2\lambda k!} \int_0^\infty (1 + x^2)\gamma\left[k, \frac{4\varepsilon}{\pi^2(1 + x^2)^2}\right] dx,$$

$$\text{for} \quad k > 0.$$ (20.14)

The integral, which contains the incomplete gamma function γ [20.42], can be evaluated numerically.

A second important PSF is the three-dimensional Gaussian PSF, which is used extensively to describe confocal detection [20.21, 20.22],

$$\overline{\mathrm{PSF}}_{3DG}(x, y, z) = \frac{I(x, y, z)}{I_0} = \exp\left[-\frac{2(x^2 + y^2)}{\omega_0^2} - \frac{2z^2}{z_0^2}\right],$$ (20.15)

with an effective beam waist z_0 in the axial direction. The PCH of a single particle is determined in the same way as for the other PSF and we derive for $k > 0$ an expression in the form of a one-dimensional integral,

$$p_{3DG}^{(1)}(k; V_0, \varepsilon) = \frac{1}{V_0} \frac{\pi\omega_0^2 z_0}{k!} \int_0^\infty \gamma(k, \varepsilon e^{-2x^2}) dx, \quad \text{for} \quad k > 0.$$ (20.16)

For the measurement of surface processes the PSF is typically approximated by a two-dimensional Gaussian,

$$\overline{PSF}_{2DG}(x,y) = \frac{I(x,y)}{I_0} = \exp\left[-\frac{2(x^2 + y^2)}{\omega_0^2}\right], \tag{20.17}$$

and the corresponding PCH is given by

$$p_{2DG}^{(1)}(k; A_0, \varepsilon) = \frac{1}{A_0} \frac{\pi\omega_0^2}{2k!} \gamma(k, \varepsilon), \quad \text{for} \quad k > 0. \tag{20.18}$$

Here, we reference, of course, to an area A_0 instead of to a volume. Finally, if the PSF is uniform, then the PCH is simply a Poisson distribution, $p_U^{(1)}(k; V_{PSF}, \varepsilon) = \mathrm{Poi}(k, \varepsilon)$.

20.3.6 Describing PCH with the Moment Generating Function

The PCH of a single particle has just been described for photon counts $k > 0$. To evaluate the PCH for $k = 0$ one must determine the sum over all photon counts $k > 0$ and subtract it from the area of the probability distribution, which is normalized to one, $\mathrm{p}^{(1)}(0) = 1 - \sum_{k=1}^{\infty} \mathrm{p}^{(1)}(k)$. Furthermore, the computation of the PCH for each photon count requires a numerical integration.

We now introduce an alternative formalism to determine the probability distribution of photon counts using a moment generating function. This approach not only determines the photon count probability for zero photon counts directly, but also expresses the single particle PCH in the form of an analytical expression. The description is based on a formal relationship between the Laplace and Poisson transforms. The Laplace transform $\boldsymbol{L}$ and the Poisson transform $\boldsymbol{P}$ are formally defined as

$$P(n) = \boldsymbol{P}[f(x)] = \int \mathrm{d}x f(x) \frac{x^n e^{-x}}{n!},$$

$$F(s) = \boldsymbol{L}[f(x)] = \int \mathrm{d}x f(x) e^{-sx}. \tag{20.19}$$

The Laplace transform $F(s)$ of the probability function $f(x)$ is also its moment generating function [20.30],

$$\langle f^n \rangle = (-1)^n \frac{\partial^n F(s)}{\partial s^n}\bigg|_{s=0}. \tag{20.20}$$

The Poisson transform can be expressed in terms of the Laplace transform $F(s)$ through the following relationship [20.30],

$$P(n) = \frac{(-1)^n}{n!} \frac{\partial^n F(s)}{\partial s^n}\bigg|_{s=1}. \tag{20.21}$$

Often, the Poisson transform can be calculated for $n = 0$ or 1 in closed form. For example, for the squared Gaussian–Lorentzian PSF, we obtain the following result for $n = 1$,

$$
p_{2GL}^{(1)}(1) = \varepsilon r \, {}_2F_2 \left[\begin{array}{c} 1/4, \, 3/4 \\ 1/2, \quad 2 \end{array} \middle| -\frac{4}{\pi^2} s\varepsilon \right] \Bigg|_{s=1} = (-1) \frac{\partial F(s)}{\partial s} \Bigg|_{s=1} \, ,
$$

$$
\text{with} \quad r = \frac{V_{\text{PSF}}}{V_0} \, . \tag{20.22}
$$

The properties of the generalized hypergeometric function ${}_pF_q$ are described in the literature [20.43,20.44]. The zeroth moment of a probability distribution is equal to one, therefore the moment generating function satisfies the condition, $F(0) = 1$, according to (20.20). We determine the moment generating function $F(s)$ from (20.22),

$$
F(s) = 1 + r\frac{8}{3} \left\{ 1 - {}_2F_2 \left[\begin{array}{cc} -3/4, & -1/4 \\ -1/2, & 1 \end{array} \middle| -\frac{4}{\pi^2} s\varepsilon \right] \right\} . \tag{20.23}
$$

The photon counting histogram can now be determined from the moment generating function according to (20.21), using the Pochhammer function $(a)_k$ and the analytical form of the derivatives of the generalized hypergeometric function,

$$
p_{2GL}^{(1)}(k) = 1 + r\frac{8}{3} \left\{ 1 - {}_2F_2 \left[\begin{array}{cc} -3/4, & -1/4 \\ -1/2, & 1 \end{array} \middle| -\frac{4}{\pi^2} s\varepsilon \right] \right\} , \quad \text{for} \quad k = 0
$$

$$
p_{2GL}^{(1)}(k) = \left(\frac{2}{\pi} \right)^{2k} \frac{\varepsilon^k}{k!} r\frac{8}{3} \frac{(-3/4)_k(-1/4)_k}{(-1/2)_k(1)_k}
$$

$$
\times {}_2F_2 \left[\begin{array}{cc} -3/4 + k, & -1/4 + k \\ -1/2 + k, & 1 + k \end{array} \middle| -\frac{4}{\pi^2} \varepsilon \right] , \quad \text{for} \quad k > 0 . \tag{20.24}
$$

By using the moment generating function to describe the PCH of a single particle we arrive at an analytical solution for $p_{2GL}^{(1)}$ including $k = 0$.

A second example of the use of the moment generating function is illustrated for the homogeneous PSF. In this case, a Poisson distribution determines the PCH of a single particle. The PCH for an open system is described by a compound Poisson distribution,

$$
\Pi(k; \overline{N}, \varepsilon) = \sum_{N=0}^{\infty} \text{Poi}(k, \varepsilon N) \text{Poi}(N, \overline{N}) . \tag{20.25}
$$

Instead of evaluating the sum in (20.25) one can use a moment generating function to determine the PCH. Here we use the factorial moment generating function $Q(s)$ of the compound Poisson distribution,

$$
Q(s) = e^{(e^{-s\varepsilon}-1)\overline{N}} . \tag{20.26}
$$

The moment generating function $Q(s)$ allows us to calculate the PCH of the open system analytically by using the following relationship

$$\Pi(k; \overline{N}, \varepsilon) = \frac{(-1)^k}{k!} \frac{d^k}{ds^k} Q(s) \bigg|_{s=1} .$$ (20.27)

20.3.7 Two-fold PCH Statistics

The photon count distribution characterizes the amplitude fluctuations of the detected photons, but lacks kinetic information. We want to demonstrate how to expand the theory of PCH to include time dependence and show how PCH is related to the autocorrelation function of FCS experiments. The photon count distribution considered so far describes the number of detected photons in a single time interval. However, by studying the joint statistics of photon counts from two short time intervals separated by a time delay τ, we incorporate time dependence into the photon count statistics [20.30].

Here, we specifically consider the two-fold distribution of photon counts for a diffusing particle. The conditional probability that a particle at position r_0 at time t_0 will be found at a later time t_1 at position r_1 is given by the solution of the diffusion equation (see for example [20.41]),

$$p_d[r_0(t_0)|r_1(t_1)] = (4\pi D\tau)^{-3/2} \exp \frac{(r_1 - r_0)^2}{4D\tau} ,$$ (20.28)

where D is the diffusion coefficient of the particle and τ the time difference $t_1 - t_0$. In the following we only consider time lags $\tau > 0$ to avoid the complication of shot noise correlations at $\tau = 0$. The photon count probability $p(k, r_0)$ for a particle with molecular brightness ε at position r_0 is given by a Poisson distribution, $p(k, r_0) = \mathrm{Poi}(k, \varepsilon \mathrm{PSF}(r_0))$. The average photon counts $\langle k(r_0) \rangle$ at position r_0 is proportional to the fluorescence intensity $I(r_0)$, $\langle k(r_0) \rangle = \eta I(r_0)$. The proportionality constant η describes the detection efficiency and is set to 1 for simplicity. The probability $p^{(1)}(k_0, k_1; \varepsilon, \tau)$ of observing k_0 photon counts at time t_0 and k_1 photon counts at a later time t_1 for a diffusing particle with brightness ε is then determined by [20.45]

$$p^{(1)}(k_0, k_1; \varepsilon, \tau)$$
$$= \int\int dr_0 dr_1 p(r_0) p(k_0, r_0) p_d[r_0(t_0)|r_0(t_0 + \tau)] p(k_1, r_1) ,$$ (20.29)

where $p(r_0)$ describes the probability of finding the particle at position r_0. The correlation function of the photon counts for a diffusing particle is equal to the correlation function of the fluorescence intensity,

$$\langle k_0(t) k_1(t + \tau) \rangle = \sum_{k_0=0}^{\infty} \sum_{k_1=0}^{\infty} k_0 k_1 p^{(1)}(k_0, k_1; \varepsilon, \tau)$$
$$= \int\int dr_0 dr_1 p(r_0) I(r_0) p_d[r_0(t_0)|r_1(t_0 + \tau)] I(r_1)$$
$$= \langle I(t) I(t + \tau) \rangle .$$ (20.30)

The above result was derived for a single particle, but it is straightforward to generalize the result to any number of particles or to an open volume with number fluctuations.

20.4   Data Analysis

We developed an algorithm to calculate the PCH $\Pi(k)$ based on the theory and implemented it on a computer to fit experimental data. Photon counts are recorded with a home-built data acquisition card, which is interfaced to a computer. The computer calculates the histogram of the experimental data. The normalized histogram represents the experimental photon counting probability distribution $\tilde{p}(k)$. The statistical uncertainty associated with each element of $\tilde{p}(k)$ is determined by the standard deviation $\sigma_k = \sqrt{M\tilde{p}(k)[1 - \tilde{p}(k)]}$. The number of data points M collected is typically of the order of 10^6. The experimental data are fit by minimizing the reduced χ^2-function,

$$\chi^2 = \frac{\sum_{k=k_{\min}}^{k_{\max}} \left(M \frac{\tilde{p}(k) - \Pi(k)}{\sigma_k} \right)^2}{k_{\max} - k_{\min} - d}. \tag{20.31}$$

The experimental photon counts range from a minimum value $k_{\min}$, which is typically 0 for most experiments, to a maximum number $k_{\max}$, and the number of fitting parameters is given by d. The normalized residuals of the fit are determined by $r(k) = M \frac{\tilde{p}(k) - \Pi(k)}{\sigma_k}$.

20.5   Single Species PCH

We tested the theory of the photon counting statistics by comparing it to experiments. The numerical PCH algorithm was used to fit the experimental photon counting histograms to the theory, as outlined in the data analysis section. The experiments were carried out using a two-photon setup and the details of the experiment are described in Chen [20.29]. The measured photon counting distributions agree with the theoretical PCH functions calculated for the experimental setup within the statistical error. The molecular brightness ε and the average number of molecules shape the photon count distribution in characteristic ways. We first study the influence of the average number of molecules $\overline{N}$ upon the properties of the histogram.

20.5.1   Influence of the Particle Concentration

The photon count distribution of a fluorescein solution was measured at three different concentrations (Fig. 20.3). We performed a global fit of all three histograms with ε linked, while the average number of particles was allowed to

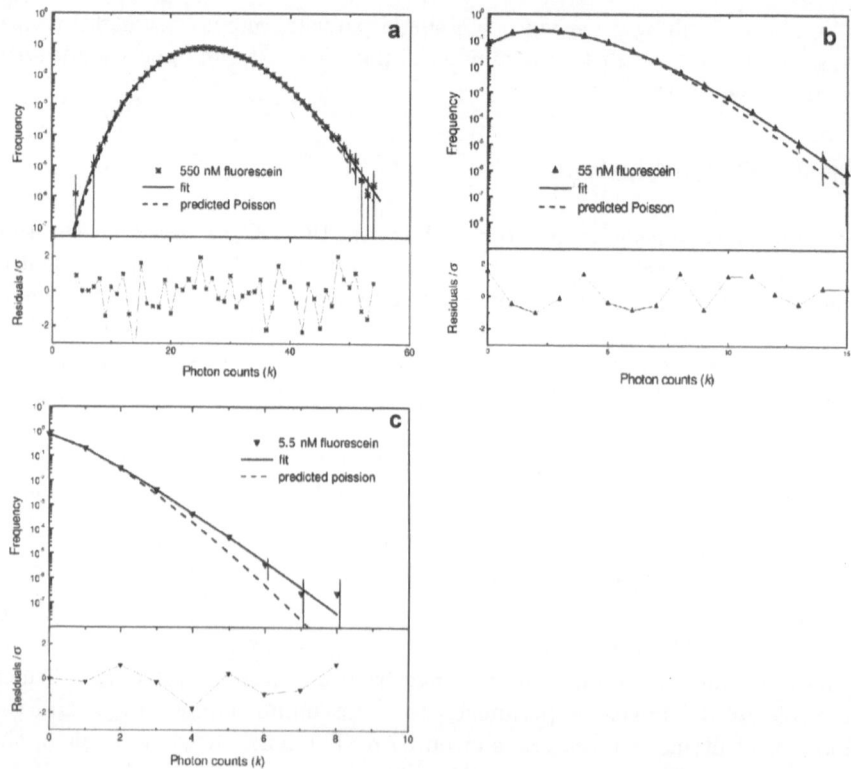

Fig. 20.3 a–c. Photon counting histogram for fluorescein at three different concentrations, **a** 550 nM, **b** 55 nM and **c** 5.5 nM. The histograms are plotted together with their Poisson distribution for a mean equal to the corresponding average photon counts $\langle k \rangle$ of the experimental histogram. For the highest concentration, only small deviations of the PCH from a Poisson distribution are observed. Lowering the concentration of the fluorescein produces stronger deviations between the histogram and the Poisson distribution as shown in **b** and **c**. Each data point is plotted together with an error bar ($\pm 3\sigma$). The data sets were fit by globally linking the molecular brightness parameter ε across the data sets, while allowing the average number of molecules $\overline{N}$ to vary. The **solid line** represents the best fit obtained by using the theoretical PCH function $\Pi(k; \overline{N}, \varepsilon)$ as explained in the text. The parameters returned by the fit are $\varepsilon_{sec} = 16\,000$ cpsm and $\overline{N} = 32.5$, 3.4 and 0.35. The lower panel displays the normalized residuals of the fit

vary. The data and the fitted histograms in Fig. 20.3 are in good agreement. The residuals between data and fit for each histogram are displayed in units of standard deviation σ. The residuals vary randomly and yield a reduced χ^2 close to 1, indicating a good description of the data by the theoretical model. The recovered number of molecules $\overline{N}$ scales with the concentration of the sample. The dashed lines in Fig. 20.3 represent the corresponding

Poisson distributions with a mean equal to the average photon counts $\langle k \rangle$ of the sample. The Poisson distribution approximates the PCH for the high fluorescein concentration case (Fig. 20.3a). However, lowering the dye concentration to 55 nM (Fig. 20.3b) already shows a broadening of the experimental PCH compared to the Poisson distribution, which is clearly visible in the tail of the distributions. The deviation of the PCH from a Poisson distribution becomes even more apparent by reducing the fluorescein concentration to 5.5 nM (Fig. 20.3c).

Thus, the photon counting histogram approaches a Poisson distribution with increasing fluorophore concentration. This behavior can be readily understood by considering the influence of the molecule concentration on the intensity fluctuations. The relative strength of the number fluctuations is given by the ratio between the standard deviation σ and the mean μ of the particle number distribution,

$$\frac{\sigma}{\mu} = \frac{\sqrt{\langle \Delta N^2 \rangle}}{\overline{N}} = \frac{1}{\sqrt{\overline{N}}}. \tag{20.32}$$

The number of molecules inside a small, open volume is Poisson distributed, and the relative strength of the particle fluctuations decreases with the inverse square root of the average number of particles $\overline{N}$. Thus, with increasing particle concentration the number distribution approaches a delta function $\delta(N - \overline{N})$. Consequently, the fluctuations in intensity become negligible, when compared to the average intensity. The second contribution to the intensity fluctuations, which is due to the diffusion in an inhomogeneous excitation profile, also vanishes at high particle concentrations. Any vacancy created by a molecule leaving a position is practically always filled by another molecule moving to that position, so that no net change in the fluorescence intensity occurs. Thus, a constant fluorescence intensity dictates a Poisson photon count distribution.

20.5.2 Influence of Molecular Brightness

We used three different fluorophores, each with its own brightness parameter ε, to illustrate the influence of the molecular brightness ε on the photon count distribution. Each fluorophore sample was made up to approximately the same concentration to facilitate the comparison of the different histograms. The count distributions are analyzed with the PCH algorithm and are shown together with the fits in Fig. 20.4. Poisson distributions with the same mean as the average photon counts are displayed as dashed lines for each histogram. The deviation between the tail of the PCH and the Poisson distribution increases with increasing ε.

To maximize the deviation between the photon count distribution and the corresponding Poisson function, one can either reduce the number of molecules within the excitation volume or increase the brightness parameter

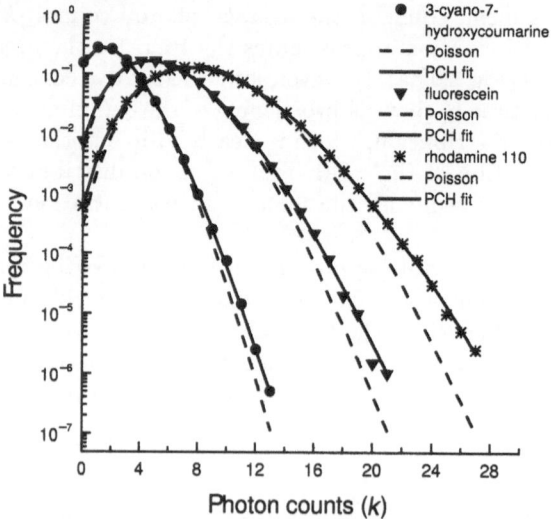

Fig. 20.4. Photon counting histograms for three dyes, each with a different molecular brightness ε. The histograms of cyano-hydroxycoumarin (•), fluorescein (▼) and rhodamine 110 (∗) taken with the same number of data points were fitted to the theoretical PCH function $\Pi(k; \overline{N}, \varepsilon)$ shown as *solid lines*. The concentration of the three samples was kept similar to facilitate the comparison between the histograms. The fit recovered the average number of molecules $\overline{N}$ as 2.6, 3.3 and 3.0 for cyano-hydroxycoumarin, fluorescein and rhodamine 110, respectively. For the molecular brightness ε, values of 0.74 for cyano-hydroxycoumarin, 1.60 for fluorescein and 2.73 for rhodamine 110 were recovered. For each histogram a Poisson distribution with a mean equal to the average number of photon counts is plotted as a **dashed line**. The deviation between the Poisson distribution and the photon counting histogram increases markedly with increased molecular brightness ε

ε as demonstrated in Fig. 20.4. The relationship between the super-Poisson character of the PCH and the molecular brightness ε can be qualitatively understood. The average fluorescence intensity of a molecule in the excitation volume is characterized by the parameter ε. A particle with a larger value of ε causes stronger intensity fluctuations as it enters and diffuses through the beam. The increase in the fluorescence intensity fluctuations leads to a further broadening of the PCH. This behavior is a consequence of the averaging of Poisson distributions over a wider intensity range as expressed by Mandel's formula. To quantify this statement, we define the fractional deviation Q, a measure of the deviation between the PCH and the Poisson distribution [20.46],

$$Q = \frac{\langle \Delta k^2 \rangle - \langle k \rangle}{\langle k \rangle} = \gamma \varepsilon, \qquad (20.33)$$

where $\langle \Delta k^2 \rangle$ and $\langle k \rangle$ are the variance and the expectation value of the photon counts, respectively, and γ is the shape factor of the PSF [20.38]. The fractional deviation Q and the $g(0)$ value of the intensity autocorrelation function [20.38],

$$g(0) = \frac{\langle \Delta I^2 \rangle}{\langle I \rangle^2} = \frac{\langle \Delta k^2 \rangle - \langle k \rangle}{\langle k \rangle^2} = \frac{\gamma}{\overline{N}} , \qquad (20.34)$$

are closely related. The $g(0)$ value is the ratio of the shape factor γ to the average number of molecules $\overline{N}$ inside the PSF. The relationship between the intensity moments and the factorial moments of the photon counts [20.33] allows us to express $g(0)$ by the variance and the average of the photon counts. Thus, the ratio of Q to $g(0)$ determines the average photon counts $\langle k \rangle$, according to (20.10).

A Poisson distribution is defined by $Q = 0$, while super-Poisson distributions require $Q > 0$ and sub-Poisson distributions mandate $Q < 0$. The molecular brightness largely determines the super-Poisson character of the PCH, since Q is directly proportional to the parameter ε, which depends on the excitation power, the detection efficiency and the molecular species.

20.5.3   Sensitivity of PCH Algorithm

To provide a quantitative description of the deviation of the photon counting histogram from a Poisson distribution, we define the following reduced χ^2_{Poi} function,

$$\chi^2_{\mathrm{Poi}}(\varepsilon, \overline{N}, M)$$

$$= \min_{\tilde{k}} \left[\sum_{k=k_{\mathrm{min}}}^{k_{\mathrm{max}}} \left(M \frac{\Pi(k, \varepsilon, \overline{N}) - \mathrm{Poi}(k, \tilde{k})}{\sigma_k} \right)^2 \frac{1}{k_{\mathrm{max}} - k_{\mathrm{min}}} \right] , \qquad (20.35)$$

where the standard deviation is given by

$$\sigma_k = \{ M \Pi(k, \varepsilon, \overline{N})[1 - \Pi(k, \varepsilon, \overline{N})] \}^{1/2} .$$

The function χ^2_{Poi} describes the statistical significance of the deviation of the PCH from a Poisson distribution. If the value of χ^2_{Poi} is less than or equal to 1 then the statistics are not sufficient to distinguish the data from a Poisson distribution. The larger the value of χ^2_{Poi}, the stronger the deviation between the PCH and the Poisson distribution. Equation (20.35) requires the minimization of the function with respect to the parameter $\tilde{k}$. Since for most cases, the value of $\tilde{k}$ is nearly identical to the average photon counts of the PCH function, $\langle k \rangle = \varepsilon \overline{N}$, matching the first moments of the PCH and Poisson function is a good approximation of the function χ^2_{Poi}.

We evaluated χ^2_{Poi} for a variety of conditions to explore the sensitivity of PCH analysis. The dependence of χ^2_{Poi} on the average number of molecules

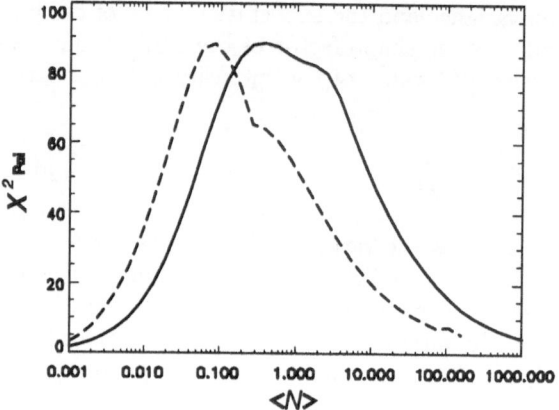

Fig. 20.5. The function χ^2_{Poi} characterizes the deviation of the PCH from a Poisson distribution and is plotted on a semi-logarithmic scale as a function of the average number of particles $\overline{N}$. The function χ^2_{Poi} was evaluated for a squared Gaussian–Lorentzian PSF with $M = 10^7$ sample points and a molecular brightness of 0.25 (**solid line**). The absolute values of χ^2_{Poi} depend strongly on the molecular brightness and χ^2_{Poi} for a molecular brightness of 12 is scaled (**dashed line**), in order to display both curves on the same graph. The curves are bell-shaped and the maximum deviation occurs near a concentration of one particle per observation volume. An increase in the molecular brightness ε shifts the maximum of χ^2_{Poi} to lower particle concentrations

$\overline{N}$ for constant molecular brightness ε is shown in Fig. 20.5. Increasing the number of molecules at low concentrations results first in a steady increase of χ^2_{Poi}, then the function reaches a maximum and decreases monotonically at high particle concentrations. Thus, an optimal concentration exists, where the photon count distribution deviates maximally from the Poisson distribution.

We can understand this result intuitively. Increasing the particle concentration leads to smaller amplitude fluctuations $g(0)$, which means that the intensity distribution is approaching a delta function, as already mentioned before. Reducing the number of molecules in the observation volume produces stronger fluctuation amplitudes. However, once the average number of molecules is less than one molecule, the probability that no molecule is found in the observation volume greatly increases. Therefore, except for the case where $k = 0$ the signal-to-noise ratio of the histogram is markedly reduced. Two effects shape the χ^2_{Poi} function and lead to a maximum around a particle concentration of one molecule in the observation volume. The function χ^2_{Poi} is shown in Fig. 20.5 for two different molecular brightness values to illustrate the influence of ε upon the peak position. For brighter particles, the maximum of the function χ^2_{Poi} occurs at a lower concentration. An increase in the molecular brightness ε increases the broadening of the PCH and shifts the maximum of χ^2_{Poi} to lower particle concentrations.

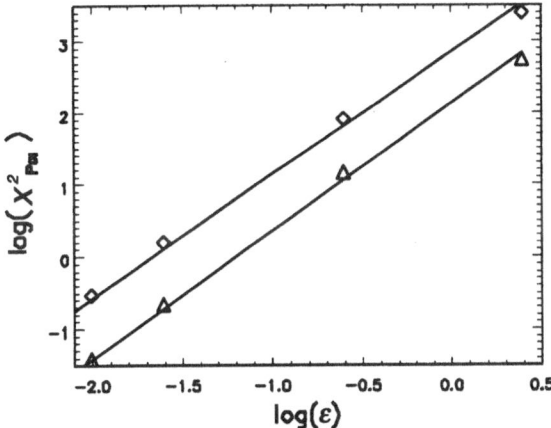

Fig. 20.6. The molecular brightness dependence of χ^2_{Poi} is shown on a doubly logarithmic scale for two different particle concentrations, $\overline{N} = 0.01$ ($\triangle$) and $\overline{N} = 1.2$ ($\diamond$). The **solid lines** with a slope of approximately 1.75 represent the best fit of each function χ^2_{Poi} to a straight line. The slope characterizes the power law dependence of χ^2_{Poi} on the molecular brightness ε

The absolute values of the χ^2_{Poi} function depend strongly on the brightness parameter. In order to show the two curves on the same graph, the corresponding functions for the larger brightness had to be scaled down to be visible (Fig. 20.5). The dependence of χ^2_{Poi} on the molecular brightness ε was evaluated for two fixed concentrations ($\overline{N} = 0.01$ and 1.2) and both were fit to a straight line that yielded a decadic logarithm of about 1.75 for both particle concentrations (Fig. 20.6).

The value of the χ^2_{Poi} function is directly proportional to the number of data points M taken, and therefore directly proportional to the data acquisition time. If we formally define the square root of the χ^2_{Poi} function to be a measure of the signal-to-noise ratio (SNR), then, we find that the SNR of PCH is proportional to the square root of the data acquisition time, and close to linear for the molecular brightness. These conclusions are similar to those of Koppel [20.47], where the SNR was derived for the autocorrelation function under the simplifying condition of Gaussian statistics.

The dependence of the SNR on the number of molecules in the excitation volume is related to another study [20.48]. There, it was shown that the SNR of the second moment is constant at high particle concentrations, but decreases at low particle concentrations. However, the SNR of the third moment displays a maximum at a concentration of about one particle per observation volume. This behavior of the SNR is also found in the SNR of the PCH, which is not surprising, since all moments are contained in the photon count distribution.

20.6   PCH for Multiple Species

Resolving multiple species is an important issue in many biological applications. Biological macromolecules interact with other molecules, and as a consequence of this network of interactions, the complex machinery of life is maintained. FCS has successfully been used to resolve multiple species [20.49]. However, it is generally recognized that resolving two species by the autocorrelation function alone requires a difference in their diffusion coefficients of the order of 2 or larger [20.50]. This poses a severe restriction on the application of the pure autocorrelation approach to many biological systems, since the differences between the diffusion coefficients of biomolecules is often less than a factor of two. The association of two monomer subunits, for example, to form a dimer is a widespread and important biological reaction mechanism [20.19]. The increase in the diffusion coefficient for a dimer is about 25% of the monomer value. Since the diffusion coefficient approximately scales with the cubic root of the molecular mass, a difference in the molecular weight of about a factor of 8 would be required to resolve two components by the autocorrelation function alone. To address this intrinsic shortcoming of the autocorrelation approach two other methods have been introduced in the literature. One is based on higher order autocorrelation functions [20.51,20.52] and the other is based on higher order moment analysis [20.53,20.54].

Here, we introduce a new approach for separating fluorescent species based on the photon counting histogram. The histogram of photon counts is sensitive to the molecular brightness and was discussed in detail for a single species. If two species differ in their molecular brightness, then a molecule of the brighter species entering the observation volume will produce a stronger fluorescent intensity change than the other species. By considering the statistics of these intensity changes, one can deduce the brightness and the average number of molecules of each species. Shot noise caused by the photodetection process is added to these intensity fluctuations. It was shown that the resulting photon counting statistics for multiple species is given by the consecutive convolution of the single species photon counting histograms. While the histogram of a single species requires two parameters, $2r$ parameters are required to describe the histogram for r species, namely the molecular brightness ε_i and the average number of particles $\overline{N}_i$ for each species. In the following, we discuss the resolution of two species in detail, consider practical limitations and demonstrate the technique experimentally.

20.6.1   Resolvability of Two Species

We discuss the most difficult case where the two species must be resolved by the histogram alone without any further knowledge. From a practical point of view one wants to know what data acquisition time and concentrations to choose in order to resolve species of a given brightness. To address this question we calculated the theoretical histograms for different conditions, in order to identify experimentally favorable concentrations and brightness conditions. The theoretically determined two-species PCH function were then fit assuming a single species model and the reduced χ^2 determined. A fit of a two-species histogram by a single species model will result in a misfit, which gives rise to systematic residuals. The magnitude of the residuals tells us whether it is feasible to distinguish between single and multiple species. A reduced χ^2 value of one or less indicates that the data statistics are not sufficient to resolve the species, while a χ^2 greater than one indicates that more than one species is present. We want to find out which brightness differences can be separated. The brightness for a given species is kept constant during the calculation of the histograms, but the concentration is varied in a systematic manner. For a fixed brightness ratio the results are best represented graphically in the form of a contour plot of the χ^2 surface as a function of the logarithmic concentration of both species. The concentration of each species is expressed in number of molecules within the PSF.

Figure 20.7 shows the result for a molecular brightness of $\varepsilon_A = 1.5$ for species A and $\varepsilon_B = 6.0$ for species B. The number of data samples was chosen as $M = 1.6 \times 10^7$. This corresponds to a data acquisition time of 13 min and

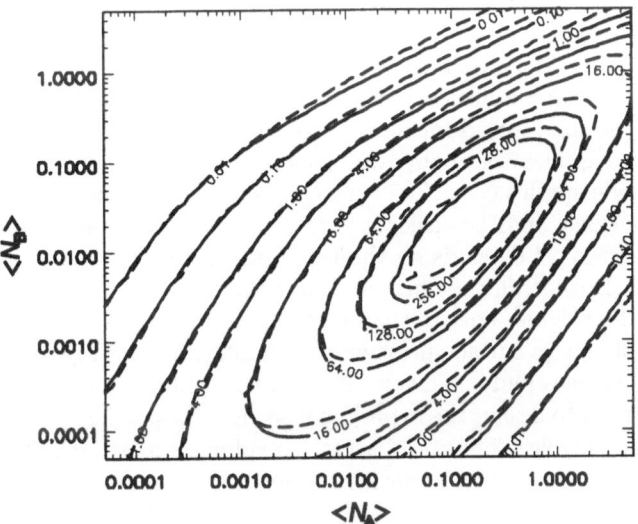

Fig. 20.7. The χ^2 contour map of the misfit between a two-species PCH by a single species PCH as a function of the logarithmic particle concentration. The **solid contour** lines represent the χ^2 surface for a brightness ratio of four with molecular brightness values of $\varepsilon_A = 1.5$ and $\varepsilon_B = 6.0$. The maximal deviation between the two- and single-species PCH functions occurs approximately at a particle concentration of 0.1 for species A and 0.02 for species B. The **dashed contour lines** represent a rescaled χ^2 surface for another brightness ratio of four, but with molecular brightness values of $\varepsilon_A = 0.25$ and $\varepsilon_B = 1.0$

a sample with a brightness of 30 000 cpsm and 120 000 cpsm, if we assume a sampling time of $\Delta t_s = 50\,\mu s$. As can be seen from Fig. 20.7 an optimal concentration for species A and species B exists where the misfit of the two-species histogram is maximal. Changing the concentration of either species in either direction results in a decrease of this deviation, thus reducing the χ^2 value. Once the reduced χ^2 value is close to 1 or lower, then the signal statistics are not sufficient to distinguish the presence of two species. The optimal average number of molecules for species A is close to 0.1 and the one for species B is about 0.02.

The behavior of the χ^2 function is similar to the χ^2_{Poi} function of the single-species case. The value of χ^2 is directly proportional to the number of data samples M and therefore to the data acquisition time. Increasing the particle concentration decreases the χ^2 value, since the fluorescence intensity fluctuations caused by the corresponding species are diminished. Decreasing the concentration to much less than 1 molecule per observation volume, decreases the χ^2 value, since the particular species is not found in the volume of the PSF during the majority of the acquisition period.

The shape of the χ^2 surface depends on the molecular brightness ratio, but is largely independent of the absolute brightness values, ε_1 and ε_2, as indicated in Fig. 20.7. The contour lines of χ^2 are plotted for the same brightness ratio, but with a difference in the absolute molecular brightness of a factor of 6. However, the amplitude of the χ^2 function depends strongly on the absolute molecular brightness values. Analyzing the dependence of χ^2 as a function of absolute brightness, but fixed ratio, yields a power-law dependence with an exponent of about two. This result is similar to the one obtained for the single species case. The brighter a molecule is, the stronger the intensity fluctuations, and the easier it is to resolve the two species.

The peak position of χ^2 depends on the brightness ratio, but the actual variations are small. In general, the optimal concentration conditions require fewer molecules of the brighter species than molecules of the dimmer species. The contour plots of the χ^2 surface provide a good guideline in judging the resolvability of two species under a given set of experimental conditions. In general, one would tend to design the experiment so that the particle concentration stays below one molecule in the observation volume while maximizing the absolute brightness for a given brightness ratio.

20.6.2   Experimental Results

To demonstrate the feasibility of using PCH to resolve species, a mixture of two dyes, rhodamine 110 and cyano-coumarin, was prepared. The mixture consisted of 80% coumarin and 20% rhodamine dye. The photon count distribution of the binary mixture was measured and then analyzed by the PCH algorithm. One of the experimental histograms of such a mixture is shown in Fig. 20.8 together with the best fit to a single-species model. The deviation between the fit and the experimental histogram is clearly visible in the tail

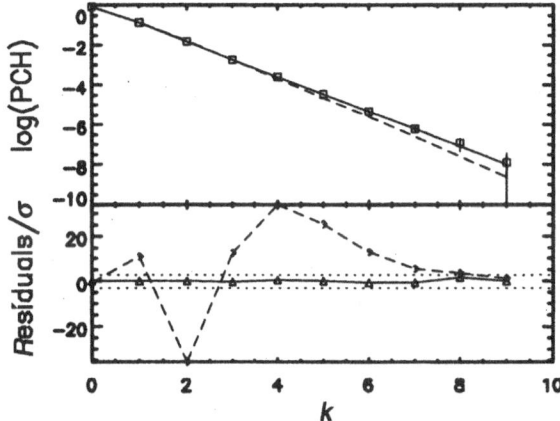

Fig. 20.8. The PCH of a binary mixture of rhodamine and coumarin (□) is plotted together with the experimental error bar (±3σ) for each data point. The **dashed line** represents a fit to a single species model. The fit and the experimental PCH deviate in the tail of the distribution. The residuals of the single species fit (**dashed line**) in the lower panel are correlated and exceed 20 standard deviations for several photon count channels. A fit of the data to a two species model (**solid line**) leads to a good description of the experimental histogram. The residuals (**solid line**) are random and the reduced χ^2 for the two species fit is 0.8. The two **dotted lines** indicate the ±3σ bounds

of the distribution. The residuals show systematic variations with standard deviations of more than $20\,\sigma$ for a few histogram values. The experimental PCH was then subject to a two-species fit. The two-species model describes the experimental PCH within statistical error. The residuals produced by the two-species fit are close to one and random (Fig. 20.8). The reduced χ^2 for the two-species fit yields a value of 0.8. After the first measurement the sample was diluted and then remeasured. This dilution experiment was performed in order to check the accuracy of the PCH analysis. The concentration of each species is reduced by the dilution, but the concentration ratio is unaffected. The particle concentrations determined by PCH analysis from each measurement are plotted in Fig. 20.9a together with the χ^2 surface representing the experimental conditions. The best concentration ratio describing the dilution experiment was then determined from the experimentally recovered number of particles and the corresponding dilution curve is shown in Fig. 20.9a. The concentration ratio was determined to be 83%/17%, in excellent agreement with the expected concentration ratio. The particle concentrations determined for each individual experiment follow the dilution curve closely and do not scatter significantly.

The fitted molecular brightness values are shown in Fig. 20.9b for the four different sample concentrations. The recovered molecular brightness values stay within ±0.1 of their mean values. For rhodamine we determine a

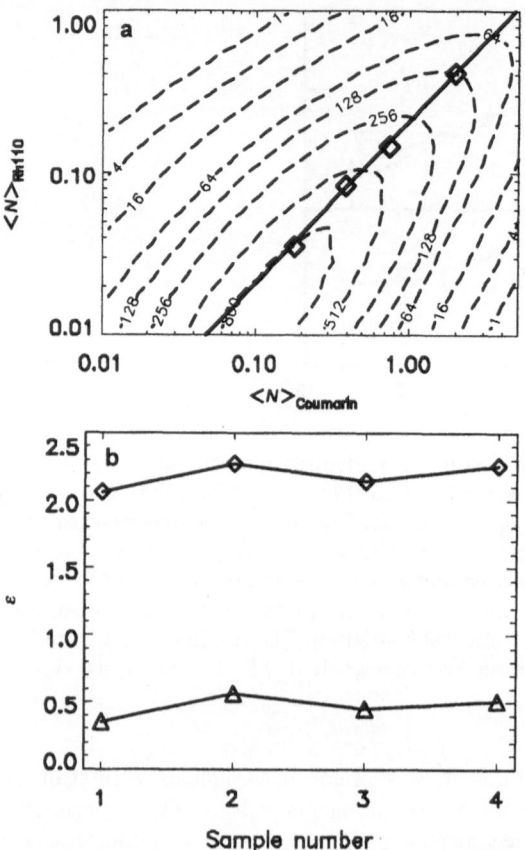

Fig. 20.9. a A binary mixture of 20% rhodamine and 80% coumarin was diluted several times and the PCH of each dilution experimentally determined. Each histogram was fit by a two-species model and the number of molecules recovered by the fit are shown in the χ^2 contour map ($\Diamond$). The average value of the fitted molecular brightness of the rhodamine and the coumarin dye were used to calculate the χ^2 contour map for our experimental conditions. The contour lines are shown as **dashed lines**. The **solid line** represents the best approximation of the fitted dye concentrations by a dilution curve and corresponds to a composition of 17% rhodamine and 83% coumarin. **b** The fitted molecular brightness values are plotted for the four samples measured. The mean of the molecular brightness is 0.47 and 2.18 for coumarin ($\triangle$) and rhodamine ($\Diamond$), respectively

molecular brightness of $\varepsilon_R = 2.18$ and for coumarin we obtain $\varepsilon_C = 0.47$. Thus, the brightness ratio of the rhodamine–coumarin pair is 4.6 under the experimental conditions. For our sampling time of $\Delta t_s = 50\,\mu s$ the molecular brightness values translate to 44 000 cpsm for rhodamine and 9 400 cpsm for coumarin.

As discussed earlier, the resolvability of two species decreases with decreasing brightness ratio. The question whether one can resolve a brightness ratio of two is of imminent biological importance. The association of two monomeric units, each labeled with a single fluorescent dye, leads, in the absence of quenching, to a dimer with a two-fold increase in the molecular brightness. Another example is the binding of a fluorescently labeled antigen to an IgG antibody. The molecular brightness of the doubly liganded antibody–antigen complex is again twice brighter than the singly liganded antibody–antigen complex. The autocorrelation function cannot distinguish between these scenarios, since the change in the diffusion coefficient is not sufficient.

To demonstrate the feasibility of distinguishing between biological samples that possess a brightness ratio of two, we labeled IgG antibodies with a fluorescent dye. For the first sample, the labeling condition was chosen so that only single labeled proteins are present. For the other sample, the labeling condition was changed, so that a small fraction of double-labeled proteins appeared. The PCH of the first sample is represented by a single species fit (Fig. 20.10a), while the PCH of the second sample required a fit to the two species model (Fig. 20.10b). The molecular brightness ratio of the two species obtained by the PCH fit of the second sample is 2.05 and the brightness ratio of the single-labeled species across the two samples equals 1.05. This demonstrates that it is feasible to resolve a brightness ratio of two. The two species differ only in the number of dyes attached to the antibody. So even lifetime measurements or dual-color FCS [20.55] cannot resolve the mixture.

We would like to point out that all experiments conducted for resolving mixtures were analyzed based on a single histogram assuming no additional information about the molecular brightness and the particle concentration of each species. This is by far the most stringent condition and is chosen to illustrate the experimental strength of the PCH analysis. One can extend the resolution power of PCH considerably by varying parameters, such as the concentration, wavelength, or temperature, to name a few. A global analysis of the combined set of experimental histograms can then be performed. For example, one species can often be isolated and measured independently in order to determine the particle concentration and its molecular brightness. This measurement serves as a calibration of the species' molecular brightness and can later be used in the analysis of a mixture containing that species.

20.7 Conclusions

We derived the theory of the photon counting histogram for fluorescence fluctuation experiments. An algorithm to calculate PCH numerically was developed, and used to compare experiment with theory. The deviation of the histogram from Poisson statistics is due to the spatially inhomogeneous excitation profile of the laser beam and the number fluctuations of the fluorescent

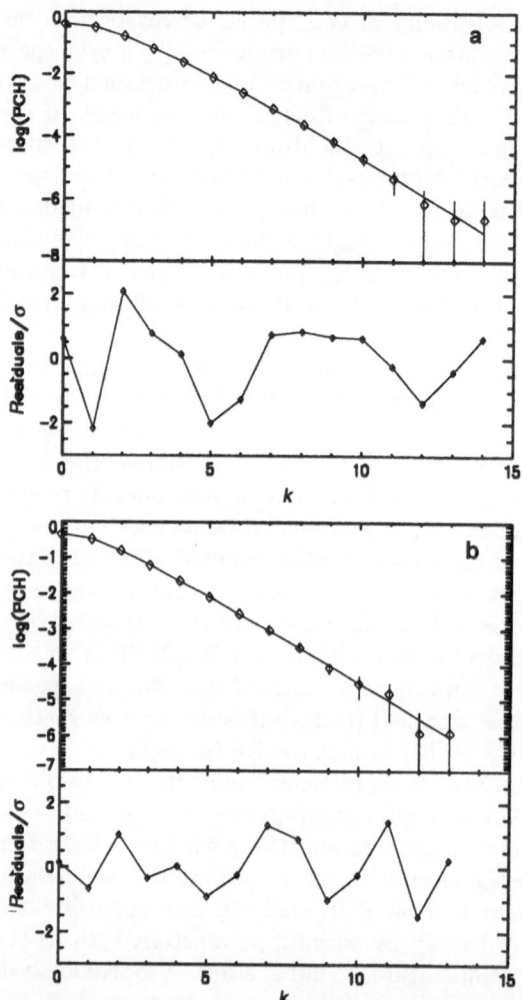

Fig. 20.10 a,b. Photon counting histogram of Alexa labeled IgG. **a** The PCH of a singly labeled IgG sample (◇) is fit to a single species model (**solid line**) and the residuals of the fit are shown in the lower panel. The fit determined the molecular brightness as 3.6 and the number of molecules as 0.28. **b** The PCH of a mixture of singly and doubly labeled IgG (◇) is fit to a two-species model (**solid line**). The residuals of the fit are shown in the lower panel. The fit determined for the singly labeled component 0.23 molecules with a brightness of 3.7 and for the doubly labeled component 0.014 molecules with a brightness of 7.6

particles inside the observation volume. Comparison between theory and experiment demonstrates that the data agree with the theoretically predicted photon counting histogram. Two parameters, the molecular brightness ε and the average number of particles $\overline{N}$ inside the observation volume uniquely specify the photon count distribution for a single species. These parameters are determined by a fit of the experimental data to the PCH model and have been used as a new tool in analyzing fluorescence fluctuation data.

PCH is sensitive to the brightness of particles; thus offering a possibility of distinguishing a mixture of species based on this feature alone. The PCH of a mixture is described by the convolution of the individual single-species photon count distribution functions. The two-species case was considered in detail and its resolvability by the PCH algorithm discussed. We demonstrated the concept experimentally by resolving a binary mixture of dyes with the histogram method. We also demonstrated that it is possible to separate a mixture of biomolecules with a brightness ratio of two, which is of considerable biological interest. These examples demonstrate the potential power of PCH in fluorescence fluctuation experiments. PCH characterizes the amplitude distribution of fluorescence intensity fluctuations and the autocorrelation function describes the time dependence of these fluctuations. Thus, PCH and FCS provide complementary information, which should prove useful in tackling biological problems with fluorescence fluctuation spectroscopy.

References

20.1 M.B. Weissman: Ann. Rev. Phys. Chem. **32**, 205–232 (1981)
20.2 J.B. Johnson: Phys. Rev. **32**, 97–109 (1928)
20.3 S.E. Luria and M. Delbrück: Genetics **28**, 491–511 (1943)
20.4 M.B. Weissman: Rev. Mod. Phys. **60**, 537–571 (1988)
20.5 D. Magde, E. Elson, and W.W. Webb: Phys. Rev. Lett. **29**, 705–708 (1972)
20.6 D. Magde, E.L. Elson, and W.W. Webb: Biopolymers **13**, 29–61 (1974)
20.7 E.L. Elson and D. Magde: Biopolymers. **13**, 1–27 (1974)
20.8 M. Ehrenberg and R. Rigler: Chem. Phys. **4**, 390–401 (1974)
20.9 J. Widengren, Ü. Mets, and R. Rigler: J. Phys. Chem. **99**, 13368–13379 (1995)
20.10 D.E. Koppel, D. Axelrod, J. Schlessinger, E.L. Elson, and W.W. Webb: Biophys. J. **16**, 1315–1329 (1976)
20.11 J. Borejdo: Biopolymers. **18**, 2807–2820 (1979)
20.12 N.L. Thompson and D. Axelrod: Biophys. J. **43**, 103–114 (1983)
20.13 S.R. Aragon and R. Pecora: Biopolymers. **14**, 119–137 (1975)
20.14 P. Kask, P. Piksarv, M. Pooga, and Ü. Mets: Biophys. J. **55**, 213–220 (1989)
20.15 M. Weissman, H. Schindler, and G. Feher: Proc. Natl. Acad. Sci. USA **73**, 2776–2780 (1976)
20.16 D. Axelrod, D.E. Koppel, J. Schlessinger, E. Elson, and W.W. Webb: Biophys. J. **16**, 1055–1069 (1976)
20.17 N.O. Petersen, D.C. Johnson, and M.J. Schlesinger: Biophys. J. **49**, 817–820 (1986)

20.18 M. Kinjo and R. Rigler: Nucleic Acids Res. **23**, 1795–1799 (1995)

20.19 K.M. Berland, P.T.C. So, Y. Chen, W.W. Mantulin, and E. Gratton: Biophys. J. **71**, 410–420 (1996)

20.20 C. Eggeling, J.R. Fries, L. Brand, R. Gunther, and C.A.M. Seidel: Proc. Natl. Acad. Sci. USA **95**, 1556–1561 (1998)

20.21 H. Qian and E.L. Elson: Appl. Opt. **30**, 1185–1195 (1991)

20.22 R. Rigler, Ü. Mets, J. Widengren, and P. Kask: Eur. Biophys. J. **22**, 169–175 (1993)

20.23 D.E. Koppel, F. Morgan, A.E. Cowan, and J.H. Carson: Biophys. J. **66**, 502–507 (1994)

20.24 K.M. Berland, P.T.C. So, and E. Gratton: Biophys. J. **68**, 694–701 (1995)

20.25 R. Rigler, J. Widengren, and Ü. Mets: Interactions and kinetics of single molecules as observed by fluorescence correlation spectroscopy. In: Fluorescence Spectroscopy: New Methods and Applications. O.S. Wolfbeis, Ed. (Springer, Berlin Heidelberg New York 1993) p.13

20.26 M. Eigen and R. Rigler: Proc. Natl. Acad. Sci. USA **91**, 5740–5747 (1994)

20.27 C.W. Gardiner: Handbook of stochastic methods. (Springer, Berlin Heidelberg New York 1985)

20.28 J.S. Bendat and A.G. Piersol: Random data: Analysis and measurement procedures. (Wiley-Interscience, New York 1971)

20.29 Y. Chen: Analysis and applications of fluorescence fluctuation spectroscopy. (University of Illinois at Urbana-Champaign, Urbana 1999)

20.30 B. Saleh: Photoelectron statistics, with applications to spectroscopy and optical communications. (Springer, Berlin Heidelberg New York 1978)

20.31 D.L. Snyder: Random point processes. (Wiley-Interscience, New York 1975)

20.32 L. Mandel: Proc. Phys. Soc. **72**, 1037–1048 (1958)

20.33 M.C. Teich and B.E.A. Saleh: Photon bunching and antibunching. In: Progress in Optics. E. Wolf, Ed. (North-Holland, Amsterdam 1988) p.1

20.34 C.L. Mehta: Theory of photoelectron counting. In: Progress in Optics. E. Wolf, Ed. (North-Holland, Amsterdam 1970) p.375

20.35 M. Bertolotti: Photon statistics. In: Photon correlation and light beating spectroscopy. H.Z. Cummins and E.R. Pike, Eds. (Plenum Press, New York 1973) p.41

20.36 E. Jakeman G. Parry, E.R. Pike, and P.N. Pusey: Contemp. Phys. **19**, 127–145 (1978)

20.37 H. Risken: Statistical properties of laser light. In: Progress in Optics. E. Wolf, Ed. (North-Holland, Amsterdam 1970) p.241

20.38 N.L. Thompson: Fluorescence correlation spectroscopy. In: Topics in fluorescence spectroscopy. J.R. Lakowicz, Ed. (Plenum, New York 1991) p.337

20.39 Y. Chen, J.D. Müller, P.T.C. So, and E. Gratton: Biophys. J. **77**, 553–567 (1999)

20.40 W. Feller: An introduction to probability theory and its applications. (John Wiley, New York 1957)

20.41 S. Chandrasekhar: Rev. Mod. Phys. **15**, 1–89 (1943)

20.42 M. Abramowitz and I.A. Stegun: Handbook of mathematical functions with formulas, graphs, and mathematical tables. (Dover Publications, New York 1965)

20.43 L.J. Slater: Generalized hypergeometric functions. (Cambridge University Press, Cambridge 1966)

20.44 Y.L. Luke: The special functions and their approximations. (Academic Press, New York 1969)

20.45 H. Qian: Biophys. Chem. **38**, 49–57 (1990)

20.46 L. Mandel: Optics Lett. **4**, 205–207 (1979)

20.47 D.E. Koppel: Phys. Rev. A **10**, 1938–1945 (1974)

20.48 P. Kask R. Günter, and P. Axhausen: Eur. Biophys. J. **25**, 163–169 (1997)

20.49 B. Rauer, E. Neumann, J. Widengren, and R. Rigler: Biophys. Chem. **58**, 3–12 (1996)

20.50 U. Meseth T. Wohland, R. Rigler, and H. Vogel: Biophys. J. **76**, 1619–1631 (1999)

20.51 A.G.D. Palmer and N.L. Thompson: Biophys J. **52**, 257–270 (1987)

20.52 A.G.D. Palmer and N.L. Thompson: Proc. Natl. Acad. Sci. USA **86**, 6148–6152 (1989)

20.53 H. Qian and E.L. Elson: Proc. Natl. Acad. Sci. USA **87**, 5479–5483 (1990)

20.54 H. Qian and E.L. Elson: Biophys. J. **57**, 375–380 (1990)

20.55 P. Schwille, F.J. Meyer-Almes, and R. Rigler: Biophys. J. **72**, 1878–1886 (1997)

21 High Order Autocorrelation in Fluorescence Correlation Spectroscopy

Nancy L. Thompson and Jennifer L. Mitchell

21.1 Introduction

Fluorescence correlation spectroscopy (FCS) was originally developed as a technique in which the temporal fluctuations in the fluorescence arising from a small volume containing fluorescent molecules are autocorrelated to obtain information about the processes that give rise to the fluorescence fluctuations. After its initial introduction [21.1–21.6], FCS was used primarily to measure translational diffusion coefficients [21.7–21.10]. It was also shown that FCS could provide information about other processes such as rotational diffusion [21.11–21.14], flow [21.15], and biochemical kinetics [21.16,21.17]. These initial applications were comprehensively reviewed a number of years ago [21.18–21.21]. FCS has become increasingly mature as a technique during the last decade, both in terms of application and development. Recent applications include characterization of receptor oligomerization on intact cells [21.22]; aggregates of prion proteins [21.23] and amyloid β-proteins [21.24]; DNA polymerization [21.25], conformational fluctuations [21.26], and interaction with transcription factors [21.27]; and phospholipid micelles [21.28]. Some of the more innovative technological advances have included the use of two-photon excitation [21.29,21.30]; cross-correlation in double-label experiments [21.31–21.35]; the combination of resonance energy transfer with FCS [21.26,21.36]; the combination of capillary electrophoresis with FCS [21.37]; and the use of evanescent illumination for examining processes at surfaces [21.38–21.42].

In the earliest studies, it was recognized that, for a sample containing only one type of fluorescent molecule, the magnitude of the normalized fluorescence fluctuation autocorrelation function was directly related to the average number of fluorescent molecules in the illuminated volume [21.1,21.2]. Thus, because FCS was sensitive to the number density of fluorescent molecules, this method could provide direct information about molecular weights [21.43]. The dependence on number densities also predicted FCS to be very sensitive to molecular aggregation and self-association. This latter feature was seen as a potential method of detecting and characterizing cell surface receptor clustering, a phenomenon of ubiquitous importance in biology and one that is not easily monitored on viable cell surfaces. The development of FCS as a technique for monitoring cell surface receptor clustering has been steadily developed in the last fifteen years [21.9,21.22,21.44–21.55] and is the subject of a separate chapter in this book. FCS is also amenable to the detection and characterization of molecular self-association in solution [21.9,21.20,21.23–21.25,21.30,21.56,21.57].

For an ergodic and monodisperse sample, the extrapolated magnitude of the normalized fluorescence fluctuation autocorrelation function is proportional to the inverse of the average number of fluorescent molecules in the sample volume. The magnitude can yield a measure of the fluorescent molecule number density provided that the proportionality constant, which is related to instrumental factors, has been calibrated with samples of known concentration [21.1,21.2]. However, for polydisperse samples, the magnitude of the normalized fluorescence fluctuation autocorrelation function depends on the number densities of the different fluorescent species as well as their relative fluorescence intensities [21.56]. One is confronted with the difficulty of having only one measured number and more than one unknown. The rate and shape of the decay of the fluorescence fluctuation autocorrelation function contain information about the temporal behavior of the different fluorescent species, but resolving the monotonically decaying autocorrelation function into the contributing characteristic rates and their amplitudes can be difficult [21.58]. If two species are present representing monomers and dimers, the characteristic rates associated with translational diffusion differ only by a factor of approximately $2^{1/3}$. For all but very high signal-to-noise ratios, resolving decay components with such small differences in characteristic times is not straightforward. One method of overcoming this difficulty is to calculate, from the same time record of fluorescence, a series of high order autocorrelation functions. For a polydisperse sample, the magnitudes of the higher order autocorrelation functions contain independent information about the number densities and relative fluorescence yields of the different fluorescent species. It is this sensitivity of high order fluorescence fluctuation autocorrelation to polydispersity and molecular self-association that is described in this chapter.

21.2 Temporal High Order FCS

21.2.1 Definitions

In temporal FCS, the fluctuating fluorescence $F(t)$ arising from a small sample volume is measured. The sample volume contains, in general, more than one fluorescent species; the number density of the i-th species at position r and time t is denoted here as $C_i(r, t)$. The sample volume is described by a function $W(r)$, which is the product of three functions describing the spatial dependence of the excitation light, the spatial dependence of the fluorescence collection efficiency, and the sample extent. The measured fluorescence may be written as [21.21]

$$F(t) = \beta \int W(\boldsymbol{r}) \sum_{i=1}^{R} \alpha_i C_i(\boldsymbol{r}, t) \mathrm{d}\Omega, \qquad (21.1)$$

where β is a proportionality constant related to, among other parameters, the fluorescence collection efficiency; R is the number of species; α_i is the product of the absorptivity and fluorescence quantum efficiency of the i-th fluorescent species; and $\mathrm{d}\Omega$ is the integral over the sample plane (for two-dimensional samples) or over all space (for three-dimensional samples). In the following descriptions, it is assumed for simplicity that the sample is two-dimensional and illuminated with a Gaussian-shaped laser beam, and that the collection optics do not alter this spatial illumination profile. In this case,

$$W(\boldsymbol{r}) = \exp\left[-\frac{2(x^2 + y^2)}{s^2}\right], \qquad (21.2)$$

where s is the $1/\mathrm{e}^2$ radius of the illuminating laser beam, $\mathrm{d}\Omega = \mathrm{d}^2 r$, and $W(\boldsymbol{r})$ has been scaled by its maximum value so that this function is unitless (21.8). Other forms for $W(\boldsymbol{r})$ have been described elsewhere [21.1,21.7,21.15, 21.21,21.38,21.59–21.65].

21.2.2 First Order Fluorescence Fluctuation Autocorrelation

For a stationary ergodic sample, the time average and the ensemble average of $F(t)$ are constant and equivalent, and are denoted here as $\langle F \rangle$. The temporal fluorescence fluctuation is defined as the deviation of $F(t)$ from its average value; i.e., $\delta F(t) = F(t) - \langle F \rangle$. The first order fluorescence fluctuation autocorrelation function is usually normalized by the square of the average fluorescence:

$$G(\tau) = \frac{\langle \delta F(t) \delta F(t + \tau) \rangle}{\langle F \rangle^2}. \qquad (21.3)$$

Equations 21.1 and 21.3 imply that [21.1,21.21,21.56]

$$G(\tau) = \frac{\sum_{i=1}^{R} \sum_{j=1}^{R} \alpha_i \alpha_j \iint W(\boldsymbol{r}_1) W(\boldsymbol{r}_2) \phi(i, j; \boldsymbol{r}_1, \boldsymbol{r}_2; \tau) \mathrm{d}^2 r_1 \mathrm{d}^2 r_2}{\left[\sum_{i=1}^{R} \alpha_i \langle C_i \rangle \int W(\boldsymbol{r}) \mathrm{d}^2 r\right]^2}. \qquad (21.4)$$

The function $\phi(i, j; \boldsymbol{r}_1, \boldsymbol{r}_2; \tau)$ is the correlation of concentration fluctuations in the i-th chemical species at position $\boldsymbol{r}_1$ and time t with concentration fluctuations in the j-th chemical species at position $\boldsymbol{r}_2$ and at time $t + \tau$; i.e.,

$$\phi(i, j; \boldsymbol{r}_1, \boldsymbol{r}_2; \tau) = \langle \delta C_i(\boldsymbol{r}_1, t) \delta C_j(\boldsymbol{r}_2, t + \tau) \rangle, \qquad (21.5)$$

where $\delta C_i(\boldsymbol{r}, t) = C_i(\boldsymbol{r}, t) - \langle C_i \rangle$ is the deviation of the number density of the i-th species, $C_i(\boldsymbol{r}, t)$, from its average value $\langle C_i \rangle$. For ideal solutions, concentration fluctuations are correlated at the same time only for the same species at the same position [21.1]; i.e.

$$\phi(i, j; \boldsymbol{r}_1, \boldsymbol{r}_2; 0) = \delta_{ij} \langle C_i \rangle \delta(\boldsymbol{r}_1 - \boldsymbol{r}_2), \tag{21.6}$$

where δ_{ij} is the Kronecker delta and $\delta(\boldsymbol{r}_1 - \boldsymbol{r}_2)$ is the Dirac delta function. This assumption implies that

$$G(0) = \gamma_2 \frac{\sum\limits_{i=1}^{R} \alpha_i^2 \langle C_i \rangle}{\left[\sum\limits_{i=1}^{R} \alpha_i \langle C_i \rangle \right]^2}, \tag{21.7}$$

where γ_2 is a constant related to the shape of the sample volume:

$$\gamma_2 = \frac{\int W^2(\boldsymbol{r}) \mathrm{d}^2 r}{\left[\int W(\boldsymbol{r}) \mathrm{d}^2 r \right]^2}. \tag{21.8}$$

For the Gaussian-shaped illumination profile in (21.2), $\gamma_2 = 1/(\pi s^2)$.

The functional form for $G(\tau)$ arising from diffusional transport of multiple species through the Gaussian-shaped sample volume described by (21.2) is [21.1,21.56]

$$\frac{G(\tau)}{G(0)} = \sum_{i=1}^{R} \frac{f_i}{1 + \tau/\tau_{\mathrm{D}i}}. \tag{21.9}$$

The characteristic diffusion time for the i-th fluorescent species and the fractional amplitude associated with the term describing decay from this species are

$$\tau_{\mathrm{D}i} = \frac{s^2}{4D_i}, \quad f_i = \frac{\alpha_i^2 \langle C_i \rangle}{\sum\limits_{i=1}^{R} \alpha_i^2 \langle C_i \rangle}, \tag{21.10}$$

where s is the characteristic distance of the sample volume and D_i is the diffusion coefficient of the i-th species. The contributions are weighted not by the relative abundance of a species but by the product of its number density and the square of its relative fluorescence yield.

Many, if not most, samples of interest contain multiple species; e.g., monomeric and oligomeric proteins in solution [21.8,21.23,21.24], monomeric and oligomeric proteins in or on membranes [21.49,21.52,21.55], macromolecular reactants and products in bimolecular or enzymatic reactions in solution [21.16,21.17,21.27,21.66–21.68], protein or DNA molecules with different conformations [21.26,21.69], DNA molecules of different lengths [21.25,21.70], or phospholipid vesicles of different sizes [21.28,21.56]. It is usually desirable to determine, from $G(\tau)$, the rates $\tau_{\mathrm{D}i}$ and amplitudes f_i for the different species. This approach can be successful if the rates τ_{Di} are sufficiently different in magnitude, but is sometimes problematic because translational diffusion coefficients are not very sensitive to molecular size and shape. The ability to distinguish different components becomes limited by the statistical accuracy of the autocorrelation function [21.58]. One method of obtaining more information about multi-component samples, as described in the next section, is to use high order autocorrelation.

21.2.3   High Order Fluorescence Fluctuation Autocorrelation

The normalized, high order, fluorescence fluctuation autocorrelation functions are defined as [21.56]

$$G_{mn}(\tau) = \frac{\langle \delta F^m(t)\delta F^n(t+\tau)\rangle - \langle \delta F^m\rangle\langle \delta F^n\rangle}{\langle F\rangle^{m+n}}, \tag{21.11}$$

where m and n are positive integers. The higher moments of the fluctuations (e.g., $\langle \delta F^2\rangle$, $\langle \delta F^3\rangle$) are in general nonzero; the subtracted product insures that $[G_{mn}(\tau)]_{\tau\to\infty} = 0$. The average fluctuation $\langle \delta F\rangle$ does equal zero, so that (21.11) reduces to (21.3) when $m = n = 1$. For systems in equilibrium, $G_{mn}(\tau) = G_{nm}(\tau)$ [21.71]. Therefore, only autocorrelation functions for which $m \leq n$ are considered.

Equations (21.1) and (21.11) imply that

$$G_{mn}(\tau) = \frac{\sum_{i_1=1}^{R}\sum_{i_2=1}^{R}\cdots\sum_{i_{m+n}=1}^{R}\alpha_{i_1}\alpha_{i_2}\ldots\alpha_{i_{m+n}}H_{mn}(\tau)}{\left[\sum_{i=1}^{R}\alpha_i\langle C_i\rangle\right]^{m+n}}, \tag{21.12}$$

where

$$H_{mn}(\tau) \tag{21.13}$$
$$= \iint\ldots\int W(\boldsymbol{r}_1)W(\boldsymbol{r}_2)\ldots W(\boldsymbol{r}_{m+n})I_{mn}(\tau)\mathrm{d}^2r_1\mathrm{d}^2r_2\ldots\mathrm{d}^2r_{m+n},$$

$$I_{mn}(\tau) = \phi_{mn}(i_1,i_2,\ldots,i_{m+n};\boldsymbol{r}_1,\boldsymbol{r}_2,\ldots,\boldsymbol{r}_{m+n};\tau)$$
$$- \phi_{m0}(i_1,i_2,\ldots,i_m;\boldsymbol{r}_1,\boldsymbol{r}_2,\ldots,\boldsymbol{r}_m)$$
$$\times \phi_{n0}(i_{m+1},i_{m+2},\ldots,i_{m+n};\boldsymbol{r}_{m+1},\boldsymbol{r}_{m+2},\ldots,\boldsymbol{r}_{m+n}), \tag{21.14}$$

and

$$\phi_{mn}(i_1,i_2,\ldots,i_{m+n};\boldsymbol{r}_1,\boldsymbol{r}_2,\ldots,\boldsymbol{r}_{m+n};\tau)$$
$$= \langle \delta C_{i_1}(\boldsymbol{r}_1,t)\delta C_{i_2}(\boldsymbol{r}_2,t)\ldots\delta C_{i_m}(\boldsymbol{r}_m,t)$$
$$\times \delta C_{i_{m+1}}(\boldsymbol{r}_{m+1},t+\tau)\delta C_{i_{m+2}}(\boldsymbol{r}_{m+2},t+\tau)\ldots\delta C_{i_{m+n}}(\boldsymbol{r}_{m+n},t+\tau)\rangle$$
$$\phi_{m0}(i_1,i_2,\ldots,i_m;\boldsymbol{r}_1,\boldsymbol{r}_2,\ldots,\boldsymbol{r}_m)$$
$$= \langle \delta C_{i_1}(\boldsymbol{r}_1,t)\delta C_{i_2}(\boldsymbol{r}_2,t)\ldots\delta C_{i_m}(\boldsymbol{r}_m,t)\rangle. \tag{21.15}$$

The magnitudes of the high order autocorrelation functions, for $m = 1$, are [21.20]

$$G_{11}(0) = \gamma_2 B_2$$
$$G_{12}(0) = \gamma_3 B_3$$
$$G_{13}(0) = \gamma_4 B_4 + 3\gamma_2^2 B_2^2$$
$$G_{14}(0) = \gamma_5 B_5 + 10\gamma_2\gamma_3 B_2 B_3$$
$$G_{15}(0) = \gamma_6 B_6 + 15\gamma_2\gamma_4 B_2 B_4 + 15\gamma_2^3 B_2^3 + 10\gamma_3^2 B_3^2 \,, \tag{21.16}$$

where

$$B_k = \frac{\sum\limits_{i=1}^{R} \alpha_i^k \langle C_i \rangle}{\left[\sum\limits_{i=1}^{R} \alpha_i \langle C_i \rangle \right]^k} \,, \tag{21.17}$$

$$\gamma_k = \frac{\int W^k(r)\mathrm{d}^2 r}{\left[\int W(r)\mathrm{d}^2 r \right]^k} \,. \tag{21.18}$$

For the Gaussian-shaped sample volume described by (21.2), $\gamma_k = k^{-1}$ $(2/\pi s)^{k-1}$. Equation (21.11) implies that the magnitudes of the high order autocorrelation functions with $m = 2$ are related to those with $m = 1$ as follows:

$$G_{22}(0) = G_{13}(0) - G_{11}^2(0)$$
$$G_{23}(0) = G_{14}(0) - G_{11}(0)G_{12}(0)$$
$$G_{24}(0) = G_{15}(0) - G_{11}(0)G_{13}(0) \,. \tag{21.19}$$

Similar relationships exist for higher values of m. Thus, all new magnitude information is contained in $G_{1n}(0)$; although it is of interest to experimentally confirm the relationships in (21.19) and those that are similar for $m \geq 3$.

Complete expressions for the time dependence of $G_{mn}(\tau)$ have been derived for transport of multiple fluorescent species by diffusion or flow through a Gaussian-shaped sample volume [21.56]. For diffusion, the $G_{mn}(\tau)$ are sums of terms which are products of the functions

$$L_{mn}(i,\tau) = \frac{1}{1 + c_{mn}\tau/\tau_{Di}} \,, \quad c_{mn} = \frac{2mn}{m+n} \,, \tag{21.20}$$

where the characteristic times τ_{Di} are given in (21.10).

The dependence of the characteristic diffusional times of the high order fluorescence fluctuation autocorrelation functions on the integers m and n suggests that global analysis of the time dependence of a series of $G_{mn}(\tau)$ would provide an improvement toward distinguishing the characteristic diffusional times and their fractional amplitudes when multiple species are present.

However, the dependence of the characteristic times on the integers m and n is weak unless either m or n is large ($c_{12} = 4/3$, $c_{13} = 3/2$, $c_{14} = 8/5$; $c_{22} = 2$, $c_{23} = 12/5$; $c_{33} = 3$). The magnitudes, not the decay times, of the high order autocorrelation functions are particularly sensitive to polydispersity.

21.2.4 Multicomponent Analysis

The behavior of high order fluorescence fluctuation autocorrelation functions is illustrated by considering the example of a diffusing, dimerizing fluorescent molecule. For simplicity, it is assumed that the relative fluorescence yield of the dimer is twice the yield of the monomer. For two species diffusing through a sample volume described by (21.2), the two lowest order autocorrelation functions can be calculated as limits of previously published general equations [21.56]. The results are

$$G_{11}(\tau) = \frac{1}{2N}\left[\frac{1-f}{1+\tau/\tau_m} + \frac{2f}{1+\tau/\tau_d}\right],$$

$$G_{12}(\tau) = \frac{1}{N^2}\left[\frac{1-f}{3+4\tau/\tau_m} + \frac{4f}{3+4\tau/\tau_d}\right], \qquad (21.21)$$

where N is the total average number of monomeric subunits in the sample volume, f is the fraction of subunits which exists in the dimer form, and τ_m and τ_d are the characteristic diffusion times for the monomeric and dimeric species. Figure 21.1 shows $G_{11}(\tau)$ and $G_{12}(\tau)$ for the case in which $N = 10$ and $\tau_d = 2^{1/3}\tau_m$. The magnitudes of these functions change significantly as a function of the fraction of the molecules that are dimerized; i.e. by factors of two and four for $n = 1$ and $n = 2$, respectively. The rates and shapes of the decays do not change as dramatically; i.e. the half-times for the decays of $G_{11}(\tau)$ and $G_{12}(\tau)$ change only by a factor of $2^{1/3}$. The values of N and

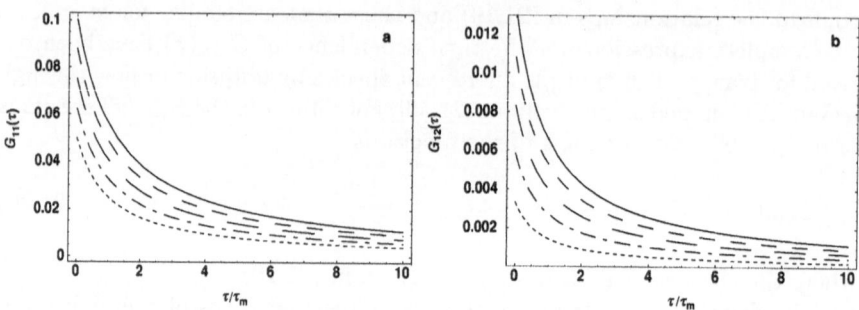

Fig. 21.1 a,b Dimerization of a diffusing species. **a** $G_{11}(\tau)$ and **b** $G_{12}(\tau)$ were calculated from (21.21). The total number of monomeric subunits in the observation volume, N, equals 10. The fraction of the monomeric subunits present as dimers, f, equals (**line**) 1, (**short dash**) 3/4, (**long dash**) 1/2, (**dash-dot**) 1/4, and (**dot**) 0

f may be algebraically retrieved from the measured, extrapolated values of $G_{11}(0)$ and $G_{12}(0)$ without the need for distinguishing the $2^{1/3}$-fold difference in the characteristic times.

It is useful to define a general method by which the magnitudes of high order fluorescence fluctuation autocorrelation functions can be used to detect the presence of polydispersity. As previously described [21.20,21.57], the parameters

$$Q_k = \frac{B_k^{1/(k-1)}}{B_2} \quad k = 3, 4, 5, \ldots \tag{21.22}$$

have the general properties that $Q_k = 1$ for all monodisperse samples and $Q_k > 1$ for all polydisperse samples. The values of Q_k can be calculated from the measured, extrapolated values of $G_{1,k-1}(0)$ (21.16) provided that the values of γ_k have been calibrated (see below).

As is evident by examining (21.17) when $R = 1$, the values of the parameters B_k equal $1/\langle C \rangle^{k-1}$ for a monodisperse sample. It is therefore convenient to define parameters

$$P_k = \langle C \rangle^{k-1} B_k, \tag{21.23}$$

which are readily calculated from data if the total concentration of monomeric, fluorescent subunits $\langle C \rangle$ is known. For a sample in which an oligomerization of order p is occurs,

$$P_k = \frac{1 - f + f\alpha^k/p}{(1 - f + f\alpha/p)^k} \quad [P_k]_{\alpha=p} = 1 - f + fp^{k-1}, \tag{21.24}$$

where f is the fraction of subunits in the oligomer form and α is the fluorescence yield of the oligomer relative to that of the monomer. The limit in the second equation describes the case in which oligomerization does not alter the fluorescence yield of the monomeric subunit. As shown in Fig. 21.2 for various oligomer sizes, the parameters P_k differ from one only in the case where oligomers are present ($f \neq 0$). In theory, curve fitting the measured values of P_k as a function of the index k to (21.24) should yield values of the oligomerization degree p, the relative fluorescence yield α, and the fractional oligomerization f. A subsequent analysis of the values of f as a function of the total concentration of monomeric subunits should yield a consistent oligomerization number p as well as a value for the equilibrium association constant describing the oligomerization.

A general method for quantitatively characterizing polydispersity has been previously described [21.56]. In this method, a number of species R is postulated, giving rise to $2R - 1$ unknowns ($\langle C_1 \rangle, \langle C_2 \rangle, \ldots, \langle C_R \rangle; \alpha_2, \alpha_3, \ldots,$ α_R). Determining the values of the unknowns requires measurement of $2R-1$ values of B_k (21.17); i.e. the extrapolated magnitudes of $G_{mn}(0)$ as well as the values of γ_{m+n} for $m + n = 2, 3, \ldots, 2R$. The situation then becomes a problem of converting the known to the unknown values. Algebraic solutions

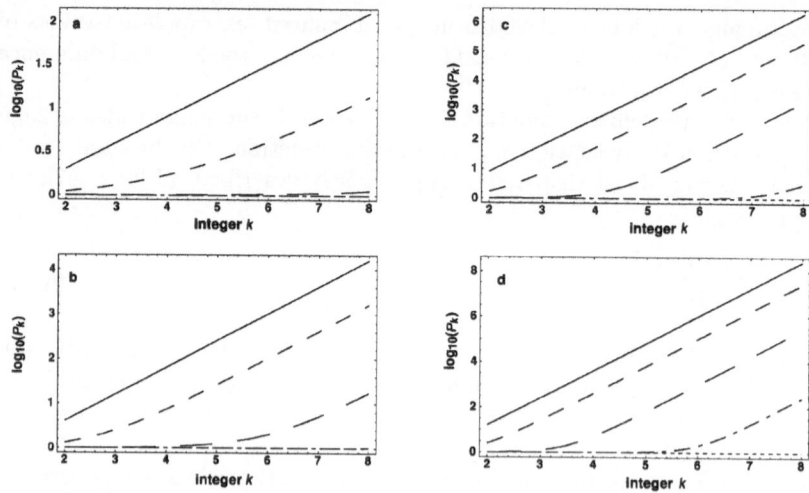

Fig. 21.2 a–d. Oligomerization detected by the parameters P_k. The parameters P_k were calculated from (21.24) with $\alpha = p$. The degree of oligomerization, p, equals **a** 2, **b** 4, **c** 8, and **d** 16. The fraction of the monomeric subunits present as dimers, f, equals (**line**) 1, (**short dash**) 0.1, (**long dash**) 10^{-3}, (**dash-dot**) 10^{-6}, and (**dot**) 0

for the concentrations $\langle C_i \rangle$ and relative fluorescence yields α_i, which are not readily condensed to simple forms, have been published for values of R up to three [21.56].

As shown in (21.16) and (21.17), the magnitudes $G_{mn}(0)$ are larger for smaller numbers of observed molecules. In practice, the number of fluorescent molecules in the sample volume must be ≤ 1000 and sample volume is usually $\geq 1\,\mu m^3$. However, many biologically significant oligomeric processes may occur at concentrations greater than $10^3/\mu m^3 \approx 2\,\mu M$. One method of decreasing the number of fluorescent molecules in the sample volume without reducing the concentration is to mix labeled and unlabeled species. For a fully oligomerized sample in which the oligomers consist of p monomers, and oligomerization does not change the fluorescence yield of the monomeric subunits,

$$
B_k = \frac{\displaystyle\sum_{s=1}^{p} s^k \langle C_s \rangle}{\left[\displaystyle\sum_{s=1}^{p} s \langle C_s \rangle\right]^k}, \tag{21.25}
$$

where $\langle C_s \rangle$ is the average concentration of oligomers containing s labeled subunits. The distribution of labeled monomers in oligomers is most simply described by a binomial distribution:

$$
\langle C_s \rangle = \langle C_o \rangle \frac{p!}{(p-s)!s!}\eta^{p-s}(1-\eta)^s, \tag{21.26}
$$

where $\langle C_o \rangle$ is the average total oligomer concentration and η is the fraction of monomeric subunits which are labeled. By using (21.26) in (21.25), one finds that [21.72]

$$B_2 = \frac{1}{\langle C_o \rangle} \left[1 + \frac{1 - \eta}{\eta p} \right], \qquad (21.27)$$

$$B_3 = \frac{1}{\langle C_o \rangle^2} \left[1 + \frac{3(1 - \eta)}{\eta p} + \frac{(1 - 2\eta)(1 - \eta)}{(\eta p)^2} \right],$$

$$B_4 = \frac{1}{\langle C_o \rangle^3} \left[1 + \frac{6(1 - \eta)}{\eta p} + \frac{(7 - 11\eta)(1 - \eta)}{(\eta p)^2} + \frac{(1 - 6\eta + 6\eta^2)(1 - \eta)}{(\eta p)^3} \right].$$

The expression for B_2 in (21.27) has been previously described [21.20]. All of the expressions in (21.27) increase with decreasing η, for $0 \le \eta \le 1$ and $p \ge 1$. When $\eta = 1$, B_k is the inverse of the $(k - 1)$-th power of the total oligomer concentration, $\langle C_o \rangle$. When $\eta \to 0$ or $p = 1$, B_k is the inverse of the $(k - 1)$-th power of the total concentration of labeled subunits, $\eta p \langle C_o \rangle$.

21.2.5 Experimental Considerations

The central components of most FCS instruments are a focused laser beam (radius $\approx 1\,\mu$m) for fluorescence excitation, an optical microscope through which fluorescence is collected (by using both a high numerical aperture objective and a confocal, limiting aperture at an intermediate image plane), and a single photon counting detector (photomultiplier or silicon avalanche diode). Pulses are counted as the number of photons detected in consecutive sample times. The laser, microscope, detector and supporting optics are mounted on a vibration-isolated air table.

The high order autocorrelation functions $U_{mn}(\tau)$ of fluctuations in the rate of detected photons may be calculated from the recorded number of pulses per sample interval as [21.20]

$$U_{mn}(k\Delta T) = \frac{1}{\langle n \rangle^{m+n}} \frac{1}{M - P} \sum_{i=1}^{M-P} (n_i - \langle n \rangle)^m (n_{i+k} - \langle n \rangle)^n$$

$$- \frac{1}{\langle n \rangle^{m+n}} \frac{1}{M^2} \sum_{i=1}^{M} (n_i - \langle n \rangle)^m \sum_{j=1}^{M} (n_j - \langle n \rangle)^n$$

$$k = 1, 2, \ldots, P, \qquad (21.28)$$

where n_i (for $i = 1, 2, \ldots, M$) is the number of pulses occurring between times $(i - 1)\Delta T$ and $i\Delta T$, P is the number of times $k\Delta T$ for which the autocorrelation function is calculated, M is the number of data points acquired, and the average number of pulses per sample time is

$$\langle n \rangle = \frac{1}{M} \sum_{i=1}^{M} n_i. \qquad (21.29)$$

The calculated pulse fluctuation autocorrelation functions $U_{mn}(\tau)$ are not identical to the photon fluctuation autocorrelation functions, denoted here as $V_{mn}(\tau)$. For values of $m + n > 2$ and $\tau > 0$, the statistical nature of photon counting affects the pulse fluctuation autocorrelation functions. The functions $V_{mn}(\tau)$ may be obtained from the calculated functions $U_{mn}(\tau)$ as previously described [21.73]. The lowest order relationships are (for $\tau > 0$)

$$V_{11}(\tau) = U_{11}(\tau)$$

$$V_{12}(\tau) = U_{12}(\tau) - \frac{1}{\langle n \rangle} U_{11}(\tau)$$

$$V_{13}(\tau) = U_{13}(\tau) - \frac{3}{\langle n \rangle}[U_{12}(\tau) + U_{11}(\tau)] + \frac{2}{\langle n \rangle^2} U_{11}(\tau)$$

$$V_{22}(\tau) = U_{22}(\tau) - \frac{2}{\langle n \rangle} U_{12}(\tau) + \frac{1}{\langle n \rangle^2} U_{11}(\tau). \tag{21.30}$$

The correction terms in (21.30) can be large if the average number of pulses detected per sample time $\langle n \rangle$ is low; for large count rates, these terms are negligible. For all values of m and n, both $U_{mn}(0)$ and $V_{mn}(0)$ contain additional contributions arising from the photon counting statistics. The values of $V_{mn}(0)$ may be obtained by extrapolating from values of $V_{mn}(\tau)$ with $\tau > 0$.

For weakly fluorescent samples, the photon signal may contain a significant contribution arising from background light. In these cases, the fluorescence fluctuation autocorrelation functions $G_{mn}(\tau)$ may be calculated from the photon fluctuation autocorrelation functions $V_{mn}(\tau)$ as [21.73]

$$G_{mn}(\tau) = \left[\frac{\langle n \rangle}{\langle n \rangle - \langle b \rangle}\right]^{m+n} V_{mn}(\tau), \tag{21.31}$$

where $\langle b \rangle$ is the average number of pulses per sample time arising from background. In the derivation of (21.31), it is assumed that fluctuations in the background are not correlated with fluctuations in the fluorescence, and that fluctuations in the background are not correlated with themselves on the time scale of interest.

Photon detectors require a finite time to register a pulse. During this dead time, additional photons cannot be counted. The dead time arises from the speed of the electronic components and from the pulse width. If the count rate is high enough so that a significant fraction of the incident photons are not detected, additional corrections to $U_{mn}(\tau)$ and $V_{mn}(\tau)$ are required [21.73].

Quantitative analysis of the $G_{mn}(\tau)$ requires knowledge of the shape of the sample volume. The extrapolated values of $G_{mn}(0)$ depend on the values of the parameters γ_{m+n} (21.18), and the rate and shape of the decays of the $G_{mn}(\tau)$ depend on the mechanism giving rise to fluorescence fluctuations (e.g., diffusion) as well as the shape of $W(\mathbf{r})$ (21.2). The values of the parameters γ_{m+n}, which are related to the norms of the function $W(\mathbf{r})$, may be found by experimentally measuring the extrapolated values of $G_{mn}(0)$ as a function of $1/\langle C\rangle^{(m+n-1)}$ for a monodisperse sample of concentration $\langle C\rangle$ [21.74]. Theoretical assumptions about the shape of $W(r)$ are not required for quantitative analysis of the $G_{mn}(0)$. Measurements of the γ_{m+n} can also be used to partially verify theoretical assumptions about the shape of $W(r)$.

21.2.6   Experimental Applications

Temporal high order FCS was first experimentally demonstrated on suspensions of the fluorescent lipid dioctadecyltetramethylindocarbo cyanine (diI) in solutions of water and ethanol [21.56]. Typical experimentally obtained $G_{mn}(\tau)$ are shown in Fig. 21.3. The characteristic decay times decrease with $m+n$ as theoretically predicted for diffusion through a focused laser beam (21.20). The values of $G_{mn}(0)$ increase with $m+n$ as theoretically predicted for observed particle numbers less than one (21.16, 21.17). The functions

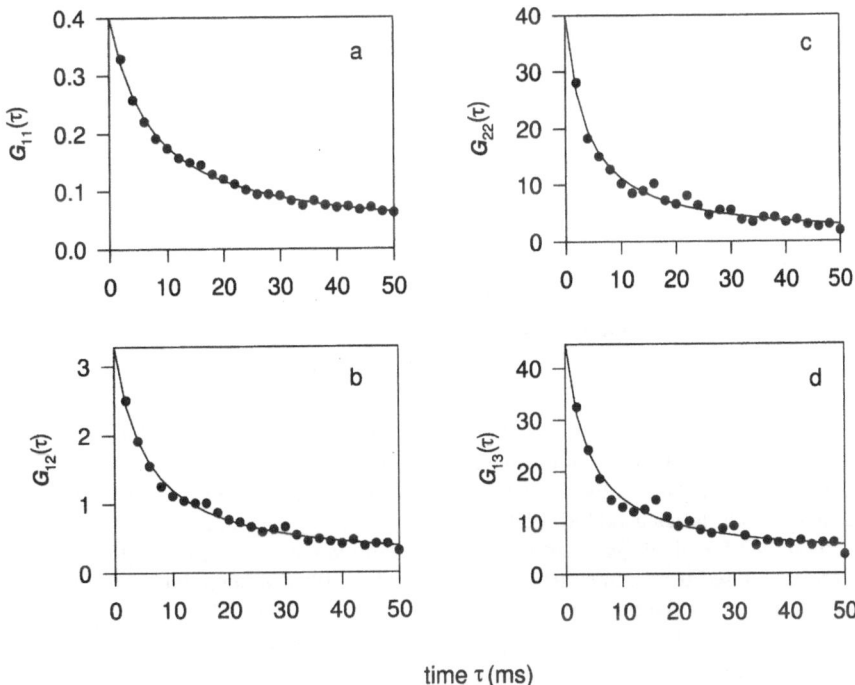

time τ (ms)

Fig. 21.3 a–d. Experimental high order fluorescence fluctuation autocorrelation functions for fluorescent lipid aggregates. Typical experimentally determined functions **a** $G_{11}(\tau)$, **b** $G_{12}(\tau)$, **c** $G_{22}(\tau)$ and **d** $G_{13}(\tau)$ are shown for 10^{-7} M diI in water/ethanol (4:1, vol:vol). The $G_{mn}(\tau)$ were calculated from a single fluorescence record of 163 840 sample points of 2 ms duration. Lines show the best fits to the theoretical forms for two-dimensional diffusion of a monodisperse suspension through a focused, Gaussian-shaped laser beam. (Adapted from the *Biophys. J.* **52**, 264 (1987), by copyright permission of the Biophyiscal Society)

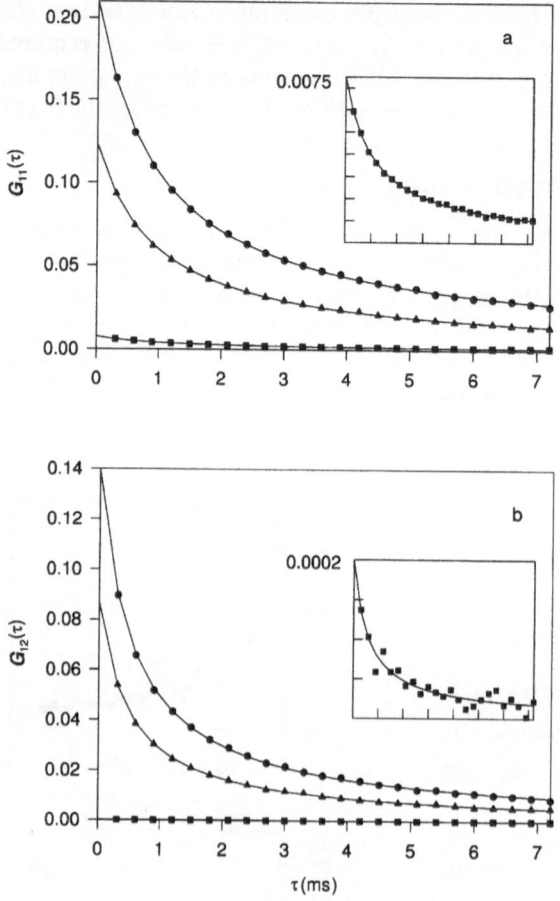

Fig. 21.4 a,b. Experimental high order fluorescence fluctuation autocorrelation functions for mixtures of fluorescently labeled IgG and B-phycoerythrin. Shown are typical experimentally determined functions **a** $G_{11}(\tau)$ and **b** $G_{12}(\tau)$. The concentrations of labeled IgG and BPE, respectively, (in units of molec/μm^3) were ($\bullet$) 0, 0.60; ($\blacktriangle$) 7.8, 0.36; ($\blacksquare$, insets) 18, 0. The sample volume was $\approx 1.2\,\mu$m^3. $G_{mn}(\tau)$ were calculated from single fluorescence records of 10^6 sample points of 0.3 ms duration. Lines show the best fits to Lorentzians in $\tau^{1/2}$. (Adapted from the *Proc. Natl. Acad. Sci. USA* **86**, 6149 (1989), by copyright permission of the National Academy of Sciences)

$G_{mn}(\tau)$ were much higher in magnitude for samples with higher water content (data not shown), where lipid aggregates are expected to be larger and more sparse. Experimentally determined functions $G_{mn}(\tau)$ were equivalent within experimental uncertainty to $G_{mn}(\tau)$, as predicted for samples in equilibrium [21.71].

In a later study [21.57], the accuracy with which high order fluorescence fluctuation autocorrelation can detect and quantify polydispersity was more rigorously evaluated. In this work, solutions containing mixtures of fluorescently labeled IgG and the highly fluorescent protein B-phycoerythrin (BPE) were examined. As shown in Fig. 21.4, the measured $G_{mn}(\tau)$ differed considerably in magnitude for solutions containing labeled IgG only, BPE only, or a mixture of the two proteins. Dramatic differences were not apparent in the records of the fluctuating, measured fluorescence intensity (data not shown). Quantitative analysis of the extrapolated magnitudes $G_{mn}(0)$, after corrections for background and for the statistics of photon counting (see above), accurately retrieved the concentrations of IgG and BPE.

21.3 Spatial High Order FCS

21.3.1 Overview

In diffusional FCS, movement of fluorescent molecules through a small illuminated region generates temporal fluctuations in the fluorescence emitted from the region. On natural or model cell membranes, difficulties sometimes arise with this approach if a significant fraction of the fluorescent molecules are translationally immobile. The focused laser beam can generate a bleached region and subsequent small mechanical fluctuations cause fluorescence fluctuations that are not related to molecular number fluctuations. In addition, the diffusion must be relatively fast so that a large enough number of data points can be acquired within an experimentally reasonable time. Scanning FCS (S-FCS) circumvents these difficulties by translating the focused laser beam through the sample or by translating the sample through the focused laser beam [21.9,21.29,21.30,21.45–21.48,21.50]. Although the fluorescence fluctuations are temporal, they arise from spatial fluctuations in the molecular number density.

An alternative to S-FCS is to use a charged-coupled device (CCD) detector to record the two-dimensional fluorescence intensity distribution on a planar model membrane or cell membrane. In this version of FCS, called imaging FCS (I-FCS), fluorescence is excited with a diffusely focused laser and detected at all pixels simultaneously [21.51,21.52,21.75] or excited with a sharply focused laser beam and detected by scanning with a confocal aperture [21.22,21.49,21.53–21.55]. The pixel-to-pixel fluorescence fluctuations from a single fluorescence image are spatially autocorrelated. A major advantage of I-FCS, compared to conventional FCS, is that fast diffusion is not required. In addition, a large number of data points rather than one is collected per sample time, which dramatically increases the signal-to-noise ratio of the autocorrelation functions. I-FCS does not require the acquisition of consecutive images, although in theory this approach would provide additional information.

21.3.2 Spatial Fluorescence Fluctuation Autocorrelation Functions

In I-FCS, the first order fluorescence fluctuation autocorrelation function is defined as

$$G(\boldsymbol{\rho}) = \frac{\langle \delta F(r) \delta F(r + \boldsymbol{\rho}) \rangle}{\langle F \rangle^2} \, , \tag{21.32}$$

where the brackets denote a spatial average. In analogy to the relationship between (21.1) and (21.11), one may define a series of high order spatial fluorescence fluctuation autocorrelation functions as

$$G_{mn}(\boldsymbol{\rho}) = \frac{\langle \delta F^m(r) \delta F^n(r + \boldsymbol{\rho}) \rangle - \langle \delta F^m(r) \rangle \langle \delta F^n(r) \rangle}{\langle F \rangle^{m+n}} \, . \tag{21.33}$$

For I-FCS with a focused, scanning laser beam, $F(r)$ is the fluorescence at position r and $\delta F(r) = F(r) - \langle F \rangle$ is the fluorescence fluctuation at position r [21.49]. For I-FCS in which a diffusely focused laser beam is employed, the measured, spatially dependent fluorescence intensity must be normalized by a function describing the spatial dependence of the excitation light after background subtraction [21.52]. In these measurements,

$$F(r) = \frac{Z(r) - B(r)}{S(r)} \, , \tag{21.34}$$

where $Z(r)$, $B(r)$ and $S(r)$ are the measured fluorescence, background, and excitation light intensities, respectively, at position r.

21.3.3 Autocorrelation Function Magnitudes and Decay Shapes

In most I-FCS instruments, the sample area corresponding to a single pixel area is less than or approximately equivalent to the optical resolution. For fluorescent monomers or oligomers much smaller than the pixel size, the extrapolated value of $G(0)$ is predicted to be proportional to that described above for temporal FCS (21.7) [21.49]. Thus, the extrapolated values of $G_{mn}(0)$ should be proportional to those shown in (21.16) after correction for the statistics of photon collection and detection [21.73]. For I-FCS with a focused, scanning laser beam, the decay shape of $G(\rho)$ is related to the beam shape [21.49]; similar results are expected for $G_{mn}(\rho)$. For I-FCS with broad illumination, the decay shapes of $G(\rho)$ and $G_{mn}(\rho)$ are related to the microscope's optical transfer function [21.52,21.75]. For clusters of fluorescent molecules that approach or are larger than the optical resolution, both the autocorrelation function magnitudes and their decay shapes deviate from the values predicted for small fluorescent entities, for both types of I-FCS [21.51, 21.52,21.65,21.75].

21.3.4 Experimental Considerations

In temporal FCS, fluorescence fluctuation autocorrelation functions are usually corrected for background by using a multiplicative factor found from the average signal intensity and the average background intensity (21.31). Similar approaches have been used in I-FCS [21.22]. The background may also be subtracted before calculating the autocorrelation functions (21.34) [21.52].

When broad illumination is employed in I-FCS, the background-corrected fluorescence, $Z(\boldsymbol{r}) - B(\boldsymbol{r})$, must be normalized by the spatial dependence of the excitation light, $S(\boldsymbol{r})$ (21.34). One way of measuring $S(\boldsymbol{r})$ would be to perfuse the sample chamber, after image acquisition, with a solution containing a high concentration of fluorescent molecules. Another method of obtaining this function is to smooth the image of interest [21.52]. For surface-based samples, it is convenient to use the evanescent field created by a totally internally reflected laser beam [21.52]. This illumination method produces a lower background and gives a smoother excitation profile because the laser does not pass through the microscope optics.

The autocorrelation functions $G_{mn}(\rho)$ may be calculated from a single fluorescence image, after correction for background and the spatial dependence of the illumination profile, at positions $k\Delta\rho$ by using (21.33) and [21.75]

$$
\langle \delta F^m(r) \delta F^n(r + k\Delta\rho) \rangle = \frac{1}{(M_x - k)M_y + (M_y - k)M_x}
$$

$$
\times \left\{ \sum_{i=1}^{M_x-k} \sum_{j=1}^{M_y} [F_{i,j} - \langle F \rangle]^m [F_{i+k,j} - \langle F \rangle]^n \right.
$$

$$
\left. + \sum_{i=1}^{M_x} \sum_{j=1}^{M_y-k} [F_{i,j} - \langle F \rangle]^m [F_{i,j+k} - \langle F \rangle]^n \right\} , \tag{21.35}
$$

$$
\langle \delta F^m(r) \rangle = \frac{1}{M_x M_y} \sum_{i=1}^{M_x} \sum_{j=1}^{M_y} [F_{i,j} - \langle F \rangle]^m , \tag{21.36}
$$

$$
\langle F \rangle = \frac{1}{M_x M_y} \sum_{i=1}^{M_x} \sum_{j=1}^{M_y} F_{i,j} , \tag{21.37}
$$

where $F_{i,j}$ is the background-corrected, normalized intensity at pixel (i,j); M_x and M_y are the numbers of rows and columns, respectively; k is the number of positions $k\Delta\rho$ for which the autocorrelation function is calculated; and $\Delta\rho$ is the length of one (square) pixel.

In temporal FCS, contributions from photon counting statistics (shot noise) are present in $G_{11}(\tau)$ only in the first channel ($\tau = 0$). However, these contributions are in general present in high order autocorrelation functions in all channels ($\tau > 0$) (21.30). The necessity of post-calculation corrections to spatial high order concentration functions for the noise associated with the statistics of photon counting has not yet been considered in I-FCS, either theoretically or experimentally.

21.3.5 Experimental Application

High order autocorrelation in I-FCS has been experimentally demonstrated on fluorescently labeled anti-trinitrophenyl (TNP) IgE specifically bound to substrate-supported planar phospholipid bilayers composed of mixtures of trinitrophenylaminocaproylphosphatidylethanolamine (TNP-cap-DPPE) and dipalmitoylphosphatidylcholine (DPPC) [21.75]. In this experimental system, the membrane-bound IgE spontaneously self-associates into clusters larger in size than optical resolution. The IgE-coated membranes were illuminated with an argon ion laser beam that was totally internally reflected at the substrate/solution interface and the evanescently excited fluorescence was measured with a slow-scan, cooled, single-photon counting CCD camera. The images were corrected for background and for the elliptically Gaussian spatial dependence of the evanescent excitation intensity. A series of high order pixel-to-pixel spatial fluorescence fluctuation autocorrelation functions were calculated from single images.

As shown in Fig. 21.5, the first order fluorescence fluctuation autocorrelation functions, $G_{11}(\rho)$, varied considerably in magnitude and shape for membranes with different densities of bound, labeled IgE, and changed significantly when IgE-coated membranes were further treated with unlabeled, polyclonal anti-IgE. Furthermore, for each sample type, the extrapolated val-

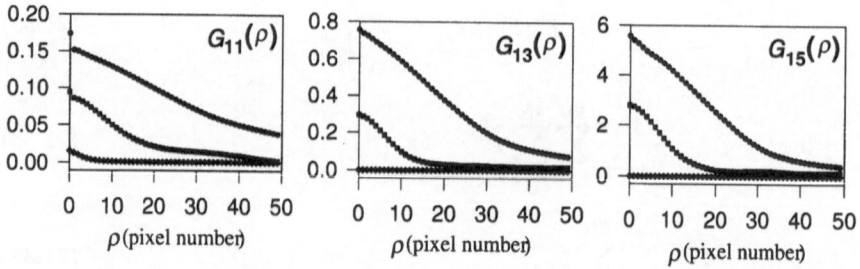

Fig. 21.5. Experimental high order spatial fluorescence fluctuation autocorrelation functions for IgE on substrate-supported planar membranes. Functions $G_{1n}(\rho)$ were calculated from images that had been corrected for background and for the spatially non-uniform illumination as described in (21.32–21.37). Values of $G_{11}(\rho)$, $G_{13}(\rho)$ and $G_{15}(\rho)$ are shown for TNP-cap-DPPE/DPPC membranes (bottom curves) 20% saturated with fluorescently labeled IgE, (middle curves) 80% saturated with fluorescently labeled IgE, and (top curves) 60% saturated with fluorescently labeled IgE and further treated with unlabeled, polyclonal anti-IgE. The pixel length was 0.11 μm. (Adapted from Vanden Broek et al. [21.75])

ues of $G_{1n}(0)$ increased dramatically with index n. The correlation distances, defined as the positions at which the $G_{1n}(\rho)$ decayed to one-half their maximum values, decreased with index n, as expected. The $G_{1n}(\rho)$ calculated from images of planar membranes containing 2 mol% fluorescently labeled lipids were much lower in magnitude than those calculated for membranes containing bound, clustered, fluorescently labeled IgE (data not shown). This work demonstrates the potential usefulness of high order fluorescence fluctuation autocorrelation in I-FCS.

21.4   Discussion

The use of high order fluorescence fluctuation autocorrelation in temporal and spatial FCS has been described in this chapter. The magnitudes and shapes of $G_{mn}(\tau)$ have been theoretically predicted for two-dimensional samples in which fluorescent molecules move through a focused Gaussian laser beam by diffusion or flow [21.56]. Initial experimental demonstrations of high order temporal FCS included studies of fluorescent lipid aggregates [21.56] and mixtures of fluorescently labeled IgG and B-phycoerythrin [21.57]. Good agreement between the theoretical predictions for $G_{mn}(\tau)$ and the experimental high order autocorrelation functions was found only after correcting the experimentally obtained functions for the noise associated with the statistics of photon counting [21.73] and calibrating the norms of the shape of the observation volume [21.74]. Recently, the feasibility of high order autocorrelation in spatial FCS has been experimentally demonstrated on fluorescently labeled anti-hapten IgE which cluster when specifically bound to haptenated, substrate-supported planar membranes [21.75].

The small body of published work in high order FCS [21.56,21.57,21.73–21.75] suggests that much more information can be obtained from the same temporal or spatial record of fluorescence fluctuations required for calculating the first order autocorrelation function, by calculating additional high order autocorrelation functions. Independent information is contained in these high order functions for samples with more than one type of fluorescent species, a characteristic of many, if not most, samples of interest in biophysics. An alternative, but theoretically equivalent, method to analyzing the magnitudes $G_{mn}(0)$ is to analyze the high order moments of the fluorescence fluctuations [21.76,21.77].

High order autocorrelation is not unique to FCS [21.78]. This technique has found application in a diverse array of fields, including light scattering [21.79–21.81], electrophysiology [21.82,21.83], and turbulence in fluids [21.84].

Numerous aspects of high order autocorrelation in FCS have not yet been explored. For temporal high order FCS, the shapes of the $G_{mn}(\tau)$ have been derived only for a two-dimensional sample in which fluorescent molecules move through a focused Gaussian laser beam by diffusion or flow. Other observation shapes have not been explicitly treated; e.g., three-dimensional samples illuminated by a focused laser beam [21.61,21.62], the evanescent illumination created by internally reflecting a laser beam at a surface/solution interface [21.38–21.42], or excitation by two-photon absorption [21.29,21.30]. It is likely that temporal high order FCS will be useful for studying the variety of mechanisms other than diffusion or flow which can give rise to fluorescence fluctuations. These mechanisms include rotational diffusion [21.11–21.14], conformational fluctuations [21.26,21.69], and biochemical kinetics [21.16,21.17,21.27,21.66–21.68]. Neither theoretical treatments nor experimental demonstrations of high order FCS for the study of these mechanisms have been presented. Finally, the use of high order autocorrelation in spatial FCS has not yet been developed in a theoretically rigorous manner [21.75]. This method holds considerable promise as an approach toward characterizing the motions and organizations of fluorescently labeled molecules on viable cell surfaces.

Acknowledgements

We thank Arthur G. Palmer for permission to reproduce Figs. 21.3 and 21.4; and Willem Vanden Broek and Zhengping Huang for permission to reproduce Fig. 21.5. This work was supported by NIH grant GM–37145 and NSF grant MCB–9728116.

References

21.1 E.L. Elson and D. Magde: Biopolymers **13**, 1–27 (1974)
21.2 D. Magde, E.L. Elson, W.W. Webb: Biopolymers **13**, 29–61 (1974)
21.3 M. Ehrenberg and R. Rigler: Chem. Phys. **4**, 390–401 (1974)
21.4 D.E. Koppel: Phys. Rev. **10**, 1938–1945 (1974)
21.5 S.R. Aragon and R. Pecora: Biopolymers **14**, 119–138 (1975)
21.6 G.D.J. Phillies: Biopolymers **14**, 499–508 (1975)
21.7 R. Rigler, P. Grasselli and M. Ehrenberg: Phys. Scr. **19**, 486–490 (1979)
21.8 C. Andries, W. Guedens, J. Clauwaert, and H. Geerts: Biophys. J. **43**, 345–354 (1983)
21.9 T. Meyer and H. Schindler: Biophys. J. **54**, 983–993 (1988)
21.10 B.A. Scalettar, J.E. Hearst, and M.P. Klein: Macromolecules **22**, 4550–4559 (1989)
21.11 M. Ehrenberg and R. Rigler: Quart. Rev. Biophys. **9**, 69–81 (1976)
21.12 J. Borejdo, S. Putnam, and M.F. Morales: Proc. Natl. Acad. Sci. USA **76**, 6346–6350 (1979)
21.13 H. Hoshikawa and H. Asai: Biophys. Chem. **22**, 167–172 (1985)
21.14 P. Kask, P. Piksarv, Ü. Mets, M. Pooga, and E. Lippmaa: Eur. Biophys. J. **14**, 257–261 (1987)
21.15 D. Magde, W.W. Webb, and E.L. Elson: Biopolymers **17**, 361–376 (1978)
21.16 R.D. Icenogle and E.L. Elson: Biopolymers **22**, 1919–1948 (1983)
21.17 R.D. Icenogle and E.L. Elson: Biopolymers **22**, 1949–1966 (1983)
21.18 E.L. Elson: Annu. Rev. Phys. Chem. **36**, 379–406 (1985)

21.19 N.O. Petersen and E.L. Elson: Meth. Enzy. **130**, 454–484 (1986)

21.20 A.G. Palmer and N.L. Thompson: Chem. Phys. Lipids. **50**, 253–270 (1989)

21.21 N.L. Thompson: Fluorescence correlation spectroscopy. In: Topics in Fluorescence Spectroscopy Vol. 1 J.R. Lakowicz, Ed. (Plenum Press, 1991) p.337

21.22 P.W. Wiseman and N.O. Petersen: Biophys. J. **76**, 963–977 (1999)

21.23 K. Post, M. Pitschke, O. Schafer, H. Wille, T.R. Appel, D. Kirsch, I. Mehlhorn, H. Serban, S.B. Prusiner, and D. Riesner: Biol. Chem. **379**, 1307–1317 (1998)

21.24 M. Pitschke, R. Prior, M. Haupt, and D. Reisner: Nat. Med. **4**, 832–834 (1998)

21.25 S. Bjorling, M. Kinjo, Z. Földes-Papp, E. Hagman, P. Thyberg, and R. Rigler: Biochem. **37**, 12971–12978 (1998)

21.26 G. Bonnet, O. Krichevsky, and A. Libchaber: Proc. Natl. Acad. Sci. USA **95**, 8602–8606 (1998)

21.27 F.W. Sevenich, J. Langowski, V. Weiss, and K. Rippe: Nucl. Acids Res. **26**, 1373–1381 (1998)

21.28 M.A. Hink, A. van Hoek, and A.J.W.G. Visser: Langmuir **15**, 992–997 (1999)

21.29 K.M. Berland, P.T.C. So, and E. Gratton: Biophys. J. **68**, 694–701 (1995)

21.30 K.M. Berland, P.T.C. So, Y. Chen, W.W. Mantulin, and E. Gratton: Biophys. J. **71**, 410–420 (1996)

21.31 P. Schwille, F.J. Meyer-Almes, and R. Rigler: Biophys. J. **72**, 1878–1886 (1997)

21.32 A. Koltermann, U. Kettling, J. Bieschke, T. Winkler, and M. Eigen: Proc. Natl. Acad. Sci. USA **95**, 1421–1426 (1998)

21.33 R. Rigler, Z. Földes-Papp, F.J. Meyer-Almes, C. Sammet, M. Volcker, and A. Schnetz: J. Biotech. **63**, 97–109 (1998)

21.34 T. Winkler, U. Kettling, A. Koltermann, and M. Eigen: Proc. Natl. Acad. Sci. USA **96**, 1375–1378 (1999)

21.35 M. Brinkmeier, K. Dorre, J. Stephan and M. Eigen: Anal. Chem. **71**, 609–616 (1999)

21.36 E. Haas and I.Z. Steinberg: Biophys. J. **46**, 429–437 (1984)

21.37 A. Van Orden and R.A. Keller: Anal. Chem. **70**, 4463–4471 (1998)

21.38 N.L. Thompson, T.P. Burghardt, and D. Axelrod: Biophys. J. **33**, 435–454 (1981)

21.39 N.L. Thompson: Biophys. J. **38**, 327–329 (1982)

21.40 N.L. Thompson and D. Axelrod: Biophys. J. **43**, 103–114 (1983)

21.41 R.L. Hansen and J.M. Harris: Anal. Chem. **70**, 2565–2575 (1998)

21.42 R.L. Hansen and J.M. Harris: Anal. Chem. **70**, 4247–4256 (1998)

21.43 M.B. Weissman, H. Schindler, and G. Feher: Proc. Natl. Acad. Sci. USA **73**, 2776–2780 (1976)

21.44 N.O. Petersen: Can. J. Biochem. Cell. Biol. **62**, 1158–1166 (1984)

21.45 N.O. Petersen: Biophys. J. **49**, 809–815 (1986)

21.46 N.O. Petersen, D.C. Johnson, and M.J. Schlesinger: Biophys. J. **49**, 817–820 (1986)

21.47 P.R. St.-Pierre and N.O. Petersen: Biophys. J. **58**, 503–511 (1990)

21.48 P.R. St.-Pierre and N.O. Petersen: Biochem. **31**, 2459–2463 (1992)

21.49 N.O. Petersen, P.L. Hoddelius, P.W. Wiseman, O. Seger, and K.E. Magnusson: Biophys. J. **65**, 1135–1146 (1993)

21.50 D.E. Koppel, F. Morgan, A.E. Cowan, and J.H. Carson: Biophys. J. **66**, 502–507 (1994)

21.51 M.D. Wang and D. Axelrod: Bioimaging **2**, 22–35 (1994)
21.52 Z.P. Huang and N.L. Thompson: Biophys. J. **70**, 2001–2007 (1996)
21.53 E. Fire, C.M. Brown, M.G. Roth, Y.I. Henis, and N.O. Petersen: J. Biol. Chem. **272**, 29538–29545 (1997)
21.54 B.J. Rasmusson, T.D. Flanagan, S.J. Turco, R.M. Epand, and N.O. Petersen: Biochim. Biophys. Acta. Molec. Cell. Res. **1404**, 338–352 (1998)
21.55 C.M. Brown and N.O. Petersen: J. Cell. Sci. **111**, 271–281 (1998)
21.56 A.G. Palmer and N.L. Thompson: Biophys. J. **52** 257–270 (1987)
21.57 A.G. Palmer and N.L. Thompson: Proc. Natl. Acad. Sci. USA **86**, 6148–6152 (1989)
21.58 U. Meseth, T. Wohland, R. Rigler, and H. Vogel: Biophys. J. **76**, 1619–1631 (1999)
21.59 H. Asai: Jpn. J. Appl. Phys. **19**, 2279–2282 (1980)
21.60 S.M. Sorscher and M.P. Klein: Rev. Sci. Instr. **51**, 98–102 (1980)
21.61 H. Qian and E.L. Elson: Proc. SPIE **909**, 352–359 (1988)
21.62 H. Qian and E.L. Elson: Appl. Optics **30**, 1185–1195 (1991)
21.63 M. Hattori, H. Shimizu, and H. Yokoyama: Rev. Sci. Instr. **67**, 4064–4071 (1996)
21.64 R.L. Hansen, X.R. Zhu, and J.M. Harris: Anal. Chem. **70**, 1281–1287 (1998)
21.65 K. Starchev, J.W. Zhang, and J. Buffle: J. Coll. Int. Sci. **203**, 189–196 (1998)
21.66 B. Rauer, E. Neumann, J. Widengren, and R. Rigler: Biophys. Chem. **58**, 3–12 (1996)
21.67 P. Schwille, F. Oehlenschlager, and N.G. Walter: Biochem. **35**, 10182–10193 (1996)
21.68 J. Klingler and T. Friedrich: Biophys. J. **73**, 2195–2200 (1997)
21.69 S. Wennmalm, L. Edman, and R. Rigler: Proc. Natl. Acad. Sci. USA **94**, 10641–10646 (1997)
21.70 Z. Földes-Papp, P. Thyberg, S. Bjorling, A. Holmgren, and R. Rigler: Nucleosides Nucleotides **16**, 781–787 (1997)
21.71 I.Z. Steinberg: Biophys. J. **50**, 171–179 (1986)
21.72 M. Zelen and N.C. Severo: Probability functions. In: Handbook of Mathematical Tables. M. Abramowitz and I.A. Stegun, Eds. (Dover Publications, 1972) p. 928
21.73 A.G. Palmer and N.L. Thompson: Rev. Sci. Instr. **60**, 624–633 (1989)
21.74 A.G. Palmer and N.L. Thompson: Appl. Optics **28**, 1214–1220 (1989)
21.75 W. Vanden Broek, Z. Huang and N.L. Thompson: (1999) High order autocorrelation with imaging fluorescence correlation spectroscopy: Application to IgE on supported planar membranes. submitted
21.76 H. Qian and E.L. Elson: Proc. Natl. Acad. Sci. USA **87**, 5479–5483 (1990)
21.77 H. Qian and E.L. Elson: Biophys. J. **57**, 375–380 (1990)
21.78 J.B. Suck, D. Quitmann, and B. Maier: J. Physique **46**, 1–164 (1985)
21.79 M.M. Hurley and P. Harrowell: J. Chem. Phys. **105**, 10521–10526 (1996)
21.80 C.J.Oliver: Recent developments in photon correlation and spectrum analysis techniques. I. Instrumentation for photo detection spectroscopy. In: Scattering techniques applied to supramolecular and nonequilibrium systems. S. Chen, B. Chu, and R. Nossal, Eds. (Plenum Press, New York 1981) p.87
21.81 P.R. Smith and D.A. Green: Appl. Optics **34**, 8475–8482 (1995)
21.82 L.S. Leibovitch and J. Fischbarg: J. Theor. Biol. **119**, 287–297 (1986)
21.83 H. Mino: IEEE T Bio-Med. Eng. **40**, 970–980 (1993)
21.84 V.S. L'vov and I. Procaccia: Physica A **257**, 165–196 (1998)

22 FCS in Single Molecule Analysis

R. Rigler, S. Wennmalm, and L. Edman

22.1 Introduction

Fluorescence correlation spectroscopy (FCS) was introduced in the early 1970s for the analysis of thermodynamic fluctuations of chemical systems [22.1–22.3] in an attempt to complement chemical relaxation spectroscopy as introduced by Eigen and de Meyer for the analysis of ultrafast kinetics. Chemical relaxation refers to the adjustment of chemical reactions into a new equilbrium state after an instantaneous change of intensive parameters such as temperature, pressure or electric field [22.4]. Chemical fluctuations depend on the spontaneous change in number density of chemical systems due to processes involving transitions into the excited state [22.3], Brownian motion [22.1,22.3] as well as chemical kinetics [22.2,22.5].

Fluctuation amplitudes are linearly related to the inverse number of molecules or particles defining the system. In the early stages of FCS correlation amplitudes of about 10^{-4} to 10^{-6} could be obtained due to limitations in the signal-to-background ratios (S/B 1:1000) attainable at that time.

A decisive change occurred with the discovery that using confocal volume elements the signal-to-background ratio could be increased to 1000:1 and above [22.6,22.7]. It also provided the possibility of single molecule detection and spectroscopy for chemical and biochemical systems. With the sensitivity in recording molecular fluctuations increased by more than 6 orders of magnitude FCS finally became the powerful routine [22.8–22.10] as envisaged at its start.

22.2 Single Molecule Detection in Solution and Correlation Functions

The first successful experiments in detecting single molecules were carried out by Moerner and Orrit [22.11]. They used the inhomogenous spectral broadening of the emission of organic molecules at cryotemperature to detect the emission of a single terylene molecule in pentacene matrices. The absence of photodynamic processes at cryotemperatures and high absorption cross-sections (10^{-20}/cm^2) were important ingredients in the sucessful detection by absorpion (Moerner) and fluorescence (Orrit) spectroscopy. With the introduction of confocal volume elements and the suppression of the solvent background consisting of Raman scattering the observaton of dye molecules such as Rhodamin in solution and at room temperature became possible. [22.6,22.7,22.12].

Similar developments were carried out by Richard Keller and collaborators in their attempt to sequence DNA at the level of single molecules [22.13–22.15]. Background suppression is achieved by using differences in the lifetime of the excited state of dye and solvent and time gating [22.13]. This technology was used earlier to record emission spectra of nucleic acids with short excited state lifetimes and low quantum yields [22.16]. Compared to time gating using pulsed excitation of confocal single molecule, detection with continuous laser exitation provides superior sensitivity since the excitation frequency is only limited by the lifetime of the excited state and not by the pulse frequency of the ecitation source.

In original experiments carried out using avalanche photodiodes (APD) as detectors counting rates for single Rhodamin G molecule of 100 000 cps where achieved when 50 mW of the Ar-laser line at 514.5 nm was exciting a Gaussian confocal volume element (VE) of radii $w_{xy} = 0.25\,\mu m$ and $w_z = 1\,\mu m$ [22.6, 22.8,22.9]. At this excitation level about 10% of the molecules are in their triplet state which has a lifetime of around $1\,\mu s$ [22.17] as compared to the singlet state with a lifetime around 2 ns. The burst size spectrum of the emission of single Rh6G molecules was analyzed by a multichannel scaler in bin sizes which are compatible with the charcteristic diffusion time of Rh6G of $40\,\mu s$. (Fig. 22.1).

The question whether the emission bursts are due to single molecules can be answered by calculating the autocorrelation function for diffusion $G(t)$:

$$G(t) = \frac{1}{N} \frac{1}{1 + \frac{\tau}{\tau_D}} \left(\frac{1}{1 + \frac{\tau w_{xy}^2}{\tau_D w_z^2}} \right)^{\frac{1}{2}}, \tag{22.1}$$

with

$$G(t) = \frac{\langle I(t)\, I(t+t) \rangle}{I^2},$$

and

$$\tau_D = \frac{w_{xy}^2}{4D},$$

where N equals the number of molecules in the VE. The value of N after correction for the background is found to be 0.005 indicating that on average 5 Rh6G molecules can be found in 1000 volume elements, i.e. the average probability P of finding Rh6G molecules in the VE. Hence the probability for detecting 2 molecules occurring simultaneously in the VE as well as its relative contribution can be calculated [22.8,22.20].

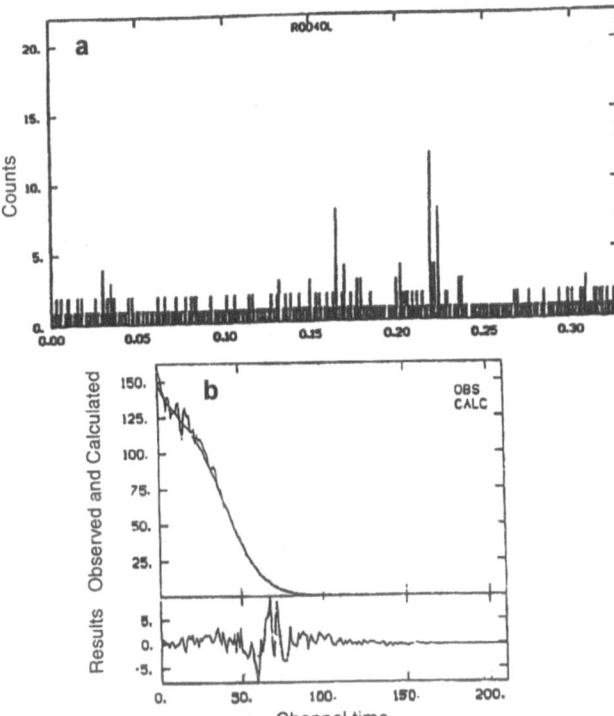

Fig. 21.1 a,b. Single molecule traces: **a** multichannel scaling traces for Single Rhodamin 6G molecules diffusing through the volume element, bin time = diffusion time (40 μs), **b** autocorrelation function of a), N (corrected for background) equals 5×10^{-3} corresponding to a concentration of 10^{-11} M [22.6,22.7]

Given a time averaged Poisson distribution of Rh molecules in the VE we calculate for the probability of finding one (p_1) and two (p_2) molecules in the VE from the average detection probability P:

$$p_1 = \frac{P}{1}e^{-1}, \quad p_2 = \frac{P^2}{2}e^{-2}, \tag{22.2}$$

and for

$$\frac{p_2}{p_1} = \frac{P}{2}e^{-1}, \tag{22.3}$$

one obtains for $p_1 = 1.8 \times 10^{-3}$, $p_2 = 1.7 \times 10^{-6}$ and for $p_2/p_1 = 9.2 \times 10^{-4}$. This means that more than 1000 single molecule events have to be registered before a double molecule event is detected given the concentrations used (Fig. 22.1).

22.3   Confocal Single Molecule Imaging

A direct consequence of confocal single molecule (SM) detection by confocal VE was the idea to use the confocal principle to image single molecules [22.18] (Fig. 22.2 a) as an alternative to near field imaging (e.g. [22.19]). Combining confocal SM detection with scanning over the area of interest reveals the two-dimensional position of single molecules adsorbed or attached specifically to surfaces. A typical image obtained from single DNA fragments mounted on a strepatvidinized glass surface by a biotin group linked to the 5'-end on one strand and visualized by a fluorescent tag (tetramethylrhodamin, TMR) attached by a linker to the 5'-end of the other strand is demostrated in Fig.22.2 b. This started the analysis of conformational kinetics of single biomolecules in our laboratory [22.20,22.21] and examplified the importance of FCS in the analysis of single molecule events.

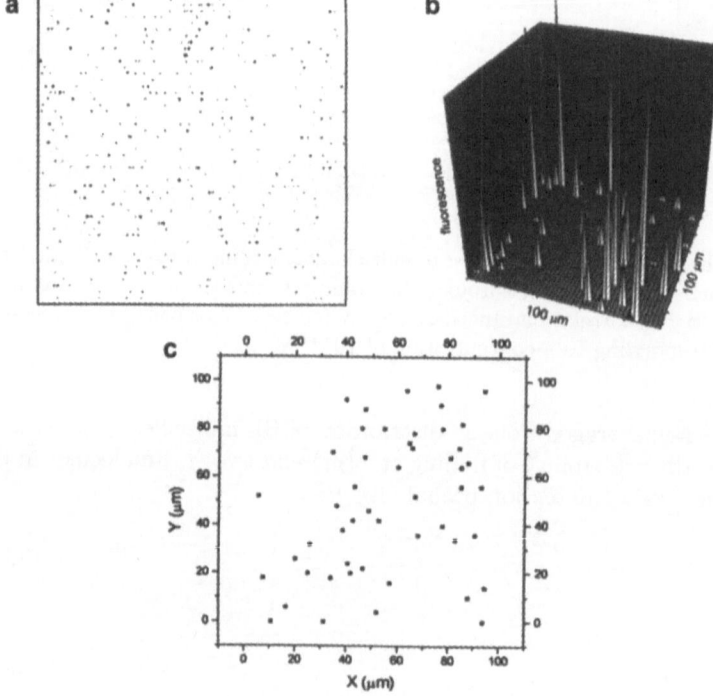

Fig. 22.2 a–c. Confocal Single molecule imaging; **a** Rhodamine 6G in agarose gel, **b** Single Rho dUTP labelled Ml3 DNA molecules in a 100 × 100 μm agarose gel matrix, **c** 217 basepair DNA synthesized by PCR with a primer containing biotin on the 5'end and a primer containing tetramethylrhodamin on the other 5'end. The DNA is bound by the biotin to the strepatavidinized glass surface [22.18,22.21]

22.4   Conformatial Transitions in Single DNA Molecules

From the determination of the excited states lifetime of 217 bp long DNA fragments tagged by TMR it became evident that the dye-DNA complex must
exist in two different states which exchange slower than the characteristic
time of the experiment, i.e. the diffusion time of the DNA fragment through
the volume element (5–30 ms) [22.20,22.22]. It also became clear from FCS
measurement (Fig. 22.3 a) that the correlation function contained in addition to the diffusion term was best described by a transition with stretched

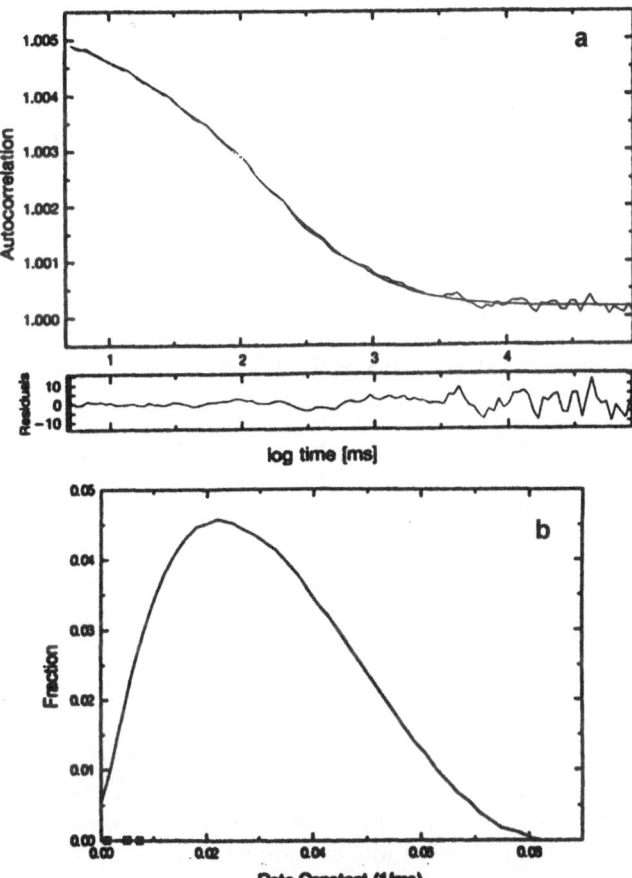

Fig. 22.3 a,b. Conformational transitions: **a** autocorrelation curve of the conformational transition in DNA. The volume element has been enlarged to yield a diffusion
time of 250 ms for the M 13-DNA molecule. The fit corresponds to a stretched exponential with $\beta = 0.44$ and $k = 43\,\mathrm{s}^{-1}$. **b** distribution function of the relaxation
rate k as evaluated form the autocorrelation function [22.20,22.21]

exponent $\exp -(kt)^\beta$. The correlation function could be described by [22.20, 22.23]:

$$G(\tau) = 1 + \text{DIFF} \left\{ A \exp[-(k\tau)^\beta] + 1 \right\} , \qquad (22.4)$$

with $A = \frac{K(1-Q)^2}{(1+KQ)^2}$

Q = quantum yield of state 2/ quantum yield of state 1
K = equilibrium constant between state 2 and state 1
k = relaxation rate of the reversible transitions between state 1 and state 2
β = stretch parameter equal to or smaller than unity
DIFF = diffusion term (22.1).
Values were found for $k = 20\,\mathrm{s}^{-1}$ and for $b = 0.44$, indicating a distribution of rates over several orders of magnitude [22.20]. This was also shown later by a direct evaluation of the Correlation function using distributed exponentials [22.21]:

$$G(\tau) = 1 + \left[\sum_i A_i \exp(k_i\tau) \right] \text{DIFF} . \qquad (22.5)$$

The meaning of β is a short notation of distributed events where $\beta = 1$ represents a Dirac distribution (single exponential) and values below one an increasing skewness of the distribution of k.

The relaxation rate of $20\,\mathrm{s}^{-1}$ represents the mean rate around which k is distributed to higher and lower values ($100\,\mathrm{s}^{-1}$ to $1\,\mathrm{s}^{-1}$) (Fig. 22.3 b). The explanation is a conformational transition matrix in which the vertical

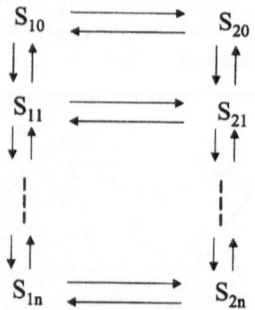

transitions equilibrate on a time scale fast compared to the horizontal transitions. A likely scenario is a multitude of interactions of the linker-bound TMR with Guanosin bases in its vicinity preceeding further steps which may involve insertion into the large groove as well as intercalation.

22.5   Single Molecule Traces

The technique developed in this laboratory is a "stochastic experiment" where the surface covered with single molecules is scanned point by point until a single molecule trace and its correlation function is observed. At this moment the scan is stopped and traces and their correlation functions are recorded [22.21].

A typical trace obtained from the tagged DNA fragments demonstrates the spontaneous occurrence of the transition between the two different fluorescescence states the intensity of which is proportional to the lifetime of their excited states [22.20,22.21] (Fig. 22.4 a). While in the ensemble transition rates are averages, in the single molecule case the transition is a spontaneous, non-predictable event, which, however, follows a distribution of exponential character [22.21,22.24]. From the time inervals between the transition times of S1 to S2 (τ_1) and vice versa (τ_2) the relaxation rates $k = 1/\tau_1 + 1/\tau_2$ can be determined. Equivalent to this evaluation, where the transition times must be determined individually, is the recording of the correlation function for the same trace (Fig. 22.4 b). The relaxation rate can then be determined directly from the correlation trace.

22.6   Homogeneous and Heterogeneous Behavior

The ergodic principle states the equivalence of the ensemble average and the time average. Hence the behavior of a single molecule observed over a sufficiently long time period should show all molecular facets as demonstrated by the molecular ensemble incorporating a vast number of individual molecules. From a kinetic point of view the distribution of transition rates, as observed in the ensemble of DNA fragments, should be observable for a single molecule if the system is ergodic and homogeneous.

In reality, however, the relaxation rates of individual DNA fragments constitute subfractions of the distribution of the ensemble (Fig. 22.4) and a heterogeneous behavior. The most likely explanation is that during the time under which single molecules can be observed (approximately 10 s) not all transitions can be performed which are possible in an ensemble with a long time history. A different description is a scenario with a high-dimensional landscape of activation barriers of different height which limit transitions to certain areas given a limited time span for transitions to take place. A particular reason for the heterogeneous behavior is, in this case, the chemical lifetime of the fluorescent molecule, which has been determined in great detail [22.25].

22.7   Time Resolution of Single Molecule Behaviour

The availability of avalanche photodiodes with substantially increased quantum yield of photon detection was an important reason for successful single molecule detection and significant reduction of the collection times in FCS. Their time resolution is limited by their recovery time which, in older versions is around 200 ns, and in modern versions is around 40 ns. Their linearity in

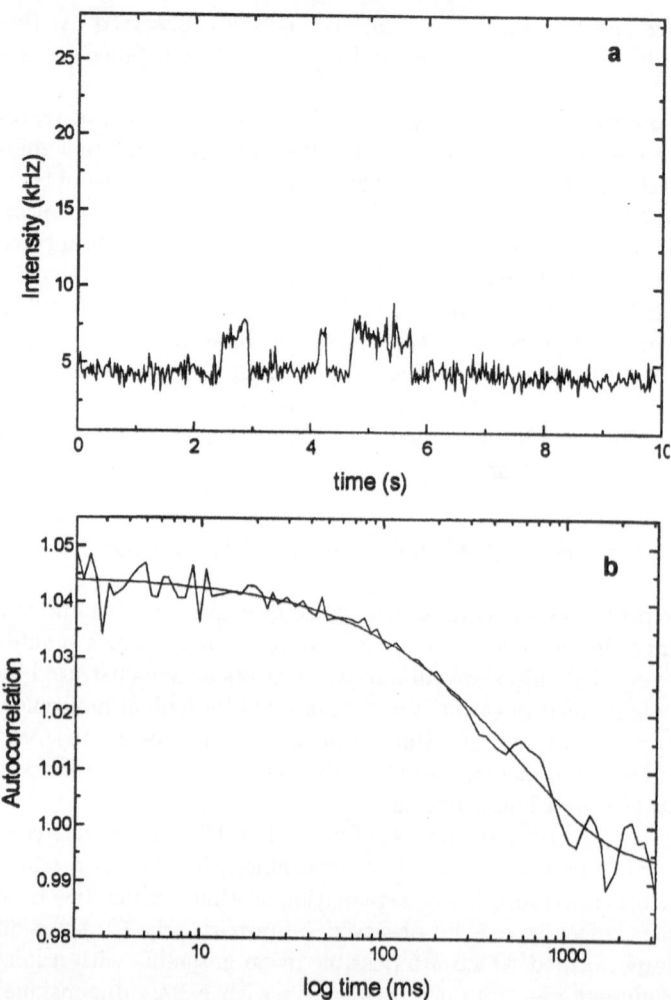

Fig. 22.4 a.b. Single molecule traces: **a** conformational transitions in a DNA molecule labeled with tetramethylrhodamin, **b** autocorrelation function [22.21]

response to impinging photons is strongly dependent on the recovery time which constitutes a time interval during which the APD is not active. Compared to our first single molecule measurements of Rh6G where 100 000 cps could be recorded with a 200 ns dead time APD a factor of two or more can be gained using short dead time versions. The important feature of APDs as detectors is their high quantum yield in combination with a fast response time. Since the recovery process is a stochastic phenomenon the time limitation can be surmounted by using two APDs and cross-correlating the signal from the same source. Time resolution up to several ns could be obtained in

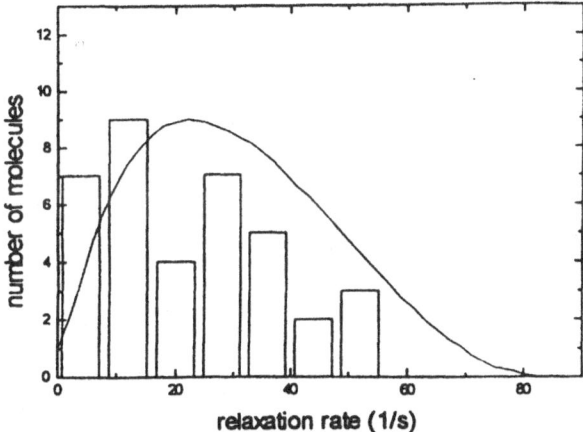

Fig. 22.5. Ensemble and Single molecule behavior: distribution functions (full line) of the relaxation rates for conformational transitions in M13-DNA molecules in Solution (ensemble), distribution function (histogram) of Single 217 bp DNA molecules attached to a glass surface by a streptavidin-biotin linkage [22.24]

this way and was applied in the analysis of triplet states [22.17,22.26] anti-bunching and rotational correlation [22.27] (see also Chap. 16) as well as in kinetic studies of dye-nucleotide interactions [22.28] and of conformational transitions in GFP [22.29].

A different way of following single molecule events is the use of 2D arrays of charged coupled devices (CCD). In this situation each pixel constitutes an individual volume element and images of single molecules can be recorded as well as their fluctuations. Compared to APDs the time resolution of CCDs is rather limited by their readout time and relatively slow processes can be observed in the second time domain. A striking example is GFP where single molecule transitions were found in the second time regime [22.30], while conformational relaxations linked to a protonization processes could be recorded in the μs to ms time regime by FCS [22.29,22.31,22.32]. In summary, confocal scanning anlaysis using APDs as introduced by Wennmalm et al. [22.21] offers an unsurpassed time resolution for the analysis of single molecules.

22.8 Kinetic Analysis, Death Numbers, and Survival Times

Important parameters for the anaysis of single molecules are the knowledge of the number of photons which can be turned over until a molecule breaks down (death number) as well as the time interval during which this takes place (survival time). For TMR, a detailed study at the level of single molecules was performed [22.25] and has given evidence for the existence of a population with a turnover (death) number of 15 000 detected photons as well as a second population with a higher value. Depending on the excitation intensity and the counting rate the break downtime (survival time) will assume different values which can be calculated from death numbers and excitation rates. Thus for recording fast processes high excitation intensities are needed in order to obtain adequate S/N ratios while for the analysis of slow processes low excitation intensities are required.

22.9   The Fluctuating Enzyme

Single molecule analysis in combination with FCS has turned out to be particularly important for the analysis of enzymatic catalysis. In a study where we analyzed the catalytic turnover of horse radish peroxidase (HRP) using a non-fluorescent Rhodamine molecule as substrate which by the action of HRP is oxidized to a fluorescent product. Based on the experience from the study of DNA fragments biotinilated HRP was attached to strepatavidin-coated glass surface at concentrations which assure the attachment as single molecules (Fig. 22.6). By the turnover of the substrate into fluorescent product, photons are generated at each single HRP molecule. Intensity fluctuations recorded show a large variation in the activity of product formation (Fig. 22.7). In comparison to the time limitation imposed by the survival time of TMR linked to the DNA fragments no such limitations exist since new fluorescent substrates are always generated which leave their point of generation by dissociation from the enzyme.

22.10   Evidence for Multiple Conformational Transition and Catalysis

Analysis of the intensity fluctuation traces shows the diffusive part of the correlation function in the range of $100\,\mu s$ and an extended (stretched) part from about 1 ms to the 1 s range (Fig. 22.8).

Horse radish peroxidase interacts with the substrate in its oxidized form (5+ and 4+) and can in either form exchange an electron with the substrate which is turned into a fluorescent product. The reduced enzyme form (3+) can be shifted into the oxidized form by hydrogen peroxide (H_2O_2) acting as cosubstrate. A simple reaction scheme is shown below, which indicates the involvement of the heme iron in the oxidation-reduction process.

(I)

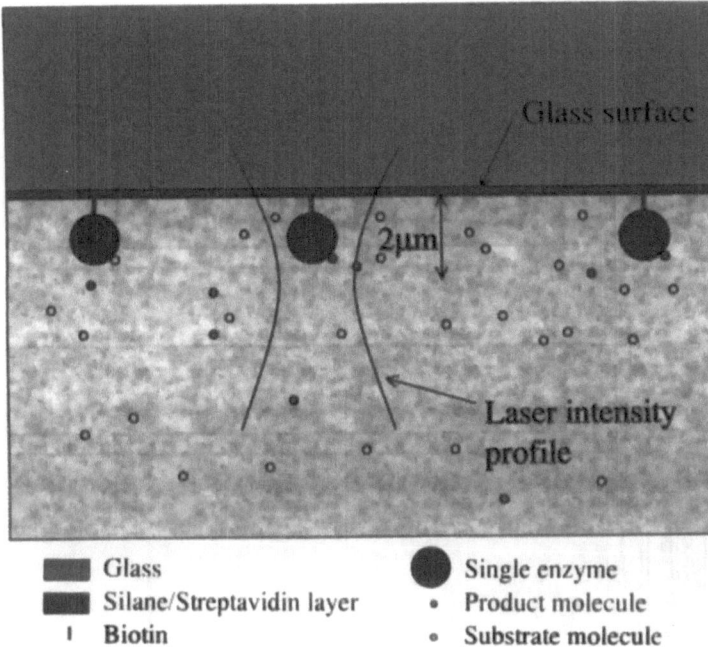

Glass
Silane/Streptavidin layer
I   Biotin

Single enzyme
• Product molecule
• Substrate molecule

Fig. 22.6. Single horse radish peroxidase molecules linked to a steptavidinized surface by biotin. The oxidized product is excited to fluorescence in the volume element [22.33]

Using a simplistic model for the analysis of correlation traces we assume as a minimal model the formation of a substrate enzyme complex (ES complex) which by catalysis will be transformed into an EP complex. The EP-complex finally dissociates into the product and the reduced enzyme. The only state which is observed is the EP complex and the product which, however, diffuses rapidly out of the VE.

The model starts from the fact that the enzyme E is transferred to the visible EP complex in the presence of the substrate S and returns to its starting position in the presence of the co-substrate which is present under saturating conditions.

$$
\begin{array}{ccc}
\text{ES} & \longleftrightarrow & \text{EP} \\
| & & | \\
\text{S} + \text{E}^{5+} & \longleftarrow & \text{E}^{3+} + \text{P}
\end{array}
\tag{II}
$$

Calculating the correlation function $G(\tau)$ for the formation of the ES and the EP complex, which is the only observable form, gives rise to two exponential processes. We consider the case of a single immobilized HRP molecule:

$$G(\tau) = 1 + [A\exp(-k_1\tau) + B\exp(-k_2\tau) + 1], \tag{22.6}$$

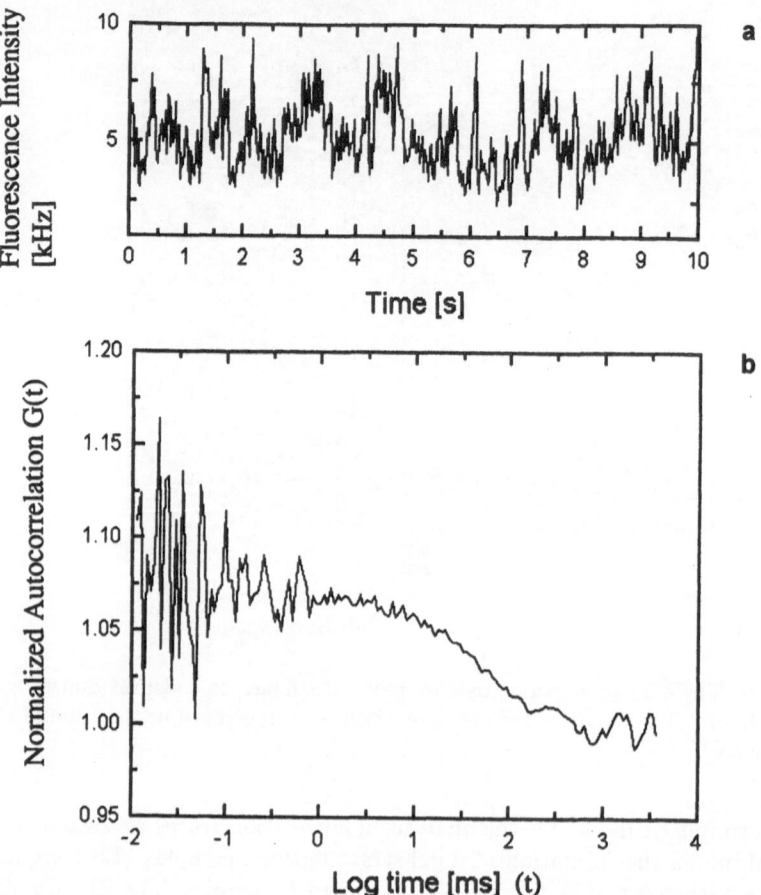

Fig. 22.7 a,b. Activity fluctuations in horse radish peroxidase: **a** fluctuations in product formation due to different catalytic activities of a Single horse radish peroxidase molecule, **b** autocorrelation function of **a** [22.33]

with

$$A = \frac{K_1(1 + K_1 + K_1 K_2)[Q_S - Q_{ES}]^2}{(K_1 + 1)(Q_S + Q_{ES}K_1 + Q_{EP}K_1 K_2)^2}$$

$$k_1 = k_{on}(E + S) + k_{off}$$

$$B = \frac{K_1 K_2 [Q_S + K_1 Q_{ES} - (1 + K_1)Q_{EP}]^2}{(K_1 + 1)(Q_S + Q_{ES}K_1 + Q_{EP}K_1 K_2)^2}$$

$$k_2 = \frac{k_{SP} k_{on}(E + S)}{k_1} + k_{PS}$$

with K_1 and K_2 = equilibrium constants for first and second step
Q_S = quantum yield of S

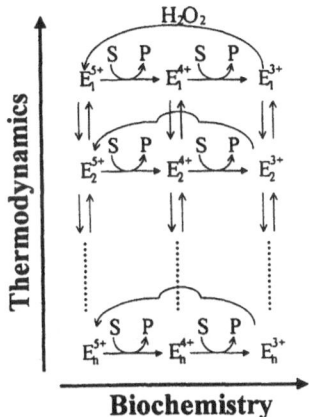

Biochemistry

Fig. 22.8. Grand scheme of the enzymatic activity of horse radish peroxidase. The existente of various conformational substates generates different levels of catalytic activity. [22.33]

Q_{ES} = quantum yield of ES
Q_{EP} = quantum yield of EP.

For the realistic case that $Q_S = Q_{ES} = 0$, $A = 0$ and the first process is not visible. Only the formation of EP can be observed: with $Q_{EP} \gg Q_S$, Q_{ES}, B reduces to

$$B = \frac{(K_1 + 1)}{K_1 K_2}.$$

The amplitude of the visible step increases with increasing substrate concentration and reaches a maximum determined by $1/K_2$.

Equation (22.6) is valid for the case that the formation of the ES complex is fast compared to the EP complex, i.e. k_{on}, $k_{off} \gg k_{SP}$, k_{PS}. It is also readily seen that (22.4) is a special case of (22.6) for the case that $K_2 = 0$.

Considering the whole catalytic cycle we have reduced the reaction scheme (Scheme 4) even further indicating that the oxidized enzyme is transformed into the visible EP complex and vice versa [22.33].

$$E \underset{k_b}{\overset{k_f}{\rightleftharpoons}} EP \tag{III}$$

In this scheme k_f describes the formation of the EP complex including all intermediary steps while k_b describes the formation of the product, the essential catalytic step as well as the reoxidation of the enzyme. The correlation function $G(\tau)$ under saturating conditions is:

$$G(\tau) = K^{-1} \exp[-(k_f + k_b)\tau] + 1, \tag{22.7}$$

with $K = k_f/k_b$ and $k_f = k_{EP} k_{on}(E + S)/k_1$.

If the formation of the EP complex can proceed via many pathways k_f will be distributed, then

$$G(\tau) = \int p(k_f)K^{-1} \exp[-(k_f + k_b)\tau]dk_f + 1,$$

with $\int p(k_f)dk_f = 1$ representing a normalized distribution function.

We assume two cases:

1. the catalytic step is much faster then the formation of the EP complex, $k_b \gg k_f$, then a single exponential correlation function will be observed:

    $$G(\tau) = b\exp[-(k_b)\tau] + 1,$$

2. the EP formation step is much faster then the catalytic step, i.e. $k_f \gg k_b$:

    $$G(\tau) = a\exp[-(k_f\tau)^\beta] + 1,$$

    with $\exp[-(k_f\tau)^\beta] = \int p(\lambda) \exp[-(\lambda)\tau]d\lambda$.

Thus, for the case that $k_f \gg k_b$ and k_f will be distributed $G(\tau)$ will have a stretched appearance. It will assume a single exponential behavior which is dominated by $\exp(-k_b\tau)$ the more the catalytic rate exceeds the rate for the formation of the EP complex.

The normalized intensity autocorrelation function for a single immobilized HRP molecule can then be described by

$$G(\tau) = a\exp[-(k_f\tau)^\beta] + b\exp[-(k_b\tau)^\gamma] + c.$$

In this description the first term represents the formation of the EP complex, while the second term represents the catalytic step including generation of the product. Stretch parameters have been introduced for both steps.

The activity of a single enzyme molecule is characterized by a landscape of conformational barriers leading to the catalytic state. In the evaluation for high substrate concentrations (130 nM) k_f varied between 510–$4100\,\mathrm{s}^{-1}$ with a β value between 0.66–0.16 while k_b was found to be between 5–$11\,\mathrm{s}^{-1}$ with a γ value 1.2–0.8. From a comparison of the measured turnover rate in solution ($50\,\mathrm{s}^{-1}$) and the dependence of the substrate concentration we assign the process characterized by a stretch parameter close to unity to the catalytic step of the reaction.

The formation of the enzyme product complex proceeds over many transitions as indicated by the low value of the stretch parameter. An interesting behavior is the observation that the stretch paramter of the EP formation decreases with increasing measurement time (Table 22.1). This is best explained by the fact that given limited observation time the catalysis proceeds

Table 22.1. Rate parameters of substrate turnovers into products HRP molecules measured for different times and at different substrate concentrations

Molecule	Time (s)	S (nM)	k_b (s^{-1})	γ	k_f (s^{-1})	β
1	10	130	5	1.2	860	0.66
2	70	130	10	0.83	2200	0.26
3	170	130	11	0.80	920	0.16
4	300	65	21	0.74	380	0.20
5	300	5	41	0.53	850	0.1

in a limited range of a landscape of activation barriers. This landscape is explored more extensively the longer the observation time, leading to a broader distribution of the transition rates and a decrease of the stretch parameter.

The analysis of the enzyme catalysis at the single molecule level has revealed properties which would not have been observed form an ensemble measurement. The results very clearly support Frauenfelder's model of conformational substates [22.34] as a basis for the functional behavior of proteins such as enzymes. The scheme in Fig. 22.8 is an example of the fact that HRP exists in many conformations of which only a few will lead to catalysis. The obvious behavior of times where the enzyme is active and other times where the enzyme is inactive as seen from the single molecule traces is generated by the large variety of activation barriers which have to be crossed to lead to catalysis.

22.11   Higher Order Correlations and Non-Markovian Behavior

Given that enzymatic action apparently is linked to a complex network of conformatonal transitions which might be involved in hysteresis and other properties which depend on the previous history we suggest strategies which are able to discriminate Markovian from non-Markovian behavior [22.35]. They are based on the comparison between correlation functions of different order. It can be shown that for Markov processes, such as exponential functions, higher and lower order correlation functions are identical. Thus the function

$$\text{NMF}(\tau_1, \tau_2) = \frac{G(\tau_1, \tau_2)}{G(\tau_1)} - G(\tau_1), \tag{22.8}$$

will be zero for Markovian processes, for which only the immediate past is known. It is therefore appropriate to call the plot of NMF a memory landscape

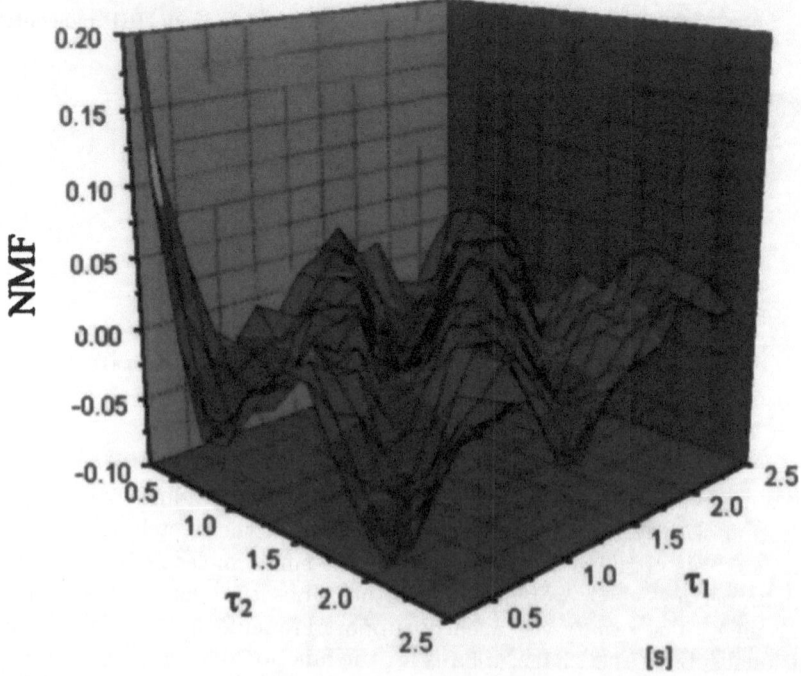

Fig. 22.9. Non-Markovian time trace of the catalysis of horse radish peroxidase as calculated according to (22.8). The System relaxes into the Markovian plane (0 plane) undergoing periodic oscillations as a result of the catalytic cycle [22.35]

(ML). Analysis of the trajectory of product fluctuation from HRP (Fig. 22.7) gives a very reproducible picture indicating a Non-Markovian relaxation on the time scale 20 ms to 2.5 s.

As can be shown from a detailed analysis of the spectrum of eigenvalues as well as from stochastic simulations the NMF relaxation as well as the periodic oscillations (Fig. 22.9) are related to the catalytic cycle embedded in the network of state transitions. It is also likely that catalysis is dependent on its past history in the sense that given the time window of NM behavior catalysis can proceed with each substrate molecule, otherwise the process of conformational transitions preceeding catalysis has to start anew. This is also supported by the observation that with increasing substrate concentrations the stretch parameter for the conformational transitions has a tendency to increase, indicating that the starting point in the energy landscape is closer to the catalytic state.

22.12   Conclusions

Single molecule sensitivity is a prequisite for sucessful FCS. On the other hand, single molecule behavior cannot be understood without an appropriate analysis of its trajectories. We demonstrate in this chapter the power of FCS for this purpose.

Acknowledgements

My thanks go to Manfred Eigen for introducing me to chemical relaxation kinetics and to Leo de Maeyer for supporting my enthusiasm for carrying out the first FCS experiments in his laboratory in 1969. My thanks go to my collaborators in the development of FCS at the Karolinska Institutet: in the early phase to Måns Ehrenberg and Pietro Graselli in the later phase to Jerker Widengren, Ülo Mets, Peet Kask, Michael Brinkmeier, Petra Schwille, Masataka Kinjo, Zeno Földes-Papp, Aladdin Pramanik as well as to the coauthors of the present chapter.

This endeavor was supported from its start in 1970 by the Swedish Natural Science Research Council and the Knut and Alice Wallenberg Foundation. The later part of the project was substantially supported by the Swedish Technical Science Research Council. The Alexander Humboldt Foundation supported a most delightful sabbatical year with Manfred Eigen during 1992/1993.

References

22.1 D. Magde, E. Elson, and W.W. Webb, Phys. Rev. Lett. **29**, 705 (1972)

22.2 E. Elson and D. Magde: Biopylymers **13**, 1–27 (1974)

22.3 M. Ehrenberg and R. Rigler: Chem. Phys. **4**, 390 (1974)

22.4 M. Eigen and L. DeMeyer: *Chemcial Relaxations* In: Techniques in Org. Chemistry, ed. by L. Weissberger, (Academic Press 1967)

22.5 D. Magde, E. Elson,W.W. Webb. Biopylymers **13**, 29–61 (1974)

22.6 R. Rigler and Ü. Mets: Soc. Photo-Opt. Instrum. Eng. **1921**, 239–248 (1993)

22.7 Ü. Mets and R. Rigler: J. Fluorescence **4.3**, 259–264 (1994)

22.8 R. Rigler, J. Widengren, and Ü. Mets: *Interactions and kinetics of single molecules as observed by fluorescence correlation spectroscopy.* In: Fluorescence spectroscopy, ed. by O.S. Wolfbeis (Springer Berlin Heidelberg New York, 1992) pp. 13–24

22.9 R. Rigler, Ü. Mets, J. Widengren, and P. Kask: Eur. Biophys J. **22**, 169–175 (1993)

22.10 M. Eigen and R. Rigler: Proc. Natl. Acad. Sci. USA **91** 5740–5747 (1994)

22.11 W.E. Moerner and M. Orrit: Science **283**, 1670–1676 (1999)

22.12 R. Rigler and J. Widengren: BioScience 180–183 (1990)

22.13 E.B. Sheras, N.K. Seitzinger, L.M. Davies, R.A. Keller, and S.A. Soper: Chem. Phys. Lett. **174**, 553–557 (1990)

22.14 W.P. Ambrose, P.M. Goodwin, J.C. Martin, and R.A. Keller: Science **265**, 364–367 (1994)

22.15 P.M. Goodwin, W.R. Ambrose, and R.A. Keller: Acc. Chem. Res. **29**, 607–613

22.16 R. Rigler, F. Claesens, and G. Lomakka: *Picosecond single photon fluorescence spectroscopy of nucleic acids.* In "Ultrafast Pnenomena IV", ed. by D.H. Auston and K.B. Eisenthal (Springer Series in Chemical Physics, 38), (Springer Berlin Heidelberg New York, 1984) pp. 472–476

22.17 J. Widengren, R. Rigler, and Ü. Mets: J. Fluorescence **4(3)**, 255–258 (1994)

22.18 J. Dapprich, Ü. Mets, W. Simm, M. Eigen, and R. Rigler: Exp. Tech. Phys. **46**, 259 (1995)

22.19 J.K. Trautman, J.J. Macklin, L.E. Bruss, and E. Betzig: Nature **369**, 40–42 (1994)

22.20 L. Edman, Ü. Mets, and R. Rigler: Proc. Natl. Acad. Sci. USA **93**, 6710–6715 (1996)

22.21 S. Wennmalm, L. Edman, and R. Rigler: Proc. Natl. Acad. Sci. USA **94**, 10641–10646 (1997)

22.22 L. Edman, Ü. Mets, and R. Rigler: Exp. Tech. Phys. **41**, 259–264 (1995)

22.23 J. Widengren and R. Rigler: Bioimaging **4**, 149–157 (1996)

22.24 S. Wennmalm, L. Edman, and R. Rigler: Chem. Phys. **247**, 61–67 (1999)

22.25 S. Wennmalm and R. Rigler: J. Phys. Chem. B **103(13)**, 2516–2519 (1999)

22.26 J. Widengren, Ü. Mets, and R. Rigler: J. Phys. Chem. **99**, 13368–13379 (1995)

22.27 Ü. Mets, J. Widengren, and R. Rigler: Chem. Phys. **218**, 191–198 (1997)

22.28 J. Widengren and R. Rigler: J. Fluorescence **7**, 211–213 (1996)

22.29 J. Widengren, T. Terry, and R. Rigler: Chem. Phys. **249**, 259–271 (1999)

22.30 R.M. Dicksom, A.B. Cubitt, R.Y. Tsien, and W.E. Moerner: Nature, 355–358 (1997)

22.31 J. Widengren and R. Rigler: Cell. Mol. Biol. **44(5)**, 857–879 (1998)

22.32 U. Haupts, S. Maiti, P. Schwille, and W.W. Webb: Proc. Natl. Acad. Sci. 13573–13578 (1998)

22.33 L. Edman, Z. Földes-Papp, S. Wennmalm, and R. Rigler: Chem. Phys. **247**, 11–22 (1999)

22.34 H. Frauenfelder, S.G. Sligar, and P.G. Wolynes: Science **254**, 1598 (1991)

22.35 L. Edman and R. Rigler: Proc. Nat. Acca. Sci. USA **97**, 8266–8271 (2000)

Subject Index

Springer Series in Chemical Physics

Editors: Vitalii I. Goldanskii   Fritz P. Schäfer   J. Peter Toennies

Springer Series in Chemical Physics

Editors: Vitalii I. Goldanskii   Fritz P. Schäfer   J. Peter Toennies

Managing Editor: H. K. V. Lotsch